Satellite Remote Sensing

Past, Present and Future

Satellite Remote Sensing

Past, Present and Future

William J. Emery
Smead Aerospace Engineering Sciences Department, University of Colorado at Boulder, Boulder, CO, United States

ELSEVIER

Elsevier
Radarweg 29, PO Box 211, 1000 AE Amsterdam, Netherlands
125 London Wall, London EC2Y 5AS, United Kingdom
50 Hampshire Street, 5th Floor, Cambridge, MA 02139, United States

ISBN: 978-0-443-22086-9

For Information on all Elsevier publications
visit our website at https://www.elsevier.com/books-and-journals

Publisher: Peter Llewellyn
Acquisitions Editor: Menka Mehta
Editorial Project Manager: Harshita Manchanda
Production Project Manager: Sruthi Satheesh
Cover Designer: Raman Kumar

Printed in England by CPI Group (UK) Ltd
Croydon CR0 4YY
United Kingdom

Typeset by MPS Limited, Chennai, India

Dedication

To my wife Dora Emery for putting up with me constantly writing books.

Contents

Preface

I have lectured on the history of satellite remote sensing for many years and formalized that into the first chapter of the book "An Introduction to Satellite Remote Sensing." I have always wanted to expand on this chapter and give a more complete view into the history of the field. I decided to write a book that would review this history in depth, discuss the present use of satellite remote sensing, and give some insight into what we can expect the future to hold for this field. This book is the product of this desire and is intended to provide a reader with an appreciation of the technical developments that led to the present use of satellite data. We have come to accept that when we see weather forecasts on television, we will see some version of the satellite view of our area of the globe. We also accept that the technologies that make this possible are readily available, and we give little thought as to what developments were needed to make these observations possible. I hope that by reading or at least reading part of this book, the reader will gain a deeper appreciation for the things that are needed to be developed to bring us to this level of routine satellite observations. The dramatic recent explosion of the use of small satellites and their technology greatly increases the use of this type of information, and it is difficult to accurately forecast where the next leap in technology will be.

Chapter 1

Early developments in remote measurement systems and aerial photographs

1.1 History of balloons (https://www.nationalballoonmuseum.com/about/history-of-ballooning/)

The origins of remote sensing of the Earth are closely linked to man's ability to leave the surface and rise to an altitude where the Earth can be viewed from above. The earliest that man was able to view the Earth from above was the first hot air balloon flight, when Joseph and Etienne Montgolfier from Ardèche, France, experimented with a fire of wool and damp straw with some success. They were invited to repeat their experiment in Paris by the Academie Royale des Sciences. In 1783 Etienne carried out an initial tethered deployment, which was successful. He repeated this effort 7 days later in front of King Louis XVI and the royal family.

The balloon was made of cotton canvas with paper glued onto both sides. It measured 18.47 m tall by 13.28 m wide and weighed 400 kg. It was named Le Reveillon after Etienne's friend Jean-Baptiste Reveillon, who was the Director of the Royal Manufacture of printed paper, and had designed a motif for the balloon. As a matter of safety, it was decided that animals and not humans would ride in the balloon, so at the blast of a cannon at 1 pm, a sheep, a duck, and a cockerel were placed in the wicker basket tied to the balloon. Eleven minutes later, a second cannon shot marked the liftoff of the balloon, which, amidst applause, rose to 600 m in the air. At this point a rip in the fabric caused it to descend slowly 8 minutes later and hit the ground after traveling 3.5 km. Pilâtre de Rozier, a physician and future aeronaut, hurried to examine the animals, which he found to be very much alive and well. It was now time for the first human to become borne aloft, and so on 21 November of that same year, Pilâtre de Rozier became the first person ever to be borne aloft in a hot air balloon.

Shortly thereafter, on December 1, 1783, the physicists Jacques Alexander Charles and Nicolas Louis Robert took to the air in a gas balloon, also in Paris, France. This flight lasted 2.5 hours and covered a distance of 25 mi. The gas used was hydrogen, a lighter than air gas that had been developed by an Englishman, Henry Cavendish, in 1776, using a combination of sulfuric acid and iron filings. Gas balloons became the mode of air travel until the invention of fixed-wing aircraft. Gas balloons were expensive to operate, and in the early days, portable heat sources were not available, so balloon travel was restricted to those who could afford gas balloons.

The first attempt to cross the English Channel in a balloon was in 1785 by Pilâtre de Rozier, along with a man named Romain. This was an experimental system using both hydrogen gas and hot air compartments. Unfortunately, this mixture of hot air and highly flammable hydrogen gas led to an explosion, and both men were killed. The first successful crossing of the channel was accomplished by the French balloonist Jean-Pierre Blanchard, along with an American, John Jeffries, using a hydrogen gas balloon.

The first manned flight of a balloon in America was in January 9, 1793, in a hydrogen gas balloon piloted by the same Frenchman who first crossed the English Channel, Jean-Pierre Blanchard. He flew from a prison yard in Philadelphia, Pennsylvania, and after ascending to 5800 ft he made a successful landing in Gloucester County, New Jersey. George Washington observed the flight.

Satellite Remote Sensing. DOI: https://doi.org/10.1016/B978-0-443-22086-9.00010-1

1.2 History of photography (https://en.wikipedia.org/wiki/History_of_photography)

Before someone could take a picture of the Earth from a balloon, we had to have cameras portable enough to take on board a balloon. The development of photography depended on two critical principles: (1) camera obscura or image projection and (2) the observation that some materials are visibly altered by exposure to light. The first consists of a dark room with a small hole in one wall. When it is bright outside, light enters through the hole and projects an upside-down image of the outside world opposite the hole. It is believed that Renaissance artists such as Michelangelo and Davinci used this device to get a sense of proportion. It can be traced in Chinese writings to about 1000 BCE, while other Arab and European historians attribute its invention to an Egyptian astronomer and mathematician, Ibn Yunus, at around 1000 AD. There is no evidence of efforts to capture images with light-sensitive materials before the 18th century.

Around 1800, Thomas Wedgwood made the first reliably documented, although unsuccessful, attempt at capturing an image in permanent form. His experiments did produce what he called "photograms," but Wedgwood and his associate Humphry Davy found no method to fix and retain these images. In 1826 Nicéphore Niépce first managed to fix an image, but it took at least eight hours or even several days of exposure in the camera. In this case, it was the back of the camera that the image was fixed on. These images were very crude.

Niepce's associate, Louis Daguerre, developed the daguerreotype process that became the first publicly announced and commercially viable photographic process. The daguerreotype process only required minutes of exposure and produced very clear and detailed images as a result. This process was introduced to the world in 1839, a date that is accepted as the birthday of practical photography. The metal-based daguerreotype process soon had competition from the paper-based calotype negative and salt print process invented by William Henry Fox Talbot and demonstrated in 1839. Subsequent developments made photography easier for the user. New materials reduced the required camera exposure time from minutes to seconds and then to a fraction of a second. Starting in 1850 the collodion process with its glass-based photographic plates combined the high resolution from the daguerreotype with the multiple print options from the calotype process. Roll films popularized use by amateurs, and it was soon possible to take pictures in full color as well as black and white.

The introduction of computer-based electronic digital cameras in the 1990s revolutionized photography, making it possible for every amateur to take properly exposed pictures. These computer-driven cameras incorporated lots of automated features that ensure proper focus and capture moving subjects. Being able to store high-quality images on small digital storage cards that could be edited and printed later greatly reduced the cost of photographs. The next big step in photography came not with the development of new camera technology but with the incorporation of a camera into the cell phone. Soon, people were buying portable cameras, which could be used to make phone calls. Everyone now carries a camera in their phone, and the ability to pass photos from your phone to a computer and also transfer them over the internet changed how everyone thinks about photography. It was now a given that everyone was able to take pictures and send them wherever they wanted.

1.3 Early balloon photography and kite photography (aerial photography, Wikipedia; file:///Users/williamjemery/Desktop/RS%20past%20pres%20futu/background%20off%20the%20web/history%20of%20aerial%20photography/History%20of%20Aerial%20Photography%20-%20Professional%20Aerial%20Photographers%20Association%20Intl.html) (https://en.wikipedia.org/wiki/Kite_aerial_photography)

The first known aerial photograph was taken from a tethered balloon at an altitude of sixteen hundred feet in 1858 by a Frenchman, Gaspard Félix Tournachon, who went by the pseudonym of Nadar (Fig. 1.1). This rather cartoon-like image shows how a box camera was mounted in the bucket under the balloon.

His subject was the French village of Petit-Bicêtre, and his pictures have since been lost. This was a major accomplishment from a tethered hot-air balloon, 80 m above the ground. The early collodion photographic process in use at that time required a complete darkroom to be carried in the basket of the balloon. Nadar's earliest surviving aerial image was taken above Paris in 1866 (Fig. 1.2).

The earliest aerial photograph over the United States was Wallace Black's image of Boston from a hydrogen balloon, taken in 1860 from an altitude of 1,200 ft in a balloon commanded by Professor Sam King. Black took eight exposures, only one of which resulted in a reasonable picture (Fig. 1.3).

FIGURE 1.1 Nadar taking a picture from a balloon.

FIGURE 1.2 Paris from a balloon taken by Nadar in 1866.

FIGURE 1.3 Boston from a balloon in 1860.

Black labored under considerable difficulty, as the balloon, even though it was tethered, was constantly moving. This motion, combined with the slow speed of the photographic materials that Black was working with, made it difficult to get an exposure without movement of the image. He used wet plates and had to prepare them in the balloon before each exposure. After descending to the ground and refreshing their supplies, they ascended again, planning on covering more of Boston and the surrounding countryside. As they rose, the hydrogen expanded, causing the neck of the balloon to open more, allowing the gas to spill on their equipment and turning the plates black and useless. In addition, the tether broke and they eventually landed in Marshfield, Massachusetts, about thirty miles away from their original point.

As photographic technology improved, it became necessary to carry less equipment on the balloon, and the first "free flight" balloon photo was taken by Triboulet over Paris in 1879. The continued improvements in camera technology enabled people to use other platforms to take cameras aloft and collect pictures of the Earth's surface. One candidate was the kite, and M. Arthur Batut took the first kite photograph over Labruguière, France, in the late 1880s. The camera was attached directly to the kite (Fig. 1.4) and had an altimeter that recorded the exposure altitude directly on the film, making it possible to scale the image later.

A slow-burning fuse connected to a rubber band–driven device, actuated the shutter a few minutes after the kite was launched. A small flag dropped after the shutter was released to indicate that it was time to pull in the kite. Batut's first aerial photograph in May 1888 was not very clear due to the slowness of the shutter. This called for some modifications to the thickness of the rubber band, which then resulted in a much clearer photograph.

One of the most famous kite photographs was taken by George R. Lawrence in 1906 after the San Francisco earthquake that devastated the city. He used between nine and seventeen large kites to lift a very large (49 lb) camera to take some very large (48×122 cm or 18×48 in) format aerial pictures (Fig. 1.5).

The camera was lifted to a height of about 2,000 ft, and it had an electric wire switch to control the shutter to produce a negative. Lawrence designed his own large-format camera especially for aerial photography. He had started with balloon photography and then graduated into kite photography. He developed a method of flying Conyne kites in trains to keep the camera steady under varying wind conditions (Fig. 1.6), which he called the "captive airship." It is still difficult to comprehend how he managed to keep 17 kites flying to lift a 49-lb camera up to 2000 ft and take a clear, large format exposure. That he was also able to operate in a most difficult time when the normal city infrastructure was compromised is also amazing.

FIGURE 1.4 Batut's kite-mounted camera.

FIGURE 1.5 Lawrence's kite photography of San Francisco after the 1906 earthquake.
George Lawrence, Library of Congress.

1.4 Pigeon photography (pidgeon photography, Wikipedia)

A most innovative aerial photography effort was created in 1903 by Julius Neubronner who wanted to take aerial photos at the 1909 Dresden International Photography Exhibition. He designed a small (70 g) camera that could be mounted on a

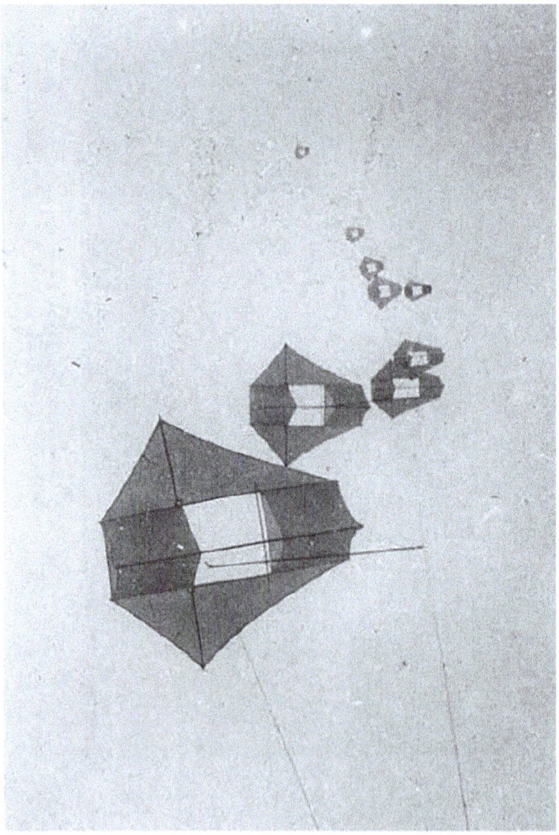

FIGURE 1.6 Lawrence's captive airship.
George Lawerence, Library of Congress.

FIGURE 1.7 Neubranners pigeon camera.

Pidgeon (Fig. 1.7). He worked as an "apothecary" and had used pigeons to deliver medications. Neubranner's initial patent application for a pigeon camera was rejected but was later granted after he produced authentic photographs taken by his pigeons. Not only did his pigeons collect photographs at the Dresden show, but they also operated at the Frankfurt International Aviation Exhibition in 1910 and at the Paris Air Show in 1911.

At the beginning of World War 1 there was some interest in using pigeon photography for mapping enemy positions. But the rapid advances in aircraft technology soon replaced the pigeons with photographs taken from aircraft, as will be discussed later.

This camera took exposures at 30-second intervals along the flight line flown by the pigeon. Picture postcards were created from the pigeon pictures, and people at the exhibition would buy them to see if they could find themselves.

In order for these cameras to be so small and to take repeat exposures, a new type of film needed to be developed. In 1870 George Eastman from Waterville in upstate New York was an accountant in Rochester. After working for years in a bank, George became bored with the monotony of the job. He decided in 1878 to take an extended vacation to the island of Santo Domingo to reevaluate his life. To record his trip, he acquired a wet-plate camera outfit. Finding the camera and associated darkroom equipment to be cumbersome and bulky, he decided he could do better and soon forgot about his trip to Santo Domingo and began working on a better camera system.

Eastman developed in 1879 a formula for making a gelatin emulsion-covered dry plate and built a machine for coating dry plates with this emulsion. This later led to the development of rolled paper film. The resulting pictures were clear, sharp, and free from paper grain distortion. George then formed his company Kodak in 1889, and introduced the flexible celluloid film, and the popularity of photography exploded. George then created a camera that took advantage of this new celluloid film called the Brownie. At the time, Kodak and photography were almost synonymous. His developments also revolutionized aerial photography shortly before the flight in 1903 of the Wright brothers.

1.5 Rocket and aerial photography (history of aerial photog; https://en.wikipedia.org/wiki/Aerial_photography#:~:text=Aerial%20photography%20(or%20airborne%20imagery,also%20known%20as%20aerial%20videography)

Another platform that offered the opportunity to take photographs of the Earth was rockets, and surprisingly enough, the first rocket photograph was taken by Alfred Nobel, the Swedish millionaire who discovered dynamite and created the world's most prestigious science prizes. Nobel unfortunately did not take the picture, but in 1896 he filed a patent for "an improved mode of obtaining photographic maps" from a rocket. Each rocket firing would produce one picture, which the camera would take while parachuting back to the ground. To control the shutter release, he used a fuse rather than a clock. Nobel died in December of that year, but engineers in his lab in Karkskoga, Sweden, carried the idea forward, and on April 26, 1897, their rocket took two photographs of the town as the camera parachuted to Earth (Fig. 1.8).

A few years later, in 1903, a German engineer, Alfred Maul, used gunpowder rockets to reach an altitude of 2,600 ft in just 8 seconds when it released a parachute-equipped camera that took photos during its descent. Around the same time, Dr. Robert Goddard of Auburn, Massachusetts, was experimenting with rockets and also released a camera equipped with a parachute to take a picture of the Earth. One such photo is of Auburn, Massachusetts (Fig. 1.9).

FIGURE 1.8 Rocket photo of Karlskoga, Sweden.

FIGURE 1.9 Rocket photo of Auburn, Massachusetts.

FIGURE 1.10 Wright brothers flying over Europe and taking the first aerial photos.
Wikipedia.

It is not surprising that the first photos and motion pictures that were collected from heavier than air aircraft were from the Wright brothers' aircraft (Fig. 1.10) flying over Rome, Italy, where Wilbur Wright made five flights. On one flight, his passenger was the photographer Frederico Valle, who shot the footage. This footage was included in a silent film titled "Wilbur Wright und seine Flugmaschine" by the French company Société Générale des Cinématographes Eclipse, which is still available on YouTube today.

Chapter 2

Aerial photography through the two World Wars

2.1 World War 1 aerial photography (https://en.wikipedia.org/wiki/Aerial_reconnaissance_in_World_War_I)

Later developments of aerial photography were driven by war. It is a normal consequence of war that many technological advances are made. In this case, both sides of the war saw the advantage of aerial views, but initially, human observers were used more than photography. Human observers frequently reported wrong results, as it was difficult to tell one type of soldier from the air from another. In one case, a German observer reported that soldiers in an English unit were running around in great disarray and appeared to be in a state of panic. As it turned out, the English soldiers were playing soccer.

Some English observers started using cameras to record enemy positions and found aerial photography easier and more accurate than directly observing and sketching ground positions. The aerial observer soon became the aerial photographer (Fig. 2.1).

The maps used by both sides at the Battle of Neuve-Chappelle in 1915 were all made from aerial photographs. By the end of the war, both the British and Germans were mapping the entire front from the air at least twice per day, which gave them detailed information on the opposing army's trench construction. England estimates that its reconnaissance planes took half a million photographs during the war, and Germany estimates that if their aerial photographs were placed side-by-side, they would cover the country six times. The war motivated dramatic improvements in camera technology, and photographs taken at 15,000 ft (4,572 m) could be blown up to the point where you could see footprints in the mud.

The first camera designed for airborne application was invented by Captain John Moore-Brabazon in 1915. He was aided by the Thorton-Pickard company who wanted to enhance the efficiency of aerial photography. This camera was mounted on the floor of the aircraft and could be triggered by the pilot at desired times. Moore-Brabazon also pioneered the use of stereoscopic methods in aerial photography, where the height of objects in the landscape is discerned by comparing photographs at different angles.

2.2 Post–World War I aerial photography

In the post war period, the first books on aerial photography were published. Reeves's Aerial Photographs Characteristics and Military Applications was published in 1927, while Winchester's and Wills' Aerial Photography, a Comprehensive Survey of its Practice and Development in 1928, and McKinley's Applied Aerial Photography came out in 1929. The first patent from Mannes and Godowsky for multilayer color film came out in 1924, which enabled Kodak to start producing Kodachrome in 1935. One of the first users of this color film was Samuel Goddard for his rocket photographs in 1937. Meanwhile Aerial photography had caught the interest of three federal agencies, namely, the Soil Conservation Service, the Agricultural Adjustment Administration, and the Forest Service.

The first commercial aerial photography company in the United Kingdom was Aerofilms, Ltd, which was actually founded during WW1 by veterans Francis Wills and Glaude Graham White in 1919. The company rapidly expanded with contracts in Africa and Asia as well as the United Kingdom. From 1921, Aerofilms carried out vertical photography for survey and mapping purposes. During the 1930s, the company pioneered the science of photogrammetry, which is cartographic mapping from aerial photographs. As early as 1920, an Australian, Milton Kent, used a "half-plate oblique aero camera" purchased from Carl Zeiss AG in his aerial photography business.

An American pioneer of this period, Sherman Fairchild, started his own aircraft company that became Fairchild Aircraft to develop and build airplanes that were specifically designed for high-altitude aerial survey missions. One such aircraft in

Satellite Remote Sensing. DOI: https://doi.org/10.1016/B978-0-443-22086-9.00003-4

FIGURE 2.1 An aerial photographer in World War 1. https://en.wikipedia.org/wiki/Aerial_reconnaissance_in_World_War_I Wikipedia.

FIGURE 2.2 Spitfire PR. https://www.nationalmuseum.af.mil/Visit/Museum-Exhibits/Fact-Sheets/Display/Article/196278/supermarine-spitfire-prxi/ Wikipedia Aerial Photography.

1935 was equipped with a camera that had five 6-in lenses along with a 10-in lens that took photos from 23,000 ft. Each photo covered 225 mi^2. A year later, Fairchild came out with an even better high-altitude camera with nine lenses in one unit that could take a photo of 600 mi^2 from 30,000 ft.

2.3 World War II·(aerial reconnaissance in World War II—Wikipedia.pdf)

Sidney Cotton and Maurice Longbottom of the Royal Air Force (RAF) were among the first in 1939 to suggest that airborne reconnaissance might be better done with small, fast airplanes that could use their speed and high ceiling to avoid detection and interception. While this seems to be accepted now as doctrine, at the time, this was considered rather radical thinking. They proposed to use a Spitfire with their armament and radios removed so that they could carry extra fuel and cameras. Thus the Spitfire PR was born (Fig. 2.2), which became very successful in this reconnaissance role.

The RAF developed an electric heating system, which was very important for taking pictures at very high altitudes without the camera parts freezing. Cotton also pioneered the collection of high-altitude, high-speed stereoscopic photography that was very useful in locating military targets.

While Germany held a considerable technological head start in aerial reconnaissance, it gradually neglected it at least relative to the emphasis placed on it by Britain. The primary reason was that Germany had no strategic bombing doctrine and instead concentrated on land armies. They did create many *Aufklärungsgruppe* (reconnaissance units) for both marine

FIGURE 2.3 The Blohm & Voss BV 141. https://en.wikipedia.org/wiki/Blohm_%26_Voss_BV_141 Wikipedia Aerial Photography.

and ground support tactical purposes, but the intellectual investment in interpretation, analysis, and strategic estimation lagged. This was justified by the German high command since most of the conflict was in land battles in the east, and long-range air capabilities were not going to change things.

Germany's emphasis on tactical aerial reconnaissance led to the creation of one rather strange aircraft, the Blohm and Voss BV 141 (Fig. 2.3).

This was an asymmetric aircraft designed to carry its camera in a pod independent of the main fuselage of the aircraft. Only 20 of these airplanes were built.

Among the Allies, the RAF continued to lead and took on the job of mentor to the Americans. British Supermarine Spitfires and Mosquitos turned out to be the best reconnaissance platforms, as it was now very clear that speed, range, and altitude were essential to the survival of a reconnaissance aircraft and the production of good aerial photographs. Other aircraft, such as the Douglas Boston, the Bristol Blenheims, and the Martin Marylands, were used in less congested areas where the platforms could be slower. A large part of the RAF reconnaissance effort was taken up in tracking German capital ships.

Aerial reconnaissance was the first to reveal the testing of German jet fighters. The RAF developed a three-phase interpretation procedure: First phase was something that required immediate response (such as advancing columns of armor); second phase needed to be handled within 24 hours (such as concentrations of landing craft in ports), and third phase was for long-term analysis (such as industrial targets and storage facilities). Moreover, the distinction between strategic and tactical aerial reconnaissance became clear. Subspecialties like weather reconnaissance, infrared and radar photography, and bomb damage assessment were identified, and groups set up to handle them. At the peak of the war, the British flew over 100 reconnaissance flights a day, collecting 50,000 images per day that had to be interpreted.

Chapter 3

Modern aerial imaging and early space imaging

3.1 Modern aerial imaging

Vertical (picture taken straight down from the aircraft) aerial imaging is used in cartography (the study of making and using maps), particularly in creating topographic maps. It is also used in land-use planning and aerial archeology. Oblique (pictures taken at an angle) aerial imaging is used more for movie production, environmental studies, power line inspection, surveillance, construction progress, commercial advertising, conveyancing, and artistic projects. In the United States, aerial imaging is often used in many Phase 1 Environmental Assessments. An example of the use in archeology is the mapping project at the site of Angkor Borei in Cambodia from 1995 to 1996. From the aerial photographs, archeologists were able to identify 112 water features such as reservoirs, fishponds, and other man-made water structures.

Since governmental regulations prohibit manned aircraft from operating at low altitudes under 1,000 ft, aerial photographers have turned to radio-controlled model aircraft (drones) to collect their photographs. Drone images became very popular in advertising, both for residential and commercial buildings. In 2014, the US Federal Aviation Administration (FAA) banned the use of drones in real estate advertisements. That ban was lifted in 2018, and both amateur and commercial use of drone imaging is regulated by the FAA.

Drones offer an attractive alternative to larger manned aircraft, but they do not fully replace them. Manned aircraft are capable of longer flights at higher altitudes than the drones. For the drones, gyroscopically stabilized camera systems have been developed that will produce excellent photographs from a drone (Fig. 3.1) even under windy conditions.

In addition to the camera technologies to improve drone images, some of them have been instrumented so that the photograph can be live-streamed to the ground to help the drone pilot to improve the image collection.

3.2 Types of aerial photographs (https://en.wikipedia.org/wiki/Aerial_imaging)

Vertical photographs were those taken looking straight down from the aircraft. These types of aerial photographs are used primarily in cartography and photogrammetry. These pictures are usually taken with large-format cameras having calibrated and well-known geometric properties. Oblique photographs, as stated earlier, are taken at an angle (Fig. 3.2).

Aerial photographs are frequently combined. Panoramas can be made by stitching several photographs taken with different angles from the same spot or from different spots at the same angle. Stereo imaging creates 3D images from several photographs of the same target taken from different spots and angles.

3.3 Orthophotomap

Vertical aerial photographs are frequently used to create orthophotos, also called orthophotomaps. These pictures have been geometrically "corrected" so that they can be used as a map. Thus, an orthophoto is a photograph taken from an "infinite" distance, looking straight down to nadir. Perspective must be removed, but variations in terrain should also be corrected for. These geometric corrections are applied to the image to make it conformal to the selected map grid. Such orthophotos are used in geographic information systems. Large sets of orthophotos are used in online mapping systems such as Google Maps and OpenStreetMap, where they are used to simulate 3D landscapes.

Satellite Remote Sensing. DOI: https://doi.org/10.1016/B978-0-443-22086-9.00005-8

FIGURE 3.1 A drone carrying a camera.
Wikipedia Aerial Imaging.

FIGURE 3.2 An oblique aerial photograph.
Wikipedia Aerial Imaging.

3.4 Leaf-on or leaf-off

The date of collection will determine whether or not an aerial photograph is leaf-on or leaf-off in terms of the deciduous foliage in the target area. This condition is related to the phenology of the plant and suggests the relative health of the vegetation. It can also be used to indicate the species of tree that is being imaged.

3.5 Aerial videographpy

Aerial video is becoming more popular and is widely used in monitoring applications such as pipelines and crop fields. Now that GPS locations can be embedded in the video frames along with accurate time/date stamps, the video sequences can easily be merged together with other data. It is common in aerial film production to use unmanned aerial vehicles in conditions where photos need to be taken that would be hazardous for a manned aircraft (too low, city buildings, electrical wires, etc.).

3.6 First picture from space (https://spacecenter.org/first-photo-taken-from-space/)

The first real view from space was taken from a captured German V2 rocket launched from New Mexico on October 24, 1946, on a suborbital flight of the rocket (Fig. 3.3).

FIGURE 3.3 One of the first photographs of the Earth was taken from a V2 rocket on October 24, 1946, with a camera provided by the Johns Hopkins Applied Physics Laboratory.

3.7 Sputnik (https://en.wikipedia.org/wiki/Sputnik_(disambiguation))

Everything changed when the Russians launched the first orbital spacecraft known as Sputnik 1. This proved that man could send a system to space and have it orbit the Earth, providing a platform from which a variety of images could be collected. Sputnik 1 had its origin on December 17, 1954 when the chief Soviet rocket scientist Sergei Korolev proposed an artificial satellite to the Minister of the Defense Industry. Korolev forwarded an earlier report by Mikhail Tikhonravov that emphasized that the launch of an orbital satellite was an inevitable stage in the development of rocket technology.

In the United States, President Dwight D. Eisenhower announced on July 29, 1955, that during the coming International Geophysical Year (IGY), the United States would launch an artificial satellite. Four days later, a leading Soviet physicist also announced that they would launch an artificial satellite. On August 8, 1955, the Politburo of the Communist Party of the Soviet Union approved the proposal to create an artificial satellite. On 30 August, Korolev presented plans for a three-stage version of his R-7 rocket to be used for satellite launches.

On January 30, 1956, the Council of Ministers of the USSR approved work on the first Earth-orbiting satellite. This first satellite was named Object D and was to be launched in 1957–58. It would have a mass of 1,000–1,400 kg (2,200–3,100 lb) and would have a payload of 200–300 kg (440–66 lb) of scientific instruments. The following groups were assigned contributions to the effort.

The USSR Academy of Sciences was responsible for the scientific leadership and was to supply the scientific instruments. The Ministry of Defense was assigned the task of building the satellite.

- The Ministry of Radio Technical Industry would develop the control system, radio/technical instruments, and telemetry system.
- The Ministry of Ship Building would develop the gyroscope.
- The Ministry of Machine Building would develop the ground launch, rocket fueling, and transportation systems.
- The Ministry of Defense was responsible for carrying out the launches.

Preliminary design work was completed in July 1955, and the scientific goals of the satellite were clearly defined. These were to measure the density of the atmosphere and its ion composition, measure the solar wind/magnetic fields, and cosmic rays.

By the end of 1956, it became clear that Object D was too ambitious, and it would not be ready on time. It would later fly as Sputnik 3. The Soviets feared that the United States would launch a satellite first, so they decided to scale back their plans and focus instead on launching a smaller, less ambitious satellite before the start of the IGY in July 1957. The new satellite would be only 100 kg and would be easy to construct, giving up the complex instrumentation in favor of a simple radio transmitter. The Council of Ministers of the USSR approved the plan for this simple satellite on February 15, 1957, and the satellite was called Object PS. They approved the launch of PS-1 and PS-2 on two R-7 rockets, provided that the R-7 would complete at least two successful test flights. The R-7 was originally designed as an intercontinental ballistic missile. At the time, it was the most powerful rocket in the world. It was designed with surplus thrust as they were not sure of how heavy the hydrogen bomb payload would be. Korolev decided to adapt the R-7 for the launch of Object PS on June 14, 1956.

The first attempt at testing the R-7 for a launch vehicle ended in failure when one of the boosters caught fire immediately after liftoff, and the rocket crashed some 400 km downrange. This was characteristic of the next two launch tests. Finally, the fourth rocket was successfully launched on August 21, 1957. The rocket reached an altitude of 10 km (6.2 mi) and traveled 6,000 km downrange. The launch of the fifth R-7 ON 7 September was again successful, and the R7 was qualified to launch satellites. On 22 September, a modified R-7 rocket arrived at the launch site to take PS-1 (later called Sputnik 1) into space.

PS-1 (Fig. 3.4) was not designed to be controlled from the ground but rather just orbit without any intervention with ground stations listening to its radio transmissions. Stations to track Sputnik were setup at six separate measurement sites clustered around the launch site, and the data were transmitted back to Moscow. The closest of these was 1 km from the launch pad. Another nationwide observation complex, called the Command-Measurement Complex, was a series of 7 measurement stations located along a line of the subsatellite track. These stations were equipped with radar, optical instruments, and communication systems. These data were transmitted to Moscow, where orbital parameters were calculated and the satellite tracked for its orbital position.

The design was led by Mikhail S. Khomyakov, an employee of OKB-1. The satellite itself was a 585-mm diameter sphere made of two hemispheres and containing an instrument mass of 83.6 kg. The metal spheres, which were 2 mm thin, were covered by a highly polished 1 mm thin "heat shield" made of aluminum–magnesium–titanium alloy (Fig. 3.5).

One of the most prominent features of the satellite is the two pairs of antennas that were designed by Mikhail V. Krayushkin of OKB-1. In each pair, one whip antenna is 2.4 m long, while the other is 2.9 m, and both have almost spherical radiation patterns. The power supply was three silver-zinc batteries. Two of the batteries powered the radio transmitter while the other powered the temperature regulation system. The batteries were expected to last 2 weeks, and they operated for 22 days.

The radio transmitter was a 1-W, 3.5 kg unit by the Moscow Electronics Research Institute, that worked on two frequencies, 20.005 and 40.002 MHz. It transmitted on the first frequency in 0.3 second pulses (~3 Hz) with pauses of the same duration filled by pulses of the second frequency. These different radio signals were used to gather information about the electron density of the ionosphere. Temperature and pressure were encoded in the duration of the radio pulses. The temperature control system contained a fan, a dual thermal switch, and a control thermal switch. If the temperature inside the satellite sphere exceeded 36°C, the fan would turn on; when it was below 20°C, the fan would be turned off by the dual thermal switch. If the temperature exceeded 50°C or fell below 0°C, another control thermal switch was activated, changing the duration of the signal pulses. The satellite was filled with dry nitrogen pressurized to 1.3 atm (130 kPa). The satellite also had a barometric pressure switch that, if activated, indicated that the pressure in the satellite was below 130 kPa, which indicated that the pressure vessel had failed, and the duration of the radio pulse would have been changed. This could be the result of being hit by space debris.

Sputnik 1 was launched on October 4, 1957, at 19:28:34 UTC, which would be October 5 at the launch site. A fuel regulator in a booster rocket failed around 16 seconds into the flight, which resulted in increased fuel consumption for

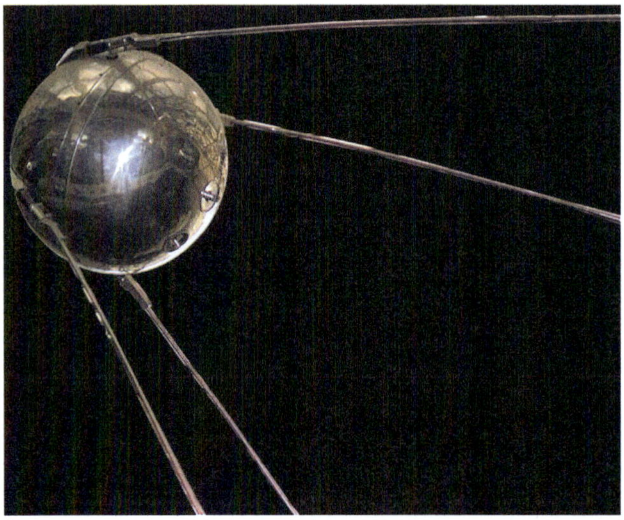

FIGURE 3.4 Sputnik 1.
Wikipedia.

FIGURES 3.5 Exploded view of Sputnik 1 (Wikipedia).

most of the powered flight, and the rocket engine was 4% above normal. This fuel consumption led to a premature thrust cessation one second earlier than planned when a sensor detected overspeed of the empty fuel pump. At 19.9 seconds after engine cut-off, the satellite separated from the second-stage rocket, and the satellite's radio transmitters were activated. The rocket had achieved a 223-km altitude where the satellite was deployed into an elliptical orbit with a perigee of 223 km and an apogee of 950 km, which was 500 km short of the intended apogee of 1,450 km. As a result, the satellite orbit was inclined at 65.10 degrees and had a period of 96.2 minutes.

3.8 A race for the utilization of satellites to benefit the Earth (https://en.wikipedia.org/wiki/Space_Race)

Sputnik 1 and later Sputniks demonstrated that orbiting platforms could be used to view the Earth. This ushered in a new space race to see who could make the greatest use of satellites as platforms for Earth observations and Earth activities. The early phases of this race were marked with repeated failures on both the Soviet and American sides of the conflict. Following the success of Sputnik 1, the USSR launched a bigger Sputnik 1 November 3, 1957, and in December of that same year, the USSR launched a much bigger Sputnik 2 (503 kg) with a dog Laïka as a passenger. This was in the same month as the United States launched Vanguard TV-3, which exploded on the launch pad.

On January 31, 1958, the United States successfully launched Explorer 1 from Cape Canaveral. On February 3, 1958, the USSR tried to launch Sputnik 3 but failed. On February 5, 1958, the United States tried to launch Vanguard 2 but failed. On March 5, 1958, the United States launched Explorer 2, which subsequently failed to achieve orbit. On March 17, the United States launched Vanguard 1 (1.47 kg), which successfully orbits and establishes the pear-shaped nature of the Earth. Another success for the United States was the launch of Explorer 3 on March 26, 1958, which orbited the Earth and collected radiation and micrometeoroid data. On April 28, 1958, another Vanguard failed to orbit. On May 15, 1958, the USSR launched Sputnik 3 (1327 kg), which orbited carrying a large number of scientific instruments, but its tape recorder failed, so it could not map the Van Allen belts.

Continuing in 1958, a Vanguard launch fails for the fourth time on May 27, followed by yet another failure of Vanguard on June 26 (fifth failure). On July 26, a successful launch of Explorer 4 yielded for the first time maps of the Van Allen belts from its 4 orbits. Explorer 5 failed to orbit from its launch on 24 August 1958, followed by a sixth failure of Vanguard on September 26, 1958. On October 1, 1958, the National Aeronautics and Space Administration (NASA) was created in the United States, replacing the National Advisory Committee on Aeronautics. In the middle of December 1958, the United States launched SCORE, the world's first communication satellite. It caught the world's attention by broadcasting a prerecorded Christmas message from US President Eisenhower, making it the first broadcast from space.

On January 2, 1959, the USSR launched Luna 1, which became known as the first "cosmic rocket" as it accidentally escaped the orbit of the moon due to its high speed, becoming the first man-made object to leave the orbit of the Earth and orbit the sun instead. The first weather satellite orbiting the Earth was Explorer 6, launched on August 2, 1959, which then obtained the first weather satellite photos of the Earth from space. On September 12, 1959, the USSR launched Luna 2, which accomplished its mission of being the first spacecraft to reach the surface of the moon. It was quickly followed by Luna 3, launched on 4 October to become the first satellite to orbit the moon and take photographs of the dark side of the moon.

In the next year, the USSR launched Sputnik 5 on August 19, 1960, with two dogs on board (Belka and Strelka) and a number of plants and returned them all alive from space. Six months later, on 31 January 1961, the United States sent Ham, a chimpanzee, into space and returned him to Earth. The USSR then achieved a clear triumph on April 12, 1961 by putting the first man in space on board the Vostok 1 spacecraft. Yuri Gagarin made a single orbit and remained in space for 1 hour and 48 minutes. On May 5, 1961, the United States achieved the first pilot-controlled space flight and the first American in space with Alan Shepard aboard the Mercury-Redstone 3. Shepard did not orbit the Earth, and his flight was only about 15 minutes in duration.

The first woman in space was Valentina Tereshkova, who orbited the Earth for three days after her launch on June 16, 1963. Another first for the USSR was when Alexei Leonov exited his spacecraft, the Voskhod 2, wearing a specialized spacesuit and carried out a 12-minute spacewalk. Voskhod 2 launched on March 18, 1965. Launched on July 14, 1965, the USA satellite Mariner 4 made its way to the surface of the planet Mars, returning the first close-up images of the Martian surface.

The year 1967 proved to be deadly for both the United States and the USSR. In January, American astronauts Ed White, Gus Grissom, and Roger Chaffee died when a fire ignited in their Apollo 1 capsule on the launch pad. A few months later, the Soviet cosmonaut Vladimir Komarov was also killed when the parachute on his Soyuz 1 capsule failed to deploy.

At the end of the next year (December 21, 1968 the United States sent the human crewed Apollo 8 to orbit the moon and return safely to Earth. It was followed-up on July 20, 1969 with the first landing on the moon of Neil Armstrong and Buzz Aldrin, who became the first men to walk on the moon, while their crewmate Michael Collins orbited the moon. This was an event seen on TV by 723 million people the world over. This was clearly a major achievement for the United States.

This was followed by many other countries getting into the satellite business. A list of first successful satellite launches by country is given in Table 3.1 stretching from 1957 for the Soviet Union up to 2012 for North Korea.

The earliest satellites came from the more developed countries led by the Soviet Union, followed by the United States, France, Japan, China, and the United Kingdom. India, Israel, Russia, and Ukraine all launched in the 20th century. Iran and North Korea had their first launches in the 21st century. Many other countries have now successfully launched their own satellites. With the advent of small sats and CubeSats, it is now possible for almost any group to build and launch

TABLE 3.1 First successful satellite launches by country.

Order	Country	Date	Rocket	Satellite
1	Soviet Union	October 4, 1957	Sputnik-PS	Sputnik-1
2	United States	February 1, 1958	Juno I	Explorer 1
3	France	November 26, 1965	Diamant-A	Astérix
4	Japan	February 11, 1970	Lambda-4S	Osumi
5	China	April 24, 1970	Long March 1	Dong Fang Hong I
6	United Kingdom	October 28, 1971	Black Arrow	Prospero
7	India	July 18, 1980	SLV	Rohini D1
8	Isreal	September 19, 1988	Shavit	Ofeg 1
9	Russia	January 21, 1992	Soyuz-U	Kosmos 2175
10	Ukraine	July 13, 1992	Tsyklon-3	Strela
11	Iran	February 2, 2009	Safir-1	Omid
12	North Korea	December 12, 2012	Unha-3	Kwangmyongsong-3, unit 2

their own satellites or series of satellites. This, coupled with the advent of commercial launch vehicles, greatly increases the opportunity to launch a satellite or constellation of satellites. The primary purpose of many of today's satellites is global communications and the internet.

3.9 Progression of satellite images of the Earth (https://en.wikipedia.org/wiki/Satellite_imagery)

Satellites provide a unique view of the Earth that eventually found many applications. Images of the Earth progressed from the 1946 image taken from the V2 rocket presented in Fig. 3.3. The first color image from space was collected on October 5, 1954, from the Aerobee rocket. Early efforts to deploy a scanner on Vanguard in 1959 produced very poor results. The first image from an orbiting satellite was from Explorer 6, and it shows a sunlit portion of the Central Pacific Ocean (Fig. 3.6).

One of the most important applications of satellite imagery was for meteorology and weather forecasting. This led to the United States launching the first of a series of Television Infrared Observing Satellites (TIROS) and the first TIROS-1, which produced the first television image (Fig. 3.7) of the Earth. Interestingly enough, TIROS-1 was launched on April 1, 1960, but it was not an April Fool's Day joke as the satellite (Fig. 3.8) provided the first series of satellite imagery that could be used for meteorological purposes. TIROS-1 was launched on a Thor-Able rocket into a 400-mi circular orbit with an inclination of 50 degrees.

TIROS-1 was a spin-stabilized satellite, as evidenced by the solar panels surrounding the outside of the satellite. The camera was pointing at the Earth from beneath the satellite, and it was surrounded by antennas that would beam the image signal to Earth (Fig. 3.9).

Note that the satellite was equipped with tape recorders to record image data when the satellite was not in communication with a ground station. As the satellite was spin stabilized, it could only point at Earth over a fixed range of

FIGURE 3.6 Explorer 6 image over Mexico.
https://www.bbc.co.uk/newsround/49251998.

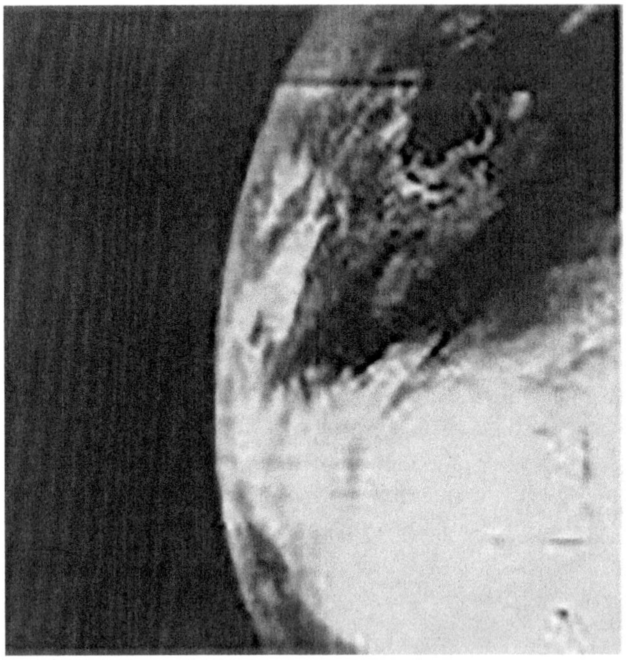

FIGURE 3.7 First television image of the Earth.
TIROS-1; image credit https://airandspace.si.edu/multimedia-gallery/136061mainbmtiros3jpg#:~:text=First%20Television%20Picture%20of%20Earth, National%20Air%20and%20Space%20Museum.

latitude, which was the latitudes of North America. During much of its polar orbit, the TIROS-1 satellite was pointed out at space. Five pairs of solid fuel thrusters on the edge of the baseplate maintained the satellite's spin rate between 8 and 12 rpm; after the launch initial spin of 125 rpm was reduced by a pair of yo-yo weights.

Both wide-angle (104 degrees field of view) and a narrow-angle (12.7 degrees field of view) vidicon cameras collect images that are then recorded on their respective tape recorders and played back when the satellite is in communication with a ground station. The wide-angle covered an area of about 1,200 km when looking straight down, while the latter covered a 120-km area at nadir. Both cameras were fitted with yellow filters to reduce the effects of light scattered by the atmosphere and to improve the detection of clouds. The cameras could be operated sequentially, alternately, or independently by signals from the ground stations. The primary contractor for TIROS-1 was the Radio Corporation of America (RCA) and its Astro division. The significant technical achievement of TIROS-1 was the integration of a simple set of electronics, with electrical and mechanical systems having built-in redundancies and able to fit into a 270-lb satellite. Another major breakthrough for TIROS-1 was the demonstration that television images could be sent to Earth in real time with no need to drop film back to Earth for processing and subsequent analysis. TIROS-1 also demonstrated that cloud patterns clearly reveal weather conditions. During its 89-day lifetime, TIROS-1 transmitted 23,000 TV images, of which forecast meteorologists found 19,000 to be useful in their weather forecast activities.

The other set of instruments carried by TIROS was a trio of radiometers to measure the Earth's infrared emissions. One of these was built by Verner Soumi at the University of Wisconsin, Space Science and Engineering Research Center. Nine days after launch, TIROS-1 discovered its first cyclone north of New Zealand. The first results from TIROS-1 were presented later in April at a joint meeting of the American Meteorological Society and the American Geophysical Union in Washington, DC. TIROS-1 operated until June 15, when a power failure permanently knocked out the cameras. During its 77-day lifetime, TIROS-1 collected a total of 22,952 images, which provided meteorologists a plethora of information to wade through.

Following this success, the 126-kg TIROS-2 was launched in November of the same year on the third flight of the new Delta A launch vehicle. TIROS-2 carried a full complement of instruments, including IR radiometers and magnetic torque rods to stabilize the attitude of the spinning satellite once it was in orbit. TIROS-2 returned 36,154 television images primarily during its first 76 days of operations. It continued to supply IR imagery for a total of 141 days, and it operated in some fashion for a total of 375 days in orbit.

FIGURE 3.8 The TIROS-1 satellite.
https://www.thisdayinaviation.com/tag/tiros-1/.

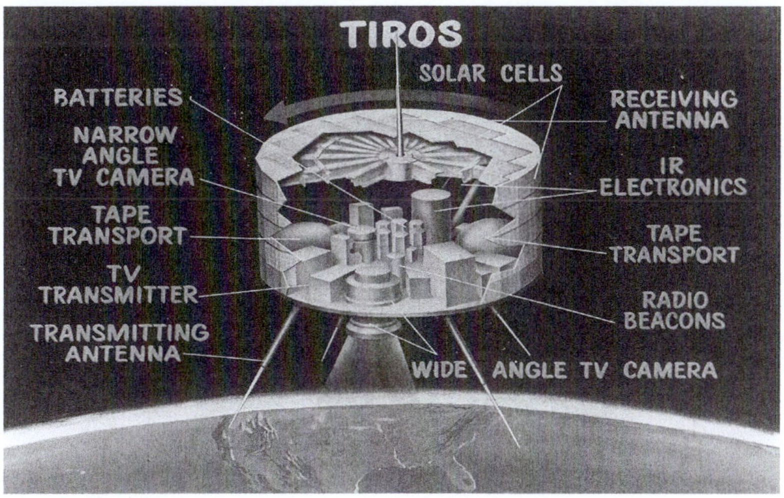

FIGURE 3.9 Cutaway diagram of TIROS-1.
https://nssdc.gsfc.nasa.gov/nmc/spacecraft/display.action?id=1960-002B.

TIROS-3 came next, and it was launched on July 12, 1961, in time for the Atlantic hurricane season, making it possible for the satellite to image tropical storms when they were forming out at sea. To increase the area covered by TIROS-3, the narrow-angle camera was replaced by another wide-angle camera aimed in a different direction than the first wide-angle camera. During its 230-day life, TIROS-3 collected 35,000 images, which allowed meteorologists to issue 70 storm bulletins. This was a first for the weather satellite program. This record continued in a series of TIROS weather satellites.

TIROS-1 (A): launched on April 1, 1960, suffered an electrical system failure on June 15, 1960.

TIROS-2 (B): launched on November 23, 1960, failed on January 22, 1961.

TIROS-3 (C): launched on July 12, 1961, deactivated on February 28, 1962.

TIROS-4 (D): launched on February 8, 1962, failed on June 30, 1962 (both cameras failed earlier).

TIROS-5 (E): launched on June 19, 1962, failed on May 13, 1963.

TIROS-6 (F): launched on September 18, 1962, failed on October 21, 1963.

TIROS-7 (G): launched on June 19, 1963, deactivated on June 3, 1968.

TIROS-8 (H): launched on December 23, 1963, deactivated on July 1, 1967.

TIROS-9 (1): launched on January 22, 1965, failed on February 15, 1967. First TIROS satellite in near-polar orbit.

TIROS-10 (OT-1): launched on July 2, 1965, deactivated on July 1, 1967.

By the time TIROS-10 was launched, it was considered operational as a weather forecast satellite, and it took on the name of the responsible agency, which was the Environmental Sciences Services Administration (ESSA). Thus the subsequent polar orbiting satellites were known as ESSA. ESSA was a predecessor of the National Oceanic and Atmospheric Administration, or NOAA, which is now the prefix used in all of the weather satellite names.

Chapter 4

Early satellite remote sensing and the TIROS program

4.1 Satellite image processing in the early TIROS era

As these were some of the first images of the Earth from orbit, people soon realized that to get information from the images, they would have to geolocate the cloud features viewed in the images. To do so, they would have to compensate for the curvature of the Earth and the rotation of the Earth that took place during the passage of the satellite in polar orbit. The first approach was to distort the latitude–longitude grid to fit the image (Fig. 4.1). The digital computer capability needed to quickly resample the satellite image to fit a selected geographic map was not available till later.

This was a very laborious process, as the analyst had to find the distorted grid that fit the image the best, and the results were always approximate. In addition, this was an analog process that did not yield a digital version of the image. Any specific analysis of the image would require a human image analysis. Fig. 4.2 demonstrates another problem with this type of image analysis. This figure is an image of Lake Erie using a number of satellite passes, each of which has a slightly different exposure. All of these images were processed with traditional photographic printing methods (Fig. 4.3), which resulted in the stripes in Fig. 4.2.

To overcome the sampling limitation of the TIROS satellite, imposed by the spin stabilization of the satellites, the satellite axis was turned by 90 degrees, and the cameras were now pointed out to the circumference of the satellite as the satellite still spun for stabilization. These became known as the "wheel" satellites and took on the designation of ESSA as explained earlier. In this configuration, the satellite would collect images of the entire Earth during successive orbits (Fig. 4.4). As the Earth rotated beneath the wheel satellite, it collected a series of circular images of the Earth, producing the first global image (Fig. 4.5) of the Earth's weather. While it is clear that this is not a truly synoptic view of the weather, due to differences in sampling times, it must have been a real breakthrough to be able to see the global cloud cover of the Earth.

This capability gave rise to the ESSA series of weather satellites. ESSA-1 was an 18-sided polygon, 107 cm in diameter, 56 cm high, and weighed 138 kg. It carried 2 cameras mounted 180 degrees opposite of each other along the axis of the satellite. The satellite maintained its orientation by its spin and using the magnetic torque rods in the satellite. It would view the weather of each area of the globe at the same local time each day. It was launched on February 3, 1966 and was deactivated 861 days later on June 12, 1968.

The next satellite was ESSA-2 of essentially the same design as ESSA-1. It was launched on February 28, 1966 and was operated for 1692 days until deactivated on October 16, 1970. ESSA-3 was launched on October 2, 1966 and was deactivated on December 2, 1968. It moved the launch site from Cape Canaveral, Florida, to Vandenberg Air Force Base in California for all polar orbiting satellites. This eliminated the need for such launches traveling across the continental US, risking potential damage if the launch vehicle failed. Thus all polar orbiter launches were moved to the west coast, shifting the risk of a launch failure to be over the ocean. ESSA-4 was launched on January 26, 1967 and replaced ESSA-2 to provide cloud-cover photography to ground stations using the applied picture transmission (APT) system. This was an analog very high frequency (VHF) broadcast system, which was later used to send down reduced-resolution imagery to global ground stations. ESSA-4 was deactivated on May 5, 1968. Launched April 20, 1967, ESSA-5 replaced ESSA-3 and lasted till February 20, 1970. ESSA-6 replaced ESSA-4 and was launched on November 10, 1967. It increased the image transmission volume to eight images daily per ground station. It was terminated on December 3, 1969. ESSA-7 was launched on August 16, 1968, to replace ESSA-5 and was deactivated on March 10, 1970. Likewise, ESSA-8 was launched on December 15, 1968, to replace ESSA-6, and was deactivated on March 12, 1978. The last in the ESSA series was ESSA-9, and it was launched on February 26, 1969, to replace ESSA-7. It lasted till November 15, 1972.

Satellite Remote Sensing. DOI: https://doi.org/10.1016/B978-0-443-22086-9.00014-9

FIGURE 4.1 Gridded TIROS image of an Atlantic Storm at 17°W, 45°S. (From Mont. Weat. Rev. July 1961, p. 235). https://www.google.com/search?q=Gridded+TIROS+image+of+the+Atlantic&sca_esv=88e799ce3e6bffdb&udm=2&biw=1270&bih=795&ei=-7ULaLP1Ccfn5NoPjNW44AY&ved=0ahUKEwjz3r6My_OMAxXHM1kFHYwqDmwQ4dUDCBI&uact=5&oq=Gridded+TIROS+image+of+the+Atlantic&gs_lp=EgNpbWciI0dyaWRkZWQgVElST1MgaW1hZ2Ugb2YgdGhlIEF0bGFudGljSM5iUKASWJtZcAN4AJABAJgBTKAB2AeqAQIxNbgBA8gBAPgBAZgCAKACAJgDAIgGAZIHAKAHowWyBwC4BwA&sclient=img#vhid=xrx8eh48h4YifM&vssid=mosaic.

FIGURE 4.2 A TIROS image of Lake Erie. https://www.google.com/search?q=first+tiros+image+of+lake+erie&sca_esv=23368486d762886d&ei=cbHmaa_xMJa-0PEPttjvwQ8&biw=970&bih=762&ved=0ahUKEwjv7drIxf2TAxUWHzQIHTbsO_gQ4dUDCBE&uact=5&oq=first+tiros+image+of+lake+erie&gs_lp=Egxnd3Mtd2l6LXNlcnAiHmZpcnN0IHRpcm9zIGltYWdlIG9mIGxha2UgZXJpZTIFECEYoAEyBRAhGKABMgUQIRigATIFECEYqwJIozZQmw5Y0RhwAXgBkAEAmAGKAaABtgaqAQMwLje4AQPIAQD4AQGYAgigAu8GwgIKEAAYRxjWBBiwA8ICCBAAGIkFGKIEwgIIEAAYgAQYogTCAgUQABjvBZgDAIgGAZAGCJIHAzEuN6AH3heyBwMwLje4B-gGwgcDMi04yAckgAgB&sclient=gws-wiz-serp.

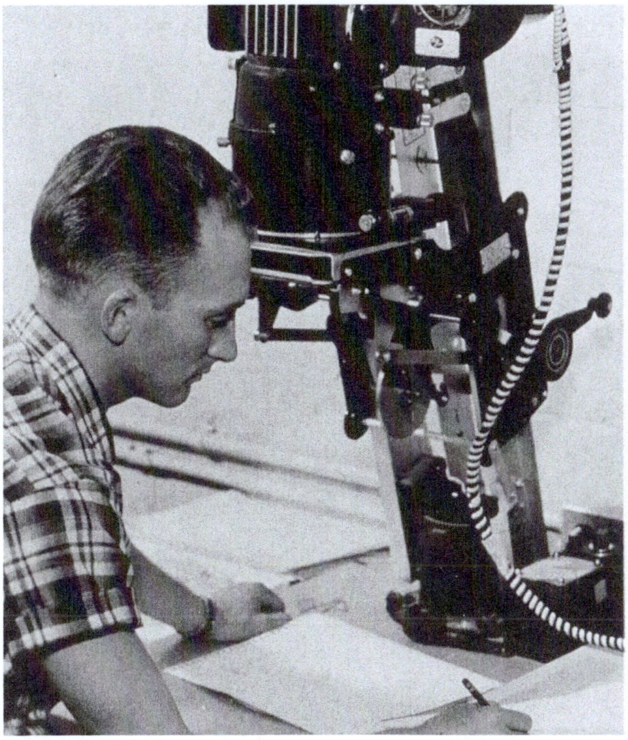

FIGURE 4.3 Image processing in the TIROS satellite era. https://science.nasa.gov/mission/tiros/.

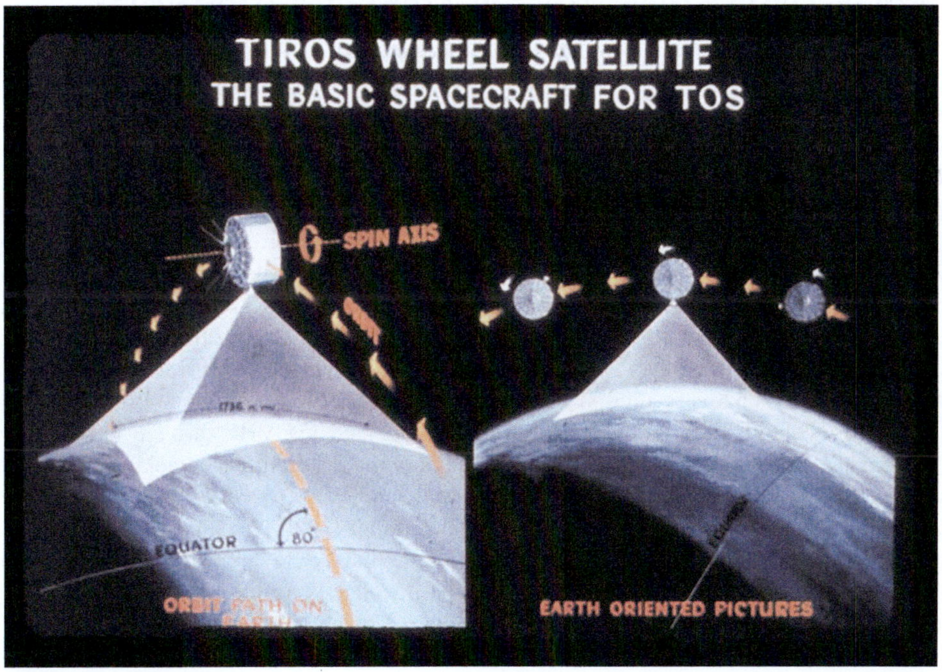

FIGURE 4.4 The TIROS wheel satellite scanning.
Image courtesy of NOAA.

It should be noted that these ESSA satellites operated in pairs, which was the method of deployment for all future polar orbiting weather satellites. It is also important to note that all of these satellites were developed, launched, and later deactivated by NASA, even though the primary user of the information was NOAA. This mode of succession had an interesting tangent, and that was the NASA Nimbus program, which will be treated separately in this book.

FIGURE 4.5 First image of the world's weather.
https://airandspace.si.edu/multimedia-gallery/migratedfirst-photomosaicjpg.

4.2 The improved TIROS satellite (https://airandspace.si.edu/collection-objects/meteorological-satellite-itos/nasm_A19761828000#:~:text=The%20Improved%20TIROS%20Operational%20Satellite,of%20the%20Earth%20weather%20systems)

The next major development in weather satellites took advantage of the three-axis stabilization technology developed for ballistic missile guidance. With three-axis control technology, it was no longer necessary to keep the satellite spinning for stabilization, and the Improved TIROS (ITOS) weather satellite (Fig. 4.6) could keep the observation side of the satellite continuously pointing at the Earth. This side is covered with ports through which the instruments view the Earth. Since the satellite no longer rotates, solar panels do not cover the outside but instead are deployed in "wing-like" structures that can be tiled to best take advantage of the solar angle to the spacecraft.

Another big change that came with ITOS was the introduction of the radiometer. Unlike the video cameras that preceded, them the radiometers collected their imagery one line at a time. Thus the image was built up as the spacecraft traveled along its orbit. The scan was just wide enough in the cross-path direction that they would overlap slightly on subsequent satellite passes. Images were stored on the spacecraft for later playback when the satellite was in contact with a ground station. To provide continuity, these satellites also carried four cameras: two television cameras for APT and two advanced vidicon cameras. It also carried a low-resolution flat-plate radiometer, a solar proton monitor, and two scanning radiometers (SRs). The reduced-resolution analog TV images were downlinked via VHF through the APT system to global ground stations. The higher resolution imagery collected by the radiometers required a tracking antenna to follow the satellite as it passed over a ground station. The radiometers were also able to collect radiation in the visible (0.52–0.73 μm), the near-infrared (NIR) (0.7–1.1 μm), and the thermal infrared (10.5–12.5 μm) bands.

The satellite was equipped with a dynamical attitude control system that could maintain the desired spacecraft orientation to keep the instruments pointing at the Earth as the satellite traveled along its orbit. Adjustments in attitude were made by means of magnetic coils and by varying the speed of the momentum flywheel.

ESSA had transitioned into NOAA, and the series of ITOS satellites were all designated as NOAA 1–8. The practice of NASA developing, launching, and operating the spacecraft was continued with this series. NASA would give these satellites alphabetic identifiers (A–H), and when they were on orbit NOAA switched to numeric identifiers NOAA 1–8. The first of the series was considered ITOS-1 or TIROS-M. After that, it was the series of NOAA 1–8. The first ITOS satellite clearly demonstrated the utility of these new satellite and sensor technologies, which were continued and improved throughout this series.

The payload of the ITOS satellite is shown in Fig. 4.7. The APT cameras continue the data from the earlier TIROS satellites. The other two cameras are the advanced vidicon camera system. There are two SRs and a flat plate radiometer for lower spatial resolution imagery. The higher resolution SR imagery required a complex and expensive tracking

FIGURE 4.6 The Improved TIROS satellite.
https://airandspace.si.edu/collection-objects/meteorological-satellite-itos/nasm_A19761828000.

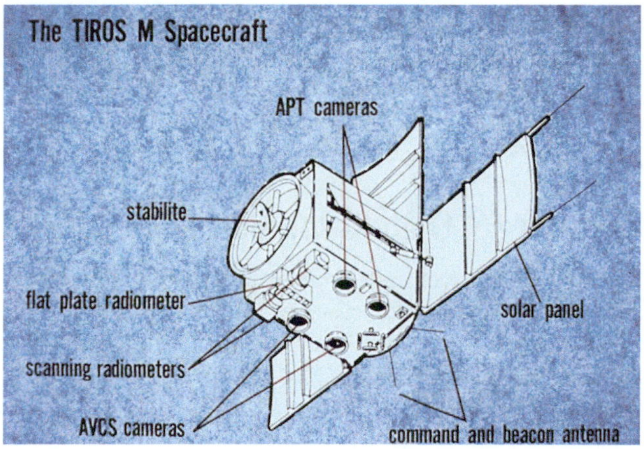

FIGURE 4.7 The Improved TIROS satellite with its sensors and other equipment; note the introduction of solar panels. APT, applied picture transmission; AVCS, advanced vidicon camera system.

antenna system, so the APT was a lower resolution analog image broadcast on a VHF frequency that could be picked up by simple whip or Yagi antennas at weather stations and ships at sea.

The details of the ITOS orbit are shown here in Fig. 4.8 that shows the satellite polar orbit relative to daylight and also exhibits all of the sensor details. The orbital altitude is about 1271 km, and the orbit is sun-synchronous with an 80 degrees inclination in a retrograde orbit with the period of 111 minutes. A big change that occurred in the evolution of the TIROS

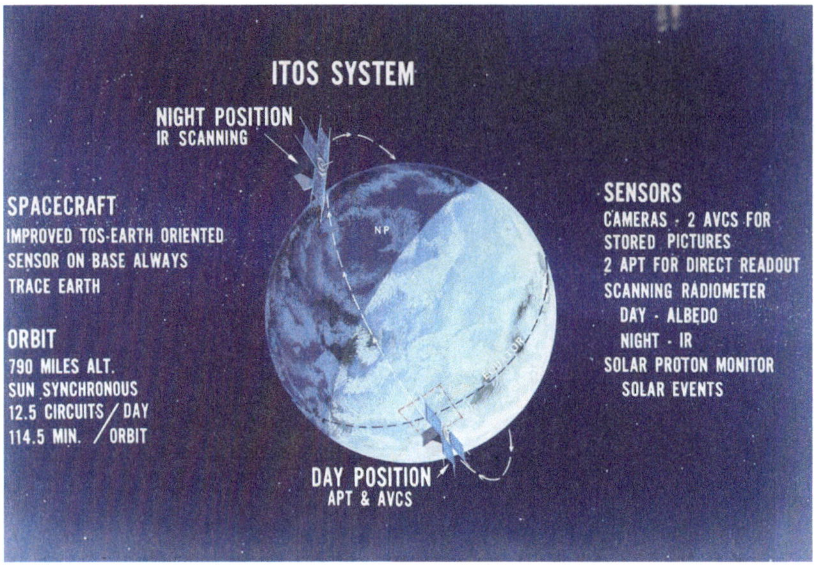

FIGURE 4.8 Improved TIROS (ITOS) satellite on orbit.

satellite was the dramatic change in size and weight of the spacecraft. The original TIROS satellites weighted about 50 kg, which increased to 250 kg with the ESSA wheel satellite. The shift to three-axis stabilized increased the ITOS satellites up to 400 kg, which later increased by a factor of three up to the later TIROS-N satellites.

Modern NOAA and Defense Meteorological Satellite Program (DMSP) satellites weigh about 1500 kg.

The SR analog images of the NOAA (ITOS) satellites were digitized on the ground to enable digital processing of the images to correct for image distortions due to Earth curvature and rotation. The digital images could also be enhanced to bring out certain features of the cloud cover that indicated weather patterns. While these satellites were initially designed to observe cloud patterns, they were also later used to map sea surface temperature (SST) using the thermal infrared channel. A system was developed that used a histogram method to filter out pixels dominated by clouds to produce SST over large 50 km grid cells. This system was found to introduce a lot of errors by letting some cloudy pixels slip through the filter. Areas that were filtered out due to cloud cover were filled in using an objective mapping method (Cressman, 1959).

A second radiometer flown experimentally on the NOAA satellites was the Very High Resolution Radiometer (VHRR), which was the first instrument to demonstrate a real capability for being able to map SST. It had channels in the visible, the near-IR, the mid-range IR, and the thermal infrared (again centered on 11 μm). Using the visible and near-IR channels for cloud clearing, the VHRR data were then used to produce a 1 km spatial resolution SST, which was the native resolution of the radiometer. The image in Fig. 4.9 is from the VHRR, and it shows a warm (dark) intrusion of the "loop current" into the Gulf of Mexico. Note that the image has been geometrically resampled, and Florida appears relatively undistorted.

4.3 TIROS-N and the Advanced Very High Resolution Radiometer

A big change in polar orbiting weather satellites came about in 1978 when TIROS-N was first launched. Again, the TIROS name was carried forward, but N was added to indicate that it was a new TIROS satellite. Like the ITOS that preceded it, TIROS-N was fully three-axis stabilized, and its sensors always pointed at the Earth. Unlike its predecessors, TIROS-N weighed 734 kg (1618 lbs) and carried a large number of sensors (Fig. 4.10):

Argos data collection system

AVCS	advanced vidicon camera system (past heritage instrument)
APT	automatic picture transmission for low resolution imagery
FPR	flat plate radiometer (another heritage instrument)
SPM	solar proton monitor (space weather instrument)
VTPR	vertical temperature profile radiometer
VHRR	Very High Resolution Radiometer (heritage from ITOS)
AVHRR	Advanced Very High Resolution Radiometer (becomes the backbone in the future)
ASS	atmospheric sounding system
SEM	Space Environment Monitor
HIRS	high-resolution infrared sounder

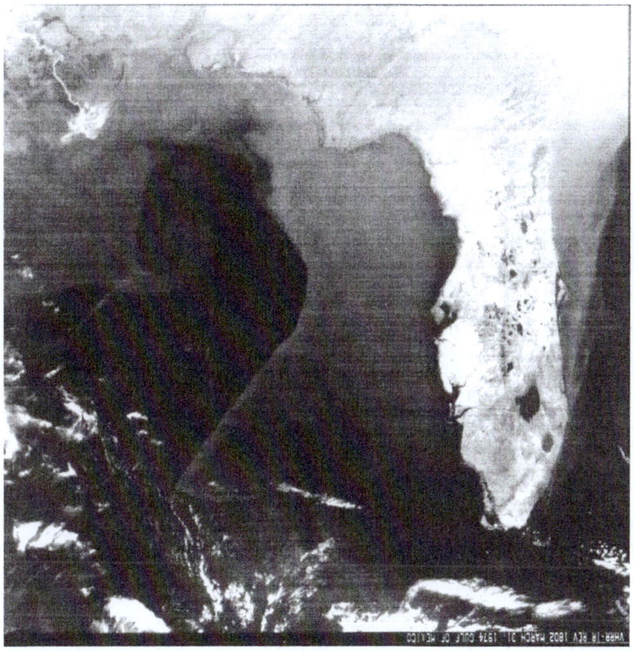

FIGURE 4.9 Very High Resolution Radiometer thermal infrared image of the Gulf of Mexico, March 31, 1974 (NOAA Photo Library). Image courtesy of NOAA.

FIGURE 4.10 TIROS-N.
https://www.google.com/imgres?q=TIROS-N&imgurl=http%3A%2F%2Fwww.astronautix.com%2Fgraphics%2Ft%2Ftiros95.jpg&imgrefur)l=http%3A%2F%2Fwww.astronautix.com%2Fa%2Fadvancedtirosn.html&docid=dI1Yfptd4zRVmM&tbnid=IrOSj0T8t_1CPM&vet=1&w=329&h=194&hcb=2&ved=2ahUKEwiahfOEnpiNAxWxGDQIHY-FLl0Q9AF6BAgZEAA.

MSU	microwave sounding unit
SSU	stratospheric sounding unit
TOVS TIROS	operational vertical sounder (combines other instruments, a combination of the HIRS, the MSU and the SSU. It was never actually the operational sounding instrument.)
TIP	TIROS information processor
MIR	Manipulated Information Rate processor
CSU	Cross-Strap Unit

These were all just in the first satellite, and the payload would be reduced in the future follow-on satellites. The Advanced VHRR (AVHRR) became one of the fundamental instruments with applications in land, ocean, and atmospheric science. It would go through a number of different versions to optimize these applications.

4.4 Operational NOAA polar orbiting weather satellites

NOAA-6 was initially known as NOAA-A before launch. This was the first in an operational series of TIROS-N satellites. The satellite used the same basic satellite bus as the DMSP satellites, which were built by General Electric. It was capable of maintaining an Earth-pointing accuracy of better than ±0.1 degree with a motion rate of less than 0.035 degree/s. It was launched from the Vandenberg Air Force Base in California on June 27, 1979. The primary imager was the AVHRR for day and night cloud cover.

On NOAA-6, the AVHRR was a four-channel SR capable of providing global daytime and nighttime cloud imagery and SST (Fig. 4.11).

In addition, it could collect information about ice, snow, and forest fires. The radiometer measured both reflected and Earth emitted radiation in the following channels: channel 1 (visible) 0.58–0.68 μm, channel 2 (NIR) 0.725–1.1 μm, channel 3 (short wave infrared) 3.55–3.93 μm, and channel 4 (thermal infrared) 10.5–11.5 μm. All four channels had a spatial resolution of 1.1 km at nadir, and the two infrared channels had a thermal resolution of 0.12 K at 300 K. The AVHRR broadcasts its data in real time as high resolution picture transmission (HRPT) and also had a limited onboard recording capability known as local area coverage, which was scheduled in advance of the acquisition. In addition, a subsampled version of the imagery (~4 km resolution) known as global area coverage (GAC) was recorded for playback at NOAA ground stations.

The TIROS operational vertical sounder (TOVS) carried by these satellites was a real misnomer, as this was never really an operational system. Moreover, it was not a single instrument but rather the combination of other instruments on the satellite. The primary instrument was the high-resolution infrared radiation sounder (HIRS), which has been modified after operation on the initial TIROS-N satellite. Thus it was known as HIRS/2. This was coupled with the stratospheric sounding unit (SSU), which was an instrument supplied by the British Meteorological Office for inclusion in the TOVS package. It measured infrared radiance in the 15 μm CO_2 band. Each SSU scan line had 8 field of views (FOVs) in the 1500 km wide scan line. The final instrument that was merged into TOVS was the microwave sounding unit (MSU), which was designed with rather large microwave vertical weighting functions to measure the lower part of the atmospheric temperature and water vapor profiles.

The HIRS/2 instrument had 20 channels: channels 1 through 5 in the 15 μm CO_2 band (15.0, 14.7, 14.5, and 14.0 μm); channels 6 and 7 in the 13.7 and 13.4 μm CO_2/H_2O bands; channel 8 in the 11.1 μm band, channel 9 in the 9.7 μm ozone band; channels 10, 11, and 12 in the 6 μm water vapor bands (8.3, 7.3, and 6.7 μm); channels 13 and 14 in the 4.57 and 4.52 μm N_2O bands; channels

FIGURE 4.11 Hurricane Allen in the Yucatan Channel.
https://en.wikipedia.org/wiki/Hurricane_Allen.

15 and 16 in the 4.46 and 4.40 μm CO_2/N_2O bands; channel 17 in the 4.24 μm CO_2 band; channels 18 and 19 in the 4.0 and 3.7 μm windows; and channel 20 in the 0.70 μm visible region. The HIRS/2 had a FOV of 30 km in diameter at nadir and sampled 56 FOVs in each scan line over the width of 2250 km. The MSU had one channel in the 50.31 GHz window region and three channels in the 55.0 GHz oxygen band (53.73, 54.96, and 57.95 GHz) and was designed to retrieve temperature profiles that were free of cloud interference. The MSU had an FOV of 110 km in diameter and sampled 11 FOVs along its 2250 swath width.

One system that was unique to these NOAA polar orbiting satellites was the data collection and platform location system, which had been developed on the ITOS series of NOAA satellites. In the TIROS-N version, the system was known as Argos and was designed and built in France to meet the meteorological data collection needs of the US and to support the Global Atmospheric Research Program (GARP). This system received low duty cycle transmissions from free-floating balloons, ocean buoys, moving animals, and remote ground stations. Using the radio signal Doppler shift, the system was able to locate the transmitter with an accuracy between 3 and 5 km and a velocity accuracy between 1.0 and 1.6 m/s. These data were recorded onboard the satellite and downlinked when the satellite was in view of a NOAA ground station. The system was capable of collecting data from up to 4000 platforms per day.

NOAA-6 also began a series of NOAA space weather observations with the Space Environment Monitor. This was an extension of the solar proton monitoring experiment flown on the ITOS satellite series. The goal was to measure the proton flux, electron flux density, and energy spectrum in the upper atmosphere. The package consisted of three detector systems and a data processing unit. The Medium Energy Proton and Electron Detector measured protons in five energy ranges from 30 keV to >2.5 MeV; electrons above 30, 100, and 300 keV, protons and electrons above 6 MeV, and omnidirectional protons above 16, 36, and 80 MeV. The High-Energy Proton Alpha Telescope had a 48 degrees view cone and looked outward from Earth to measure protons in four energy ranges above 370 MeV, and alpha particles in two energy ranges above 850 MeV. The Total Energy Detector measured electrons and protons between 300 eV and 20 keV.

4.5 Advanced Very High Resolution Radiometer evolution and applications

The AVHRR (Fig. 4.12) was the imager on the TIROS-N series of spacecraft, and it was used for a great many applications in geoscience.

It also evolved over its years of operation in response to improving these applications. It evolved from the four-channel radiometer first deployed on TIROS-N to a five-channel radiometer when the thermal infrared channel was split into two adjacent windows to improve the retrieval of SST. NOAA-6, 8, and 10 satellites carried the four-channel AVHRR instruments, while NOAA-7, 9, 11, 12, and 14 used the 5-channel AVHRR (Table 4.1).

The series of satellites began with TIROS-N in 1978 and continued up to the launch of NOAA-14 in 1994 (Table 4.2).

NOAA-B was launched on May 29, 1980 but failed to achieve orbit. NOAA-13 launched on August 9, 1993 and failed on orbit due to an electrical short in the solar array.

A graphical presentation of this information gives a good idea of the length of time that this TIROS-N series of satellites operated and how it transitioned into the European MetOp series (Fig. 4.13).

FIGURE 4.12 The Advanced Very High Resolution Radiometer (AVHRR). https://www.google.com/search?q=picture+of+The+Advanced+Very+High+Resolution+Radiometer+%28AVHRR%29&sca_esv=9bb75c165dc77b23&ei=J-8eaMWAN_W00PEP0az3wAQ&ved=0ahUKEwiFsOqloJiNAxV1GjQIHVHWHUgQ4dUDCBA&uact=5&oq=picture+of+The+Advanced+Very+High+Resolution+Radiometer+%28AVHRR%29&gs_lp=Egxnd3Mtd2l6LXNlcnAiP3BpY3R1cmUgb2YgVGhlIEFkdmFuY2VkIFZlcnkgSGlnaCBSZXNvbHV0aW9uIFJhZGlvbWV0ZXIgKEFWSFJSKUiLNFDXDFjQKnABeACQAQCYAeoBoAGsD6oBBjAuMTEuMbgBA8gBAPgBAZgCCqACygvCAgoQABiwAxjWBBhHwgIIEAAYgAQYogTCAgUQABjvBcICChAhGKABGMMEGArCAggQIRigARjDBMICBBAhGAqYAwCIBgGQBgiSBwMxLjmgB4wqsgcDMC45uAe-Cw&sclient=gws-wiz-serp#vhid=iE3XPUsxwpjX2M&vssid=_bO8eaNe1LZjh0PEPieyYUQ_77.

TABLE 4.1 Advanced Very High Resolution Radiometer (AVHRR) channel distributions and satellites equipped with them.

Band	NOAA-6, 8, 10	NOAA-7, 9, 11, 12, 14	IFOV
1	0.58–0.68	0.58–0.68	1.39
2	0.725–1.10	0.725–1.10	1.41
3	3.55–3.93	3.55–3.93	1.51
4	10.50–11.50	10.3–11.3	1.41
5	Band 4 repeat	11.5–12.5	1.30
	μm	μm	Milliradians

TABLE 4.2 The series of TIROS-N satellites.

Satellites	Launch	Dates of operation
TIROS-N	10/13/78	10/19/78–01/30/80
NOAA-6	06/27/79	06/27/79–11/16/86
NOAA-7	06/23/81	08/24/81–06/07/86
NOAA-8	03/28/83	05/03/83–10/31/85
NOAA-9	12/12/84	02/25/85–02/13/98
NOAA-10	09/17/86	11/17/86–08/30/2001
NOAA 11	09/24/88	11/08/88–09/13/94
NOAA-12	05/14/91	05/14/91–08/10/2007
NOAA-14	12/30/94	12/30/94–05/28/2007

This series of satellites carrying the AVHRR produced a long continuous record that was used for a great many applications. This abundance of useful data prompted NOAA to create the Comprehensive Large Array-Data Stewardship System (CLASS), which initially held all of the AVHRR data for free online access to interested users. Today, CLASS holds a wide variety of satellite data and makes it all available to users online at no cost. CLASS has a user-friendly search interface so that users can locate the data they are interested in and determine if it merits being downloaded.

4.6 Cloud mapping and monitoring

The imager on the NOAA satellites always had the responsibility for imaging cloud cover to detect weather patterns and their evolution over time. Cloud properties derived from the AVHRR are used to generate the NASA climate data record (CDR) that identifies global cloud cover and cloud macro and microphysical properties (Fig. 4.14). This CDR algorithm also addresses AVHRR sensor noise, erroneous image geolocation, and AVHRR shortwave channel calibration. Daytime and nighttime cloud properties are available, which represents an advantage over many other cloud climatology which are daytime only. Most of the CDR products have been validated using NASA A-Train (to be introduced later) observations, which have not been used very often. A highly evolved version of the AVHRR known as AVHRR/3 flies on the European MetOp-C satellite. Unlike its predecessors, AVHRR/3 has six channels. More will be said about this in the section of this book that treats European polar orbiting weather satellites.

A study by Tag et al. (2000) introduces a multiple cloud-type classification package based on AVHRR data. It is a neural network cloud classifier valid over the world's maritime regions. To extend this classifier to land regions, they created a database of nearly 8000 16X16 pixel cloud samples (from 13 Northern Hemisphere land regions) independently classified by three experts. This database provided training and testing samples for developing a cloud classifier valid over land. Approximately 200 features, calculated from visible and one infrared channel, form the basis for the computer

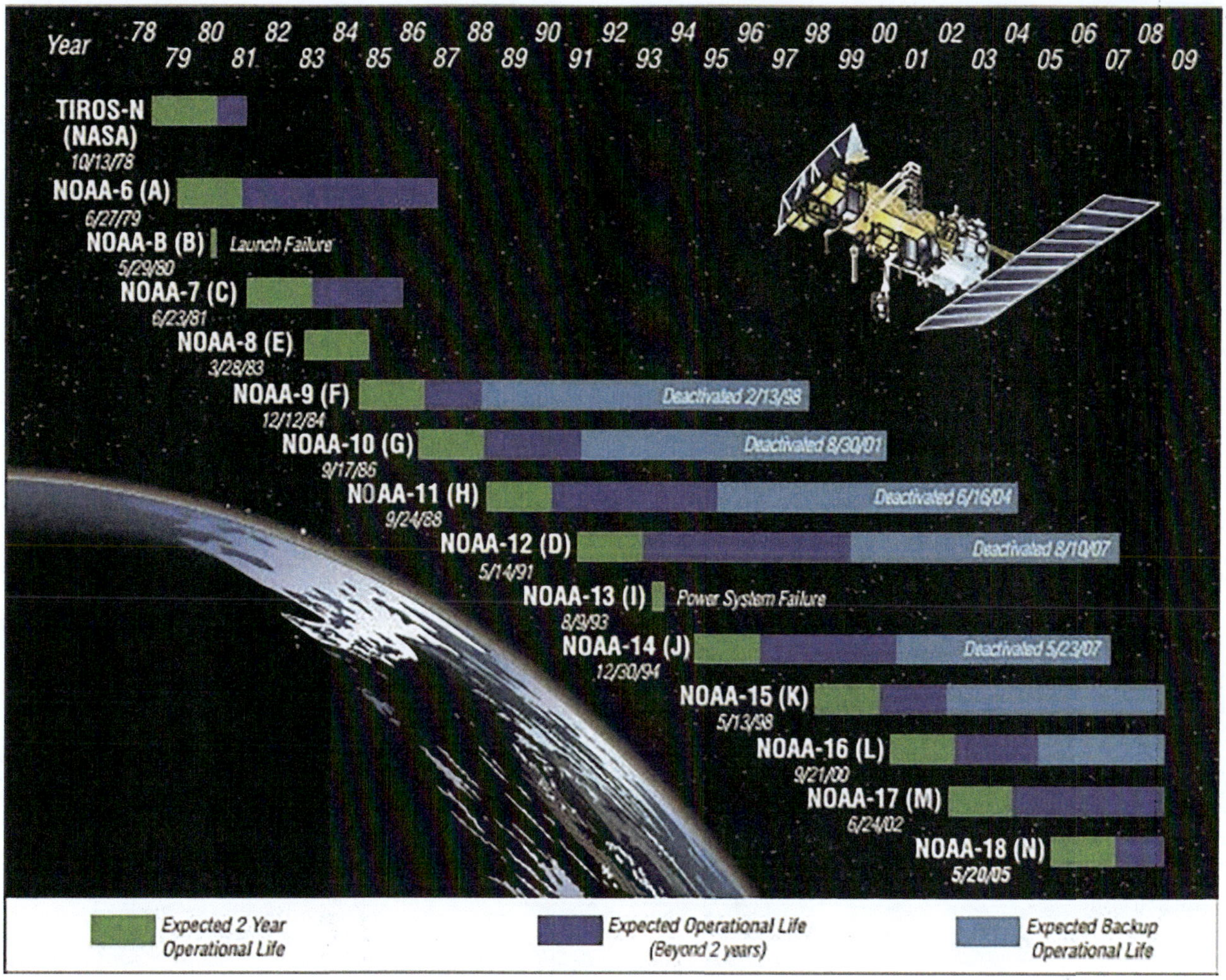

FIGURES 4.13 Time series of NOAA operational polar orbiting (POES) satellites. The green and blue color codes indicate midmorning or afternoon equatorial passes.
https://www.nesdis.noaa.gov/our-satellites/related-information/history-of-noaa-satellites/poes-history.

network analysis. They used a nearest neighbor classifier, meshed with a feature selection method using backward sequential selection. From this approach, the authors selected the fewest features that maximized the classification accuracy. Overall, accuracies ranged from 86% to 78% for the water and land classifiers. An example of their classifier is given here in Fig. 4.15.

4.7 Atmospheric corrections

One of the primary concerns for Earth studies using AVHRR data was the ability to correct for atmospheric corruption of the signal reaching the satellite from the Earth. This was different depending on the target background, with ocean applications being very different from land applications. A study by Tanre et al. (1992) addressed the correction for the visible and NIR bands of the AVHRR used for studies of vegetated surfaces. They showed that invariant effects (ozone absorption and molecular scattering) and variant effects (water vapor absorption and aerosol scattering) dominate the atmospheric effects on the transmission of the AVHRR solar band transmissions. Since these components depend on view angle (VA) and illumination geometry, their effects on atmospheric transmission are highly variable and at time dependent.

Atmospheric water vapor is the major contributor to attenuation of the thermal infrared signal from the Earth's surface and affects land and SST observations the most, reducing the IR transmission by as much as 10%–30%. This range depends upon the amount of atmospheric moisture and the transmission direction of the emitted radiation. Ozone

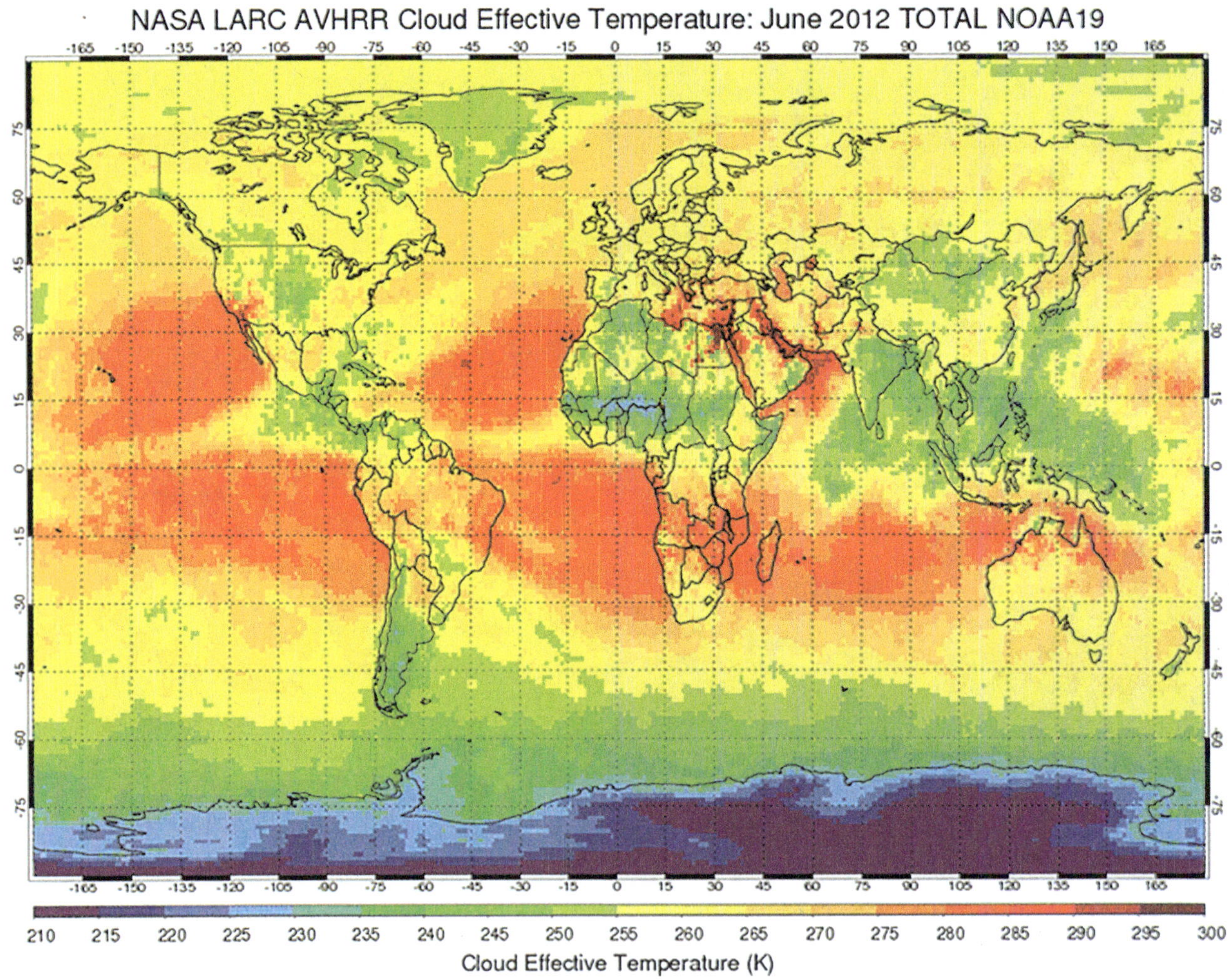

FIGURE 4.14 NOAA Advanced Very High Resolution Radiometer (AVHRR) cloud properties climate data record (CDR). Map lines delineate study areas and do not necessarily depict accepted national boundaries.
https://www.ncei.noaa.gov/products/climate-data-records/avhrr-cloud-properties-nasa.

absorption can reduce the apparent reflectance in AVJHRR channel 1 by 5%–15%. Molecular scattering can increase the apparent channel 1 reflectance from 0.02 to 0.07. In the red-centered channel 1, the effect of moderate haze in the atmosphere is of the same magnitude as the reflectance of vegetated surfaces. Dense haze, dust storms, or smoke can completely overwhelm the reflectance of the surface. Both molecular scattering and ozone can easily be corrected for when using AVHRR to study land surface reflectance. When computing SST from thermal infrared emissions, atmospheric water vapor should be compensated for. The effect of aerosols in the atmosphere is highly variable as a function of time and location, and while its correction should be made when using visible and NIR AVHRR channels, it often requires the use of rather complex models.

4.8 Sea surface temperature

The primary application of the AVHRR to the ocean was the computation of SST from the thermal infrared channels on the AVHRR. This was preceded by similar calculations of SST using the earlier SR and the VHRR. There have been a great many studies on the computation of SST from AVHRR data. Minnett et al. (2019) provide a review of half century of satellite remote sensing of SST. They define that it became possible to correct for the atmospheric water vapor attenuation of the infrared signal when radiometers with multiple infrared AVHRR channels were deployed in 1979. About the same time, microwave radiometers with SST sensing capabilities were also first launched. Since that time, SST

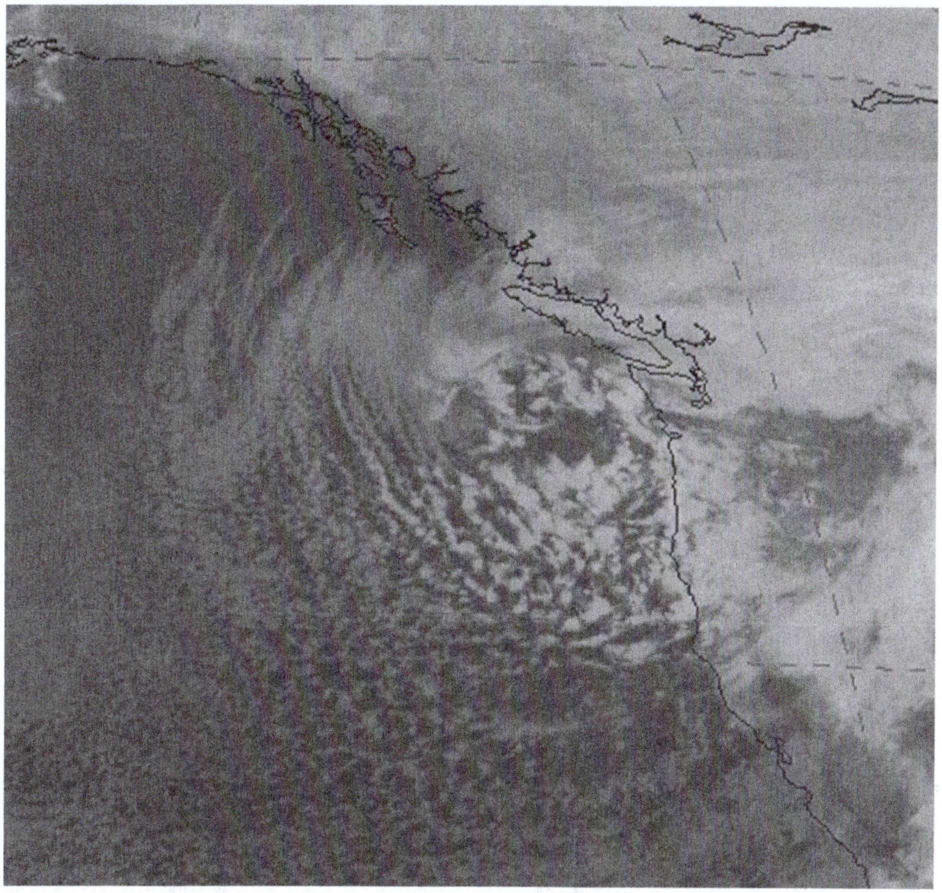

FIGURE 4.15 Wintertime frontal band approaching the US West Coast in the visible image (Warner et al, 2012).

has been derived from combinations of infrared and microwave radiometers on polar orbiting spacecraft and from infrared radiometers on geostationary spacecraft.

Reynolds et al. (2007) introduced a blended SST analysis that produced daily high-resolution SST fields. They use optimum interpolation to merge and map AVHRR imagery with microwave data from the advanced microwave SR with SST observations from ships and buoys. To do so, adjustments were made to compensate for the fundamental differences in these SST measurements. These differences can best be described using a formal definition created by the Group for High-Resolution Sea-Surface Temperature to better separate the various types of SST (Fig. 4.16).

Here, five different SSTs are defined: (1) The interface SST (SST_{int}) which is the temperature of the surface that interacts with the atmosphere on a molecular scale; (2) the skin SST (SST_{skin}) which is the temperature measured by an infrared radiometer typically operating at wavelength 3.7–12 μm and represents the temperature within the conductive diffusion-dominated sublayer at a depth of 10–20 μm; (3) the subskin temperature ($SST_{subskin}$) which is the temperature at the base of the conductive laminar sublayer of the ocean (which for practical purposes can be approximated by the measurement of SST by a microwave radiometer operating over the 6–11 GHz frequency range); (4) the surface temperature at depth (SST_2 or SST_{depth}) which covers all water temperature measurements beneath the $SST_{subskin}$; and (5) the foundation temperature (SST_{fnd}) which is the temperature that is not exposed to diurnal variability, it is defined as the temperature at the first daylight when the heat gain from solar radiation first exceeds the heat loss at the sea surface.

It should be acknowledged that historically SST was considered a measure of upper layer temperature, and now new sensing systems have introduced this variety of definitions of SST. A person new to this field will be confused by the many different definitions, but they represent potential new research areas for people interested in SST and its space/time variations. As this text is focused on satellite remote sensing of the Earth, we will restrict our discussion to those SST definitions most appropriate for satellite sensing of SST. As discussed by Minnett et al. (2019), the upper layer of the ocean is nearly everywhere/always warmer than the overlying atmosphere, and there is, in general, a heat flow from the ocean to the atmosphere. This upwards flow of heat is accomplished in the top hundred micrometers or so of the water (Wong and Minnett, 2018) by molecular conduction as the air–sea

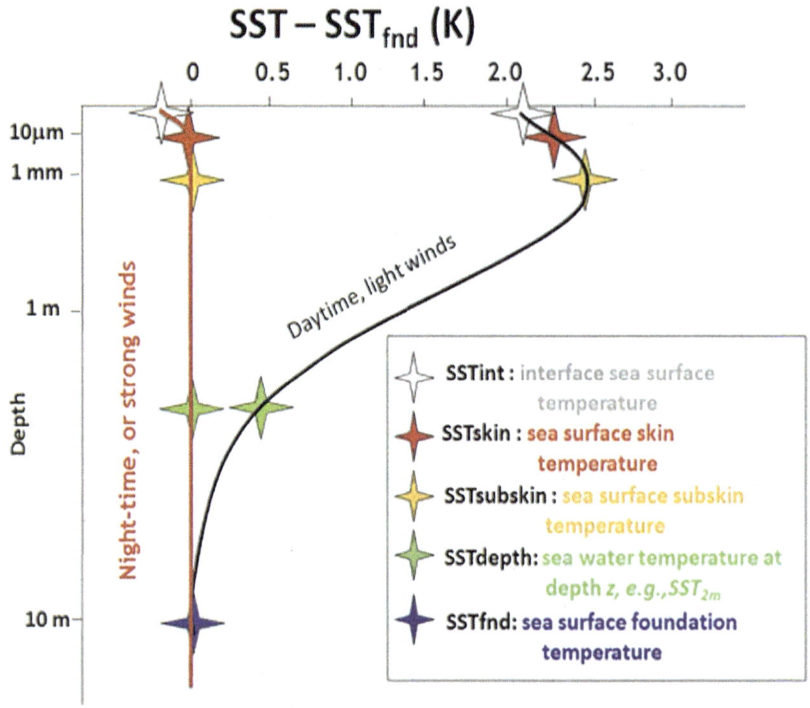

FIGURE 4.16 Group for High-Resolution Sea-Surface Temperature definitions sea surface temperature (SST). Minnett et al. (2019).

density difference at the skin layer suppresses turbulent heat transfer (Soloviev and Lukas, 2014). The skin layer is, therefore, the interface between a turbulent ocean and a turbulent atmosphere. As a consequence, the skin temperature is always colder than the temperature just below it. Seasonal changes in the upper layer temperature profile may make the skin SST appear warmer, but that is only because the entire upper layer has warmed up.

The first satellite to carry an infrared radiometer was NASA's Nimbus-1 satellite, which carried the high resolution infrared radiometer (HRIR). This instrument had a single IR channel in the 3.4–4.2 μm wavelength range (Minnett et al., 2019). It had a spatial resolution of 8.6 km at the nadir point with a swath that extended to the horizon. Since this wavelength band responds to both infrared emitted radiation and solar reflected radiation, it could only collect SST observations at night when there was no solar radiation present. A study by Allison and Kennedy (1967) focused on SST variations under clear sky conditions. Nimbus-1 only lasted a month, but Nimbus-2, launched in May 1966, operated for 6 months. The data were transmitted in analog form and were displayed as photo-facsimiles in gray tones or were printed on a line printer and contoured by hand. These data were used to identify boundaries between currents or to delineate fronts. LaViolette and Chabot (1968) studied the response of SST to changes in wind forcing in the Arabian Sea, while Szeieidlda et al. (1972) did a similar analysis in the Persian Gulf. A study of coastal upwelling was carried out by Warnecke et al. (1971). A third HRIR, which included a visible channel, was launched in 1969 on Nimbus-3. Nimbus was used as a testing facility for new instruments.

Following these early effort a two-channel radiometer called the SR was developed from the HRIR. Now the infrared channel was located in the 10.5–12.5 μm atmospheric window (Fig. 4.17) and the instrument could collect SST measurements both night and day.

As stated earlier, this SR flew on the NOAA TIROS-M launched in Jan of 1970 into a sun-synchronous orbit with an altitude of 1470 km. The spatial resolution was 3.6 km at nadir. The TIROS-N satellite was followed by five operational satellites, NOAA 1-5, launched between Dec 1970 and July 1976 (https://earthobservatory.nasa.gov/features/RemoteSensing/remote_04.php).

The SR led to the development of the AVHRR that first flew on NOAA-2 launched in Oct 1972. The VHRR had channels similar to the SR but with a spatial resolution of 0.9 km at nadir and a much reduced (noise equivalent temperature change) NEΔT of ~0.5 K. The VHRR provided a much improved observation of SST in cloud-free areas. Another improvement was the introduction of onboard satellite recording of the VHRR data, which allowed data to be collected in areas out of range of the ground stations. A great many investigations into ocean thermal fronts were carried out with VHRR data (Legeckis, 1978, Fig. 4.18). This figure shows the thermal structure of the Gulf Stream off Cape Cod, USA, collected by the VHRR on NOAA-4 in November 1975 (Legeckis, 1978).

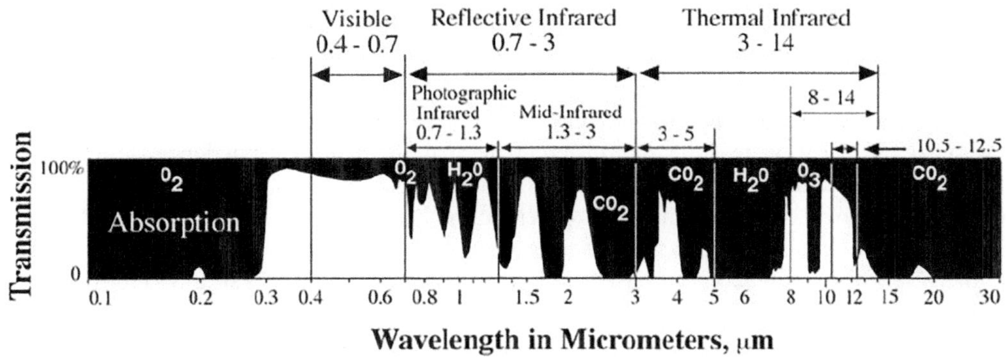

FIGURE 4.17 Transmission in the atmosphere showing the gases that cause the windows (https://www.researchgate.net/figure/Atmospheric-windows-in-the-electromagnetic-spectrum_fig1_267397779).
Image courtesy NASA.

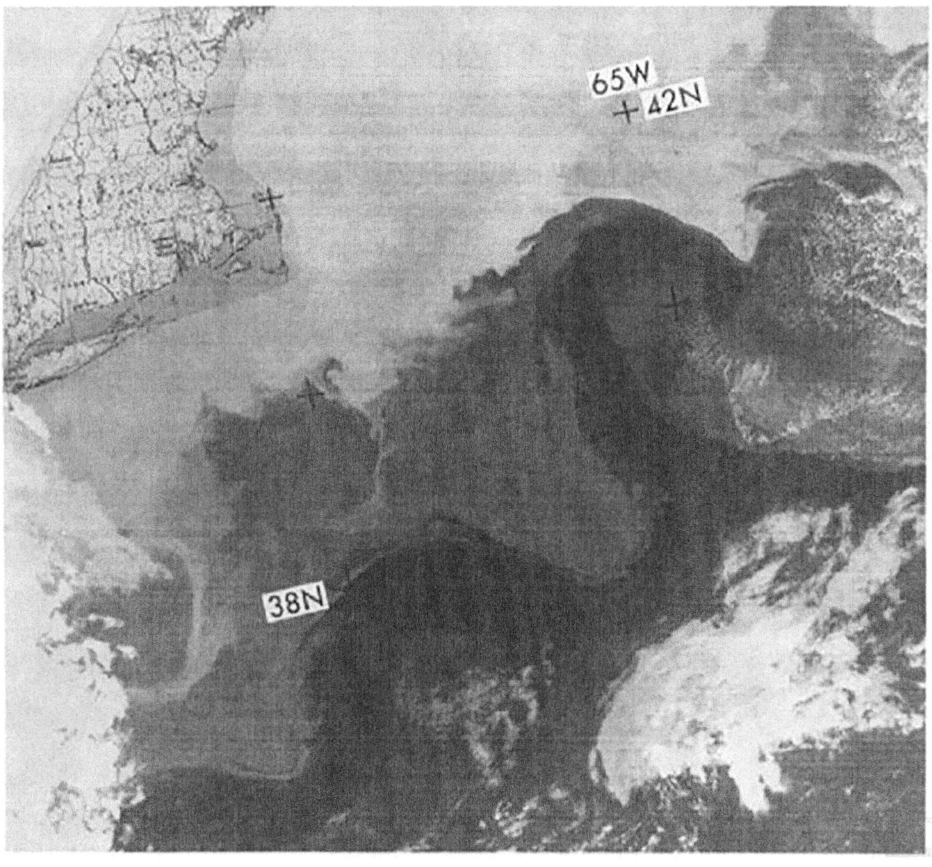

FIGURE 4.18 VHRR image of the Gulf Stream off Cape Cod November, 19755 (Legeckis, 1978).

This image revealed the complexity of the Gulf Stream and explained why it had been so difficult to capture the fronts associated with the Gulf Stream by ship-based surveys that could not capture the mesoscale frontal structure properly (Fuglister, 1955).

The instrument that followed the VHRR to observe SST was the AVHRR that flew on the NOAA polar orbiting satellites, as discussed earlier. The first flight of the AVHRR was on TIROS-N that was launched in late 1978 (Cracknell, 1997). In terms of its SST measurement capabilities, the AVHRR evolved over time. The AVHRR on TIROS-N and NOAA-6 carried an AVHRR with only four channels, which then lacked the "split-window" (later channels 4 and 5, which were 10.8 and 12.0 μm channels respectively) to "correct" SST for the influence of atmospheric moisture. Notice the difference in atmospheric transmission between these two wavelengths in Fig. 4.17. The SST retrievals from the earlier AVHRRs with only four channels were carried out using a linear relationship to the "coincident" in situ SST measurements and were uncorrected for atmospheric variability. This introduced a lot of scatter into this relationship (Barton, 1985). It was suggested early on that the two different thermal IR channels experience different moisture attenuation and that this difference could be used to correct the SST for atmospheric water vapor.

Early infrared satellite measurements of SST really only observed the skin SST (Fig. 4.16). In the early studies of satellite infrared SST, the only available ground truth measurements of SST that could be used for calibration/validation of the satellite SSTs were ship and land-based SSTs, which do not sense the skin SST observed by the satellite. Even after these validation measurements switched to drifting buoys, the buoy SST does not measure the skin SST. Still, in the absence of in situ measurements of skin SST, the community had to accept in situ measurements of the deeper drifting buoy SSTs. Thus there was an accepted offset that came with this type of SST calibration/validation (McClain, 1980).

The attention of the SST research community turned to doing the best possible atmospheric corrections as a way to reduce this variation between the satellite and in situ SST estimates. The first approach, as mentioned earlier, was to use the two thermal infrared channels of the AVHRR to infer the atmospheric moisture content from their differential water vapor absorption profiles (McClain et al., 1985). This approach can be mathematically expressed as follows:

$$\mathrm{SST} = \mathrm{aT_i} + \gamma(T_i - T_j) + c \tag{4.1}$$

where T_i and T_j are the brightness temperatures (BTs) measured in the two thermal infrared channels of the AVHRR and c is an offset. The term γ is the differential absorption coefficient of the equation (McMillin, 1975), and this equation is called the multichannel SST (MCSST). The coefficients depend on the state of the atmosphere at the time of observation, particularly the atmospheric water vapor content. These coefficients were generally determined by regression with a set of nearly coincident in situ SST measurements (called matchups). An alternative approach that was later used was to perform radiative transfer simulations of the atmosphere, which would more closely correspond to the satellite-observed skin SST. The statistical approach required a large number of matchups to be able to sample the global atmospheric variability (McClain et al., 1985). It was important in selecting these matchups to try to reduce the spatial variability of the SST field, and therefore, matchups were generally limited to within 10 km of the satellite pixel and 1 hour of data collection. The radiative transfer approach used a representative collection of atmospheric radiosonde profiles to simulate atmospheric conditions over the satellite observation.

Later improvements on the MCSST included the effects of varying path length for satellite data collection. To do this, a (sec(θ)-1) term was included in the calculation, where θ is the satellite zenith angle measured at the sea surface.

$$\mathrm{SST} = a_0 + a_1 T_{11} + a_2(T_{11} - T_{12})T_{\mathrm{sfc}} + a_3(\sec(\theta) - 1)(T_{11} - T_{12}) \tag{4.2}$$

Here, the T_n are the BTs measured at the n μm wavelength channels, and T_{sfc} is a "first guess" SST often taken as the simple MCSST in (4.1). The a_n are constants that are found using the matchups or by radiative transfer. This formulation was first called the "nonlinear" SST (Walton et al., 1998, Walton, 2016).

This nonlinear equation could be used both day and night as the 11 and 12 μm channels are not affected by sunlight. A slightly different formulation was developed for night SST computations, which used the shorter 3.5 μm IR channel.

$$\mathrm{SST_{triple}} = a_0 + a_1 T_{11} + a_2(T_{3.7} - T_{12})T_{\mathrm{sfc}} + a_3 \tag{4.3}$$

where in 1.2 a_n are constants and T_{sfc} is a first guess SST usually taken as the MCSST. This "triple" equation has the advantage of using all the thermal infrared information contained in an AVHRR sampling. Again, coefficients are found with statistical matchups with in situ SST measurements or using radiative transfer simulations employing a suite of global radiosonde profiles.

As discussed by Minnett et al. (2019), the statistical approach generally does well when the atmospheric conditions are close to the average conditions, such as occur in the midlatitudes. In areas of dramatically different atmospheric moisture content, such as the wet tropics or the dry high latitudes, the statistical approach does not do as well.

A very different approach to this atmospheric water vapor correction was taken by the Along-Track SR (ATSR) built by Rutherford Appleton Labs in the United Kingdom for the European Space Agency (ESA). The ATSR takes nearly simultaneous measurements of SST through different path lengths using a conical scanner which samples a forward view at 55 degrees to nadir, along with the usual nadir view (Fig. 4.19). These samples are just minutes apart but view the SST through very different atmospheric paths. The difference in these paths can be used to derive an atmospheric correction for the individual SST measurement. The ATSR had channels at 1.6, 3.7, 10.8, and 12.0 μm that are spatially collocated and

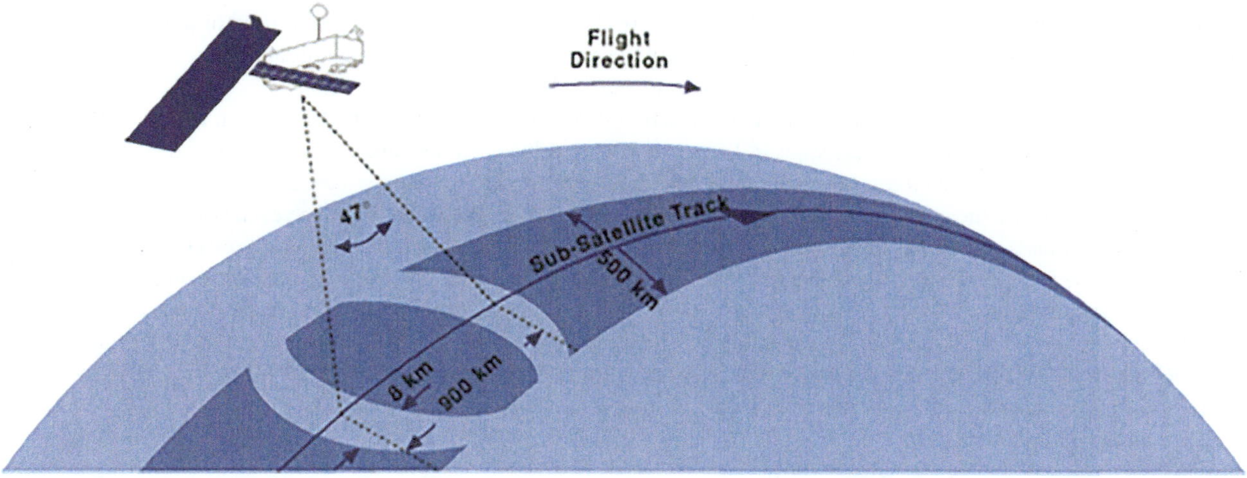

FIGURE 4.19 AATSR scan geometry (https://earth.esa.int/eogateway/instruments/aatsr/description).

experience the same differential atmospheric path lengths. In a later version of this instrument, known as the Sea and Land Surface Temperature Radiometer, the conical scan was shifted so that the nadir surface sample came first (Donlon et al., 2012).

The differences in the BTs between the forward and nadir samples differ by the effects of the atmosphere. Hence the atmospheric correction takes the form:

$$SST = c_0 \sum_i c_{n,i} T_{n,i} + \sum_i c_{f,i} T_{f,i} \tag{4.4}$$

where the n and f indicate the nadir and forward scans, respectively. Here, i indicates the atmospheric window channel, and c are coefficients (Zavody et al., 1995). The practice for ATSR SST estimates was to use the radiative transfer approach as outlined in this chapter. These coefficients vary with latitude in the radiative transfer solution.

Later, an optimal estimation approach was developed for the dual-view radiometers (Embury and Merchant, 2012; Embury et al., 2012), which has become the basis of a 20-year time series of SST measurements from the ATSR intended to be of climate quality (Merchant et al., 2012).

We will return to further developments in SST remote sensing in later chapters of this book.

4.9 The normalized difference vegetation index (https://en.wikipedia.org/wiki/File:Par_action_spectrum.gif)

One of the earliest efforts to monitor vegetation from space used the multispectral scanner on the early Landsat satellites. A first experiment was to study the spring vegetation green-up and subsequent summer and fall dry-down over the area known as the Great Plains region of the central United States. This area covered from the southern tip of Texas to the United States–Canada border, which resulted in a wide range of zenith angles for the satellite observations.

The investigators for this study were PhD student Donald Deering and his advisor, Dr. Robert Hass. They found that it was difficult to map the biophysical characteristics of the rangeland vegetation because of the strong differences introduced by these differences in zenith angle introduced by the significant changes in latitude. Working with a colleague in mathematics (Dr. John Schell), they found that a ratio of the difference between the red and infrared radiances (note in terms of the channels on the AVHRR, this is really the NIR) divided by their sum, was able to compensate (normalize) for the contribution of the zenith angle variations. They called this ratio the "vegetation index." Other researchers were identifying the simple red/infrared ratio as the "vegetation index," so that the Great Plains researchers decided to call their index the normalized difference vegetation index or NDVI.

The earliest recorded use of the NDVI was in 1973 by Dr. John Rouse, then the Director of the Remote Sensing Center of Texas A&M University, which then housed the Great Plains study. They were, however, preceded in publishing on the NDVI, which first appeared in a paper by Kriegler et al. in 1969.

Soon thereafter, Dr. Compton (Jim) Tucker of NASA's Goddard Space Flight Center produced a series of early scientific journal articles describing uses of the NDVI. It quickly became one of the most successful of various attempts to monitor vegetation from space. The NDVI can be tho of various attempts to monitor vegetation from ught of as a measure of the health of the vegetation rather than a measure of the amount of the vegetation. NDVI has been used to quantify the photosynthetic capacity of plant canopies. It has been used in a wide variety of conditions and situations.

The rationale behind the success of this index is shown here in Fig. 4.20.

Healthy plants absorb solar radiation in the photosynthetically active (PAR) spectral region, which they then use as energy in the process of photosynthesis. Hence, the peaks are in the spectral region between 400 and 500 nm. Due to their water content, leaf cells also reflect solar radiation in the NIR portion of the spectrum. Photon energy longer than 700 nm cannot synthesize organic molecules and thus, solar radiation above 700 nm is strongly reflected by healthy vegetation. Therefore, healthy green plants appear dark in the PAR bands and relatively bright in the NIR. In contrast, cloud and snow appear bright in the red (and other visible wavelengths) and quite dark in the NIR. The green pigment in the plant leaves, due to chlorophyll, strongly absorbs visible light with wavelengths from 400 to 700 nm. The healthier a plant is, the more leaves it has and the more it absorbs visible light. On the other hand, the cell structure of the leaves strongly reflects NIR radiation (700–1100 nm). Since the AVHRR has both of these channels, it is an ideal instrument from which to calculate the NDVI. The long series of operational AVHRR instruments makes it possible to construct a long time series of the NDVI from it.

The NDVI is calculated as follows:

$$\mathrm{NDVI} = \frac{(\mathrm{NIR} - \mathrm{Red})}{(\mathrm{NIR} + \mathrm{Red})} (\text{when NIR} > 0 \text{ and/or Red} > 0) \tag{4.5}$$

$$\mathrm{NDVI} = 0 \ (\text{when NIR, Red} = 0)$$

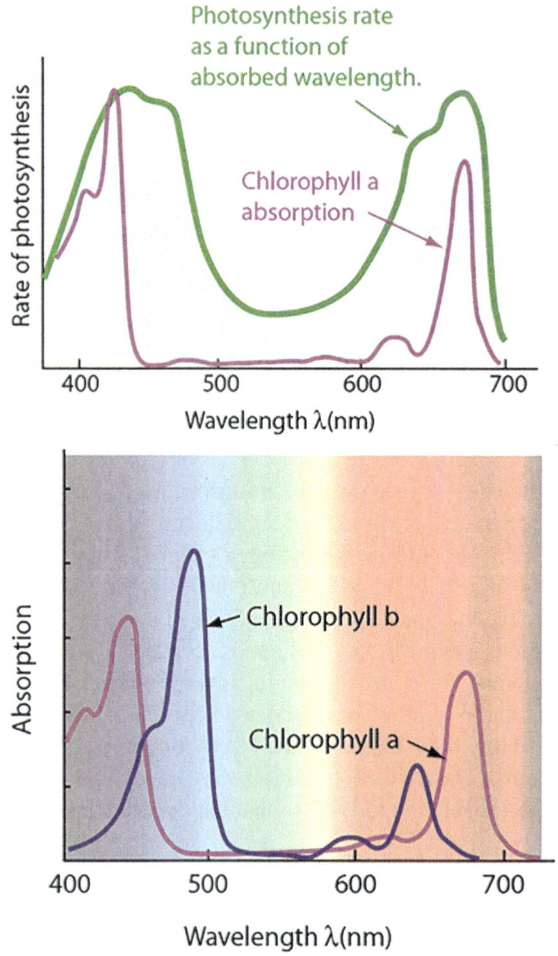

FIGURE 4.20 Leaf absorption spectra (chlorophyll-A, chlorophyll-B, and carotenoids; top) and photosynthetically active radiation (PAR) spectrum (bottom) http://hyperphysics.phy-astr.gsu.edu/hbase/Biology/ligabs.html.

Healthy vegetation (chlorophyll) reflects more NIR and green light compared to other wavelengths. It also absorbs more red and blue light. Thus healthy vegetation would increase the numerator in 1.4 while the denominator normalizes the ratio to account for the differences in zenith angles due to latitude variations. A couple of examples are presented here in Fig. 4.21.

In Fig. 4.21, the higher reflection in the NIR combined with the absorption in the red yields a higher NDVI for the healthy green tree on the left versus the tree on the right, which has experienced wilting with the season. You can see in the calculation how the NIR and visible channels combine to give you the NDVI. The NDVI ranges from −1 to +1 but is usually on the positive side.

The NDVI can be composited and mapped to produce maps of global vegetation health (Fig. 4.22). It needs to be emphasized here that this is a map of the health of the vegetations and not a measure of the amount of vegetation present. It is also important to realize that the colors assigned to the NDVI are completely subjective, and there is no standard for the presentation of such maps.

Another global NDVI map is provided by Wikipedia (Fig. 4.23), which chose to display the NDVI on a map including the polar regions, which changes the range of the NDVI. Here, negative values (values approaching −1) correspond to water. Values close to zero (−0.1 to 0.1) correspond to barren areas and the polar regions. Lastly, low positive values indicate shrub and grassland (0.2 to 0.4), and high values correspond to temperature and tropical rainforests (values approach +1).

When compositing NDVI values, some of the same problems occur as with SST. Cloud cover and residual cloud elements left in a pixel will bias the composite toward low NDVI values. Thus it is important to filter out all of the cloudy pixels before compositing the NDVI. Again, as with SST, one should take the highest NDVI in the composite. In addition, the presence of water, such as in lakes and rivers, will bias the NDVI low, so they should also be removed before compositing.

Another global map exhibits a very different color scheme for the NDVI (Fig. 4.24).

Produced by the German Center for Earth System Research and Sustainability, this global NDVI map is for a single date and uses a very different color code, which is indexed at the right of the map. High productivity is indicated in the green shades, while low productivity is in the blue and purple shades. Note that the only NDVI that goes below zero is the dark blue that represents the ice-covered polar regions. The data set from which this data is taken is based on recalibrated AVHRR/2 and AVHRR/3 observations from the NOAA-7 to NOAA-19 satellites. These data have been processed to reduce artifacts in the data and recovered snow-covered NDVI for regions in winter in Northern latitudes. Due to the use of different AVHRR sensors

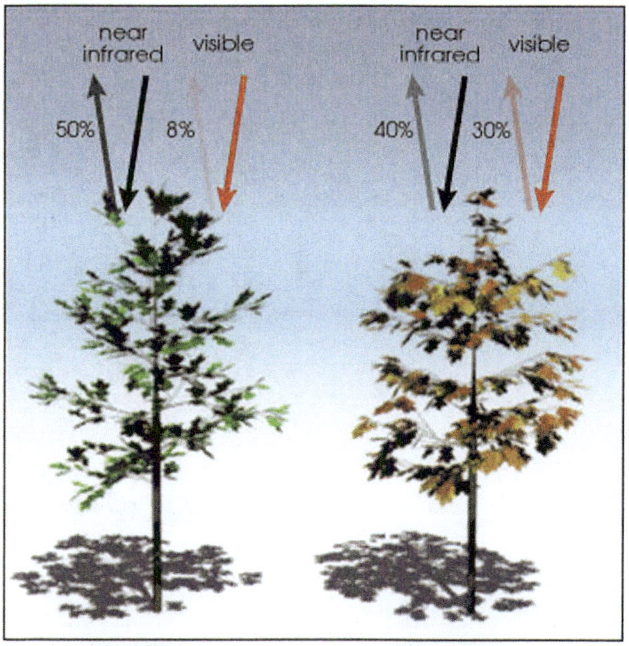

$$\frac{(0.50 - 0.08)}{(0.50 + 0.08)} = 0.72 \qquad \frac{(0.4 - 0.30)}{(0.4 + 0.30)} = 0.14$$

FIGURE 4.21 Two different cases of radiation response to vegetation health and the corresponding normalized difference vegetation index value. On the left is a case of healthy vegetation where 50% of the near-infrared (NIR) radiation is reflected and only 8% of the visible is reflected. On the right is the same tree after senescence has turned the leaves brown, and only 40% of the NIR is reflected, while now 30% of the visible is reflected. Courtesy NASA.

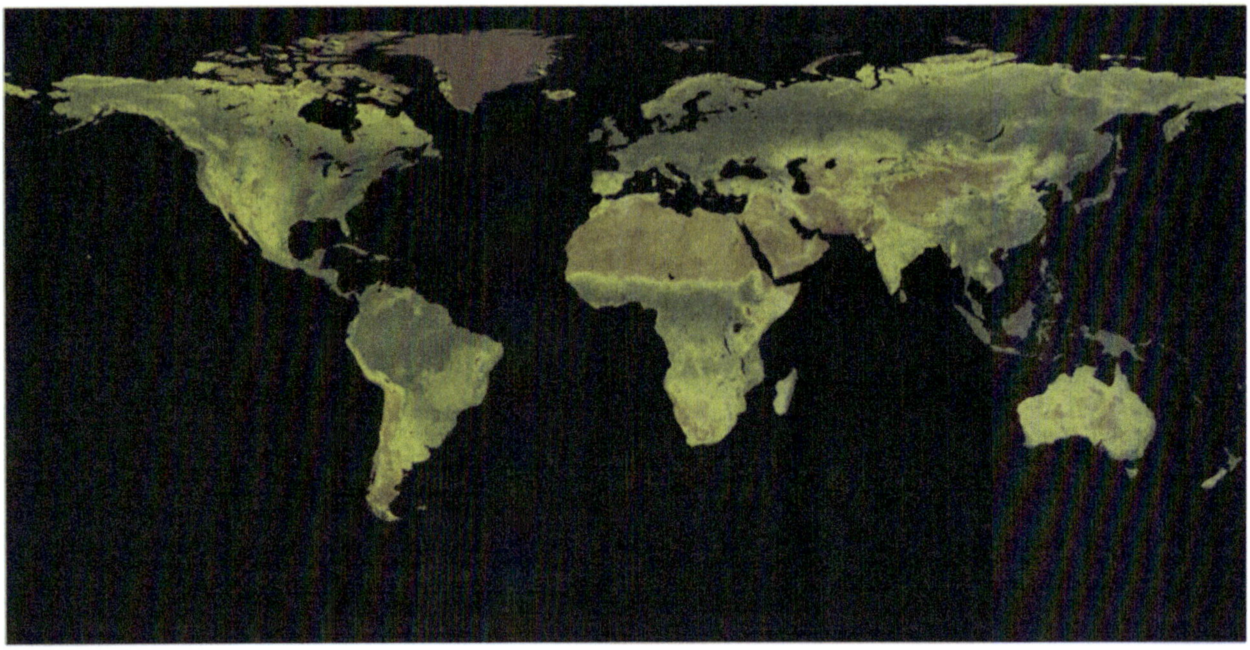

FIGURE 4.22 Normalized difference vegetation index derived from the NOAA climate data record (CDR) of Advanced Very High Resolution Radiometer (AVHRR) surface reflectance. The data record goes from 1981 to 2018 and is derived from 8 AVHRR instruments flying on NOAA polar-orbiting spacecraft. The data are projected onto a 0.05 degree X 0.05 degree global grid. This is one of the Land Surface CDR Version 5 products produced by NASA Goddard Space Flight Center and the University of Maryland (Vermote, 2019). Map lines delineate study areas and do not necessarily depict accepted national boundaries. Courtesy NASA.

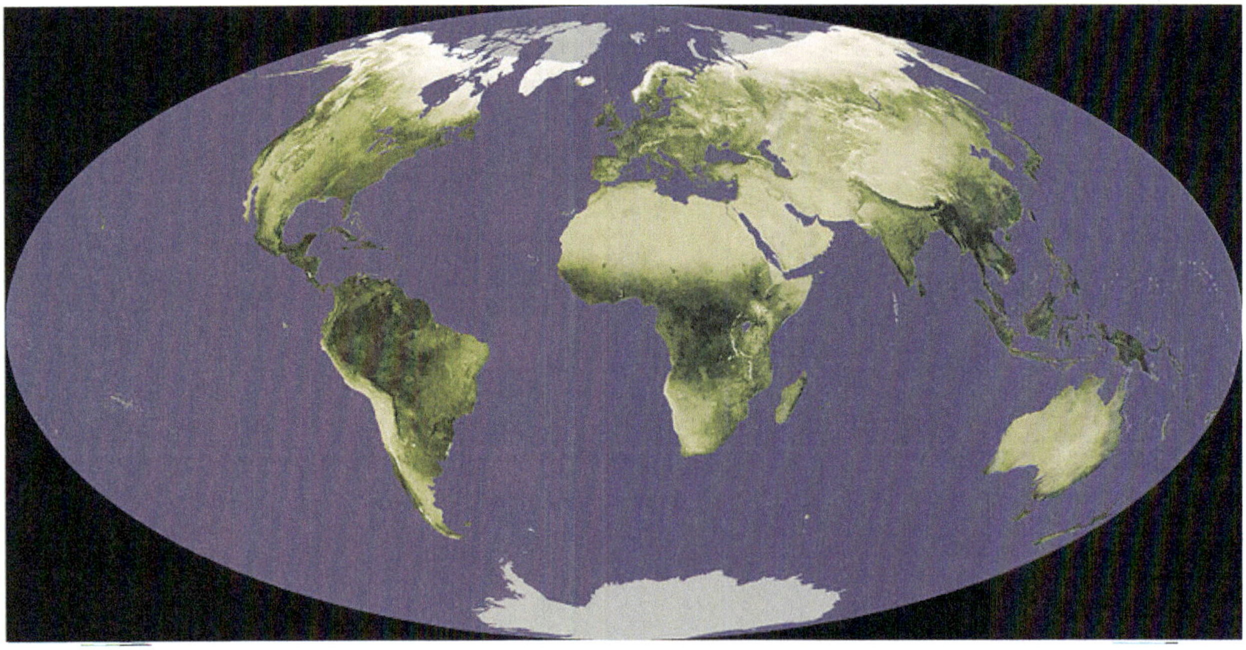

FIGURE 4.23 Global normalized difference vegetation index (NDVI) map (NDVI from Wikipedia). Map lines delineate study areas and do not necessarily depict accepted national boundaries.

and the fact that cloud masking is somewhat limited for AVHRR data, these NDVIs are not of as high a quality as those that came later from NASA satellites.

The NDVI is popular for its accuracy (https://en.wikipedia.org/wiki/Normalized_difference_vegetation_index). It has a very high correlation with the true state of the vegetation on the ground. The precision of the NDVI calculated from AVHRR observations was studied by Roderick et al. (1996) using data collected over Australia. They found that the NDVI was

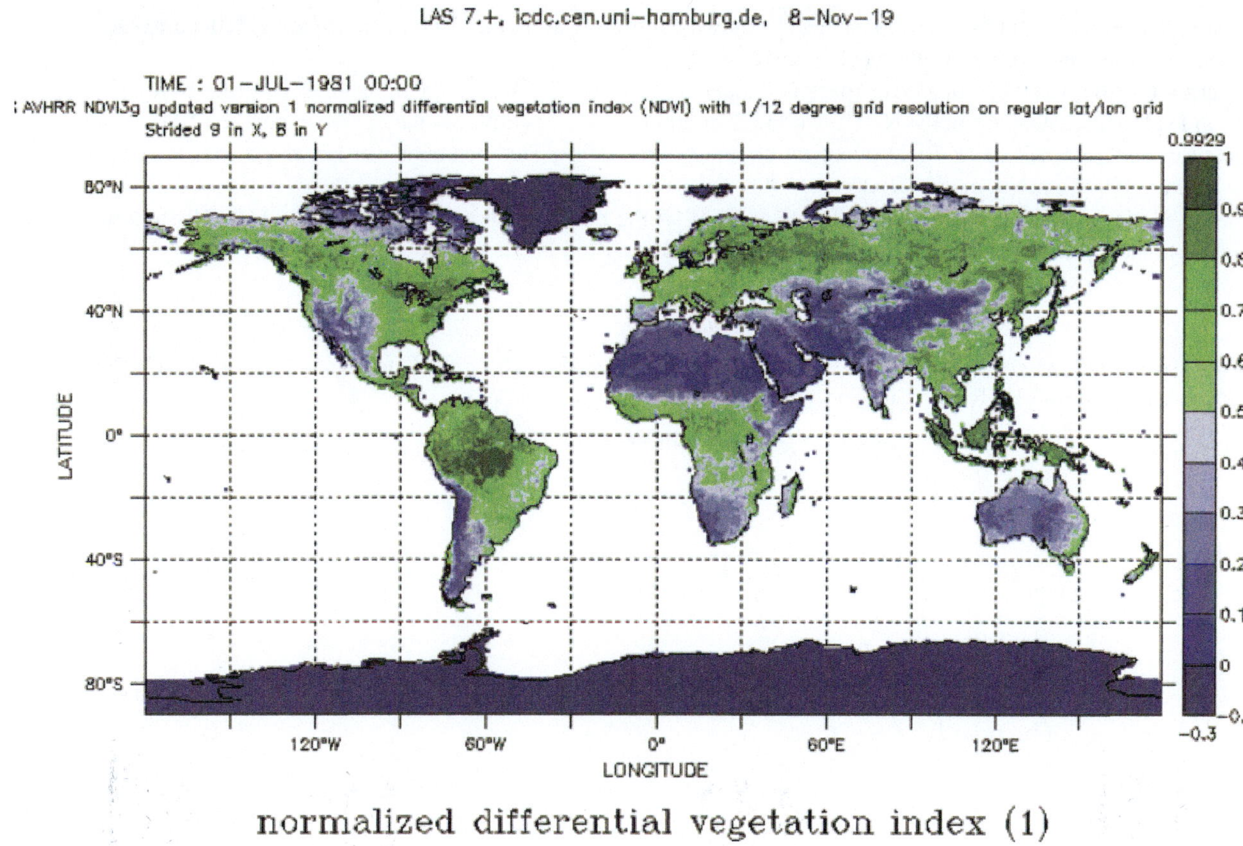

FIGURE 4.24 Global normalized difference vegetation index (NDVI) values for Nov. 8, 2019. Map lines delineate study areas and do not necessarily depict accepted national boundaries.
Courtesy NASA.

accurate to within ±0.01, which degrades significantly over dark targets and at large solar zenith angles. They present a framework whereby the precision of other vegetation indices can be estimated. In their study, Roderick et al. (1996) state that "the accuracy of the NDVI derived from NOAA–AVHRR data is a complex function of the entire data acquisition and processing system" (Goward et al., 1993). This system includes atmospheric corrections (Holben, 1986), sensor calibration (Price, 1987, 1988; Goward et al., 1991, Kaufman and Holben, 1993), variations in soil background color (Huete and Tucker, 1991), and the impacts of bidirectional reflectance (Gutman, 1991). In addition, for time series comparisons, the accuracy of spatial positioning is very important (Townshend et al., 1992). Finally, the transformation to biophysical parameters from the NDVI introduces additional errors (Price, 1992; Huemmrich and Goward, 1992).

All of these studies involve the effects of systematic errors on the AVHRR and the derived NDVI. None of them touch on the precision of the NDVI itself. There is no use making any of the aforementioned corrections if the variability of an NDVI estimate itself is greater than the correction. The precision of the digitization of the satellite-received radiance is determined by the sensitivity of the instrument and the number of levels used for quantization. This quantization process defines an absolute limit on the observational precision, which in turn sets a limit for the derived vegetation index computed from these observations.

Tucker (1980) investigated the link between this quantization and the subsequent interpretation of multispectral satellite data using simulation techniques to regress biophysical information (leaf water content) against simulated satellite observations. The satellite data were then quantized at various levels, and he found that a precision of at least 128 levels was needed to estimate biophysical properties from satellite data. He also observed that the precision of the reflectance measurement was inversely related to the cosine of the solar zenith angle. This dependence on solar zenith angle has also been observed by Graetz and Gentle (1982) in their study of Australian arid rangelands.

Looking only at NOAA–AVHRR data, Goward et al. (1991) studied the relationship between reflectance precision and solar zenith angle. They found that the NDVI was unreliable for solar zenith angles greater than 80 degrees. Errors in NDVI were also shown to be dependent on the reflectance of the target. For areas of low reflectance, the NDVI was found to be extremely sensitive to quantization error.

In their study, Roderick et al. (1996) developed a well-known diagram relating the red reflectance to the NIR reflectance (Fig. 4.25) that describes the NDVI in graphical form.

We can learn a lot from the theoretical data space represented in Fig. 4.25. Clearly, small changes in either NIR or red reflectance can lead to large changes in the NDVI, while for brighter targets, the reverse is true. We expect the NDVI over darker targets to be more sensitive to signal quantization and system noise. This behavior was observed over dark soils (Huete and Tucker, 1991) and the ocean (Roderick, 1994).

In their study, Roderick et al. (1994) do a careful analysis of all the possible sources of error in the AVHRR sampling system. They concluded that the reflectance resolution of the NIR and red bands can be approximated by

$$\Delta p \sim \frac{0.1}{\cos\theta}\% \tag{4.6}$$

where θ is the solar zenith angle, and p represents reflectance. Thus Δp gives the reflectance resolution of the NIR and reflectance channels of the AVHRR as a function of solar zenith angle (Fig. 4.26).

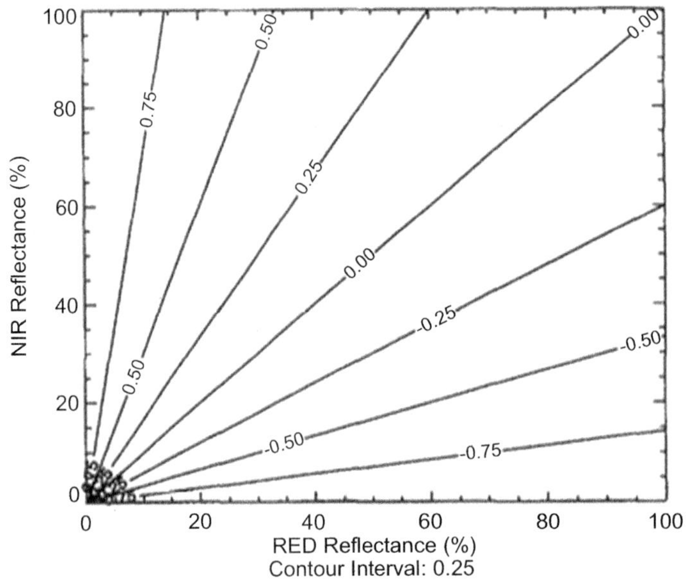

FIGURE 4.25 Contours of the normalized difference vegetation index (NDVI) function (Roderick et al., 1996). *NIR*, Near-infrared.

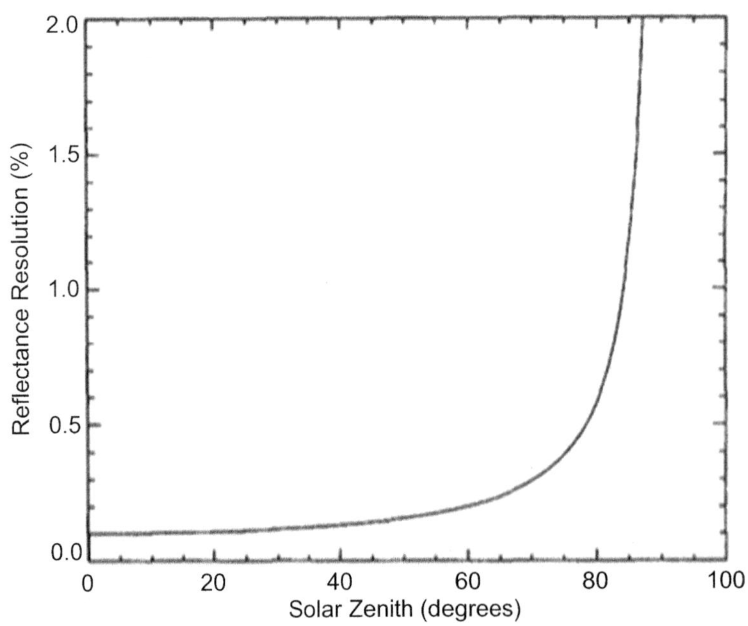

FIGURE 4.26 Reflectance resolution (%) for the Advanced Very High Resolution Radiometer channels used in the normalized difference vegetation index calculation (Cracknel, 1997).

The NOAA satellites carrying the AVHRR are all in sun-synchronous orbits designed to maintain a constant local solar time at each equatorial crossing. This does not, however, lead to constant solar zenith angles throughout the year. These variations in solar zenith angle will cause variability in the reflectance used to calculate the NDVI. In addition, the equatorial crossing times of these satellites have been found to shift later by about 20 minutes each year after launch (Brest and Rossow, 1992). There is also a seasonal variation of the solar zenith angle (Fig. 4.27) as simulated by Roderick (1993) for the nominal equatorial crossing time of 14:30 hours in local standard time (LST) or land surface temperature.

Due to the dependence of reflectance on the solar zenith angle and the seasonal variation of that angle, it is safe to say that the reflectance components of the NDVI calculation will vary seasonally. In most cases, however, the solar zenith angle will be less that 70 degrees for midlatitudes locations. Using Eq. (4.6), we find a reflectance resolution of about 0.3%.

In their paper, Roderick et al. (1996) further examined the quantization error and found that this error depends on where in the NIR versus red space you are operating. Thus the error PDF of a specific NDVI value will vary with the position in the NIR versus red space. Two NDVIs might be equal, but their error statistics will vary according to their position in the NIR versus red space (Figure 1.44). For typical land surface reflectance in the range of 5%–45% and a reflectance resolution of 0.3%, the maximum error in the NDVI varies from ±0.03 for dark targets and ±0.003 for bright targets.

The dependence of the NDVI on atmospheric correction methods also depends on the position in the NIR versus red space. For bright targets, the NDVI is rather insensitive to atmospheric and other external effects, while for dark targets, these corrections may be very important in assessing the accuracy of the NDVI. Roderick et al. (1996) found that for Australian conditions, the best possible precision estimate for AVHRR was ±0.01 NDVI units.

One of the real advantages of using the AVHRR to calculate NDVI was the ability to produce a long time series of NDVI by combining data from a temporal series of AVHRR instruments. In doing so, it is necessary to develop a robust method of compositing the NDVI calculated from a series of AVHRR instruments. In their study, Asam et al. (2023) examine thirteen different compositing methods for AVHRR data over Europe and North America in 2007. They examined daily, decadal, and monthly composites over Europe and North America and evaluated each composite relative to the others. They felt that long time series of climate-sensitive variables is needed to measure, understand, and adapt to global environment changes. The AVHRR is one sensor that provides this type of time series, and its emphasis was on the NDVI as an indication of vegetation variability.

To provide a good indicator of climate change, the NDVI computed from AVHRR data needs to be corrected for variations in viewing and illumination geometries. We have previously discussed the changes in AVHRR viewing due to the fact that the orbit degrades, making the AVHRR not repeat the same sampling geometry. It is, thus, very important to take these changes into consideration when compositing the NDVI data. The rules that are applied for averaging NDVI over time or space need to consider all viewing effects that will influence the individual NDVI values. It is generally the maximum NDVI for the compositing, which is much like compositing the maximum for SST compositing. Unfortunately, compositing on the maximum NDVI value does not provide a representative value for regions with low NDVI, such as water, snow, and bare ground.

Cloud filtering is critical to ensure that the NDVI will represent the vegetation. Even the best cloud filter, however, may still leave some residual cloud in the AVHRR pixel. In their study, Wang et al. (2014) developed an effective compositing method to produced cloud-free AVHRR images. Their method of cloud removal consists of three steps: (1) compositing daily AVHRR images using the maximum BT in channel 4 (thermal infrared), (2) replacing water pixels with pixels that have maximum channel 1/channel 2 ratios (red/NIR), and (3) replacing vegetation pixels with pixels that have maximum NDVI values. They tested this method on the Land Long-Term Data Record AVHRR data set in the region of 37°E–180°E, 3°S–73°N for the year 1993, where they generated monthly minimal cloud-effected images. Their method takes into account the radiometric changes of land surfaces that take place due to topography and temporal changes. They looked at the AVHRR red reflectance and the channel 4 BT for their target area for July 1993 (Fig. 4.28).

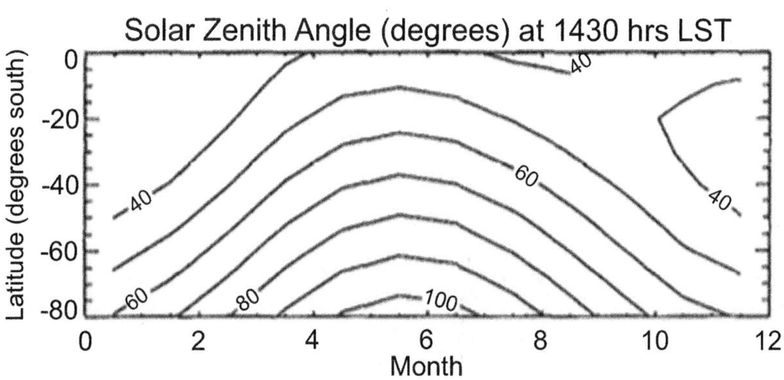

FIGURE 4.27 Solar zenith angles as a function of latitude and time (month of the year) for the southern hemisphere.

Their area in central Asia included the Taklimakan Desert, Tianshan Mountain, Junggar Basin, and the Qinghai–Tibet Plateau in China. It is interesting to note that in the vegetation panel (Fig. 4.28C), there is more variability in both Refl1 and T4 in the middle of the month. Moreover, in general, the T4 line shows a maximum where the Refl1 exhibits minimum. In Fig. 4.28B, the time series is relatively flat as would be expected for snow.

Their flow chart to produce the "cloud-free" composite image is shown here in Fig. 4.29. The first step is to composite on the maximum T4 value, which is commonly used in many cloud identification and removal approaches. Pixels with the

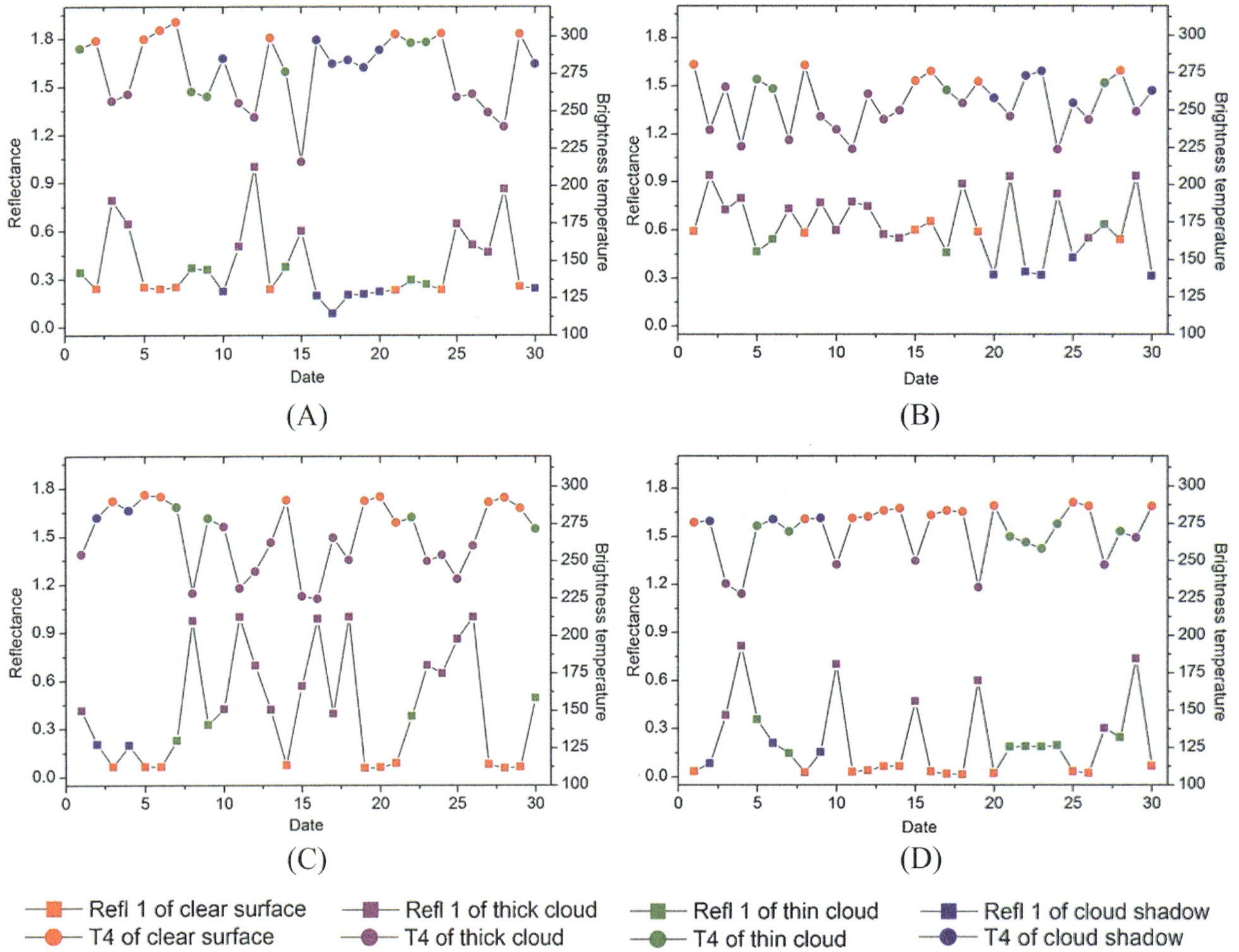

FIGURE 4.28 The reflectance in channel 1 (Refl 1 top) and the brightness temperature from channel 4 (T4 bottom) for (A) barren soil, (B) snow, (C) vegetation, and (D) water (Wang et al., 2014).

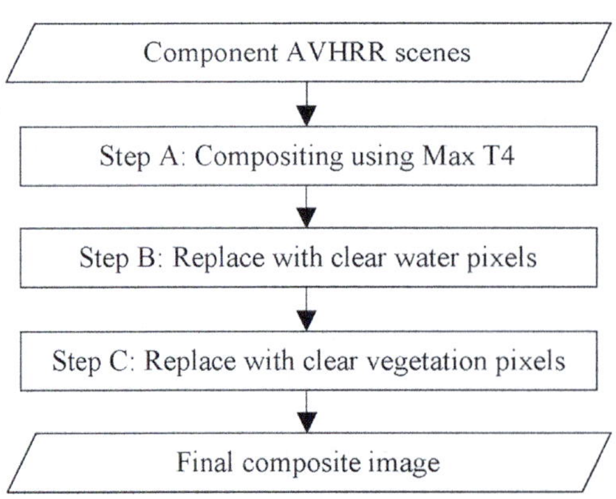

FIGURE 4.29 Flowchart of the process stream to produce "cloud-free" Advanced Very High Resolution Radiometer (AVHRR) images (Wang et al., 2014).

highest T4 value are the least affected by clouds and cloud shadows. They found that T4 maximum cloud filtering works well over large areas, but it fails to properly identify clear pixels when the BT is low. Clear pixels at high latitude have lower T4 values due to thin clouds, haze, cloud shadows, or smoke.

These pixels will be missed when compositing only on the maximum T4 values and cloudy pixels will be incorrectly identified. To rectify these problems, Wang et al. (2014) introduced additional filtering criteria.

Their next step (Fig. 4.29) is to replace water pixels (that may be contaminated by T4 maximum compositing) with pixels having maxima in the ratio of channel 1/channel 2 in the AVHRR (Red/NIR). Clear water pixels have a higher reflectance in channel 1 than in channel 2, while cloud over water has the opposite effects. Hence, the largest channel 1/channel 2 values are best for identifying "cloud-free" pixels in an AVHRR scene. Wang et al. (2014) found that the threshold value for this ratio was reflectance channel 1<0.2 and the reflectance in channel 2<0.1. They tested this threshold with 500 water samples and found 90% of the pixels meet this cutoff. Next, they wanted to correct their "cloud-free" pixels over vegetation at high latitudes. In these cases, they should be composite based on the maximum NDVI. It is a well-accepted practice that compositing on maximum NDVI suppresses clouds because clear vegetation has a higher NDVI than does a pixel contaminated with cloud or cloud shadows. It should be remarked here that quotation marks are frequently used with the phrase "cloud free" since no filtering on compositing algorithm will actually remove all of the clouds, and some subpixel cloud remains undetected and unfiltered.

Using this compositing scheme, Wang et al. (2014) computed 30-day composite AVHRR images in their area of interest. Their target area contains various land cover types such as vegetations, snow, water, and bare ground. It also contains mountains and basins (Fig. 4.30). The places are labeled in Figure 1.49A. A comparison of their final composite with one done using the maximum NDVI makes it apparent that the lake areas and some of the desert are covered with clouds because water and bare soil have NDVI values similar to or even smaller than clouds. Compositing on the maximum NDVI works best when the ground area is well vegetated.

Using a CLAVR-1 (clouds from AVHRR Phase 1) cloud mask to identify clouds in the AVHRR composite Wang et al. (2014) found that cloudy, mixed cloudy, and cloud-shadowed pixels are all classified as cloud-contaminated. In Fig. 4.30E, the image contains many gaps (black pixels) due to the overclassification of cloudy pixels. This is a weakness of the CLAVR-1 cloud classification method.

To quantify these differences further, Wang et al. (2014) computed the cloud cover percentages for their method versus the two other methods (Table 4.3) as a function of month. Here, you can see that, overall, the Wang et al. (2014) proposed method had a smaller percentage of cloud cover. The annual average residual cloud cover percentage is 18.74% lower

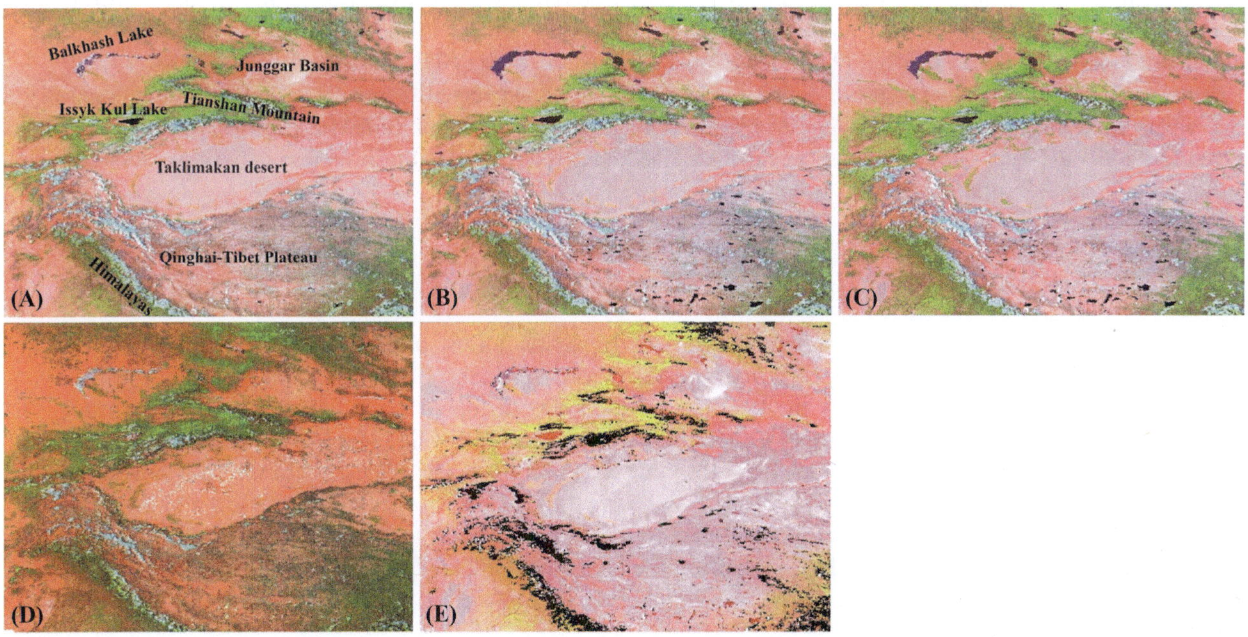

FIGURES 4.30 Advanced Very High Resolution Radiometer images in Central Asia for July 1993. Channel 1 is in *blue*; Channel 2 is in *green*, and Channel 4 is in *red*: (A) Composite image after the first step on the method in Figure 1.48, (B) Composite image after the second step (C) Composite image after the final imaging, (D) Composite image using only maximum normalized difference vegetation index, and (E) Composite image using maximum T4 cloud masks (Wang et al. (2014)).

TABLE 4.3 Cloud cover percentages for three different compositing methods.

Months	Composite using proposed method (%)	Composite using maximum NDVI (%)	Composite using cloud masks (%)
March	39.38	54.87	50.59
April	21.82	42.17	36.26
May	15,15	29.05	27.25
June	13.79	19.94	15.32
July	11.57	18.53	17.42
August	11.47	17.85	15.74
September	13.21	17.94	15,45
October	23.54	34.57	36.97
Average	18.74	29.37	26.87

NDVI, Normalized difference vegetation index.

using the Wang et al. (2014) method and lower than the other two methods. The cloud percentage also depends on the length of the compositing period. Wang et al. (2014) looked at the cloud percentages by month for 10-, 15-, and 20-day compositing periods instead of the month-long average used in Table 4.3. As would be expected, the percentages are higher than those in Table 4.3. The average cloud percentage for 10-day composites is 23.53%, which is higher than the average in Table 4.3, which demonstrates that more than a 10-day composite is needed to remove clouds from an AVHRR image. Previous studies (Tucker et al, 2005, James and Kalluri, 1994) have suggested that a 10-day composite was adequate to suppress the clouds in an NDVI image field.

Returning to the compositing of Asam et al. (2023), they found that while compositing can be used to generate gap-free, equidistant, and spatially comprehensive time series, considerable consideration must be given to the solar and scan geometries. Algorithms that include the viewing and illumination conditions often overcorrect and produce NDVI artifacts. Conversely, methods that rely only on the NDVI value frequently led to NDVI overestimations and saturated pixels. They found that the MED (median NDVI), which is the median value of the NDVI observations in the target area, yields an improvement. It is thought to reduce signal attenuation due to undetected clouds and BRDF (bidirectional reflectance) effects while retaining the prefiltering of physical observations. Asam et al. (2023) chose MED as the best-performing compositing algorithm of the 13 they investigated. The AVHRR MED NDVI time series is now 40 years long and provides a unique monitor of vegetated surface conditions.

4.10 Land surface temperature (LST)

One would think that a section on land surface temperature (LST) should follow after SST, but it is important to recognize the relative difficulty in computing LST versus SST (Vazquez et al., 1997). Unlike the ocean with its radiatively neutral composition and distribution, the land has a wide range of topographic differences, in addition to composition and spatial structural elements that must be accounted for in computing LST. Many of these considerations needed to be accounted for in the computation of the NDVI, so this section follows our discussion of the NDVI. In their study, Vazquez et al. (1997) identify the two main problems in computing LST as the atmospheric effect and the land emissivity effect. It should be stressed here that atmospheric corrections developed for SST will not be the same as those for LST again due to the influence of the land on them and the fact that the land has a much larger diurnal change than does the ocean. They consider an accurate LST to be in the range of ±1°C–2°C, which is a much lower accuracy than is generally accepted for SST.

Vazquez et al. (1997) review a variety of proposed methods to compute LST. They first discuss a variety of techniques that have been used for SST and argue that these are not appropriate for LST. They point out that to derive an accurate LST algorithm, one needs nearly coincident field and satellite measurements over land. They also mention that there is a lack of high-quality in situ surface temperature measurements over land. They identify as primary problems the difficulty in synchronizing both land surface and satellite temperature measurements, as well as the impossibility of covering the wide range of atmospheric conditions, land surface temperatures, and emissivity contained in a satellite image in a single field experiment. Added to that is the fact that the AVHRR samples a 1 km^2 area while the scale of land variability is much smaller. Different factors, such as topography and vegetative cover, can result in these smaller scale changes. In their study, Vazquez et al. (1997) used a flat soil surface with uniform land cover over various seasons and a wide range of land surface temperatures. In addition, the area is known to have relatively low atmospheric moisture content and a low probability of cloudiness. Their selected area covered a 2X2 AVHRR pixel region, which showed a distinct contrast with the surrounding area during the growing season. This area is covered with natural grassland and patches of bare soil. They examined data for this area from spring to fall to be able to include a wide variety of atmospheric and land surface conditions.

Vazquez et al. (1997) used HRPT AVHRR data that were collected by the receiving station at the University of Granada. They only used AVHRR data with VAs of less than 30 degrees from nadir, and the images were georegistered using coastal and river features. The radiometric resolution of the data was reduced from 10 to 8 bits in such a way that the temperature resolution was better than 0.3 K. They used 5 cm deep in situ temperature measurements collected by the Spanish Meteorological Office following World Meteorological Organization (WMO) guidelines. These temperature values were integrated over 30 minutes to reduce noise in the observations. The 30 minutes average closest to the time of the NOAA satellite overpass was selected for comparison. To assess the differences between the 5 cm temperatures and the skin temperatures that would be measured by the satellite, they made additional measurements over their study area with an Everest infrared thermometer and found the differences with the 5 m temperatures to be almost zero at a time of 7:00–8:00 GMT.

A list of the satellite passes, and their dates and times are given here in Table 4.4. Vazquez et al. (1997) used only morning NOAA satellite passes, which they found to be the time of minimum difference between the surface skin temperature and that measured at 5 cm. Thus they felt that it was adequate to use the 5 cm temperature as a surrogate for surface skin temperature. They cautioned that this assumption is not valid all year.

TABLE 4.4 List of Advanced Very High Resolution Radiometer (AVHRR) images used in the Vazquez et al. (1997) study.

Year	Day/Month	Hour/Minute (GMT)
1995	09/03	08:01
1995	14/03	07:53
1995	15/03	07:30
1995	18/05	07:42
1995	19/05	07:20
1995	27/05	07:46
1995	28/05	07:24
1994	14/06	08:10
1994	15/06	07:48
1994	23/06	08:45
1994	24/06	07:55
1994	29/06	07:45
1994	20/07	08:20
1994	07/07	08:10
1994	12/07	08:03
1994	20/07	08:30
1994	27/08	08:05

(Continued)

TABLE 4.4 (Continued)

Year	Day/Month	Hour/Minute (GMT)
1994	31/08	08:17
1994	01/09	07:56
1994	02/09	07:34
1994	05/09	08:10
1994	06/09	07:48
1994	07/09	07:26
1994	09/09	08:23
1994	11/09	07:40
1994	27/09	08:33
1994	21/10	08:13
1994	29/10	08:41
1994	01/11	07:35
1994	26/11	08:34
1994	01/12	08:25
1994	03/12	07:12
1994	10/12	08:30

Vazquez et al. (1997) analyzed five different LST algorithms as shown below:
Becker and LI (1990)

$$T = 1.274 + \frac{T_4 + T_5}{2}\left[1 + 0.15616\frac{1-\epsilon}{\epsilon} - 0.482\frac{\nabla\epsilon}{\epsilon^2}\right] + \frac{T_4 - T_5}{2}\left[6.26 + 3.98\frac{1-\epsilon}{\epsilon} + 38.33\frac{\nabla\epsilon}{\epsilon^2}\right] \quad (4.7)$$

Prata and Platt (1991)

$$T = 3.45\frac{T_4 - T_0}{\epsilon_4} - 2.45\frac{T_5 - T_0}{\epsilon_5} + 40\frac{1-\epsilon_4}{\epsilon_4} + T_0 \quad (4.8)$$

Price (1984)

$$T = [T_4 + 3.33(T_4 - T_5)]*\left[\frac{5.5 - \varepsilon_4}{4.5}\right] + 0.75T_5\nabla\varepsilon_4 \quad (4.9)$$

Ulivieri et al. (1992)

$$T = T_4 + 1.8(T_4 - T_5) + 48(1 - \varepsilon) - 75\nabla\varepsilon. \quad (4.10)$$

Kerr et al. (1992)

$$T = CT_s + (1 - C)T_{ive} \quad (4.11)$$

$$T_s = -2.4 + 3.6T_4 - 2.6T_5$$

$$T_{ive} = 3.1 + 3.1T_4 - 2.1T_5$$

$$C = \frac{NDVI - NDVI_{ive}}{NDVI_s - NDVI_{ive}}$$

where $\varepsilon = \frac{\varepsilon_4 + \varepsilon_5}{2}$, $\nabla\varepsilon = (\varepsilon_4 - \varepsilon_5)$, and $T_0 = 273.15K$

TABLE 4.5 Statistical results for the various LST algorithms.

	R^2	RMSD (K)	MBD (K)	MD (K)
B & L	0.97	3.8	3.3	7.5
P & P	0.97	3.0	1.8	6.8
PRICE	0.97	2.3	0.1	4.2
ULIVIERI	0.97	1.9	0.7	4.7

Some of these algorithms require the choice of surface emissivity. In the Kerr et al. (1992) study, the NDVI is used as a surrogate for the emissivity. For the other algorithms, we must first choose a surface emissivity. To allow for seasonal variability in emissivity, Vazquez et al. (1997) divided their data set into three seasonal subsets.

The results of this comparison for these algorithms are presented here in Table 4.5.

Note R^2 is the coefficient of determination, RMSD is the root mean square error, MBD is the mean bias difference, and MD is the maximum difference for each algorithm. The Ulivieri and Price algorithms have the lowest RMSDs with a correspondingly low MBDs and MDs. The Becker and Li algorithm has the largest RMSD with an equally high MBD and MD, followed by the Prata and Platt algorithm.

These results are shown graphically in Fig. 4.31.

While it is not overwhelmingly obvious, one can see the wider spread of the dots in C and D versus the scatter in A and B.

Vazquez et al. (1997) treated the Kerr et al. algorithm separately due to its use of the NDVI as a surrogate for emissivity. They decided to use a number of vegetation indices with the Kerr et al. algorithm but found that the NDVI performed the best. They then compared the Kerr et al. NDVI algorithm with the Ulivieri algorithm (Fig. 4.32) and found that they are nearly equivalent.

That they are nearly equivalent confirms that they perform best against the ground truth measurements. It also demonstrates that the choice of emissivity is not critical to the computation of LST.

Vazquez et al. (1997) concluded that land surface temperature (LST) can be inferred from infrared satellite data provided that at least two infrared measurements are available and that there is some *a priori* information on the spectral emissivity of the target land surface. They find that the use of a surrogate such as NDVI proposed by Kerr et al. (1992) provides a simpler way to deal with land surface effects. This latter method produced results very close to those with more complex methods of estimating the emissivity. They found that the Kerr et al. and the Ulivieri et al. LST algorithms are statistically indistinguishable. In addition, both of these algorithms have equivalent similarities to the ground measurements of LST. These authors also conclude that errors in the emissivity propagate through each LST algorithm in a different manner which contributes to the errors in LST estimates. An error of about 1% in emissivity will result in an LST error between 0.6 and 2.0 K.

One of the primary advantages of using AVHRR data is the long time series provided by this series of instruments and satellites. Reiners et al. (2021) describe the TIMELINE project to derive LST at a 1 km spatial resolution from AVHRR over Europe for the entire operating period of this sensor. They validate the LST time series derived from AVHRR data with in situ LST measurements and also by considering the consistency of the AVHRR LST time series. They also want to evaluate the effects of total column water vapor (TCWV) and VA on the LST product. They consider only daytime AVHRR imagery since the first two channels are needed to estimate surface emissivity. The ground LST measurements used for validation were 10 stations located in Europe, North America, and Southern Africa (Table 4.6). These sites all carried out thermal infrared measurements of LST. The sites are shown on Fig. 4.33 along with the overall TIMELINE study area.

Reiners et al. (2021) quote Frey et al. (2017) as finding that of the many algorithms for LST, the Becker and Li (1990) did best of the split-window algorithms. Thus the split window algorithm that they tested was

$$\alpha = 1 + p_1 \frac{1 - \varepsilon}{\varepsilon} + p_2 \frac{d\varepsilon}{\varepsilon^2} \tag{4.12}$$

$$\beta = p_3 + p_4 \frac{1 - \varepsilon}{\varepsilon} + p_5 \frac{d\varepsilon}{\varepsilon^2}$$

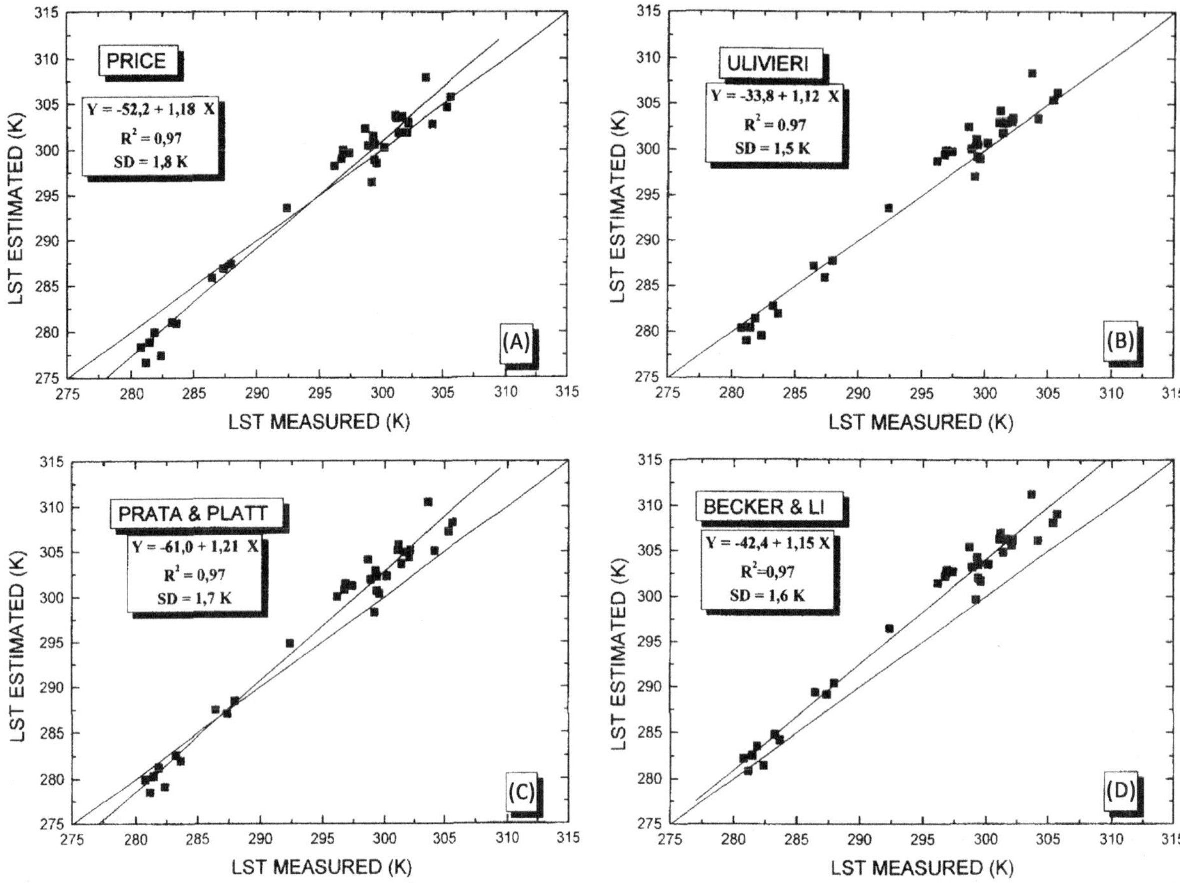

FIGURE 4.31 Scatter plots of the in situ measured ground temperatures against the LST estimate from the Advanced Very High Resolution Radiometer (AVHRR) overpass: (A) Price algorithm, (B) Ulivieri algorithm, (C) Prata and Platt algorithm, and (D) Becker and Li algorithm (Becker and Li, 1990).

$$\mathrm{LST} = p_0 + \alpha\frac{t_4 - t_5}{2} + \beta\frac{t_4 - t_5}{2}$$

where t_4 and t_5 are the BTs in AVHRR channels 4 and 5, ε is the mean emissivity for both thermal infrared channels, and $p_0 - p_5$ are the coefficients added by Frey et al. (2017). These authors also examined a single-channel algorithm, but we will restrict our discussion to the split-window approach. These authors used AVHRR data with a nominal 1 km resolution but acknowledged that this resolution increases to 6.5 km near the edges of the 55 degrees swath. For their analysis, they used AVHRR Level 1B data, which had been calibrated, geolocated, and quality checked for errors. The data were available from the DIMS archive at the German Aerospace Institution (DLR). DLR processing included the calculation of sun and satellite zenith angles as described in Bachman et al. (2015).

The additional data needed for the LST computation consisted of total column water vapor (TCWV) and land cover characterization to derive emissivity. Reiners et al. (2021) used data from the ERA-archive of the European Center for Medium-Range Weather Forecasts (ECMWF). These data are a reanalysis of 3-hourly measurements and have a spatial resolution of about 79 km. As for emissivity, the land cover estimates used the method derived by Caselles et al. (2012), which combines spectral information for different land cover types with the seasonal fluctuation of vegetation cover by estimating the fraction of vegetative cover of each scene from the AVHRR reflective bands. The derived land cover type for each pixel is then assigned to an emissivity class. For TIMELINE, the GlobeCover classification for 2009 was used (Bontemps et al., 2011).

These authors also calculated the uncertainty with all of their LST values. They followed the method used in Frey et al. (2017) to estimate the uncertainty, which includes the uncertainty of the emissivity calculation, the noise equivalent temperature difference (NEdelT), and the uncertainty due to geolocation. For the NEdelT, they used a constant of 0.12 K taken from Trischenko and Trends (2002). The uncertainty of the geolocation is taken as the standard deviation of the LST

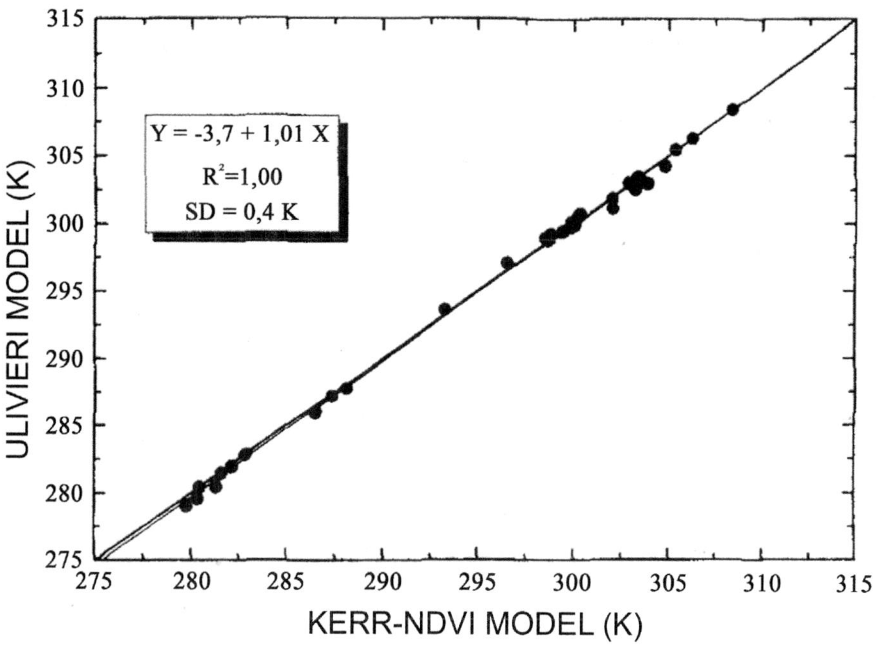

FIGURE 4.32 Scatter plot of the Kerr-normalized difference vegetation index (NDVI) LST algorithm versus the Ulivieri algorithm (Becker and Li, 1990).

TABLE 4.6 Description of the 10 stations used for in situ LST measurements.

Station name	Lat	Long	Land cover at the station	Land cover around the station	Emissivity classes	Validation period
BND	40	−88.3	Grassland	Cropland	3	2010–13
BO	40.1	−105.2	Sparse grassland	Grassland/cropland	3	2010–13
DR	36.6	−116	Arid shrubland	Arid shrubland	4	2010–13
FP	48.3	−105.1	Grassland	Grassland	3	2010–13
GC	34.2	−89.9	Grassland	Grassland	3	2010–13
PEN	40.7	−77.9	Cropland	Cropland/forest	3	2010–13
SF	43.7	−96.6	Grassland	Grassland	3	2010–13
HE	−22.9	18	Arid grassland	Arid grassland	3	2010, 2013
EV	37	−6.4	Open savannah	Savannah, 33% Tree Crown cover	3	2010
DN	38.5	−8	Grassland	Grassland	3	2011–13

EV, Essential variables.

in an *nxn* box with *n*=3, 5, 7, depending upon the quality of the level 1B data used in their calculation. The calibration error is taken as 1% of the BT. Thus the overall uncertainty of the calibration is

$$\mathrm{Unc} = \sqrt{\mathrm{alg}^2 \mathrm{emi}^2 + \mathrm{geo}^2 + \mathrm{cal}^2} \tag{4.13}$$

where alg is the uncertainty of the LST algorithm, emi is the emissivity uncertainty, geo is the geolocation uncertainty, and cal is the uncertainty of the calibration.

A problem working with a time series of satellite measurements is that the individual satellites measure at slightly different times, in addition to the orbital drift these satellites experience. To solve this, Reiners et al. (2021) introduced a daytime normalization by applying a diurnal temperature cycle (DTC) model used by Göttsche and Oleson (2001) which is

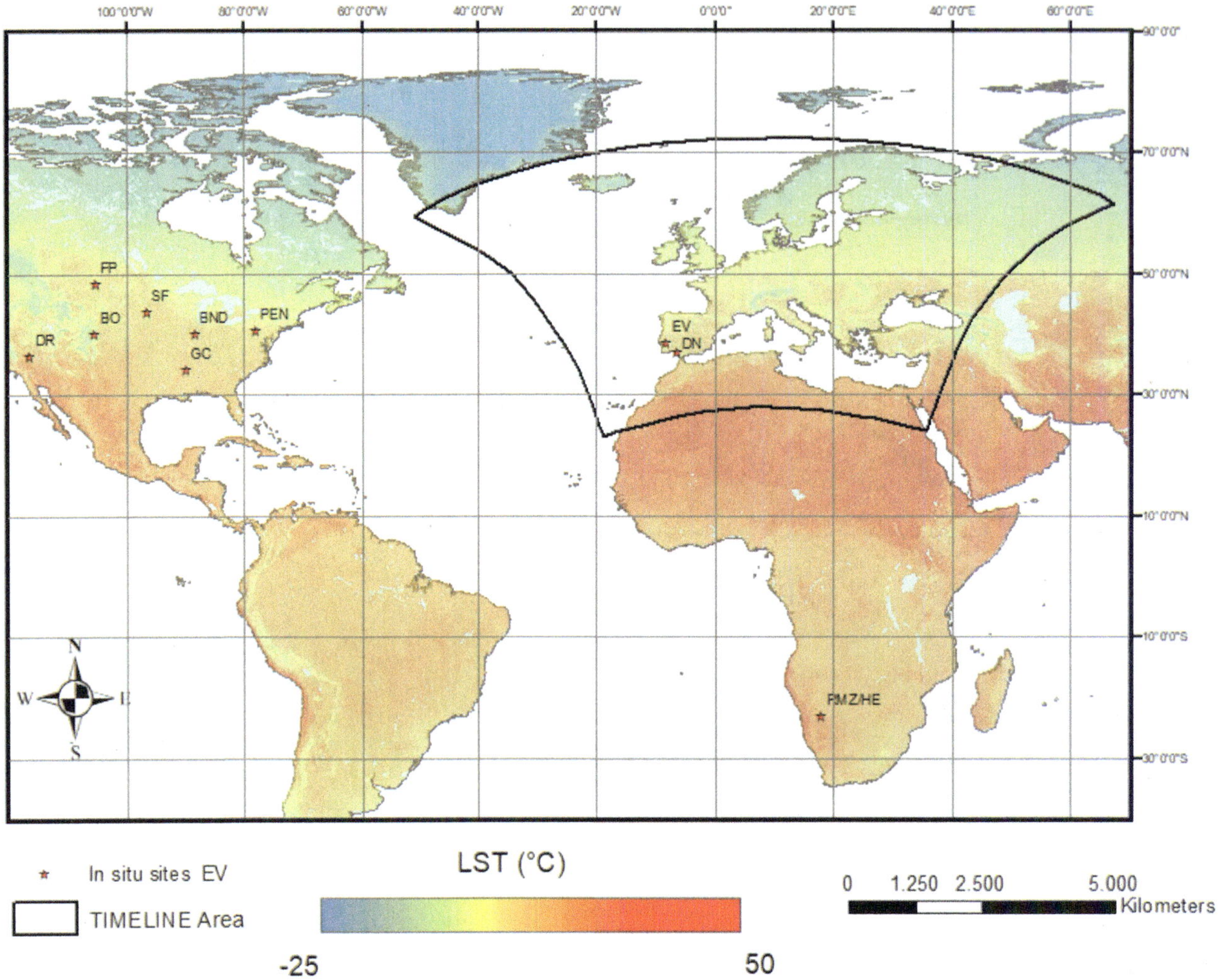

FIGURE 4.33 The TIMELINE study area and the LST ground truth sites (Reiners et al., 2021). Map lines delineate study areas and do not necessarily depict accepted national boundaries.

$$\mathrm{LST}(t) = T_0 + T_a \cos\left(\frac{\pi}{\omega}(t - t_m)\right) \tag{4.14}$$

where T_0 is the LST at sunrise, T_a is the amplitude of the diurnal temperature change, ω is the length of day, t is the observation time, and t_m is the time of the maximum temperature. This model was modified by Liu et al. (2019) to normalize AVHRR-derived LST to a certain observation time as

$$\mathrm{LST}(t_1) = \mathrm{LST}(t_2) + T_a \cos\left(\frac{\pi}{\omega}(t_1 - t_m)\right) + T_a \cos\left(\frac{\pi}{\omega}(t_2 - t_m)\right). \tag{4.15}$$

where t_1 and t_2 are two different times in the DTC model. The length of the day can be calculated by the geolocation and the day of the year.

The in situ infrared radiometers and the AVHRR sensor measure on completely different scales. Thus it must be assumed that the ground measurement is representative of the surrounding area covered by the 1 km AVHRR pixel. Since the ground truth radiometer measurements were only available for the period between 2010 and 2013, the validation with the ground truth data can only be carried out in these years. For the period 2010–13, the LST was compared to the near-ground radiometer measurements. The correlations and scatter of these comparisons are shown in Fig. 4.34.

The best agreements exhibited by the reduced scatter in these plots are for DR and GC, while the largest scatter occurs for BND. The stations at FP, BND, DN, and SF all exhibited marked positive MDs, while slightly weaker positive MDs were observed for essential variables (EV), PEN, and HE. Modest negative MDs occurred at GC and BO. There are considerable positive MDs at BND, FP, and EV at LSTs over 300 K.

Previous studies of the TIMELINE LST (Frey et al., 2017) estimated the error in LST to be between 1 and 3 K. Similar accuracies were found by Ma et al. (2020) in their comparison of AVHRR-derived LST with SURFRAD sites, getting RSME values between 2.25 and 3.866 K. Reiners et al. (2021) conclude that the comparisons with ground truth yielded MADs between 1.19 and 2.81 K and RMSE between 1.54 and 3.68 K, which is consistent with the estimates from earlier studies. They observed a distinct seasonal difference with an MAD of 2.07 K in summer and an MAD of 1.54 K in winter.

4.11 Forest fire detection and monitoring

Another application of AVHRR data is in the detection and monitoring of forest fires. A great many studies have been made of this problem, but the main focus has been on AVHRR channel 3, which at 3.7 μm contains both emitted radiation and reflected radiation. One would think that the thermal infrared channels would be ideal for this application, but the fact is that these channels saturate at the high temperatures of forest fires and therefore offer no valuable information. Channel 3, however, saturates at a much higher temperature and is therefore responsive to wildfire behavior. Li et al. (2001) provide a review of the many AVHRR-based fire detection and monitoring algorithms. They examine three different types of AVHRR forest fire algorithms: (1) single channel threshold, (2) multichannel threshold, and (3) spatial context. They examine the performance of these different algorithms and make suggestions to improve them.

They report that several international programs have been established to study fire activity around the world using satellite sensors. Programs such as the International Geosphere Biosphere Program data and information system (IGBP-DIS), Global Fire Product Initiative (Justice and Malingreau, 1993, 1996), and the World Fire Web. These many activities are endorsed by the Global Observation of Forest Cover (GOFC) program. The objective of GOFC is to improve our understanding of forest dynamics and forest fires impacts on the global carbon budget to provide resource managers and decision makers with the information they need about forest changes on local to global scales. GOFC has two foci: (1) near-real-time detection and monitoring of fire activity and (2) mapping burned out areas after the fire. The former goal requires close communication and coordination with the firefighting community and the second is useful to forest managers to understand the impact of the fire.

While many different satellite sensors can be used to detect and monitor forest fires the AVHRR has the longest running consistent series of measurements that can be used for this purpose. Therefore AVHRR data are the focus of the Li et al. (2001) study. They review the different AVHRR fire algorithms and provide an evaluation of the efficiency and performance of five of these algorithms in their application to Canadian boreal forests. They note that an advantage of the AVHRR is that it provides daily coverage at a moderate (~ 1 km) resolution. Second, the AVHRR channel 1 (~0.63 μm), channel 2 NIR (~0.83 μm), channel 3 mid-infrared (~3.76 μm), and its two thermal infrared channels 4, 5 (~11, 12 μm) all are sensitive to some aspect of forest fires, but each contains different information. Smoke can be identified in the visible channel, but since clouds and smoke have similar reflectance values, it is best to use other channels to discriminate smoke from clouds (Li et al., 2000a). Burned areas can be mapped using the reflectance difference between the visible and NIR channels (Li et al., 2000b). The size and temperature of subpixel fires can be determined using two thermal infrared channels (channels 3 and 4), assuming a uniform background (Dozier, 1981) if the sensor is not saturated and the fire signature is significant in both channels. This is often very difficult as, for example, in the deforestation fires in South America that when detected by the AVHRR increase the signal in the 10 μm channel by only 1.5 K, which is within the background variability of this channel.

The primary AVHRR channel used in fire detection and monitoring is channel 3 (~3.75 μm), which is less sensitive to wildfire temperatures. It is very sensitive to the presence of fires, which can also be a problem for fire detection, as a small, intense fire can easily saturate the channel. For example, a subpixel fire of 1000 K in an area with a 300 K background needs to be only 13X13 m^2 to saturate the 1 km 3.75 μm channel (Kaufman et al., 1990; Robinson, 1991). Often, to overcome this problem, channel 4 is used together with channel 3. Since the fire will have varying temperatures within the pixel, this leads to very different radiances in these two channels.

Forest fire algorithms using AVHRR data have been reviewed and compared by a number of studies (Robinson, 1991; Justice et al., 1993; Setzer and Malingreau, 1996; Martin et al., 1999). These algorithms can be grouped into three categories: (1) single channel algorithm based on channel 3, (2) multichannel algorithms, and (3) contextual algorithms that compare the potential fire pixel with the thermal properties of the background. It should be noted that the single and multichannel methods are fixed threshold methods, while the contextual method can have variable thresholds.

The single-channel approach uses only the channel 3 radiances. Since the AVHRR was not designed to detect and monitor wildfires, it saturates at about 320–331 K (Robinson, 1991), well below the fire temperatures. Li et al. (2001) give the example for a background of 300 K a fire of 1000 K that fills only 0.02% of the pixel area can raise the channel 3 pixel temperature to 324 K. Since the background seldom exceeds 300 K, the channel 3 radiance can still be very useful in

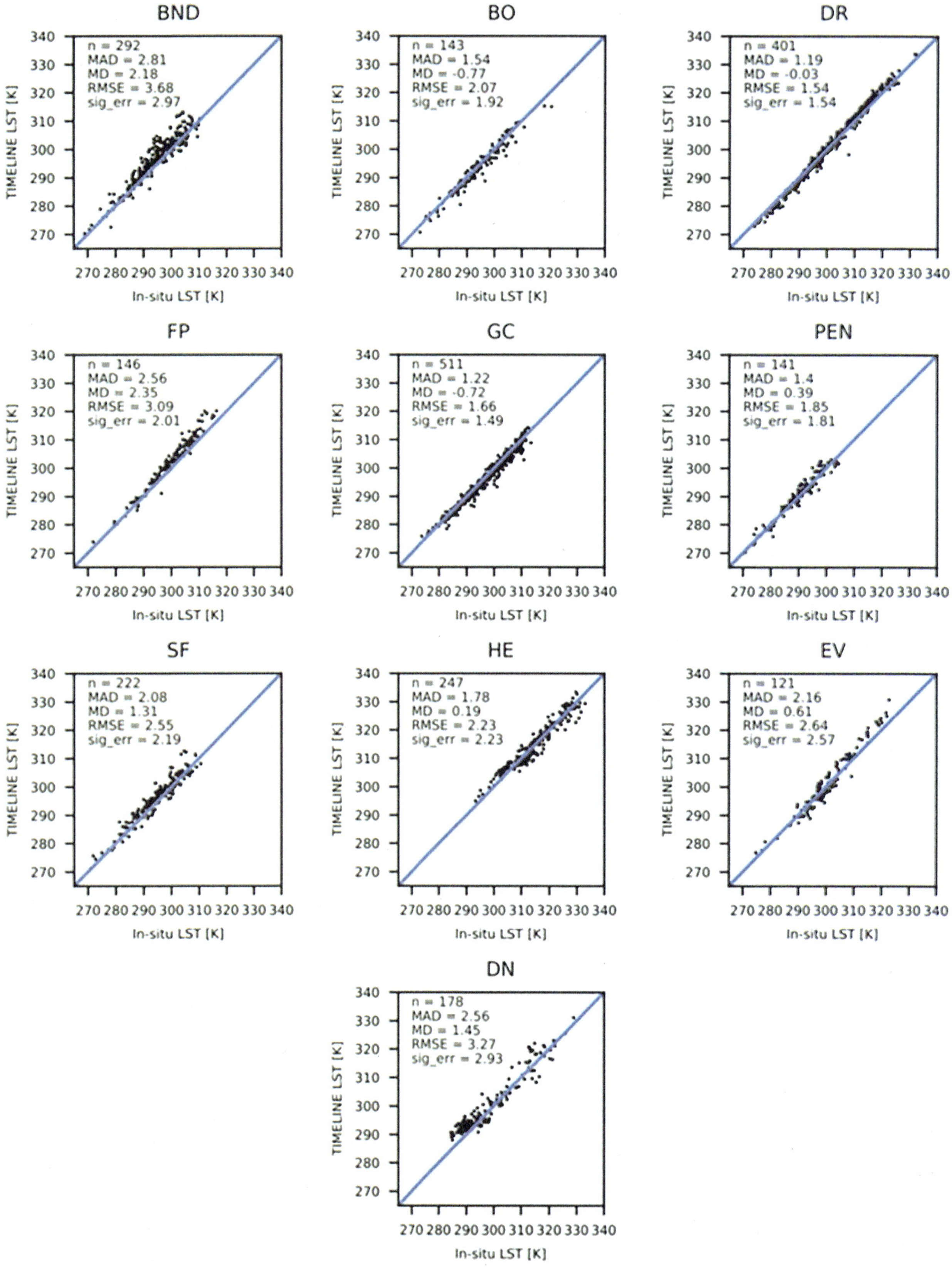

FIGURE 4.34 Scatter correlation diagrams for LST versus ground truth radiometry. Geographical locations of the ground truth spots are given in Fig. 4.33. Here, MAD is the mean absolute difference, MD is the mean difference, sig_{err} is the standard deviation of the error, and RMSE is the root mean square error (Liu et al., 2019). *EV*, Essential variables.

detecting and monitoring wildfires particularly for those areas with low background emission values (Muirhead and Cracknell, 1985; Malingreau and Tucker, 1988; Setzer and Pereira, 1992b, 1992). The single channel algorithm is usually applied to nighttime satellite data where the reflected portion of channel 3 is absent, and it is only viewing the thermal infrared (Malingreau 1990; Langaas, 1992). One of the real challenges in the single-channel fire algorithm is correcting for highly reflective clouds. Thus it is important to filter out clouds before applying the single channel 3 fire algorithm.

Multichannel fire algorithms were introduced to overcome some of the limitations of the single channel 3 algorithm. These algorithms usually have three steps: (1) use channel 3 to identify all potential fires, (2) use the thermal infrared channel 4 to eliminate clouds, and (3) use the difference between channels 3 and 4 to isolate the fires from warm backgrounds. It should be noted that these algorithms do not account for highly reflective ground surfaces. These multichannel algorithms have been used in many different biomes such as tropical forest (Kaufman et al., 1990; Belward et al., 1994), savanna fires in West Africa (Langaas, 1993; Kennedy et al., 1994; Franca et al., 1995), and boreal forests (Cahoon et al., 1991, 1994; Li et al., 1997, 2000c). In each of these applications, the algorithms were tuned to account for differences in the environment. It should be noted that while channel 4 is efficient at detecting cold high cloud decks, it is not useful in mapping low clouds. The visible channel 1 has been used to map low cloud (Kennedy et al., 1994; Arino, 1998), but it must be recognized that channel 1 is sensitive to smoke plumes and can mix clouds and smoke. Some investigators have used the NIR data from channel 2 (Li et al., 2000c), which is less affected by smoke. The difference between channel 4 and 5 is useful to remove thin cirrus clouds (Inoue, 1987; Franca et al., 1995; Li et al., 1997, 2000c). The difference between channels 3 and 4 is the most efficient at identifying false fires. The common belief behind this idea is that active fire pixels have larger T3–T4 differences than do unburned backgrounds. Unequal atmospheric effects in these two channels contribute to their difference, which reflects the different transmittance of the atmosphere (Fig. 4.35).

The atmosphere is more transparent in channel 3 than in channel 4. It must also be considered that atmospheric emission is proportional to transmission, thus contributing to the channel 3 measurements, and thus reducing the T–T4 difference. To better understand this difference, Li et al. (2001) used MODTRAN (Anderson et al., 1986) to compute the T3 and T4 top of the atmosphere BTs versus the corresponding ground surface temperatures (Fig. 4.36).

While both channels have the expected linear response to temperature, they are similar at the lower temperatures characteristic of the ground surface values but diverge at the higher temperatures. It is this difference that one wants to exploit for detecting and monitoring forest fires. For the T3–T4 difference to reach 10 K (a typical threshold in fire detection algorithms), the fire temperature needs to be around 370 K, where T3 would be about 363 K. Since channel 3 saturates at about 325 K, this compromises the use of the multichannel temperature difference in many cases of uniform "hot" pixels. Hot fires might actually result in a negative T3–T4 difference as T4 saturates at a slightly higher temperature of about 363 K. The maximum T3–T4 difference at about 325 K is 4–5 K.

The primary source of large, measurable T3–T4 differences is the inhomogeneity of the distribution of fire within a fire pixel (Dozier, 1981; Kaufman et al., 1998). Due to the nonlinear relationship between radiance and temperature, the more heterogeneous the temperature field is, the larger the T3–T4 difference will be. This is the basis for eliminating false fires when using moderate spatial resolution, such as the 1 km AVHRR imagery. The difference T3–T4 is found to reach maximum values for a very small fraction of an AVHRR pixel having an active fire signature (Li et al., 2001). The hotter the fire the smaller the fraction of a pixel is required to reach a maximum of T3–T4. For those fires whose temperature is greater than the surface temperature by more than 500 K (rather typical, Li et al., 2001), the T3–T4 difference is a maximum when the fire fraction is well below 0.1%, depending on VA. The use of AVHRR data having large VAs contributes to this issue, as the fire fraction decreases with increasing VA. At the same time, low VAs decrease the ability to detect small fires. It should be noted that if the entire 1 km AVHRR pixel is on fire the T3–T4 difference vanishes. Li et al. (2001) report that it is unusual for a forest fire to occupy a large portion of an AVHRR pixel with a uniform temperature.

Contextual fire algorithms also use the multichannel approach (Lee and Tag, 1990). This contextual method was first explored using AVHRR data by Justice and Malingreau (1993) as part of the IGBP-DIS (International Geophysical Biophysical Program, Data Information System) fire working group. This approach was extended to global application (Justice and Malingreau, 1996; Eva and Flasse, 1996; Dwyer et al., 1998) and became part of the World Fire Web initiative (Grégoire et al., 2001). Rather than having fixed thresholds for a particular region, the textual method computes varying, pixel-specific thresholds. This approach involves two steps: first, identifying potential fire pixels and second, fine-tuning the thresholds to confirm those identified in the first step (Martin et al., 1999). Those first step thresholds are set a bit wider to ensure the inclusion of all fire pixels. In the second step, the means and standard deviations of the important fire indicators (e.g., T3 and T3–T4) are computed for the nonfire pixels surrounding the potential fire pixel. The windows for these calculations can vary, ranging anywhere from 3X3 to 21X21 pixels (Flasse and Ceccato, 1996) until the number of background pixel involved reaches a preset value (usually 25%). These statistics are then used to refine the threshold and to confirm a fire. Application of a fire algorithm to a specific region may benefit from a threshold tuned to that area,

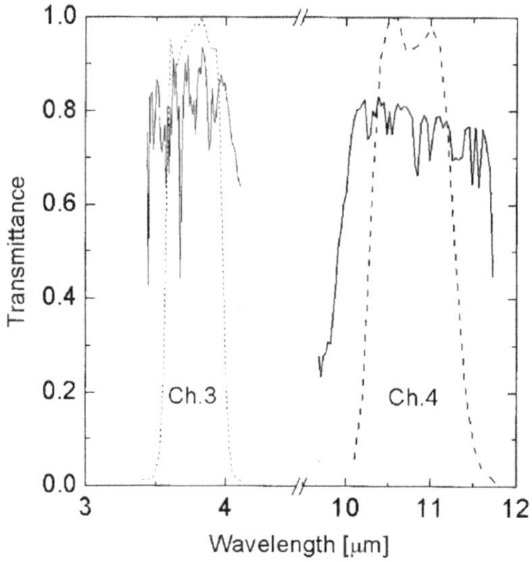

FIGURE 4.35 The atmospheric transmittance (solid lines) in the spectral regions of the channels 3 and 4 as computed by MODTRAN 3.7 for the sub-Arctic summer atmosphere. Moreover, shown are the channel 3 and 4 response functions (dashed lines, Li et al., 2001).

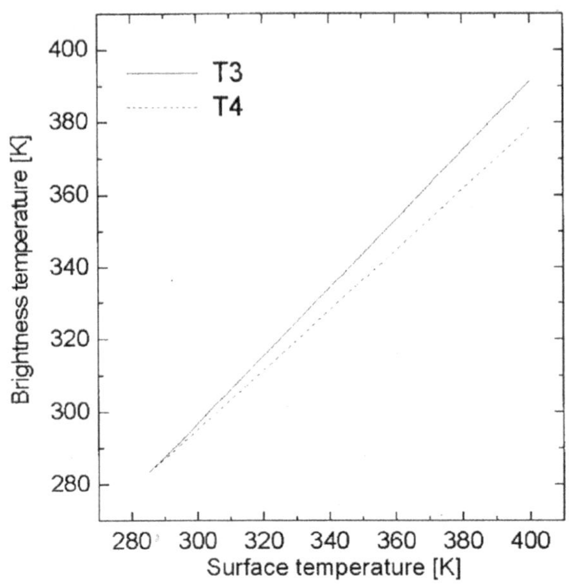

FIGURE 4.36 Top of atmosphere T3 and T4 brightness temperatures versus ground surface temperatures computed using MODTRAN 3.7, which includes atmospheric attenuation by absorption and emission (Li et al., 2001).

while global fire algorithms will have more conservative thresholds that may miss the smaller fires. These global approaches detect the larger fires, which are frequently those of greatest concern.

Before evaluating the performance of the different fire algorithms, Li et al. (2001) examined the ability of different indices to respond to fire signatures. This was an effort to get a better understanding of the contributions contributed by the different AVHRR channels. Li et al. (200c) examined the relative efficiencies of these channels and channel combinations when applied to Canadian boreal forests. This efficiency is a measure of the number of false fires eliminated (NFF) and the number of actual fires (NTF) retained (Table 4.7).

It should be noted that the real fires in this test data set consisted of data not limited to the Canadian Fire System DIS but included visual inspection of images for smoke plumes. A 1000X1000 km^2 region in western Canada was used to derive the statistics presented in Table 4.7. This table makes it clear that channel 3 alone results in a large number of false fires. While you can reduce the false fire detection by increasing the threshold, you stand a good chance of eliminating many of the real fires. Thus it is best to use a series of tests that can reduce the false fire detection by 98% (Li et al., 2001) while leaving the real fire detection only slightly reduced (~10%). The most effective test for eliminating false fires is the T3–T4 test statistic, followed by the channel 2 reflectance and the implementation of land cover information. Fortunately, the T3–T4 test is optimal, making it possible to use AVHRR data alone to detect and monitor forest fires.

To specifically test AVHRR fire algorithms, Li et al. (2001) tested five different fire algorithms applied to AVHRR data collected over Canada. The fire detection results from these algorithms are compared with each other and with ground truth data. These authors wish to note that the comparison results apply to the specific biome studied in Canada and may not apply to other very different regions. Four algorithms were designed for global fire applications, while one was specifically designed for boreal forests. Of the four global algorithms, one was created for the IGBP-DIS (Malingreau and Justice, 1997), one was used for AVHRR and the Moderate Resolution Imaging Spectrometer (MODIS) to be discussed in a later chapter (Kaufman et al., 1998a,b), one was developed for ESA (Arino and Mellinotte, 1998), and one was proposed by Giglio et al. (1999). The results of the comparisons by Li et al. (2001) are presented here in Table 4.8. The fifth algorithm was specifically designed for the Canadian boreal forest (CCRS).

Two of these algorithms used fixed thresholds (CCRS and ESA), and three are contextual methods with variable thresholds (IGBP, MODIS/AVHRR and Giglio et al., 1999). In implementing the contextual algorithms, the same cloud screening methods were applied. Some of these algorithms had unique aspects, like using smoke to eliminate false fires in the ESA algorithm. Therefore it needs to be stressed that these comparisons are limited by the application area and the general use of each algorithm.

The IGBP algorithm is the only one that has been applied on a truly global scale. The ESA algorithm was used mostly in the tropics. To adapt the MODIS algorithm to AVHRR data, it was modified to better fit the characteristics of the AVHRR channels (Li et al., 2001). The Giglio et al. (1999) fire algorithm has not been applied on a global basis, and the CCRS algorithm was used operationally in 2000 to detect fires in the boreal forest areas of Canada and the USA just a few hours after the AVHRR imagery was collected. The results had a significant impact on the Multiagency Coordination Group of the Northern Rockies Coordination Center to formulate its daily firefighting strategies.

A graphical comparison of these methods is presented in Fig. 4.37. Note that the panel labeled MODIS is the MODIS algorithm tuned to work with AVHRR, and the data used are AVHRR imagery.

The CCRS algorithm was specifically designed for boreal forest and therefore does the best job in terms of the trade-off between finding false fires (commission) and missing real fires (omission). Most of the hot spots detected by the CCRS algorithm fall within the Canadian Forest System (CAS) fire polygons, while some of those hot spots outside the polygons were verified to be real fires not mapped by the provincial and territorial fire agencies (Fraser et al., 2000). Cloud cover is the primary reason that the CCRS algorithm missed fires. While the commission errors are low, the CCRS algorithm errors are not the lowest among the five, which belong to the MODIS/AVHRR algorithm, which we again stress was the MODIS algorithm adapted to work with AVHRR data. This low commission error was at the expense of missing a large number of real fires. The ESA algorithm is quite conservative, also producing few false fires. By contrast, the IGBP algorithm is very liberal, with the highest rate of commission errors but with the lowest rate of omission errors. The Giglio et al. (1999) fire algorithm lies somewhere between the two, with moderate errors in both emission and commission. Li et al. (2001) conclude that while these leading fire algorithms do not agree well in terms of fires detected, they do generally agree in

TABLE 4.7 Summary of fire index tests and their efficiency (Li et al., 2000c).

Test no.	Description	Threshold	NTF	NFF
1	Initial test	T3>315K	12,569	168,168
2	Eliminate warm background	T3–T4>l4K	12,569	48,855
3	Eliminate nonforest scenes	Land cover	12,569	30,511
4	Eliminate bright scenes	R^2 <0.22	12,442	5665
5	Eliminate cloud edges or thin clouds	T4–T5>4.1K T3–T4<19K	11,307	2673
6	Eliminate cold clouds	T4<260K	11,307	2673
7	Eliminate single fire pixels		11,160	1828

TABLE 4.8 Five fire detection algorithms were tested the Canadian boreal forest. Note that all the parameters (av, ad, md, sd) refer to background pixels.

Algorithm/description	CCRS	ESA	IGBP	GIGLIO	MODIS/AVHRR
Algorithm category and geographic applicability	Fixed thresholds, regional (Canada)	Fixed thresholds, global/regional	Contextual, global	Contextual, global	Contextual, global
Potential fire detection			T_3>311 and T_{34}>8	T_3>310 and T_{34}>6	T_3≥315 and T_{34}≥5
Background window size			3×3–15×15	5×5–21×21	3×3–21×21
Minimum number of pixels			Max(25% of pixels tested, 3)	Max(25% of pixels tested, 6)	Max(25% of pixels tested, 3)
Background selection			T_3≤311 or T_{34}≤8	T_3≤318 or T_{34}≤12	T_3≤320 or T_{34}<20
Actual fire detection with T_3 and/or T_4	T_3>315	T_3>320	Define: ξ_3=av(T_3)+2*sd(T_3)+3 ξ_{34}=Max{8, av(T_{34})+2*sd(T_{34})}	Define: ξ_4=av(T_4)+ad(T_4)-3 ξ_{34}=av(T_{34})+Max {2.5*ad(T_{34}),4}	Define: ξ_3=Min[320,av(T_3)+4*Max {sd(T_3),2}] ξ_{34}=Min[20,md(T_{34})+4*Max{sd(T_{34}),2}]
			Then, confirm potential fires as real if: T_3>ξ_3 and T_{34}>ξ_{34}	Then, confirm potential fires as real if: T_4>ξ_4 and T_{34}>ξ_{34}	Then, confirm potential fires as real if: T_3>360 or [T_3>ξ_3 and T_{34}>ξ_{34}]
Filter hot surfaces	T_{34}≥14	T_{34}>15	Incorporated into fire detection	Incorporated into fire detection	Incorporated into fire detection
Filter clouds	T_4≥260	T_4>245	R_1+R_2≤1.2 and T_5≥265 and (R_1+R_2≤ 0.8 or T_5≥285)	*IGBP criteria applied here (no external cloud mask)*	*IGBP criteria applied here (no external cloud mask)*
Filter reflective surfaces	R_2≤ 0.22	R_1<0.25	R_2<0.20	R_2<0.25	
Filter sun glint		|R_1-R_2|>0.01			R_1≤0 3 or R_2≤ 0.3 or reflected sun angle ≥40 degrees
Other detection criteria	T_{34}*19 or T_{45}<4.l				
Postprocessing (not applied in this investigation)	Elimination of nonforest and isolated pixels	Quicklook inspection Max annual NDVI>0			

R_1: reflectance in band 1 (0.66 μm); R_2: reflectance in band 2 (0.86 μm); T_3: brightness temperature in band 3 (3.75 or 3.95 μm for MODIS); T_4: brightness temperature in band 4 (10.8 μm); T_5: brightness temperature in band 5 (11.9 μm); T_{34}: T_3–T_4; T_{45}: T_4–T_5; av(): mean; ad(): mean absolute deviation; md(): median, sd(): standard deviation. A temperature is in K. *AVHRR*, Advanced Very High Resolution Radiometer; *ESA*, European Space Agency; *IGBP*, International Geosphere Biosphere Program.

terms of the spatial distribution of fires. It is unlikely that any one algorithm will work well everywhere, and there is a need to tune the algorithm for different biomes and environments.

All of the algorithms had significant commission errors (Fig. 4.38) over an agricultural region in the western Canadian prairies. Solar reflection from bare soil is thought to be the primary cause of false fires that were detected in May and June when the snow has melted but the crops have not yet appeared.

Li et al. (2001) carried out a correlation analysis of the five algorithms as applied across Canada (Table 4.9). The ESA and CCRS algorithms are highly correlated because they are both multichannel fixed threshold algorithms. The contextual algorithms are all correlated with the CCRS algorithm about the same amount. The highest correlation is between the ESA algorithm and the MODIS/AVHRR algorithm.

These authors also note that a potential bias exists due to the validation method whereby false alarms in the Canadian fire polygons will be interpreted as active fires increasing the apparent detection rate. One source of this type of error is that subpixel areas that are burned in a previous season and are bare at the time of satellite overpass. The bare soil may contribute to the larger number of false fires being identified as active fires. This is likely the explanation for the IGBP algorithm showing clusters of false alarms in areas that had burned the previous year (Fig. 4.37).

Li et al. (2001) provide the following suggestions to improve the fire hot spot algorithms:

1. Set thresholds based on land cover type, which can be used to estimate albedo or emissivity.
2. Replace the threshold with physical models that simulate the fires.
3. Develop new and innovative methods that take the best advantage of the new sensor capabilities planned for future satellites.
4. Explore detection methods that take advantage of the time series of satellite data.

They conclude that the most challenging aspect of AVHRR-based fire algorithms is to correct for the reflection from clouds and Earth surfaces. Many of the algorithms include a test to remove cloudy pixels but there is no test for surface reflection and its contribution to commission errors over bare ground. One can either use channel 3 to screen out bright surfaces or adjust the fire threshold according to land cover type. While the T3–T4 difference made significant contributions to the fire algorithms, it is important to remember that this value is only useful for inhomogeneous fire

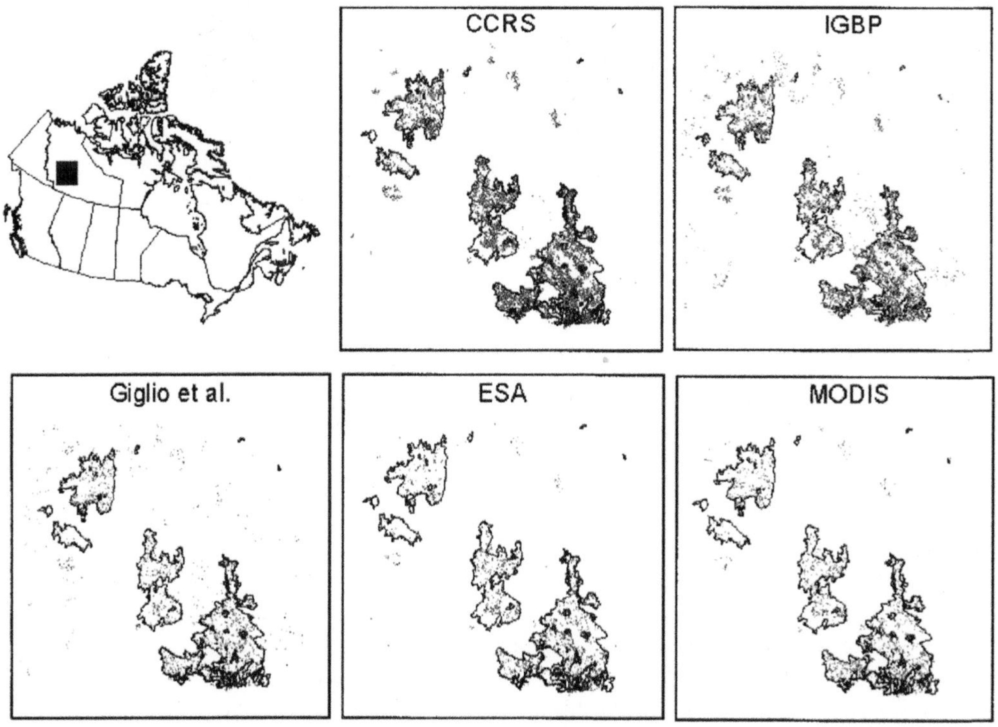

FIGURE 4.37 Comparison of hot spots detected by the five algorithms in an area of the Canadian boreal forest is indicated on the map. The satellite hot spots are compared with fire polygons. Map lines delineate study areas and do not necessarily depict accepted national boundaries. *ESA*, European Space Agency; *IGBP*, International Geosphere Biosphere Program.
From the Canadian forest agencies (Li et al., 2001).

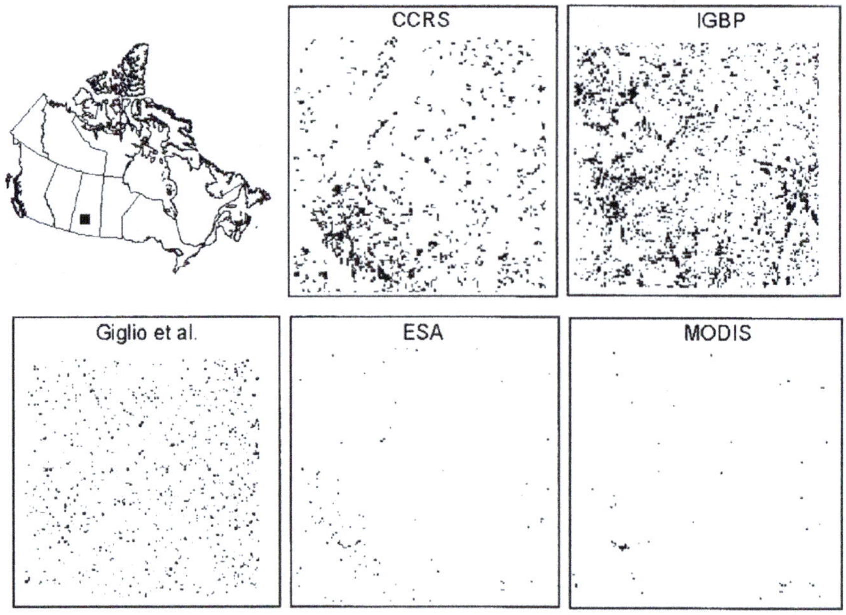

FIGURE 4.38 Fire detection results for the five algorithms over a Canadian prairie area (Li et al., 2001). Bare soil has a much larger reflectance than green vegetation. Map lines delineate study areas and do not necessarily depict accepted national boundaries. *ESA*, European Space Agency; *IGBP*, International Geosphere Biosphere Program.

TABLE 4.9 Correlations between the hot spots identified by each of the algorithms.

	CCRS	ESA	GIGLIO	IGBP	MODIS/AVHRR
CCRS	1.00				
ESA	0.91	1.00			
GIGLIO	0.80	0.81	1.00		
1GBP	0.82	0.70	0.80	1.00	
MODIS	0.81	0.94	0.77	0.62	1.00

ESA, European Space Agency; *IGBP*, International Geosphere Biosphere Program.

pixels. This can be explained as when a portion of an AVHRR pixel is burning, the radiance emitted by this hot spot leads to a much greater response in channel 3 versus channel 4. This difference will be much smaller in a nonburned area. Thus this difference between the channels is inversely proportional to the thermal homogeneity of the fire scene. Since most fires are not uniform at the scale of an AVHRR pixel, this channel difference works well for the removal of false fire detection. They also found that the contextual algorithms were very sensitive to the selection of the initial fire thresholds and recommended that more liberal thresholds be used for potential fire pixels and more conservative thresholds be used to calculate the nonfire background.

In their study, Csiszar et al. (2003) examined the interannual changes in active fire detectability in North America using the long-term records provided by the AVHRR. They analyzed 1 km AVHRR imagery provided by NOAA with a goal of producing a long-term time series of active fire detection over the continental United States and Canada. The interannual variations were driven by changes in background temperatures and by the orbital drift of the NOAA afternoon polar orbiting satellites. Two global-scale fire inventories have been derived from AVHRR data: the Global Fire Product (Strappiana et al., 2000) covering the years 1992 and 1993, and the World Fire Web system which started in 1998 (Gregoire et al., 2001). Several countries operate national AVHRR-based fire monitoring systems such as the those in Canada (Li et al., 2000a) and in Brazil (Franca and Setzer, 2001).

While prelaunch calibrations show that the channel 3 saturation temperatures vary slightly between satellites and their AVHRRs, Csiszar et al. (2003) decided to use a constant of 322.0 K for all of the satellites. Over the lifetime of the AVHRR sensor operation, this saturation temperature shows some modest increase with sensor degradation; these authors felt this was small enough that a constant value could be used for the saturation temperature of channel 3. In contrast, the saturation temperatures of channels 4 and 5 (the longer wavelength thermal infrared channels) have a 5–8 K higher nominal saturation temperature.

The primary goal of their study (Csiszar et al., 2003) was to discuss the different factors influencing the creation of a high-quality, long-term time series of active fire occurrences. There are two main factors influencing the interannual changes in the background conditions: (1) the orbital drift (Fig. 4.39) of the NOAA satellites and (2) the interannual changes in weather patterns. It is easier to quantify the first effect, but their analysis also addresses the second factor.

The interannual variation in the local time of observation (here represented by changes in equatorial crossing times) will lead to changes in the background temperatures. It should be noted that both omission and commission errors depend strongly on the discrimination of the fire signal from the background temperature. This change in viewing time also changes the amount of reflected radiation (sun glint problem), which affects the channel 3 response. This is a particular problem when subpixel water bodies are present to cause the direct specular reflection, which is the sun glint. This is a typical condition, particularly in Canadian boreal forests. These pixels need to be identified and eliminated from consideration.

The analysis of Csiszar et al. (2003) was based on the NOAA/NESDIS Pathfinder Atmosphere (PATMOS) data set (Stove el al., 2002) which is available from mid-1981 to the end of 2000 for the afternoon satellite of the NOAA polar orbiting series. While this data set was established to study clouds and aerosols its clear-sky radiance products can be used for the study of land surface processes (Gutman et al., 2000). Csiszar et al. (2003) used PATMOS data to characterize spatially homogeneous background temperature conditions and therefore the use of 3 km GAC data is acceptable. They used AVHRR channel 3 and channel 4 radiance products projected onto a 1X1 degrees grid. These radiance values were converted into BTs using satellite specific centroid wavenumbers and conversion coefficients. They further assumed that the number of fire pixels in a 1X1 degrees cell each month would be small enough so as not to affect the monthly means used to characterize the background conditions.

To identify the active fires, they used a continuous record of 1 km AVHRR data available at NOAA (http://www.saa.noaa.gov), which goes back to May 1985. In spite of the fact that in the earlier years the coverage is spotty, they decided to use the part of the 20 years covered by PATMOS for their series between 1985 and 2000, which covered the NOAA satellites 9, 11, and 14. The primary requirement for fire detection is to be able to distinguish the fire signature from the background radiance. Even small fires can often saturate the mid-infrared of channel 3 while the thermal infrared channel 4 is less sensitive to the presence of subpixel hot spots. Thus the T3–T4 difference is a good indicator of the presence of active fires. Note that if the fire hot spot fills the pixel, T4 increases substantially, and the T3–T4 difference is no longer useful. Fortunately, this is seldom the case for AVHRR pixels. If the background temperature increases, this difference becomes smaller, and when T3 reaches its saturation temperature, this difference tends to reject valid fire pixels, resulting in omission errors. If the T3–T4 threshold is decreased to account for valid fires, the likelihood of detecting false fires (commission errors) will also increase.

It should be noted that the use of contextual heterogeneity tests (e.g., Flasse and Ceccato, 1996; Justice et al., 1996) that derive dynamic thresholds from the background temperatures may reduce the ambiguity of the fire signature but not fully

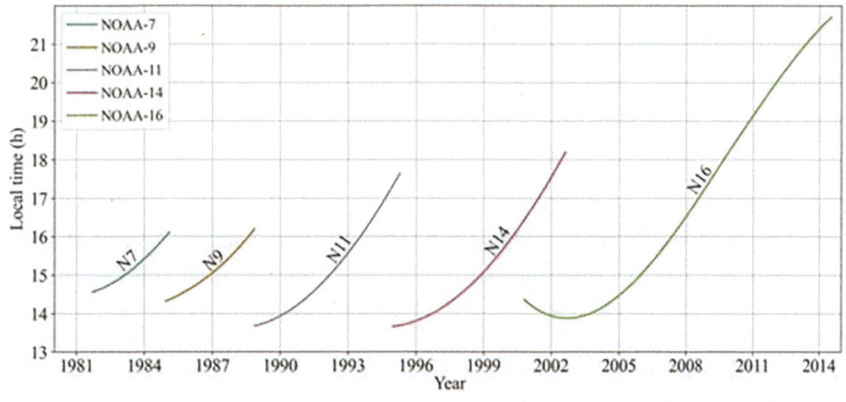

FIGURE 4.39 Equator crossing times of the afternoon NOAA satellites between 1981 and 2000.

resolve it because (1) these algorithms are still based on the assumption that the presence of fire in a pixel will result in a significant difference between the potential fire pixel and its surroundings and (2) as the background temperature increases, the differences between the fire pixel and the surroundings become increasingly ambiguous. While contextual algorithms tend to reduce false alarms, there are still commission errors at the edges of the hot spots or of clouds.

Csiszar et al. (2003) do not select a particular fire detection algorithm to test but rather just examine the characteristics of the channels that go into these algorithms. Since the T3–T4 difference is so important, they first address these two channels (Fig. 4.40). July was picked as the height of the fire season in North America, and the four areas were selected to show the case for very different vegetation ground covers. It should be noted that the higher monthly mean T3 values compared to T4 are not caused by fires but rather the solar reflection component of the channel 3 value. This causes a large number of commission errors in daytime fire detection for algorithms that use this difference.

All the plots show some amount of interannual variability but the effect on the T3–T4 difference varies considerably with the different ground covers. In Montana, the interannual variation of T3 values is significant, with relatively low values, but the background T4 values have a similar interannual variation and are so low that there is still a measurable difference between T3 and T4. The croplands in Kansas show a very marked interannual variation in both T3 and T4, but the latter is significantly lower, making the T3–T4 difference a good fire indicator. Over the shrublands of NW Utah, the interannual variation is less than in the first two areas, and T3 appears to hit an upper threshold, reducing the interannual variation and compromising the T3–T4 detection difference about half the time. Finally, the sparse vegetation cover of Arizona shows a saturated T3 value for most of the years and a widespread of standard deviations of T4, suggesting that the standard T3–T4 difference would not be very useful in detecting fires.

Csiszar et al. (2003) conclude that for long-term studies of fire detection that minimize both omission and commission errors, it is important that these errors remain constant over time. Nonforested areas are seriously impacted by the interannual changes in solar illumination and sensor VA. This type of uncertainty can only be minimized by careful analysis of the changes in land conditions that modify both emission and reflectance, changing the requisite thresholds for fire detection (Cuomo et al., 2001). Over forested areas, these variables can be assumed to be fairly constant, increasing the ability of the T3–T4 differences to detect active fires and separate them from false alarms. There is the possibility of combining hot spot identification with visible channel smoke detection to better map and monitor the active fires. Other sources of error in the interannual variability of fire detection are changes in sun-glint conditions, changes in the postlaunch calibration of the AVHRR sensor, and compensating for the orbital drift of the NOAA satellites carrying these sensors.

Data from the AVHRR are also useful in estimating the fire fuel potential. Chuvieco et al. (2004) use NDVI and surface temperatures computed from AVHRR data to study the fuel moisture content (FMC) from Mediterranean grasslands and shrubs. They had a 6-year time series of field measurements of FMC and a 4-year time series of AVHRR data. In forest fire literature, the FMC is defined as the percent of water weight over the sample dry weight. The FMC is computed from field measurements using gravimetric sampling (wet samples are weighed, the oven-dried at 60°C or 100°C, (Vegas et al., 1992) and then weighed again to get the dry weight). FMC can be calculated for both live and dead vegetation.

For the application to fire fuel mapping, FMC correlates well with NDVI, which can then be used as a surrogate to compute FMC in areas where there are no field measurements. FMC can be considered to be a function of equivalent water thickness that is usually calculated from satellite data. The estimation of FMC from satellite reflectance measurements is best carried out when a single vegetation species is being sensed. Strong empirical relations and satellite variables have been found (Alonso et al., 1996; Chladil & Nunez, 1995; Chuvieco et al., 1999, 2002, 2003a; Hardy & Burgan, 1999; Paltridge & Barber, 1988). Chuvieco et al. (2004) based their estimates on an empirical model using linear regression and the AVHRR products of NDVI and LST (both of which have been discussed earlier). The empirical fit employed long time series of FMC field measurements by the Cabaneros National Park study site in central Spain along with FMC measurements at some additional sites (Fig. 4.41).

The primary site for the FMC measurements is in the National Park in central Spain. The central area of this park is covered in grassland and shrubs on gentle slopes. As a park, the area is protected, and there is no farming or any other agricultural activity, and the damage from anthropogenic use is minimized. Three grassland plots were sampled along with two areas covered with multiple species of shrubs. Each sample plot area was 50X50 m^2 and separated by 3–5 m. In their study, Chuvieco et al. (2004) used the grasslands and only *Cistus ladanifer* one of the dominant shrub species in the Mediterranean area. Field measurements were collected every 8 days from April to September and a time series from 1996 to 1999 was available for model calibration. For validation purposes, another field campaign was done between 2001 and 2002 when the samples were collected every 16 days since analysis showed that the earlier FMC values did not exhibit changes over shorter periods. When the data from the calibration sites were examined, it was found that there were no

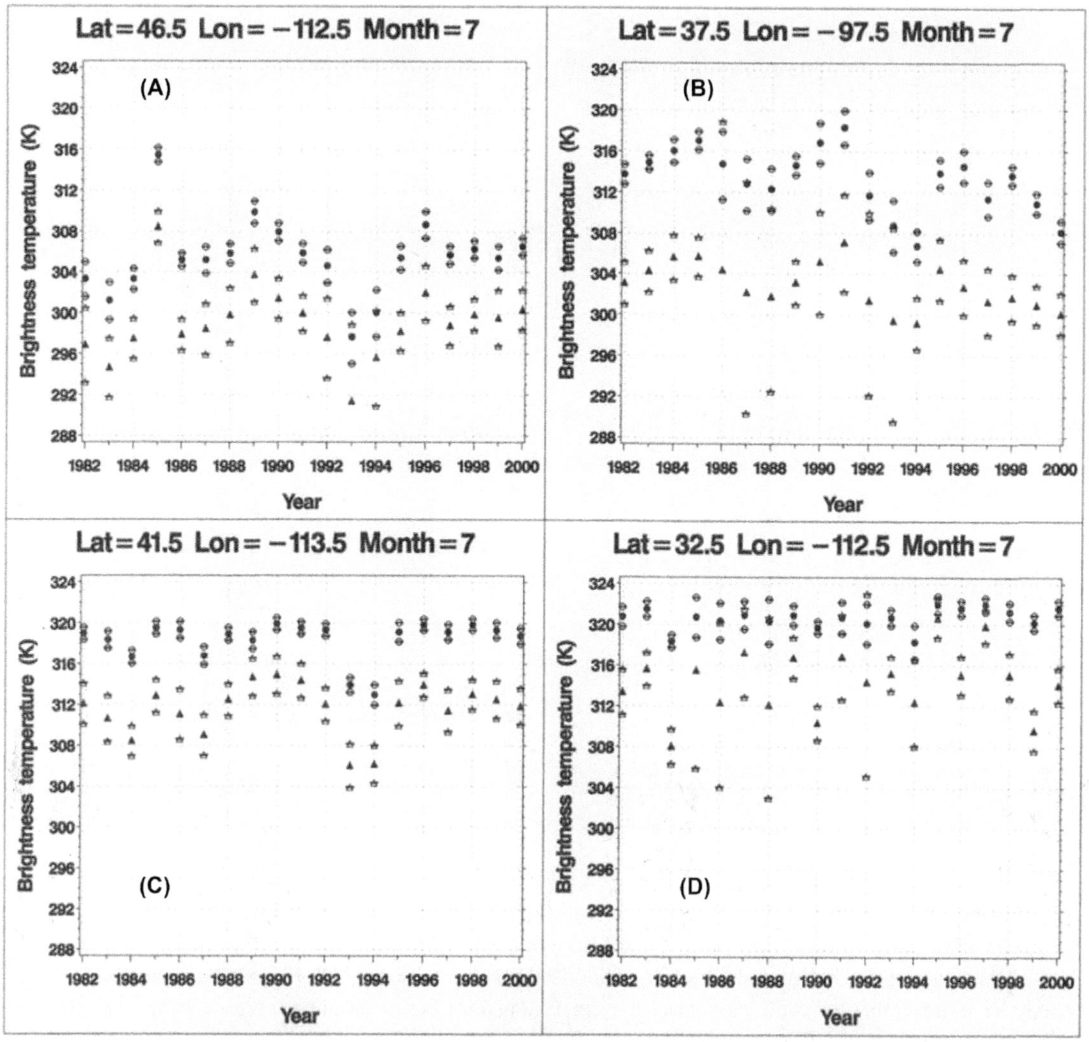

FIGURE 4.40 Time series of Advanced Very High Resolution Radiometer channels 3 (circles) and 4 (triangles) monthly mean brightness temperatures and one standard deviation (symbols with a dash) in July over a 1X1 degrees grid in (A) Montana, (B) Kansas, (C) NW Utah, and (D) Arizona.

significant differences between the sample plots in spite of the 3–5 km separations. Thus we can assume that these smaller spot samples are representative of the much larger area represented by an AVHRR pixel.

During the 2001–2002 validation phase, they used a different set of field measurement sites (also shown in Fig. 4.41) that included grassland and a mixture of *C. ladanifer* and *Rosmarinus officinalis* shrubs together with some other minor shrub species. These plots are 200–500 km away from the park calibration site and have different elevations, but they still are part of the Mediterranean ecosystem. The plots were selected to have as homogeneous ground cover as possible. While the grasslands come close to satisfying this assumption, the shrub plots have mixtures of shrub species. o keep things consistent, the sampling methods used at the calibration sites were repeated at the validation sites.

The AVHRR data were received by the University of Alcala's HRPT receiving station. Raw digital counts were converted to reflectance using procedures specified by NOAA, and LST was calculated using the method proposed by Coll and Caselles (1997). Geometric corrections were based on orbital models, and further adjustment was carried out using manual control points. Daily values were used to formulate 8-day composites of maximum NDVI values. The median value of a 3X3 pixel window was extracted for each ground location and correlated with the field measurements. These investigators used an empirical fit of the FMC to AVHRR reflectance following previous studies (Alonso et al., 1996; Chladil & Nunez, 1995; Chuvieco et al., 1999, 2002, 2003a; Hardy & Burgan, 1999; Paltridge & Barber, 1988). The statistical model was developed using 88 periods (22 periods of 8 days over 4 years, 1996 to 1999) covering spring and summer data in the National Park region. This period included some dry years (1997 and 1999), where the overall

FIGURE 4.41 Locations of the ground truth sites in Spain (Chuvieco et al., 2004). Map lines delineate study areas and do not necessarily depict accepted national boundaries.

precipitation was close to 200 mm in 6 months, as well as some wet periods with 230 to 250 mm of moisture over the same period (1996 and 1998). This empirical relationship was validated with the National Park measurements as well as the other field sites discussed earlier. This model included both NDVI and ST.

For the 6 years of field data, the 3 grassland sites and 2 shrub plots show very similar behavior (Fig. 4.42) in spite of their spatial separations. Thus it may be assumed that the calibration plots are representative of the larger areas covered by an AVHRR pixel. As stated earlier, the equations to estimate FMC from AVHRR images were derived from multiple linear regression analysis using NDVI, LST, and the seasonal trends in FMC for the 4 years of National Park calibration data (1996–99).

These equations are as follows:

$$FMC_g = -57.103 + 284.8808 \times NDVI - 0.089 \times LST + 136.75 \times FD_g \tag{4.16}$$

$$FMC_c = 70.195 + 53.520 \times NDVI - 1.435 \times LST + 122.087 \times FD_c \tag{4.17}$$

where FMC_g and FMC_c are the AVHRR estimated FMC values for the grasslands and the *C. ladanaifer* shrubland, respectively. NDVI is the NDVI as discussed earlier in this section (ranges from −1 to 1), LST is the surface temperature (in °C) and FD is a function of the day of the year. Using these equations over the time span available, the coefficient of determination (r^2) was 0.737 for grasslands and 0.672 for the *C. ladanifer* shrub. Significance levels were less than 0.1 for NDVI and FD for the grasslands, and for LST and FD in the case of *C. ladanifer* shrub. LST was not significant for grasslands ($P>.5$) since most of the discrimination power was in the FD_g variable, but it was retained in the model fit to improve spatial estimations. For the same reason, NDVI was kept for *C. ladanifer* in spite of its very low significance ($P<.187$). The contribution of NDVI to grassland FMC is substantial due to the fact that the drying process in herbaceous species is followed by a loss of chlorophyll activity and the decrease of leaf area index (LAI) that would be sensed by the NDVI. For both cases, the seasonal trends (FD) are very significant ($P<.001$), and the NDVI is highly correlated with the FMC.

The FMC estimated from the AVHRR-derived values is plotted as a time series in Fig. 4.43 for the Cabaneros site against the ground truth FMC for the 2001–2002 validation period.

The temporal changes are very well represented by the satellite-derived FMC, and the differences are quite small with no consistent bias. The worst fit occurred for the grasslands in late spring (early June) and in the middle of summer (August). The former misfit is during a sudden decrease of the FMC, which in both years changes over 30-35% in 16 days.

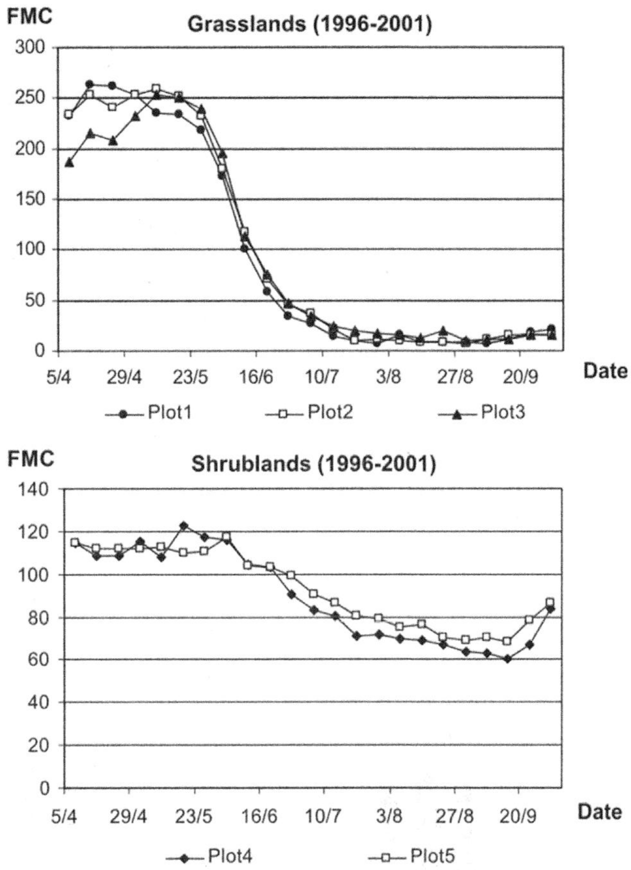

FIGURE 4.42 Time series of average fuel moisture content (FMC) values for the different ground truth sites (Coll and Caselles, 1997).

This rapid drop is seen in the decrease of the NDVI and the increase of LST, but a less steep decrease is seen in the observed values. Chuvieco et al. (2004) argue that for practical purposes these misfits are not a problem, and filters could be applied to remove them. The validation of *C. ladanifer* exhibited a better fit of the estimated FMC to that observed. The additional two validation sites for grasslands also showed very good fits with r^2 values of 0.881 and 0.905. The additional *C. ladanifer* site also had a high r^2 value of 0.7241.

This study concluded that AVHRR-derived estimates of field moisture content were reliable, at least for the Mediterranean ecosystems. This is an important factor in estimating the fire fuel potential if it can be extended to other vegetation cover. This provides a means of calculating the forest fire danger from the NOAA AVHRR. Thus this instrument is not only useful in detecting and monitoring fires but also in evaluating the fire potential.

In another study of FMC, Boyer and Dawson (2004) use a modeling approach to determine the sensitivity of satellite reflectance data at the leaf and canopy levels to changes in the biophysical variables that contribute to FMC. At the leaf level, their results indicate a sensitivity of reflectance to changes in leaf water and dry matter content that is used to calculate FMC, which is greatest in the short wave infrared (SWIR, AVHRR channel 3) and the (NIR, AVHRR channel 2). Variations in FMC have no effect at the visible wavelengths, which are centered in the red for the AVHRR. At the canopy level, they find that the reflectance sensitivity to changes in leaf water and dry matter content depends greatly on the type of model used and the variations in the canopy composition. In the longer wavelengths of the SWIR, the influences of variable LAI, fractional vegetation cover, and solar zenith angle are greater than those at the shorter wavelengths of the SWIR and the NIR. Empirical correlations between NDVI and FMC are shown to be weaker than with canopy water content.

Another aspect of forest fires where AVHRR data can be useful is in estimating the fire-burned area. Again, there is a study in Spain that uses AVHRR imagery to assess the damage caused by forest fires (Gitas et al., 2004). Their focus is to study the impact of a large forest fire that took place on the Spanish Mediterranean coast in the Crus Cape region. Many studies have investigated the use of satellite imagery to generate accurate estimates of burned areas (Caetano et al., 1994; Chuvieco & Congalton, 1988; Milne, 1986; Mitri & Gitas, 2002; Pereira, 1992; Pereira et al., 1997; Pereira & Setzer, 1993;

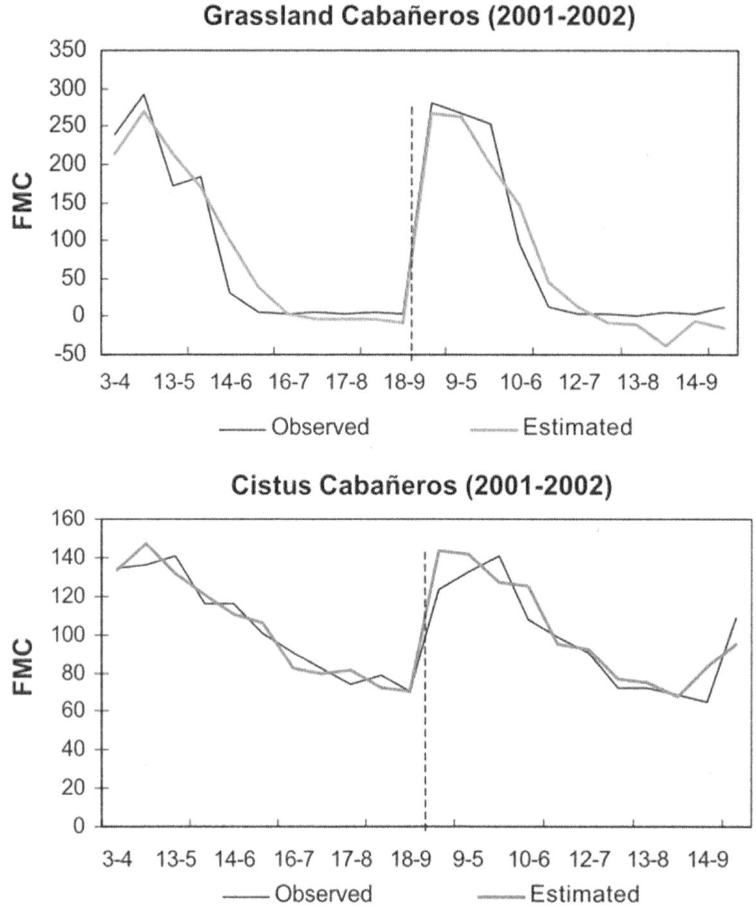

FIGURE 4.43 The observed and estimated fuel moisture content (FMC) for the grasslands (top) and the *Cistus ladanifer* (bottom) over the 2001–2002 validation period measured at the initial calibration sites (Coll and Caselles, 1997).

Siljeströ m & Moreno, 1995; Tanaka et al., 1983). The great advantage of the remote sensing estimates of fire-burned areas is the ability to quickly survey large areas and to generate long time series. While the thermal channels of the AVHRR play a significant role in separating burned areas from other types of land cover (Martin et al., 2002), there are still some significant problems to overcome. The area sampled by the radiometer changes with scan angle, changing the amount of atmospheric interference, which also influences the amount of greenness detected (Holben, 1986).

The study by Gitas et al. (2004) focused on a large fire on the coast of Spain (Fig. 4.44) that occurred in August 2000 in the eastern extremity of the Pyrenees in Creus Cape, Spain. The total burned area of 6,000 ha was covered 83% by wooded areas and 17% by non-wooded areas. The fire, which had an anthropogenic ignition, was worsened by strong prevailing winds. The fire started at 11 am and was extinguished later the following day. The AVHRR image was acquired immediately after the fire, at the HRPT receiving station in the Geography Department of Alcala de Henares University in Madrid, Spain. The images were geometrically corrected and the band reflectance and radiances calibrated using the standard NOAA procedures.

To study the burned out areas of the AVHRR image, we need to classify the image which is by segmentation. The segmentation used in this study was a bottom-up, region-merging method starting with single pixel objects. During this procedure, the whole image was segmented, and image objects were generated based on adjustable criteria of homogeneity in color and shape. The visual interpretation of different image segmentation results indicated that it was best to use the thermal infrared channel 4 (10.3–11.3 μm), which was then the sole information source in the segmentation process. The objects resulting from the segmentation were grouped into three different classes. Each class was defined by a fuzzy set which consisted of membership functions of the object features (Fig. 4.45).

Membership functions designate whether a feature is typical, atypical, or typical of a different class. Three different classes resulted: (1) nonburned area, (2) burned-a for fire-affected areas larger than 3900 ha, and (3) burned-b for areas smaller than 3900 ha. The burned thresholds were selected to match the segmentation results. The next step was to

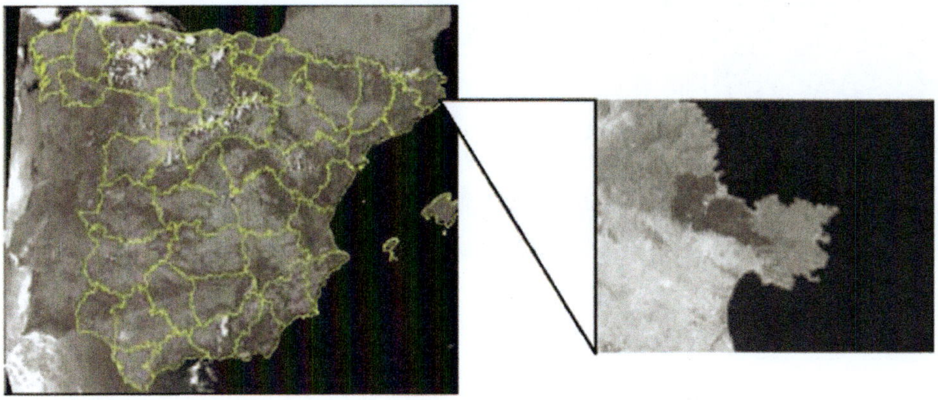

FIGURE 4.44 The study area in Spain. On the left is the Advanced Very High Resolution Radiometer image of the Iberian Peninsula, and on the right is the Landsat-TM image of the Ceus Cape (Gitas et al., 2004). Map lines delineate study areas and do not necessarily depict accepted national boundaries.

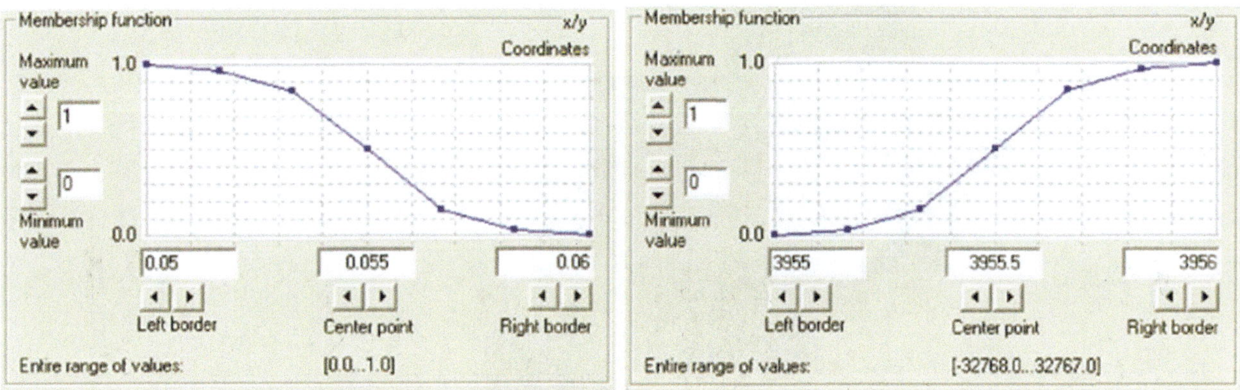

FIGURE 4.45 Two examples of membership functions. On the left is the ratio of band 1, while on the right is the mean of band 3 (Gitas et al., 2004).

classify the image using the segmentation groupings. In this process, burned areas were merged with burned-b. All of the classes were organized to form a structured group. By merging image objects at level 1 (Fig. 4.46), level 2 is created.

This final product compared well with ground surveys of the burned area (Fig. 4.47).

Here, we can clearly see how well the AVHRR-derived burn area matched the fire perimeter based on ground measurements. While this is limited to a comparison of a fairly small area in Spain, it suggests that similar methods may be applied to AVHRR data to estimate the area burned by a forest fire.

4.12 Flood monitoring and inundation estimation

Closely related to forest fire and burned scar monitoring is the monitoring of floods and estimating the areas inundated by floods. The AVHRR has again been a useful tool in these types of applications. Barton and Bathols (1989) discuss using AVHRR data to study the movement of a flood down the Darling River in central eastern Australia. They found that the thermal infrared channels collected at night yielded a better contrast between the flooded areas and the surrounding land than was possible with the visible or NIR channels. This flood progressed slowly, which is typical of floods in central Australia (Fig. 4.48). This figure also shows the dates when the flood hit a particular area. The monthly total rainfall for three sites in this catchment area for the first 6 months of 1988 is presented in Fig. 4.49. These rains resulted in considerable flooding along the entire length of the Darling River. This river falls from 107 m above sea level at Bourke to 64 m at Menindee. This drop of 43 m takes place over a distance of 440 km. The AVHRR images over this period show the movement of the flood down the river, the extent of the flooding, and the speed of the flood front.

Looking at Fig. 4.49, we can easily see how this April rainfall event led to a flood. Five AVHRR images were used in this analysis as listed in Table 4.10.

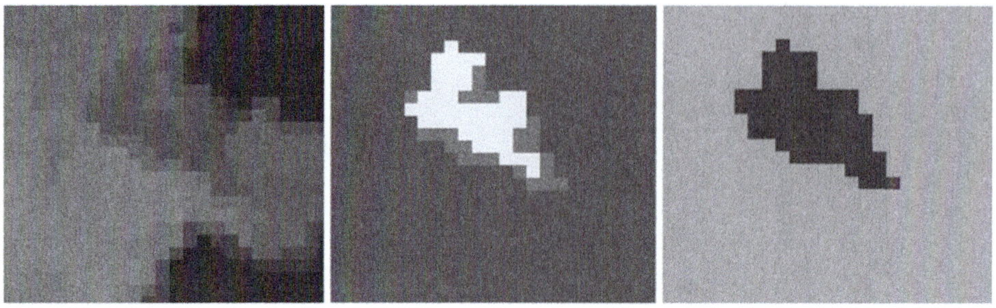

FIGURE 4.46 Burned area classifications. The final result is on the right (Gitas et al., 2004).

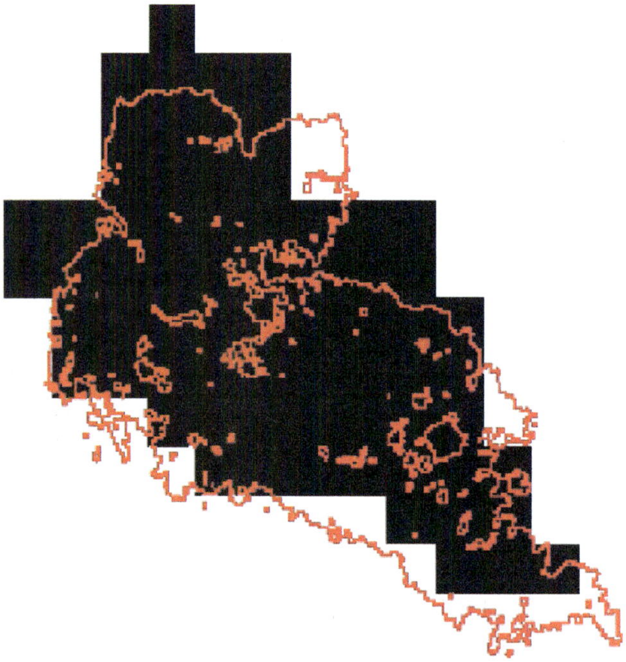

FIGURE 4.47 Advanced Very High Resolution Radiometer derived burn area (*black* area) versus the fire perimeter (*red*) (Gitas et al., 2004). Map lines delineate study areas and do not necessarily depict accepted national boundaries.

For their study of this flood, Barton and Bathols (1989) decided to use a simple vegetation index based on AVHRR channels 1 and 2 (visible and NIR). This index is given by

$$VI = (\alpha_2 - \alpha_1)/(\alpha_2 + \alpha_1) \quad (4.18)$$

where α_1 and α_2 are the reflectance in AVHRR channels 1 and 2. The daytime image of this VI on May 29, 1988, is shown here in Fig. 4.50 over the target area containing the Darling River.

Here, the river is partly obscured by clouds, and even when clear, the VI indicates a very noisy background. Barton and Bathols (1989) found instead that the nighttime channel 4 thermal infrared image (Fig. 4.51) gave a much clearer picture of the river and the flooded areas.

They then took random samples of 20 pixels from both the VI image (Fig. 4.50) and the thermal infrared image in Fig. 4.51 and computed the average values and the standard deviations, which are summarized in Table 4.11.

These statistics reveal that the nighttime thermal infrared image does a much better job of discriminating land and water boundaries than do either of the daytime images. The results show that the flood vegetation index is closer to the value of the land than for the lakes. This is likely due to silt and debris being present in the floodwaters, which raises the VI index to a level close to the land surface albedo. The thermal image during the day does a poor job of discriminating between both lakes and floods since the water has temperatures similar to the land. At night, however, the cloud cover is minimal, and the heat loss from the land cools it to a temperature below that of the water in the flood and the lakes. It is interesting to note that there is a diurnal variation in the temperatures of the flood that is larger than that for the lake.

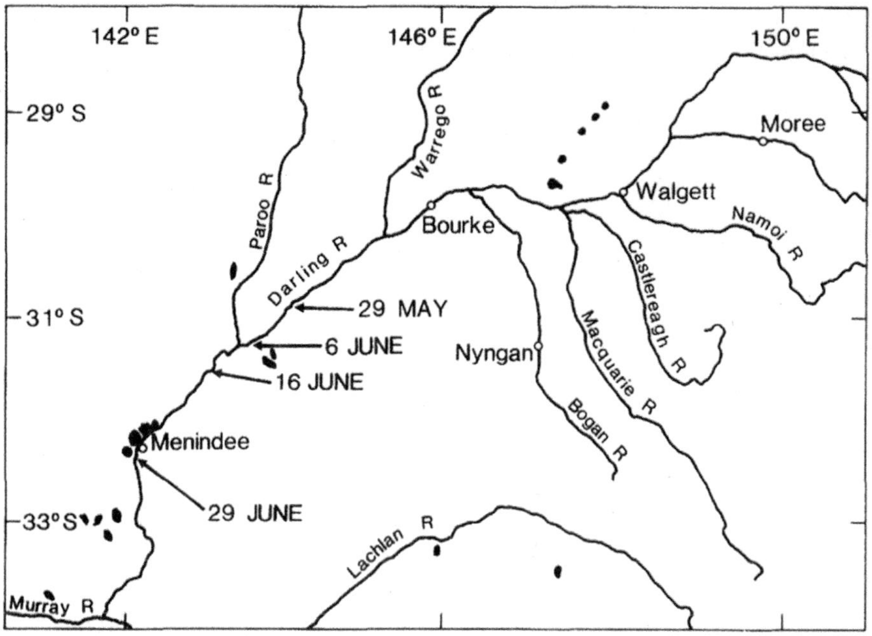

FIGURE 4.48 The Darling Rivers and its catchment area (Barton and Bathols, 1989). Map lines delineate study areas and do not necessarily depict accepted national boundaries.

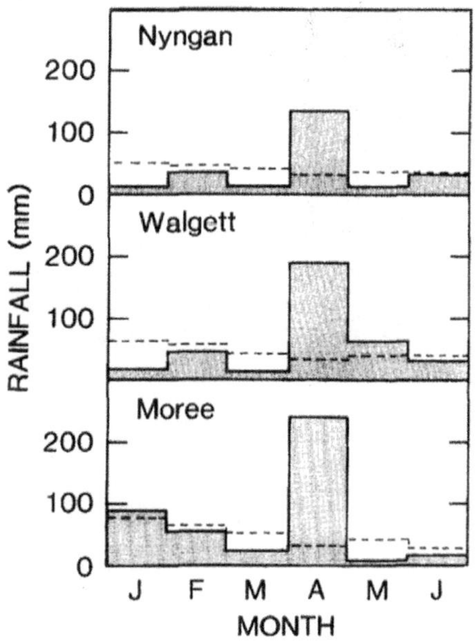

FIGURE 4.49 Rainfall for three sites in the Darling River catchment area. The dotted lines show the average values (Barton and Bathols, 1989).

The front location of the flood is given in Table 4.10 and shows that by June 29, the flood had reached the lakes near Menindee. This leading edge was observed for the nighttime thermal infrared images of channel 4. The estimate of the areal coverage of this flood was performed by dividing up the image into 64X64 pixel areas and then deriving a histogram of the BT in channel 4 (Fig. 4.52).

The peak in the histogram between 2°C and 6°C is due to the area (about 1 km^2) being totally covered by land with no water cover. The part of the histogram above 10°C is completely water, indicating the presence of lakes and the river. In between are pixels that combine land- and water-covered surfaces. Barton and Bathols (1989) assumed that the land and

TABLE 4.10 Details of the NOAA-9 Advanced Very High Resolution Radiometer (AVHRR) data use in this study.

			Flood front location		
Date	**Time (Z)**	**Orbit no.**	***Lat***	**Long**	**Distance (km)**[a]
29 May	0639	17828			
29 May	1802	17835	31°05′S	144°13′E	0
6 June	1816	17948	31°25′S	143°51′E	58
16 June	1807	18089	31°39′S	143°23′E	116
29 June	1726	18272	32°32′S	142°10′E	240

[a] *Distances are from the front location on 29 May.*

FIGURE 4.50 Image of the vegetation index for the daytime NOAA-9 image on May 29, 1988. The dotted lines indicate latitude and longitude (Barton and Bathols, 1989).

water pixels at each end of the histogram have a symmetric distribution with temperatures such that the pixels are not classified as either land or water and are likely water-covered land surfaces caused by the flood. They suggest that the pixels near 7°C are more land than water, and those near 9°C are more water than land. In this way, they use the image in Fig. 4.52 to be about 920 km^2 with an accuracy of 25%.

In a similar study, Jain et al. (2006) examined floods of the Brahmaputra River in the state of Assam in India. This river is known to flood every year, but the extent of the inundation experienced varies from year to year. These authors used AVHRR data to study the maximum spatial extent of this inundation in 2003. They state that of all the natural disasters,

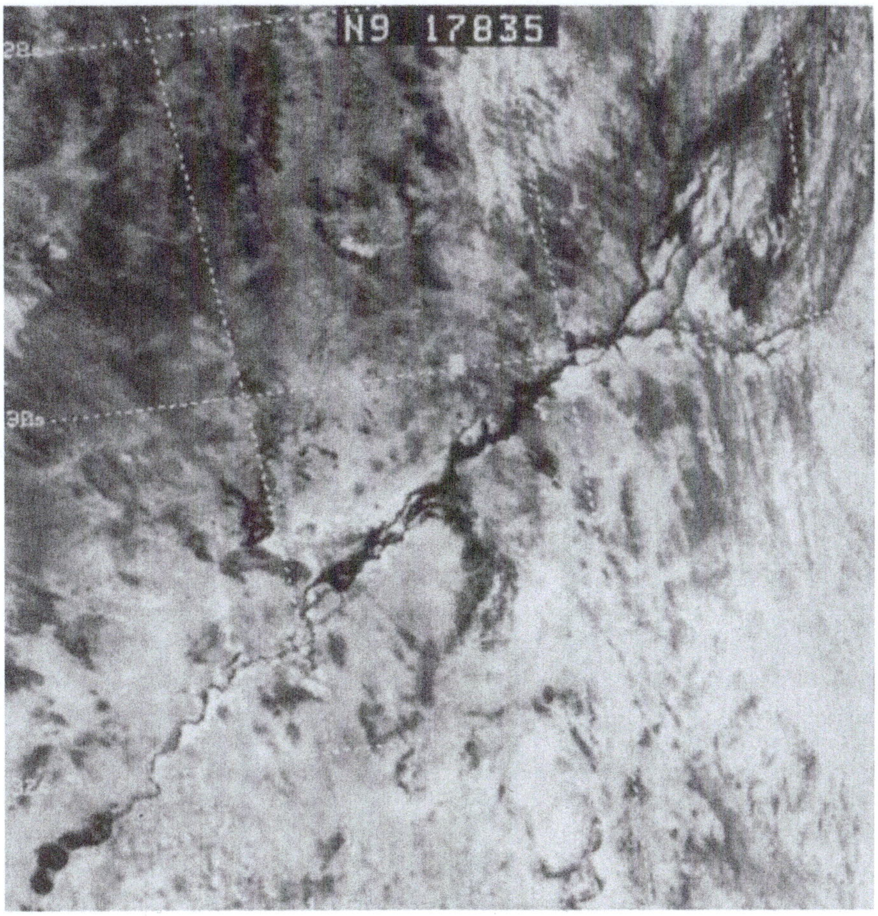

FIGURE 4.51 Advanced Very High Resolution Radiometer channel 4 brightness temperature for May 29, 1988 (Barton and Bathols, 1989).

TABLE 4.11 Averages and standard deviations for VI and the thermal infrared.

	Day				Night	
	Vegetation index		Temperature (°C)		Temperature (°C)	
Surface	**Avg.**	**Std. Dev.**	**Avg.**	**Std. Dev.**	**Avg.**	**Std. Dev.**
Land	0.26	0.05	14.0	0.6	4.1	1.5
Flood	0.12	0.08	14.3	0.4	10.3	1.1
Lakes	-0.13	0.05	13.5	1.0	12.1	0.6

[a]Values are obtained from a sample of 20 randomly selected sites for each surface type. The vegetation index is defined in the text.

floods are the most frequent to occur in India. Since all of the major rivers in India flow toward the east and empty into the Bay of Bengal, it is the eastern parts of India that are most susceptible to floods. While these floods are primarily during the monsoon season, there are still considerable temporal and spatial variations in flooding, often caused by variations in the monsoon season and geographic distribution. Jain et al. (2006) remark that damage from floods in India has been increasing every year, causing loss of lives and property, so knowledge and prediction of these floods would be useful in avoiding damage and threats to human activity. They further suggest that such knowledge would be cost-effective compared to structural methods such as dams and dikes.

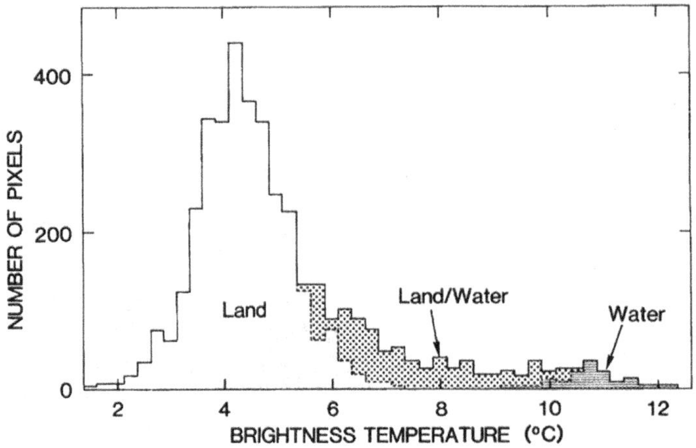

FIGURE 4.52 Histogram of the brightness temperatures for 64X64 pixel area centered on 30°33'S, 145°16'E for the channel 4 image collected on the night of May 19, 1988 (Barton and Bathols, 1989).

Many previous studies used Landsat or SPOT images to study the extent of water bodies, but their high resolution over a limited area makes it difficult to get a view of the larger area. While it has a more modest spatial resolution, the AVHRR provides a much larger spatial coverage with a rapid refresh. This higher frequency of coverage increases the potential for finding images with fewer clouds in the subject area. As reviewed earlier in this text, Barton and Bathols (1989) used AVHRR nighttime thermal infrared channel 4 images to map a flood of the Darling River. Also in 1989, Ali used AVHRR imagery to map the 1984 Bangladesh flood using visible, NIR, and thermal infrared channels. Islam et al. (2000) assessed the flood hazard in Bangladesh using AVHRR data and comparing images collected during the flood and the dry season when no flood was present. These latter authors also used elevation data to examine the correlation between elevation and the interpretation of the AVHRR images.

In 2003 the state of Assam experienced heavy rains starting June 13 that were the combination of the seasonal monsoon rains and increased local precipitations. This combination of rain events resulted in flooding of the low-lying areas, which affected 22 out of 24 districts in the Assam state (Fig. 4.53).

The flood caused widespread damage and a number of people lost their lives. It wiped out crops and destroyed the very banks that were erected to prevent flood damage. This study used AVHRR data from NOAA 15 and 17 from March to October 2003. The data were received at the Satellite Earth Station of the Indian Institute of Technology in Roorkee, India. In spite of the frequent coverage of the AVHRR, only five cloud-free images were found for March 24, May 13, July 23, August 5, and November 5. These images were processed using the ERDAS IMAGINE 8.6 software package (from Hexagon Ltd.) for radiometric correction and geometric mapping correction using 15 ground control points, which resulted in a geolocation error less than ±1 km.

To evaluate the role of topography, they used data extracted from a global digital elevation model (DEM) GTOPO30 available from the US Geological Survey's (USGS's) Earth Resources Observation System (EROS) data center in Sioux Falls, South Dakota. GTOPO30 is regularly spaced at 30-arc sec (approximately 1 km) intervals. The horizontal coordinate system is in decimal degrees of latitude and longitude referenced to WGS84. The study area in India was downloaded from the EROS data center and was presented here in Fig. 4.54.

To determine the flooded areas, Jain et al. (2006) used three different indicators computed from the AVHRR data. First, a channel 2 threshold (Liu, 1989; Sheng et al., 1998) where

$$\begin{aligned}&\text{Water}, \quad \text{if } CH_2 \le T_0\\&\text{Land}, \quad \text{if } CH_2 > T_0\end{aligned} \tag{4.19}$$

Second, a ratio of two AVHRR channels (Sheng and Xiao, 1994)

$$\begin{aligned}&\text{Water}, \quad \text{if } CH_2/Ch_1 \le T_0\\&\text{Land}, \quad \text{if } CH_2/CH_1 > T_0\end{aligned} \tag{4.20}$$

Third, a difference between channels 1 and 2 (Xiao and Chen, 1987)

$$\begin{aligned}&\text{Water}, \quad \text{if } CH_2/CH_1 \le T_0\\&\text{Land}, \quad \text{if } CH_2-CH_1 > T_0\end{aligned} \tag{4.21}$$

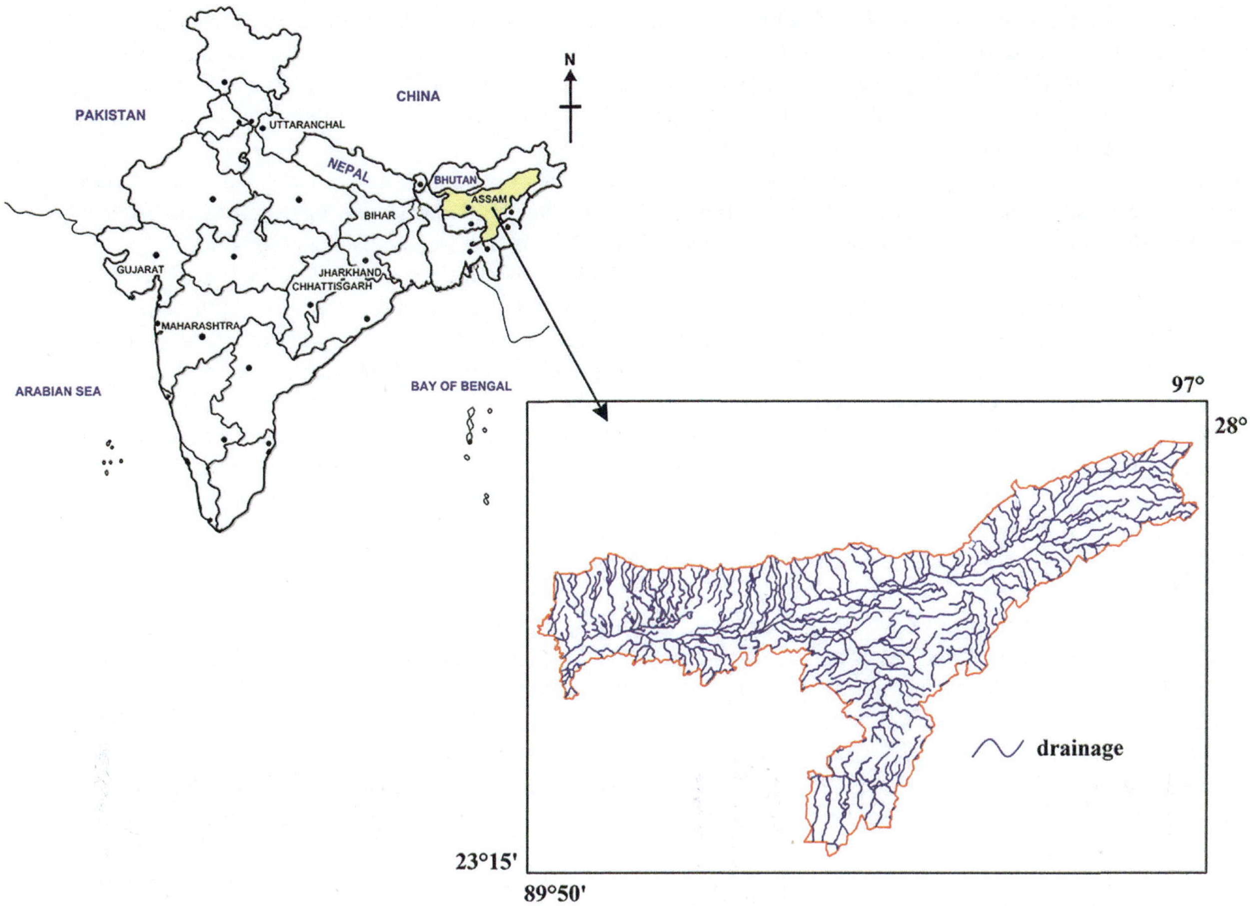

FIGURE 4.53 The study area of the 2003 flood (Jain et al., 2006). Map lines delineate study areas and do not necessarily depict accepted national boundaries.

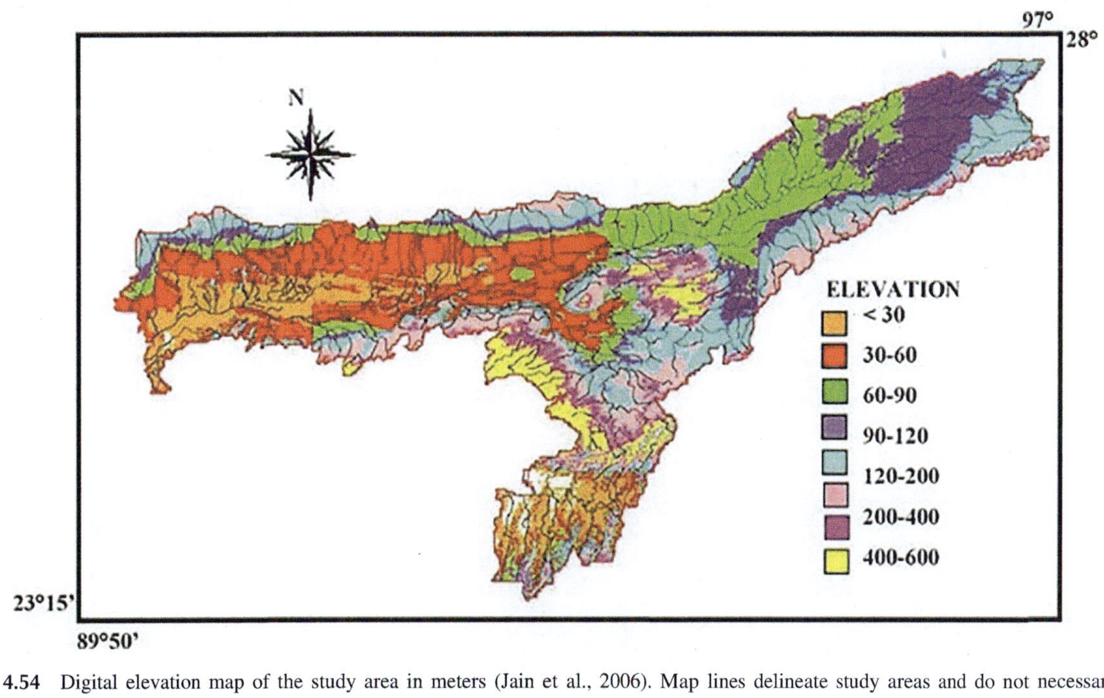

FIGURE 4.54 Digital elevation map of the study area in meters (Jain et al., 2006). Map lines delineate study areas and do not necessarily depict accepted national boundaries.

They found that for premonsoon periods the ratio method worked best due to the lower volume of water. During the postmonsoon season, the single band threshold worked the best. Once these indicators were used to define the water pixels, the area of submerged land could be easily calculated. The image from March of 2003, which was before the flood, clearly shows the normal distribution of land and water in the river. They computed the area of the water bodies and the flooded areas computed from each of the five AVHRR images (Table 4.12).

The dramatic increase in the flooded area on July 23 and August 5 shows the effects of the flood clearly, where the flooded areas increase by 25%–30%. The river swelled slightly during this period, but the flooded areas increased by an order of magnitude and more. The flood affected areas are shown for July 23 in Fig. 4.55 and for August 5 in Fig. 4.56.

It is easy to see that the severity of the flood was much greater in August when it peaked. It should be noted that there are some flooded areas that are not likely to be caused by the river overflow but are instead areas that captured the rain events into a local flood area. These areas are present in the July image but not connected to the river overflow flood. By August, the river flood dominates the area, and these independent spots are merged with it. There are some smaller flooded areas to the south that are quite distant from the river.

One persistent problem in the remote sensing of floods is that it is generally cloudy in flood events since rainfall is the primary source of water for the flood. It has already been mentioned that, in spite of its modest spatial resolution, the AVHRR has the advantage of frequent coverage, which has the potential of getting an image that is cloud-free or at least has reduced cloud cover. A study by Sheng et al. (1998) looks at the challenges associated with cloud contamination in the use of AVHRR imagery for flood monitoring. They developed a simple but practically effective method for identifying water-covered land due to a flood.

They begin by reviewing typical cloud screening methods, correctly saying that there are a variety of cloud clearing algorithms (Coakley and Bretherton, 1982; England and Hunt, 1985; Saunders, 1986; Saunders and Kriebel, 1988). They also noted that Saunders (1988) concluded that a combination of the spatial coherence at infrared wavelengths (Coakley and Bretherton, 1982) together with dynamic visible threshold methods was the most effective. The combination of algorithms became known as the Spatial Coherence Visible (SVC) algorithm, which was most effective for use with daytime imagery, with the exception of uniform thin cirrus clouds. Sheng et al. (1998) used the SVC in their analysis of cloud contamination removal for studies of AVHRR for flood monitoring. Since the spatial changes in cloud top temperatures are larger than those of the Earth's surface, they also used surface temperature in their cloud screening procedure.

They also addressed the effect of cloud shadows on the interpretation of AVHRR imagery for flood inundation. Cloud shadows reduce the amount of sunlight illuminating the Earth's surface making it more difficult to distinguish between land and water covered land. Moreover, the similar appearance of the cloud shadow and the water surface adds to the difficulty of discriminating land from water. Sheng et al. (1998) used the visible and NIR channels to screen the cloud shadow effects.

These authors note that floods generally occur when the vegetation is healthy and lush. Thus an AVHRR 1.1 km pixel contains primarily water, vegetation, and soil. The spectral characteristics of these surface elements are presented here in Fig. 4.57.

Also shown in Fig. 4.57 are the wavelengths covered by AVHRR channels 1 and 2 (shaded areas). The differences in ground cover reflectance suggest that the possibility of discriminating ground cover type from the AVHRR reflectance. The reflectance during a flood is complicated by the fact that for water, its visible signal shifts toward the red (the center of

TABLE 4.12 Areas of water bodies and flooded areas for each Advanced Very High Resolution Radiometer (AVHRR) image.

	Area of water bodies		Flood affected area		
Date	km^2	%	km^2	%	Note
March 24, 2002	6028	8.2	2304	3.2	Before monsoon
May 13,2002	6124	8.3	3129	4.2	Just before floods
July 23, 2002	6357	8.6	22,543	30.6	Flood period
August 05, 2002	6431	8.7	25,201	34.2	Flood period
November 02, 2002	6028	8.2	3744	5.1	After flood retreated

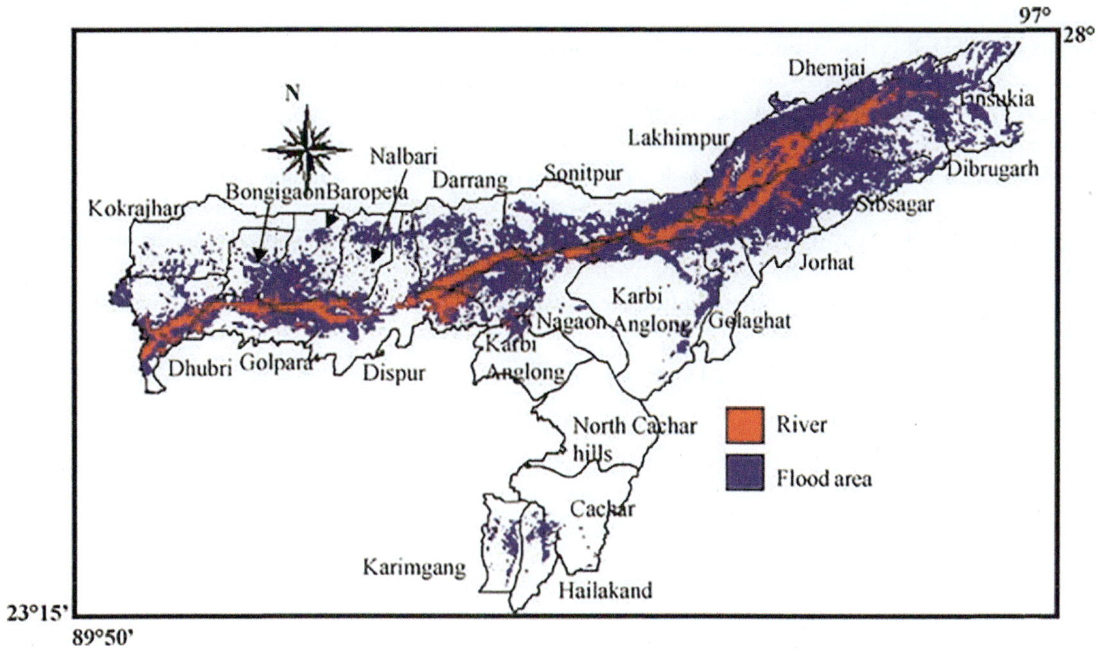

FIGURE 4.55 Flooded areas on July 23 along with the river (Jain et al., 2006). Map lines delineate study areas and do not necessarily depict accepted national boundaries.

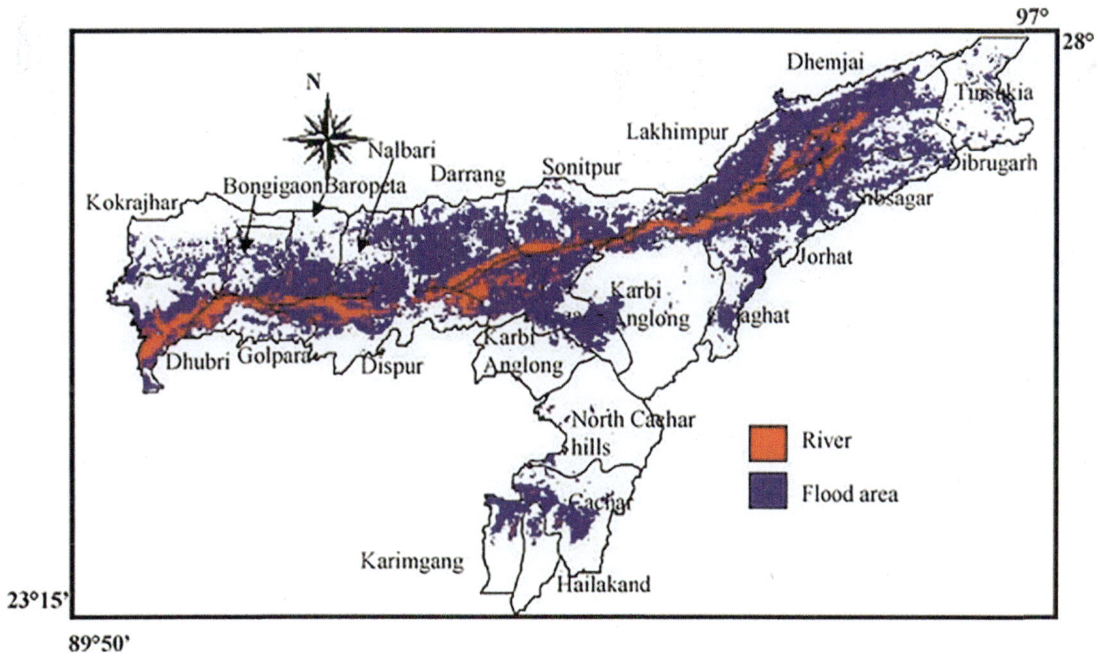

FIGURE 4.56 Flooded areas on August 5 along with the river (Jain et al., 2006). Map lines delineate study areas and do not necessarily depict accepted national boundaries.

the AVHRR channel 1) due to the presence of silt and debris in the flood waters. Pure water has a much lower albedo in the NIR compared with vegetation, which has a strong response in the same channel 2. In channel 1, vegetation has a very low albedo which jumps up in the NIR. Therefore these authors felt that a ratio between channels 2 and 1 would enhance the difference between water and land.

This conclusion is born out in the plots in Fig. 4.58, which shows the typical histograms of the AVHRR responses for channel 1, channel 2, and the ratio of 2/1. While we can see a small separate peak for the water in both channel 1 and channel 2, the ration histogram clearly delineates between the water and the land peaks. These peaks can be easily separated in the ratio diagram using a simple threshold. It is the separation of these peaks in the ratio histogram that makes it possible to explore the mapping of flood-impacted areas under partly cloudy conditions. The Earth's surface is covered by cloud 50% of the time (Paltridge and Platt, 1976) and in some regions of persistent cloud cover it may be a lot higher. As mentioned earlier, the frequent AVHRR coverage makes it possible to capture an image with reduced cloud amount as clouds move around with the wind. While AVHRR cannot sense the ground when thick clouds are present, it is also true that clouds move around and sometimes quite quickly, making it possible that the next AVHRR overflight will find the target area with minimal cloud cover. In the case of thin cirrus clouds, while the AVHRR receives less solar illumination, it may be sufficient to separate land and water-covered areas, particularly when using the ratio of the channel 2 and channel 1 reflectance.

In the thin cloud condition, the AVHRR receives radiation both from the ground surface and the cloud. We can write the radiation sensed by a channel as:

$$CH_{i\sim}C_i + G_i + P_i \tag{4.22}$$

where I is the AVHRR channel number 1 or 2, CH_i is the value measured by the sensor for channel i, C_i is the radiation contributed by the clouds, G_i is the radiation contributed by the ground, and P_i is the radiation over the atmospheric path length, which is close to zero in the NIR.

As a consequence the ratio of CH_2/CH_1 is given by

$$R = k*\frac{CH_2}{CH_1} = k*\frac{C_2 + G_2 + P_2}{C_1 + G_1 + P_1} \tag{4.23}$$

For water this becomes

$$R(\text{water}) = k*\frac{C_2 + G_{2(\text{water})} + P_2}{C_1 + G_{1(\text{water})} + P_1} \tag{4.24}$$

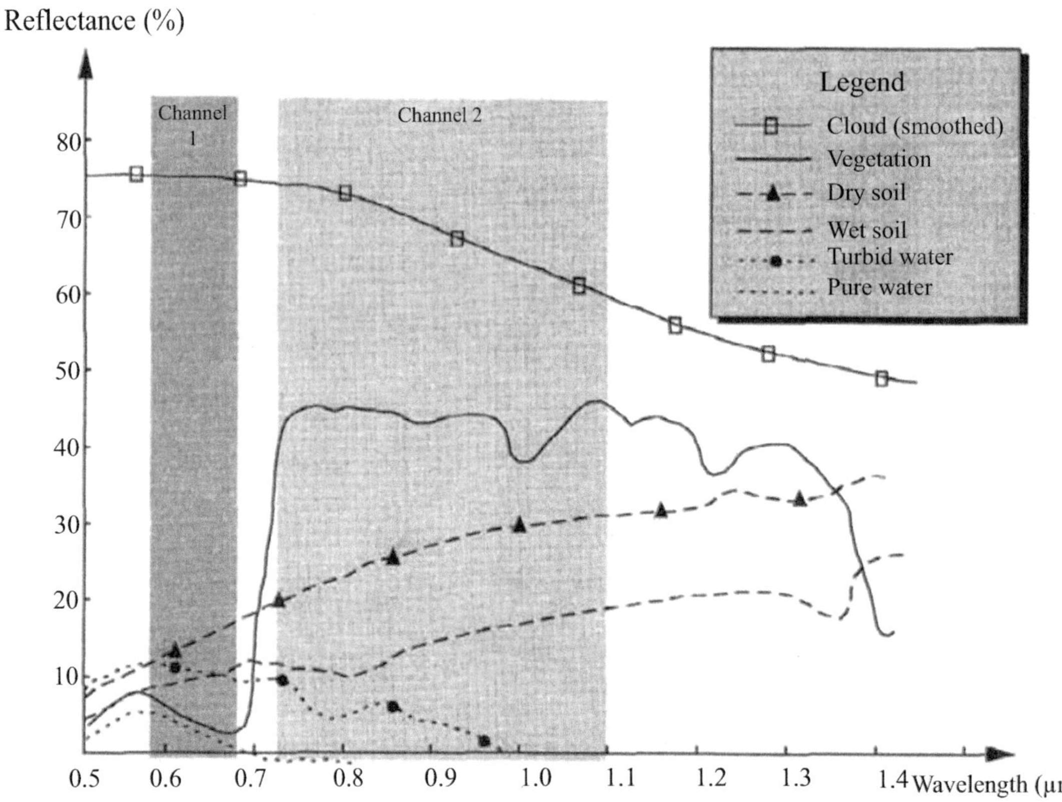

FIGURE 4.57 Typical spectral reflectance of the primary ground cover types. Shaded areas represent the Advanced Very High Resolution Radiometer channels 1 and 2 (Jain et al., 2006).

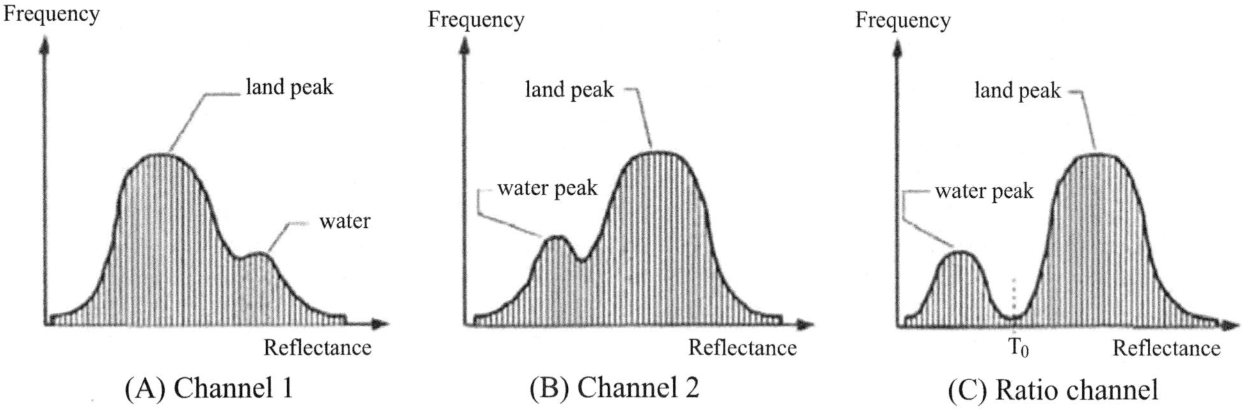

FIGURE 4.58 Advanced Very High Resolution Radiometer typical histograms in flooded areas (Jain et al., 2006).

and for land we can write

$$R(\text{land}) = k * \frac{C_2 + G_{2(\text{land})} + P_2}{C_1 + G_{1(\text{land})} + P_1} \tag{4.25}$$

The term k is an amplification factor that is set to 128 for their study.

As given by the spectral curves in Fig. 4.58, G_1 (water) is generally greater than G_1 (land) while G_2 (water) is usually less than G_2 (land). As a result, R(water) <R(land). Thus water and land under thin clouds have different values of this ratio, and it should be possible to discriminate land from water-covered surfaces. A similar approach to cloud shadow sensing eliminates the cloud reflectance term. They conclude that cloud shadow channel 1 response would be a bit larger than for land, but the channel difference is not sufficient to discriminate between the two, and once again their ratio is used to differentiate between water and land in cloud shadow conditions.

Sheng et al. (1998) conclude that while the individual AVHRR channels 1 and 2 may be strongly attenuated by even moderate cloud cover the ratio of CH_2/CH_1 is much less affected by the presence of even moderate cloud cover. This ratio enhances the differences between land and water surfaces under cloud shadow as well as moderate cloud cover. It is acknowledged that thick cloud cover eliminates this advantage and makes it impossible to differentiate water surfaces from land cover.

4.13 Measuring and monitoring snow cover

A somewhat related application of AVHRR imagery is the mapping of snow-covered land. Baumgartner et al. (1994) used AVHRR data to monitor snow cover variations in the European Alps. They coupled the AVHRR image analysis with a Snow Melt Runoff model to monitor changes in the Alpine snow cover. Recently, climate scientists have recognized the importance of understanding snow cover variations in determining climate change. Large areas of the Northern Hemisphere are covered by snow in the winter, which results in a marked increase in land surface albedo. Likewise, in summer, the absence of snow greatly decreases the albedo. These albedo changes markedly influence the climate and need to be included in climate model data assimilation. In addition, changes in snow cover influence: (1) snowmelt and runoff conditions which dictate the fresh water supply for human consumption, crop irrigation, and electricity generation, (2) the regional climate, (3) the ambient vegetation, and (4) tourism.

Experience has shown that since the mid-eighties in the mid latitudes of Northern Europe, the weather patterns and hence the snow cover have changed substantially. For example, the snow cover in the prealpine regions of the Vosges, the Black Forest, the Apenin, and the Jurra lasted at least 3–4 weeks. Now these regions are snow covered only a few days (and this was written in 1994). Baumgartner et al. (1994) state that "presently, snowfall starts in the same usual period, but the main snowfall occurs later in the long-term average between February and March when there is more solar energy to melt snow than in January." In addition, the humidity is higher in these later months, which increases the snowfall at higher elevations but results in rainfall in the lower elevations. As a consequence, snow depths in the lower elevations show a dramatic decrease. At higher elevations, snow depths are similar compared to the average, but the duration of snow accumulation and ablation is much shorter.

Historically, the primary snow accumulation takes place between October and March with sporadic periods of warmer weather. In more recent years, the snowfall starts at about the same time, but the intense snow period has shifted to the later months of February and March. During these months, there is more solar radiation to melt snow than in the earlier months of December and January. In addition, the relative humidity is higher resulting in greater snowfall at higher elevations and rainfall in the lowlands. Hence, overall snow depths at lower elevations show a dramatic decrease, while at higher elevations, the snow cover depths do not change very much.

Baumgartner et al. (1994) created their own HRPT receiving station to receive AVHRR imagery of their study region. They classified these images using supervised techniques of minimum distance and Mahalanobis distance (Duda and Hart, 1973). Their final result was a semantic map differentiating between several types of snow-covered and nonsnow-covered areas. Their procedure was repeated to generate a time series of classified maps. These maps were integrated with a geographic information system (GIS) that includes topographic information, rivers, lakes, and climate information. To model snowmelt runoff, they used the snowmelt runoff model (SRM; Martinec et al., 1994). SRM calculates the daily runoff based on the degree-day method. The inputs to the model are daily temperatures (minima and maxima), precipitation, and snow coverage (computed from the satellite imagery).

Combining the maps of snow cover with an elevation model one can derive elevation-dependent snow cover variations (Fig. 4.59).

It should be noted here that the snowline in the central Alps never went below 1,600 m, which is significantly higher than a decade earlier when even the lowlands (600–800 m) were also snow covered for at least 3–4 weeks out of the year. This type of event can be seen in the historical snow cover record. Baumgartner et al. (1994) conclude that AVHRR data enable the quantification of the areal distribution of snow. Over the past decade, snow cover patterns, snowmelt runoff, snow depths, and water equivalent show major changes compared to the long term average. They state that if these changes persist for an extended period economic and climatic consequences would result.

In their study of snow cover, Rössler and Dietz (2002) used past and present AVHRR data to generate a snow cover detection system. They state the obvious that snow cover is the largest and most transient part of Earth's cryosphere system in terms of its area coverage. On local scales, these changes can result in floods or drought, and on the planetary scales, snow cover dictates the planetary albedo. They used daily AVHRR imagery to create the TIMELINE (TIME Series

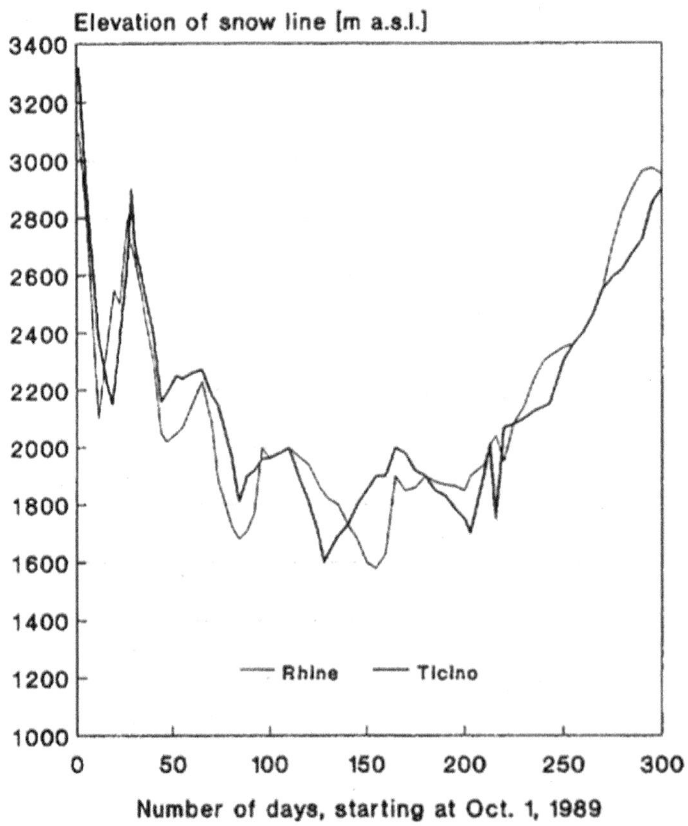

FIGURE 4.59 Elevation variation of the snowline in the Rhine-Felsberg and Ticino-Bellinzona basins (1989–90) (Baumgartner et al., 1994).

processing of Medium Resolution Earth Observation Data assessing Long-Term Dynamics in our Natural Environment) processor of the German Aerospace Center (DLR) from an HRPT station, giving them a nominal resolution of 1 km. For snow detection, they use the normalized difference snow index (NDSI) and also include information on the skin surface temperature to yield a binary snow mask. The daily results are used to calculate 10-day and monthly snow covers.

The importance of a knowledge of "area covered by snow" is emphasized by the fact that it is included as one of the 54 EV by the Global Climate Observing System (GCOS; WMO 2016). Medium-resolution optical satellite systems are ideal for studying the variability of snow cover which is a very transient variable. The primary requirement is for rapid refresh and that is characteristic of AVHRR data from the NOAA satellites with the AVHRR data from the sensor flying on MetOP (Dietz et al., 2012). The result is a time series from 1982 up to 2012. There are many studies of snow cover computed from AVHRR data often as a byproduct of cloud classification in snow covered areas. Many of these studies take place in regional studies such as the Tibetan Plateau (Zhu and Shi, 2018), in Turkey (Akyurek and Sorman, 2002) or in the Alps (Zhu and Shi, 2018; Foppa et al., 2004).

To detect snow, Rössler and Dietz (2002) took advantage of the reflective properties of snow and their difference from tree reflectance. This is demonstrated in Fig. 4.60, which shows the reflectance spectrum of snow versus conifer trees. In the reflection spectrum of snow, there is almost total reflection in the visible channel 1, while the conifer tree tops show relatively little reflection. In the NIR of channel 2, the snow falls off a bit but is still higher than the conifer line. The 3*a* channel shows the conifer increasing while the snow line has dropped almost to zero. Taking advantage of this reality they introduce the NDSI as

$$\text{NDSI} = \frac{R_1 - R_{3a}}{R_1 + R_{3a}} \tag{4.26}$$

where R_1 is the atmospherically corrected reflectance in AVHRR channel 1 and R_{3a} is the reflectance in channel 3*a*. It should be noted that 3*a* is from 1.58 to 1.64 μm, but variant 3*b* goes back to the original AVHRR channel 3 of 3.55–3.93 μm, which then contains both reflected and emitted radiation. To adjust the channel 3*b* value for the NDSI calculation, the thermal infrared of channel 5 is used to subtract the thermal radiance from channel 3*b*, leaving only the reflectance to be entered into equation 1.24. Thus we can write our corrected 3*b* reflectance as

$$R_{3b} = \frac{L_{3b} - B(\lambda_3, T_5)}{B_{\text{sun}}\mu_0 - B(\lambda_3, T_5)} \tag{4.27}$$

where L_{3b} is the measured radiance of channel 3*b*, $B(\lambda_3, T_5)$ is the blackbody radiance at wavelength λ_3 and temperature T_5 (BT of channel 5), B_{sun} is the solar radiance corrected for the Earth-sun distance and μ_0 is the cosine of the solar zenith angle.

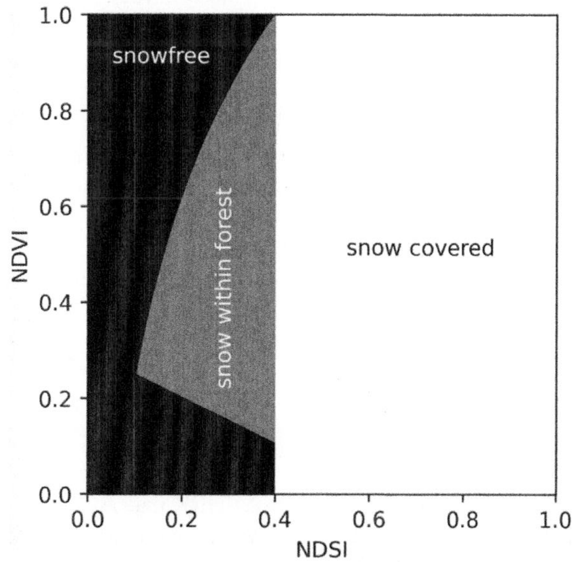

FIGURE 4.60 Normalized difference snow index (NDSI) thresholds for detecting snow and the relationship to normalized difference vegetation index (NDVI) (Baumgartner et al., 1994).

The NDSI can be used to compute the area covered with snow, and an NDSI threshold of 0.4 corresponds to a fractional snow cover of 50%, and all NDSI values above this value indicate snow cover greater than 50% (Hall et al., 2002; Salomonson and Appel, 2004; Salomonson and Appel, 2006). For NDSI values below 0.4, there is no longer a binary distinction between snow covered and partially covered areas. Klein et al. (1998) combined a canopy reflectance model with a snow reflectance model to map snow in forested areas. Thus, Rössler and Dietz (2002) added the NDVI to their snow algorithm. The NDVI is defined as

$$\mathrm{NDVI} = \frac{R_1 - R_2}{R_1 + R_2} \tag{4.28}$$

where R_1 is the atmospherically corrected reflectance in AVHRR channel 1, and R_2 is the reflectance in channel 2 as defined earlier in this text. Following Klein et al. (1998), they introduced new NDSI values in the NDSI and NDVI space. These are

$$\mathrm{NDSI}_{\mathrm{exponential}} = \mathrm{NDSI} - 0.0652X\ e^{1.8069\mathrm{XNDVI}} \tag{4.29}$$

$$\mathrm{NDSI}_{\mathrm{linear}} = \mathrm{NDSI} - (\mathrm{NDVI} - 0.2883)/{-0.428} \tag{4.30}$$

$$\mathrm{snow} = \begin{cases} \mathrm{NDSI} \geq 0.4 \\ \mathrm{NDVI} \geq 0.25 \wedge \mathrm{NDSI}_{\mathrm{exponential}} \geq 0 \\ \mathrm{NDVI} \geq 0.1 \wedge \mathrm{NDVI} < 0.25 \wedge \mathrm{NDSI}_{\mathrm{linear}} \geq 0 \end{cases}$$

This set of criteria is best understood in Fig. 4.60. Here, it is very clear that in the area with NDSI values below 0.4 the snow cover is a function of NDVI, with large portions having no snow cover, while one area is a snow-covered forest. Above 0.4, all the ground is snow covered.

Various other information was needed for the TIMELINE processor. This is all summarized in Fig. 4.61, which is a flowchart of the TIMRLINE Snow Mask Processor. The process starts by masking out cloud and water vapor. Clouds often have signatures very close to snow, so this is a challenge. Still, it is primarily the NDSI that gives the overall snow cover. The other tests mentioned above are used to identify snow-free areas in the forest. The DEM is useful in separating land and water. The cloud mask is an important part of the process. Solar zenith angles greater than 85 degrees are classified as "night." A temperature-height criterion is first usedwhere a channel 4 temperature is above 281 K and the altitude is below 1,300 m. This condition is assigned to the snow-free class. Next pixels with more than 45% reflectance in channel 3 are also classified as "snow free." Finally, the cloud mask is used, and any pixels with a cloud mask value of over 70% are assigned to the cloud class.

An example of the level 2 product in terms of snow mapping is presented in Fig. 4.62.

The RGB image of Fig. 4.62A shows the snow-covered areas as the color cyan, while the clouds appear gray or white. At the top of Fig. 4.62B, there is a sharp transition from black to gray shades introduced by the transition of channel 3*b* (night mode) to channel 3*a* (day mode). This NDSI image of Fig. 4.62B also shows high values for water and some clouds. The final result of the TIMELINE process appears to have properly separated snow from the other elements of the image. An example of the level 3 processing result is presented here in Fig. 4.63. This figure exhibits linear artifacts that are the result of inadequate cloud clearing at the edge of the image. This problem occurs when the satellite zenith angle is larger than 65 degrees, which is considered a bad satellite viewing geometry in the quality layer of the system. This may also be caused by low reflection in the visiblem whereby these areas are classified as "snow free" in spite of a large NDSI value.

The NDSI method is independent of absolute reflection and bidirectional reflectance distribution function (BRDF) effect over snow, which are well known to have a strong influence on AVHRR data (Zhonghai and Simpson 1999). The TIMELINE process created a daily cloud-free AVHRR snow product covering a time series of almost 40 years. This series enabled observations of snow cover down to the catchment area level, including snow changes in the mountains, which has not been previously possible over a long period of time.

4.14 Urban heat islands

The urban heat islands (UHIs) of three coastal cities were studied with AVHRR data by Roth et al. (1989). They used the thermal infrared channels to study the UHIs for Vancouver, BC, Seattle, Washington, and Los Angeles, California. The data were received at the HRPT station in the Department of Oceanography at the University of British Columbia. They found that heat island intensities were greatest during daytime in the warm season. They also found that daytime land intra-urban thermal patterns were most closely related to land use, where industrial areas are warmest, while vegetated, riverine, and

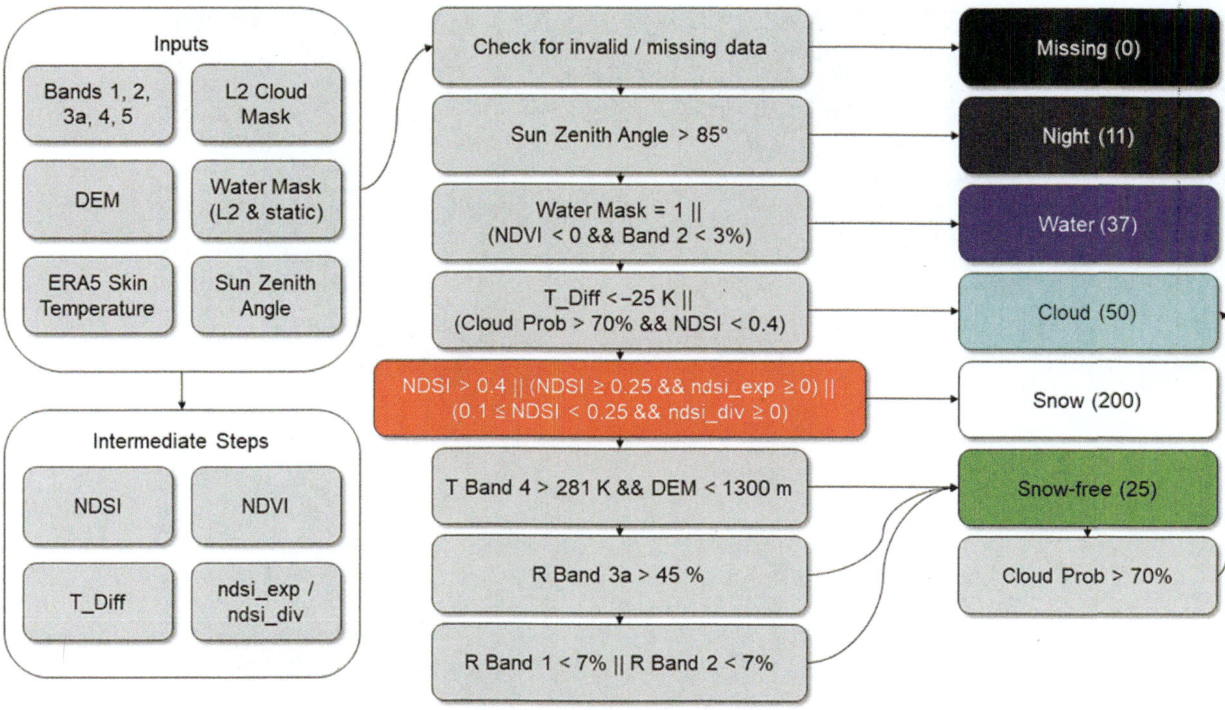

FIGURE 4.61 Flowchart of the TIMELINE snow mask processor. The central normalized difference snow index (NDSI) snow recognition is highlighted in red (Baumgartner et al., 1994). *DEM*, Digital elevation model; *NDSI*, Normalized Difference Snow Index; *NDVI*, normalized difference vegetation index.

coastal areas are the coolest (Fig. 4.64). This temperature map corresponds well with the land use map for the Vancouver area (Fig. 4.65). It is interesting how the temperature map in Fig. 4.64 captures the rail line out to the loading point on the coast. The most intense activity is in the downtown area and along the northern coast of Burrard Inlet. Other areas of industrial activity are along the Fraser River. Note that the logarithmic temperature scale needed to accommodate the wide range of temperatures, ranging from low temperatures for vegetated areas to the highest temperatures in the industrial areas.

Similar results were found for Seattle and Los Angeles. The land use areas (Fig. 4.66) are well reflected in the surface BTs of Fig. 4.67.

These comparisons allowed Roth et al. (1989) to conclude the following:

1. The location of the warmest surface areas sensed by the satellite during daytime is in industrial-commercial zones, especially those with large flat-topped buildings or extensive open areas of pavement (airports, shopping malls, and major highway intersections). The highest surface temperatures are not usually found in the central business district, where the buildings are tall.
2. These zones are also the sites of relatively warm near-surface air temperatures (found by examining a selection of air temperatures).
3. The correlation between satellite temperature and land use is much stronger than for air temperature in the daytime.
4. The daytime urban–rural surface temperature differences are considerably larger than those in the near-surface air.
5. Daytime urban–rural differences and intraurban variability of surface temperatures are largest in the warm season.
6. Nocturnal urban–rural and intraurban variability in surface temperatures are much smaller than in the daytime. This is the reverse of the case for near-surface air temperatures.
7. These results are consistent with almost all other satellite based UHI studies (e.g., Carlson et al, 1981; Price, 1979; Vukovich, 1983; Lombardo, 1985; Kidder and Wu, 1987).

Roth et al. (1989) go on to further evaluate the use of satellite BTs in the study of UHIs. They felt that the following questions needed to be answered.

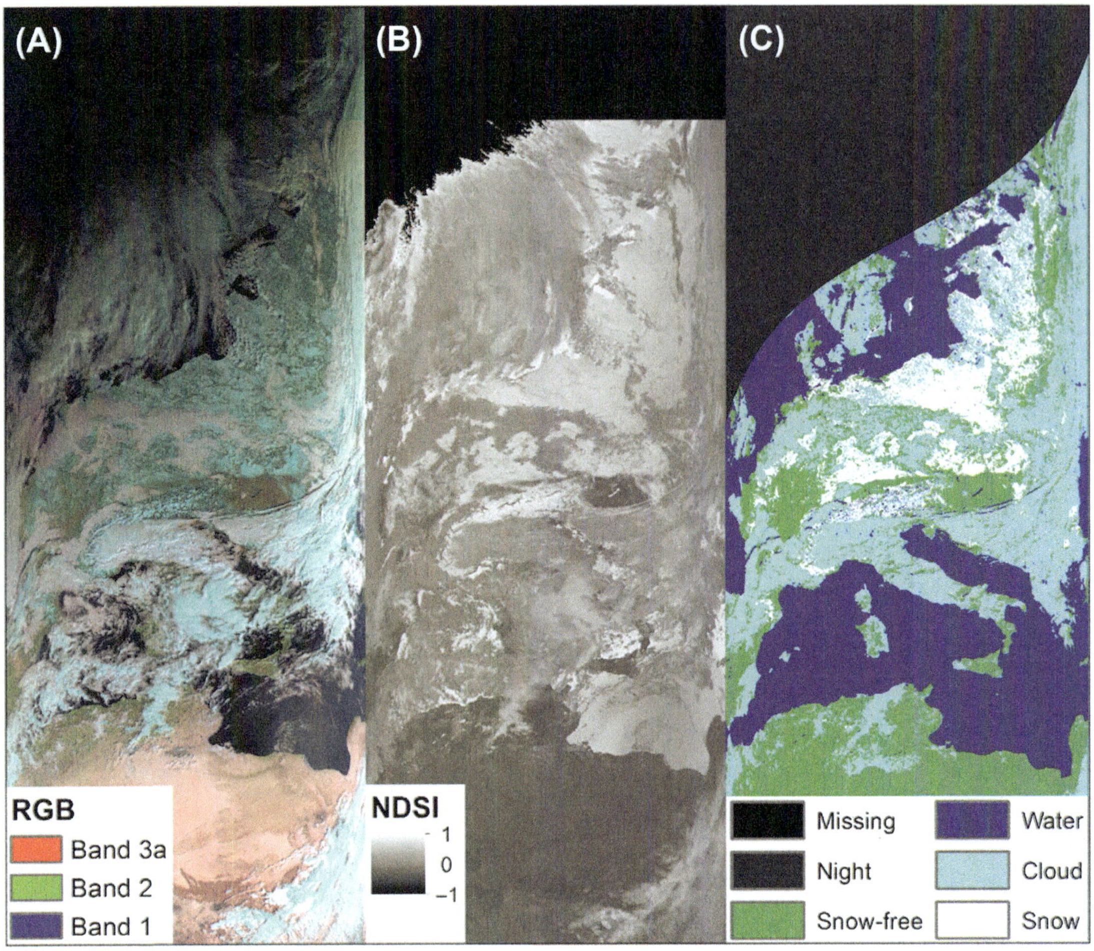

FIGURE 4.62 A level 2 snow cover product in satellite swath projection from January 8, 2009 (A) RGB composite from Advanced Very High Resolution Radiometer channels 3*a*, 2, 1; (B) normalized difference snow index (NDSI) calculated from channels 1 and 3*a* in grayscale; and (C) snow mask created according to the scheme in Fig. 4.61 (Baumgartner et al., 1994). Map lines delineate study areas and do not necessarily depict accepted national boundaries.

1. What is the nature of the surface "seen" by the satellite sensor? Many studies have confirmed that the satellite infrared is only able to see the skin of the surface.
2. How do the observed radiant temperatures relate to the true temperatures of the full urban-air interface?
3. What is the relationship between satellite derived UHIs and those measured in the air?
4. How appropriate are these data as input to urban boundary layer models?

The answer to question (1) is simple if the city is located directly under the satellite overpass. In this case, the city consists of roof-tops, roads, and open flat areas such as parking lots, parks, and gardens. There are other elements such as home gardens, trees, and driveways. For a densely built-up city, the active area seen by the satellite is much larger than the strictly planned area. As an example, cubic buildings separated by a distance equal to their characteristic length (i.e., a plan density of about 2.5) have an active area to plan ratio of about 2. A much lower density suburban area having scattered trees and shrubs may have a ratio of about 1.5, while in a downtown core, this ratio may be ≥3. The satellite is unable to view the vertical building surfaces or sense the surface under the treetops.

For off nadir satellite tracks, the surface sensed will depend on the viewing angle. Off nadir increases the response to vertical building surfaces. In addition, one must take into account the solar illumination angle when considering off-nadir satellite viewing. Many UHI studies fail to consider the influence of the viewing geometry. Roth et al. (1989) restrict their study to relatively large viewing angles (≥60 degrees), which is closer to the nadir viewing condition.

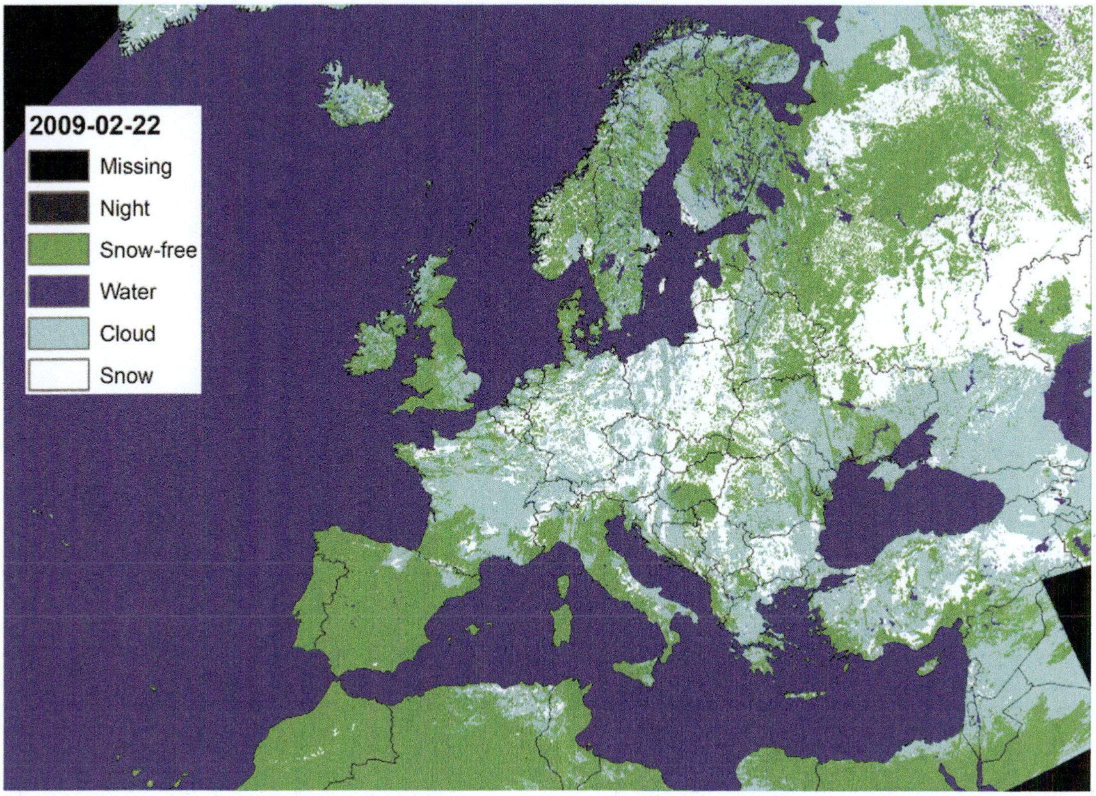

FIGURE 4.63 Example of a daily level 3 product for February 22, 2009 (maximum snow cover) (Baumgartner et al., 1994). Map lines delineate study areas and do not necessarily depict accepted national boundaries.

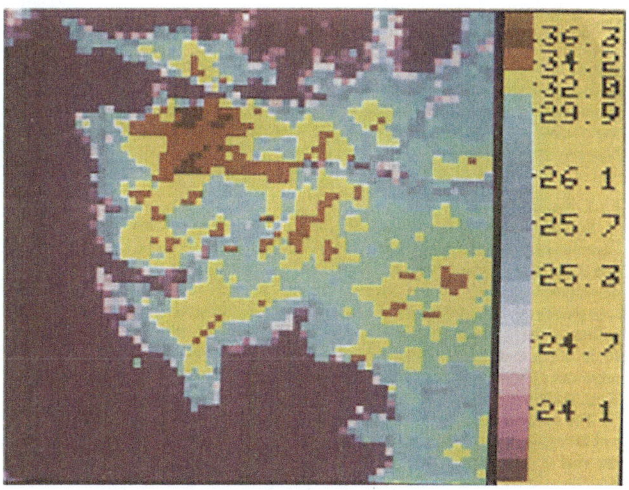

FIGURE 4.64 Brightness temperatures (°C) from the Advanced Very High Resolution Radiometer channel 4 (Roth et al., 1989).

If we ignore the errors introduced by satellite sensor limitations, atmospheric attenuation, and changes in surface emissivity question (2) is tightly related to the surface geometry. First, there is the problem of undersampling of the active target surface area. This can become a major source of error if the materials (and hence the temperature) viewed by the satellite are very different from the materials in the areas obscured from the satellite sensor. This will result in a very biased estimate of the surface temperature. Roofs, for example, have a low albedo and are built to minimize conductive heat transfer. They are therefore good absorbers of solar radiation and have low thermal inertia, which contributes an anomalously large diurnal range of surface

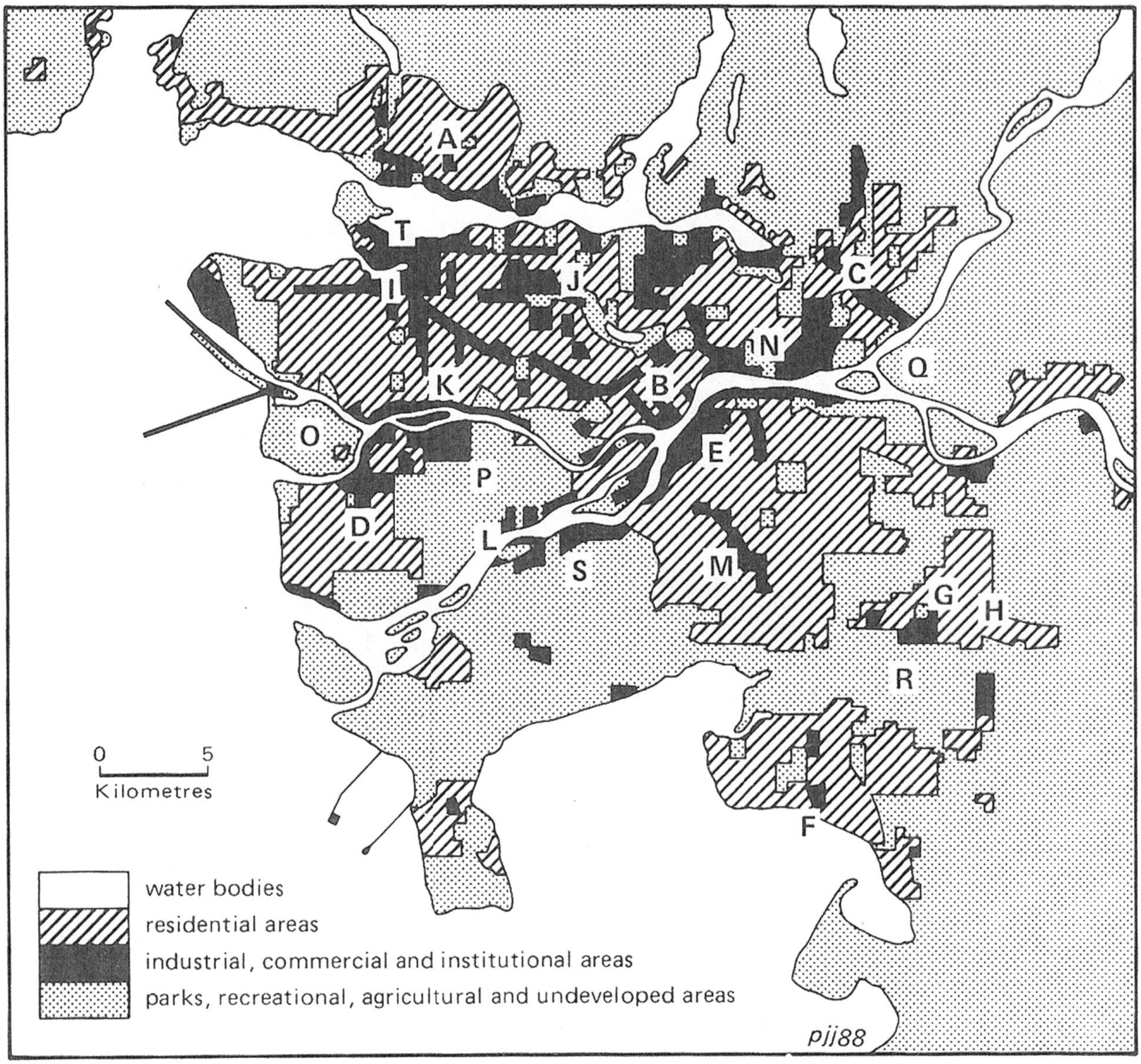

FIGURE 4.65 Land use map of the Vancouver, BC area (Roth et al., 1989). Map lines delineate study areas and do not necessarily depict accepted national boundaries.

temperatures (Goward, 1981). The importance of this can be seen in Fig. 4.68, which is an enhancement of an area of a light industrial area in south Vancouver collected by an aircraft survey taken on a summer afternoon and during the subsequent night.

The color rainbows are the same for both images, but the absolute temperatures and the temperature ranges are very different. The purpose of this figure is to demonstrate which surfaces are warmer (yellow and red) or colder (blue and black) when compared to the green background computed as the average of the entire image. The target area is composed of light industrial activity on the right (warehouses and engineering activities), while on the left, it is mostly individual houses with their gardens and trees. There are a school and a park in the lower left corner of the image. An outstanding aspect of both images is that the roofs of both the larger industrial buildings and the smaller residential structures were the hottest during the day and the coldest in the night image. The parking lots around the commercial building were also very hot during the day and cooled off dramatically at night. The yards that are next to the commercial buildings appear warm at night, which is likely due to their reduced sky view factor (Oke, 1987).

It seems clear that in all of the 3 cities studied, the daytime warmth of the industrial area is due to the combined influence of the large roof and storage yard areas, together with the relative lack of vegetation and moisture. In many cases

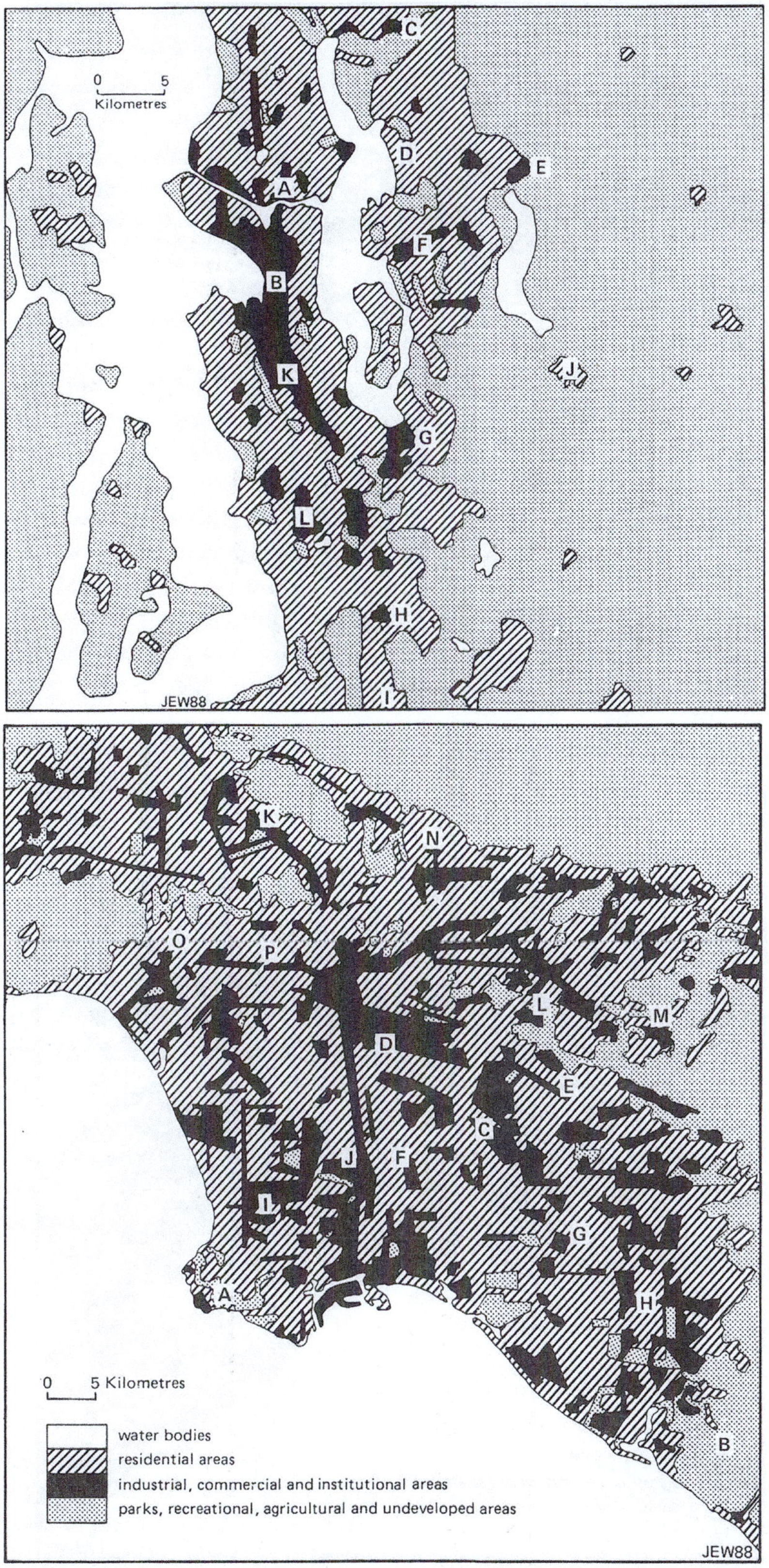

FIGURE 4.66 Land use maps; top Seattle, bottom Los Angeles (Roth et al., 1989). Map lines delineate study areas and do not necessarily depict accepted national boundaries.

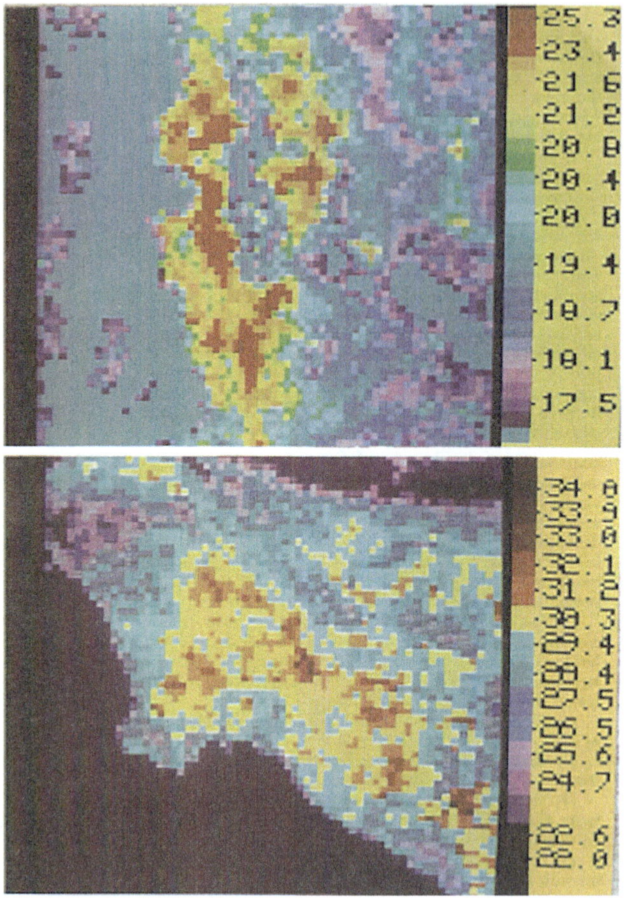

FIGURE 4.67 Advanced Very High Resolution Radiometer brightness temperatures (top) Seattle on September 24,1985 and (bottom) Los Angeles on October 11, 1985. Note that the higher temperatures for Los Angeles (Roth et al., 1989).

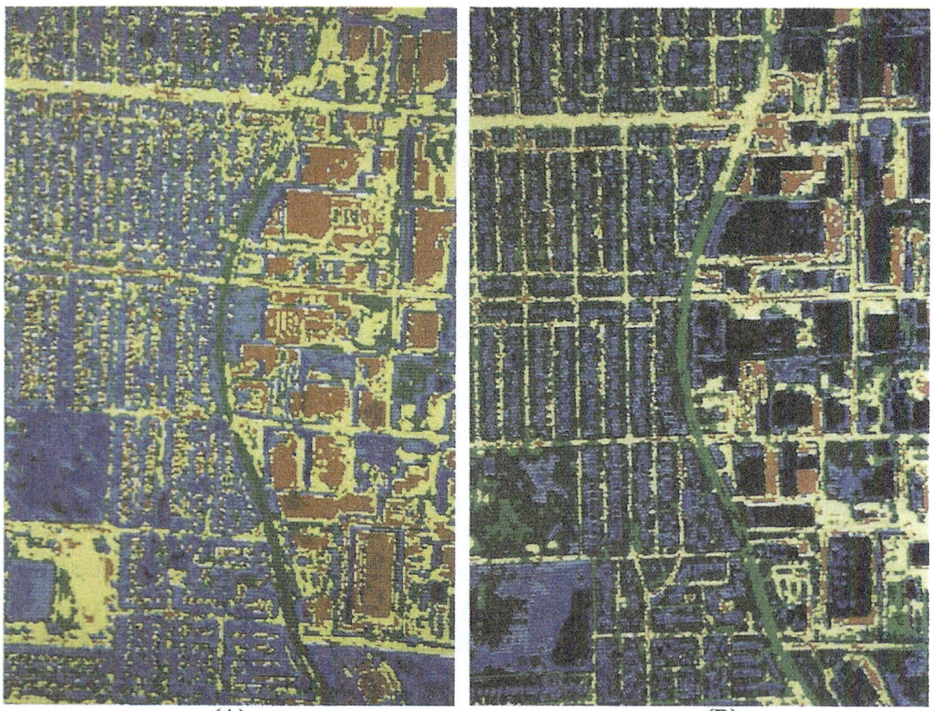

(A) **(B)**

FIGURE 4.68 Surface radiant temperature distribution for suburban and industrial areas of South Vancouver at (A) 14:15 (LST) on August 25, 1985 and (B) 05:05 (LST) on August 26, 1985. Colors range from *black* (coldest) →*blue*, →*green*, →*yellow*, →*red* (warmest) (Roth et al., 1989).

the heat is associated with the release of heat by some combustion activity. Error is introduced by the fact that the satellite views the targets as if they only have roofs and no sides. This is particularly true for higher solar zenith angles, which will result in higher roof-wall temperature differences. In addition, the low thermal inertia of the treetops contributes to the large diurnal temperature difference, with the trees providing shade and trapping the longwave radiation at the surface not viewed by the satellite.

A number of studies have analyzed the differences between remotely sensed heat islands and coincident air temperature measurements. Roth et al. (1989) found these two measurements to be relatively poorly related, independent of whether they are based on stationary sites or mobile platforms. They even found one case where the average of the daytime pixel-based surface temperature was larger, while the nighttime conventional temperatures were smaller. They felt that this difference was due to the following:

1. A lack of coupling between the surface and the air in the urban environment.
2. Biased spatial sampling of the surface temperature by the remote sensors. For example, a few very hot roofs or a source of combustion heat may result in a warm pixel.
3. Failure to recognize the different scales of urban climate phenomena and the fact that the satellite measurement is over a large area (~1.1 km pixels), while the stationary and even the mobile temperature measurements are spot samples compared with the satellite.

Roth et al. (1989) suggest that many different kinds of UHIs should be defined. These include heat islands determined by their direct air, surface, and subsurface temperature measurements. This contrasts with the radiative heat island as observed from orbiting satellites or from lower flying aircraft with infrared sensors. Even just within the conventional measurements, it should be realized that a heat island defined by 2 m temperature measurements made at fixed stations will differ greatly with automobile surveys conducted with similar instrumentation. Remotely sensed heat islands should not be confused with those determined by 2 m air temperatures or even in situ surface measurements.

They also conclude that remotely sensed UHIs can be used as inputs to a boundary-layer numerical model. These models treat the surface as a warm, moist platform of uniform surface properties. This surface is involved in heat and moisture fluxes and is located at some elevation, making it ideal for the type of UHI sensed remotely by infrared sensors on satellites or on aircraft. Unlike past boundary layer model efforts, which were tuned to conventional surface temperature measurements (mostly air temperature), the satellite view provides a pixel-averaged view of the surface temperature. They further conclude that while the influence of the mismatch between conventional temperature measurements cannot be easily evaluated, the satellite view of the UHI certainly provides a new approach to defining the temperature of the urban surface.

In a more comprehensive study of the UHI effect as sensed by the AVHRR, Gallo et al. (1993) looked at 37 cities in the continental United States and their surrounding rural areas for 28 June through August 8, 1991. They computed both a vegetation index and radiative surface temperature from AVHRR data collected at 13:30 LST. Their purpose was to evaluate the use of radiative satellite temperature measurements of urban and rural areas with direct air temperature measurements to assess their ability to observe the influence of the urban environment on the observed minimum air temperature (the heat island effect). This UHI effect has been detected even for cities with populations less than 10,000 (Karl et al., 1988). In the United States, the effect is best seen in the temperature range (maximum minus minimum temperature). Most of the earlier studies were based on air temperature measurements made at about 1.5 m off the ground (Lowry, 1977; Oke, 1978). In an effort to explain these differences, Carlson et al. (1981) suggested that the differences between the materials that are associated with urban and rural surfaces are responsible for these differences. The concrete and asphalt that dominate the urban areas will absorb heat during the day and release it at night. In rural areas, the vegetation will vary with time of year and regional differences. During the growing season, the vegetation redistributes the heat absorbed, with some of it being lost through evapotranspiration. As a consequence, the overall amount of heat that is lost at night in these rural areas is much less than in the urban areas. During the nonvegetated season, there are still measurable temperature differences between the urban and rural areas (Karl et al., 1988) due primarily to heat lost from internally heated structures in the urban areas.

Using population to predict the UHI bias in temperature records has been examined (Oke, 1973; Oke 1982; Karl et al., 1988). Oke (1973) related the rural-urban temperature difference and population statistics for North American and European cities and found different relationships for each region. Contributing to this lack of agreement was the relative paucity of reliable global population statistics. The conclusion is that population alone is not a globally applicable method to estimate UHI and remove its effect from long term land surface temperature records.

There have been many studies of satellite thermal infrared imagery to define the UHI (Rao, 1972; Carlson et al., 1977; Matson et al., 1978; Price, 1979, Kidder and Wu, 1987, Carnahan and Larson, 1990). In 1972 Rao was the first to demonstrate that the UHI could be identified by the analysis of thermal infrared satellite data. Matson et al. (1978) used

VHRR thermal infrared (10.5–12.5 μm) acquired at night to study urban and rural surface temperature differences. Price (1979) used imagery from the Heat Capacity Mapping Mission to determine the extent and intensity of urban surface heating in the northeastern US Land use applications led to higher surface temperatures in industrial areas when compared with rural vegetation regions. At night, the surface temperature differences between the urban and rural areas became minimal, suggesting that the sides of buildings, rather than just the roofs, contribute significantly to the higher minimum temperatures observed in urban areas when compared with rural regions.

The largest air temperature differences between urban and rural areas are found at night, while the greatest surface temperatures in these areas are found during the day (Roth et al., 1989). The reduced amount of heat stored in the soil and surface structures in rural vegetation-covered areas, when compared to the relatively unvegetated areas of the urban areas, has been identified as a significant contributor to the UHI effects. Vegetation indices computed from satellite data have been shown to be an effective method of estimating the leaf area indices in vegetated areas (Gallo and Daughtry, 1987; Asrar et al., 1989; Peterson and Running, 1989; Running et al., 1989; Curran et al., 1992).

Gallo et al. (1993) examined both land-based and AVHRR imagery for 37 US cities (Table 4.13).

The daily minimum air temperatures for June, July and August of 1991 were acquired from the US Cooperative Station Network that were located within a 1 degree latitude–longitude square centered on the selected cities. Temperature data from 322 stations were used in this study. Five of the stations (<2%) were located on rooftops. Data from these five stations (BNGR, CNTI, SCTO, SDGO, and WSTR) were compared with results from the other stations and concluded that the location effect of the temperature station is negligible.

Thirty-nine percent of the stations used a maximum–minimum temperature measurement system (MMTS). Quale et al. (1991) determined that during the summer months, the averaged MMTS sensors are approximately 0.3°C above the

TABLE 4.13 The 37 cities studied by Gallo et al. (1993).

ABQE	**Albuquerque, New**	**GRFL**	**Great Falls, Montana**
	Mexico	LBCK	Lubbock, Texas
ALTN	Allentown	LTRK	Little Rock, Arkansas
	Pennsylvania	LVGS	Las Vegas, Nevada
BLGS	Billings, Montana	MLKE	Milwaukee, Wisconsin
BNGR	Bangor, Maine	MTGY	Montgomery, Alabama
BOSE	Boise, Idaho	NOLS	New Orleans, Louisiana
BSMK	Bismarck, North	NSVL	Nashville, Tennessee
	Dakota	OLDO	Orlando, Florida
CL-M	Columbia, Missouri	OMHA	Omaha, Nebraska
CL-S	Columbia, South	RMND	Richmond, Virginia
	Carolina	SATO	San Antonio, Texas
CNTI	Cincinnati, Ohio	SCTO	Sacramento, California
COSP	Colorado Springs	SDGO	San Diego, California
	Colorado	SLKC	Salt Lake City, Utah
CSPR	Casper, Wyoming	SPKN	Spokane, Washington
DLAS	Dallas, Texas	STLE	Seattle, Washington
DLTH	Duluth, Minnesota	TCSN	Tucson, Arizona
EGNE	Eugene, Oregon	WSTR	Worcester, Massachusetts
FLSF	Flagstaff, Arizona	WTCA	Wichita, Kansa
FRGO	Fargo, North Dakota	WTLO	Waterloo, Iowa

minimum temperature measured with a standard "mercury in glass" thermometer. These investigators did not recommend an adjustment factor for the daily temperatures and suggested that it only be used for monthly or longer average data. Therefore Gallo et al. (1993) made no adjustment for this effect. All of the ground stations were classified manually based on the surrounding environments. An urban area was defined as one with a population of more than 50,000 people, while those areas with less than 10,000 people were considered as rural. Stations in rural areas that were close to urban areas were considered suburban.

Satellite data used for their study were the 1.1 km resolution AVHRR imagery from the NOAA-11 satellite collected between 28 June and 8 August of 1991. The data were acquired and processed to weekly and biweekly products by the USGS EROS Data Center. Daily AVHRR images were screened for cloud cover in 1991 before they were used to compute the weekly and biweekly products. These images were geometrically registered to a Lambert azimuthal equal-area map with a cell resolution of 1 km^2. They then defined a NDVI they called ND, which is equivalent to the NDVI discussed earlier in this book. They used prelaunch calibration information to calibrate the visible (0.58–0.68 μm) and NIR (0.72–1.1 μm) channels of the AVHRR (Holben et al., 1990). Note, we will use ND for vegetation index in this part of the book but in general will denote the vegetation index as the NDVI.

High vegetation index values indicate the presence of green vegetation, and the vegetation index can be used to monitor seasonal changes in vegetation activity (Goward et al., 1985; Malingreau, 1986; Gallo and Flesch, 1989). Surfaces with no vegetation, such as water and clouds, have very low vegetation index values. For this reason, after cloud removal, the weekly and biweekly products are computed as averages of high vegetation index values to suppress the attenuation of the surface vegetation signal due to the presence of residual cloud.

Calibrated thermal IR data from channels 4 (T4; 10.3–11.3 μm) and 5 (T5; 11.5–12.5 μm) were used to estimate the radiative surface temperature T_{sfc} as

$$T_{sfc} = \mathrm{T4} + 3.3(\mathrm{T4} - \mathrm{T5}) \tag{4.31}$$

for a surface emissivity of 1.0 (Price, 1990). While the surface emissivity of urban and rural areas is likely to vary considerably, and no information was available for the emissivity differences in the cities studied. Oke (1978) gave emissivity values for agricultural areas and deciduous and coniferous forests ranging from 0.9 to 0.97, while his emissivity for urban areas (concrete, asphalt, and stone) ranged from 0.71 to 0.95. Roth et al. (1989) suggested that the differences due to unaccounted for differences in emissivity could be as large as 1.5°C. In their study, Gallo et al. (1993) did not correct for area differences in atmospheric effects since relative and not absolute temperature differences were needed. Roth et al. (1989) estimated that the error introduced by not compensating for atmospheric effects could be as large as 1.0°C or greater.

The AVHRR images were sampled for three-pixel by three-pixel windows centered on each weather station. No additional filtering was performed on these windows but in the comparison with the station data any vegetation index values of T_{sfc} values that indicated cloud cover were eliminated from the comparison. Differences for urban and rural areas were computed for daily, weekly, biweekly, and monthly values of T_{sfc} and the vegetation index. Daily AVHRR images were not available for all 37 cities due to cloud contamination, and it was decided to use the composites of T_{sfc} and the vegetation index (their ND = NDVI) for comparisons with composites of T_{min} computed from the station air temperature data. The results of their regression analysis are summarized here in Table 4.14.

Only the time period between 28 June and July 4 showed the statistically significant relationship between the urban–rural differences of T_{min} and $T_{sfc.}$ They found that the elevation of the measurement station influences this temperature difference. The examination of weather station elevation less than 500 m did not improve this relationship. The amount of variation of urban-rural differences in T_{min} associated with variation in the vegetation index (ND=NDVI) were also statistically significant for four of the sex weeks analyzed if all the cities were included. Biweekly urban–rural differences indicated a significant relationship between T_{min} and T_{sfc} for stations below 500 m. Variations in T_{min} were not statistically related to ND in the 5–18 July biweekly interval (Fig. 4.69). There are a great many other detailed conclusions that can be drawn from Table 4.13 and it is left to the reader to determine which are important.

Gallo et al. (1993) conclude that the satellite-derived NDVI is linearly related to the differences in observed urban and rural minimum temperatures. Relationships between differences in the vegetation index and observed differences in temperature were improved when the analysis was restricted by elevation differences between the station locations within and around the urban areas. This relationship between urban and rural temperature differences appears to be dependent upon the length of time over which the differences are examined. Biweekly and monthly satellite composites and minimum temperatures are the recommended time periods for urban-rural analyzes. Urban and rural differences in the vegetation index explain a larger proportion of the variance in observed minimum temperature differences than was

TABLE 4.14 Results of regression analyzes of urban-rural differences of T_{min} (dependent variable), with.

	T_{sfc}				ND		
Observation interval of T_{min}	r^2	rmse	b_1	n	r^2	rmse	b_1
28 June–4 July	0.13	1.5	0.15	36	0.12	1.5	-6.3
(wx<500 m)	0.22	0.8	0.11	27	0.44	0.6	-6.5
5 July–11 July	0.04ns[a]	1.7	0.06	37	0.04ns	1.7	-4.4
(wx<500 m)	0.06ns	1.2	0.05	28	0.18	1.1	-6.4
12 July–18 July	0.01ns	1.7	0.03	37	0.09	1.6	-7.1
(wx<500 m)	0.01ns	1.1	0.03	28	0.26	0.9	-8.1
19 July–25 July	0.00ns	1.6	0.02	37	0.04ns	1.6	-3.0
(wx<500 m)	0.04ns	1.2	0.05	28	0.14	1.1	-4.6
26 July–1 August	0.07ns	1.6	0.08	32	0.10	1.6	-5.4
(wx<500 m)	0.02ns	1.1	0.04	23	0.30	0.9	-6,0
2 August–8 August	0.05ns	1.5	0.04	37	0.08	1.5	-4.3
(wx<500 m)	0.00ns	1.1	0.01	28	0.20	1.0	-5.2

[a] *The designation ns after r^2 value indicates lack of significance at P≤0,1 probability level.*

typical for past studies that used population data. A careful approach to the use of satellite data to sample urban and rural areas may prove to be a global method for estimating the UHI effect.

4.15 Biological applications

There have been a wide variety of applications of AVHRR data to biological studies. We will only review a few of them to give an idea of the way in which these data can be used to provide added insight into other biological studies.

4.15.1 Mapping the phenology preference of the yellow-billed cuckoo

The western, yellow-billed cuckoo is facing a steep population decline in its western breeding grounds primarily due to the loss of native habitat (Laymon and Halterman, 1987; Wallace et al., 2015). The favored nesting ground for the cuckoo in the southwestern United States is low-elevation riparian forests and woodlands (Johnson et al., 2008). Over the past 60 years, reductions in the breeding range of the cuckoo have been observed owing to riparian habitat alterations due to the clearing of vegetation, water management, agriculture, urbanization, overgrazing, and recreation (Laymon and Halterman, 2985). As a result, the cuckoo has been identified as a possible candidate for listing under the endangered species act.

The cuckoo breeds in the southwestern United States for a short period of time during the summer. Thus field measurements of cuckoo breeding require a rather demanding amount of personal and expense over this 1–2 month period. The cuckoo migrates from South America to the southwestern Uniterd States to breed. Such a survey was conducted by the US Fish and Wildlife service in 1998 and 1999. This makes this period ideal to explore the potential for using AVHRR data to map the cuckoo breeding habitat.

The study by Wallace et al. (2015) aims to explore the relationships between vegetation phenology patterns captured by satellites and phenometrics related to the distribution of the yellow-billed cuckoo and to use this data to map; the cuckoo's habitat. This study was conducted in Arizona, USA. Their analysis used 1 km resolution AVHRR data to compute biweekly composite NDVI for 1998 and 1999. In spite of this, relatively low spatial resolution, the frequent and continued temporal resolution made this data source ideal for this study. As has been stated earlier in this book the NDVI is a measure of the photosynthetically active vegetation whereby high values indicate dense vegetation and low values are associated with sparse vegetation, mixed soil, and negative values in areas of no vegetation such as water and snow (Price, 1992; Huete et al., 2002).

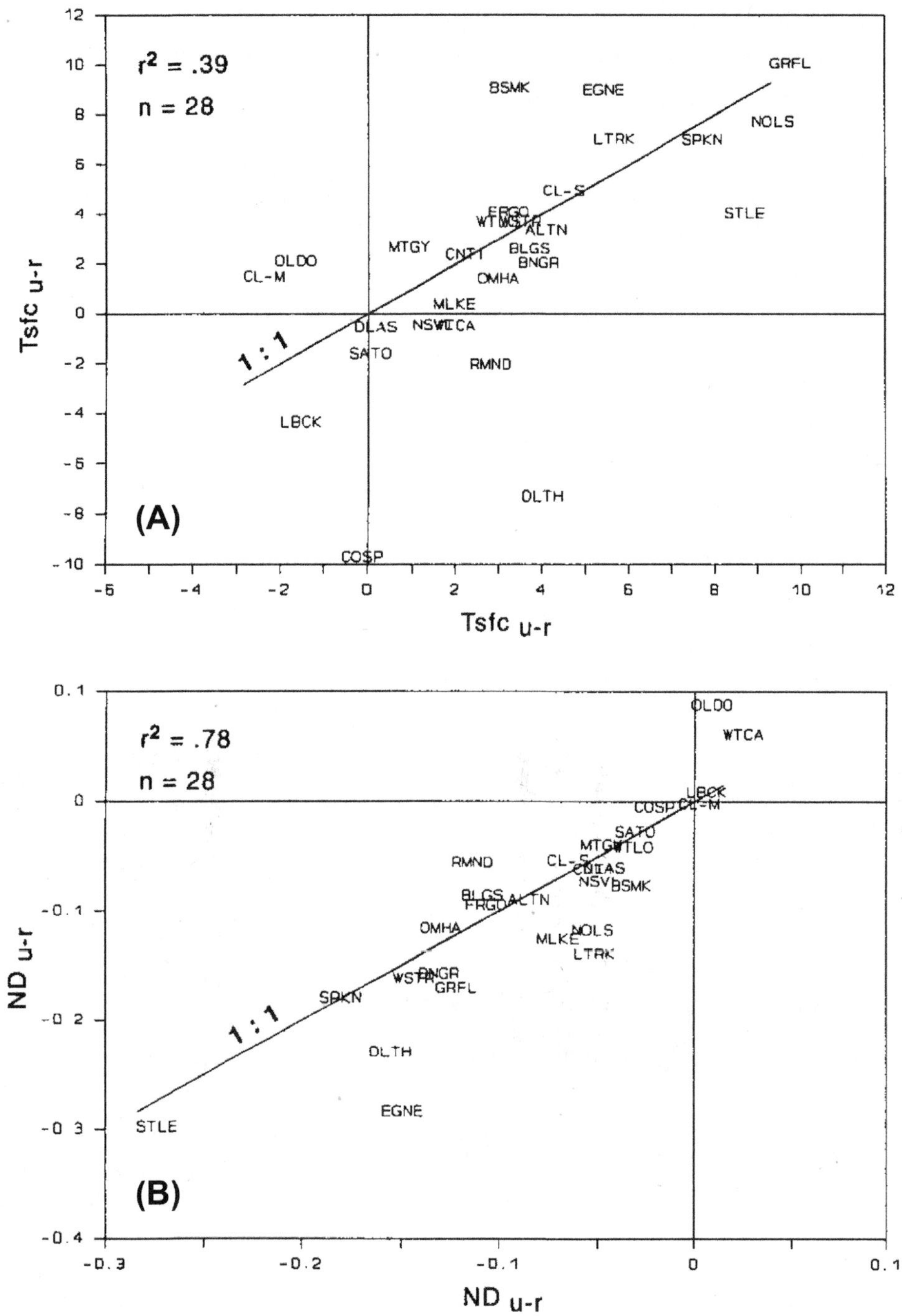

FIGURE 4.69 Relationship between urban-rural differences in (A) surface temperature and (B) vegetation index (ND) for the 5–18 July 1991 biweekly composite interval compared to the 19 July–1 August 1991 biweekly interval (Gallo et al., 1993).

Following common practice to reduce cloud effects, maximum-value composites of the daily NDVI images were computed ever 2-week periods (Eidenshink, 1992). They used these composites for the years 1998 and 1999, yielding 26 images per year.

They also used a riparian vegetation map created by the Arizona Game and Fish Dept. (AZGFD) in 1994. This polygon map provides details of the vegetation found in the area and identifies 12 classes including marsh, mesquite, tamarisk, mixed broadleaf, and agricultural land. The mapping was done with a combination of Landsat thematic mapper (TM) imagery and aerial videography using a minimum polygon size of 900 m^2. They created two binary images: (1) all riparian vegetation and (2) the subset of a single vegetation class of cottonwood-willow. They further excluded the far north and northeast of Arizona as cuckoos were not sighted in these areas. Five sets of data values were created:

1. A data set with 178 survey locations with birds present for 1998 (BP98),
2. A data set consisting of 43 survey locations with birds absent during 1998 (BA98),
3. A data set of the points that were mapped as cottonwood-willow (CAW). These values were generated using ArcGIS (ESRI, Redland, CA) by first converting the polygon coverage to a 30 m grid consistent with the spatial resolution of Landsat TM for this period. The grid was then converted to a point coverage (n=27,893), resulting in a point sample of the cottonwood-willow polygons weighted by the size of each polygon.
4. A data set of points representing the general riparian vegetation (RIP) was also converted to a 30 m grid and then to a point coverage (n=691,712). Note CW is part of RIP.
5. A data set of 359 survey locations with birds present for 1999 (BP99).

Wallace et al. (2015) then computed the response of photosynthesis (phenometrics) for the state of Arizona from the satellite greenness data. They computed NDVI biweekly composite images and extracted the time series of NDVI values for all image pixel locations in the state. They then used Fourier analysis to compute the annual curve at each location (Jakubauskaa et al., 2001; Moody and Johnson, 2001; Wallace, 2002). They then defined three new types of image products: (1) The amount of greenness which is related to the vegetation productivity, (2) The variability of the greenness which represents the vegetation green up and senescence, and (3) The timing of the peak greenness which represents the maximum greenness at the peak of the amplitude curve.

Student's t-tests were used to identify the phenometric variables that were significantly different in the survey locations with cuckoos (BP98) versus the other point locations (BA98, CW and RIP). They used two tailed tests with unequal variances and a significance level of 0.95. This t-test evaluates the null hypothesis that the difference in the means of the phenometric parameters observed for two distinct populations is zero (David, 1986). The satellite derived Fourier phenometrics were used to create cuckoo preference models that were found to be effective at separating locations with cuckoo sightings from other landscapes. The three 1998 phenometrics derived for the state of Arizona using the Fourier analysis are presented here in Fig. 4.70. Fig. 4.70A shows the highest vegetation along the Mogollon Rim that nearly bisects the state. High levels were seen along drainages and on the "sky island" mountains in the southeastern part of the state. Low vegetation areas are seen along and within the Grand Canyon.

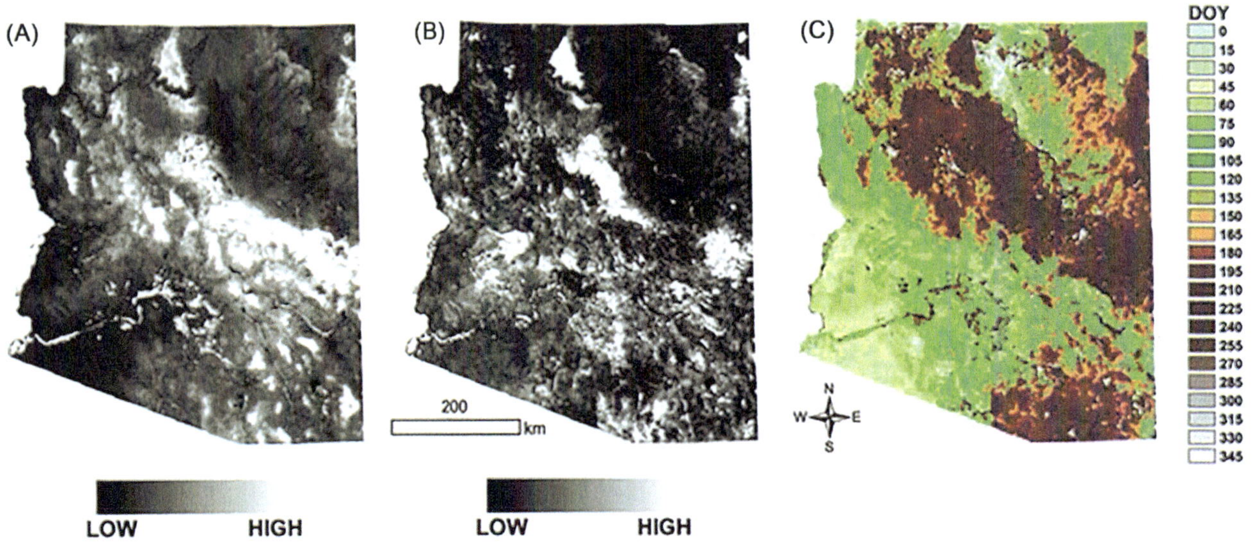

FIGURE 4.70 1998 phenometrics for the state of Arizona using a Fourier analysis: (A) productivity, (B) dynamics, and (C) timing of the peak greenness. These are the layers used to predict the distribution of the *yellow-billed* cuckoo in Arizona (Wallace et al., 2015). Map lines delineate study areas and do not necessarily depict accepted national boundaries.

The dynamics in Fig. 4.70B reveal marked changes along some of the Mongollon Rim and along the north rim of the Grand Canyon (caused by the contrast between summer pine forest and winter snow cover). The colored panel in Fig. 4.70C demonstrates that the peak greenness occurs earlier in the year for most of the southwestern Arizona deserts and on the Colorado Plateau in northeastern Arizona, while the peak greenness is later for the southeastern Arizona and the Mogollon Rim.

Histograms of the phenometrics of the 4 landscapes are shown in Fig. 4.71.

The timing histogram exhibits the greatest differences that demonstrate that sites with birds present are dominated by landscapes that reach maximum greenness in the late summer. Those sites with birds absent have a maximum greenness in the late spring. The histograms of dynamics and vegetation amount have much less obvious relationships. The *t*-tests reveal that the landscapes with cuckoos present have phenometrics that are dramatically different from the other landscapes examined. Again, the landscapes with cuckoos present have significantly later peak greenness than sites without birds. Moreover, the landscapes with cuckoos present are greener in general and are significantly more dynamic than the other landscapes that are generally less green with sites without birds such as cottonwood-willow vegetation (CW) or riparian vegetation (RIP)

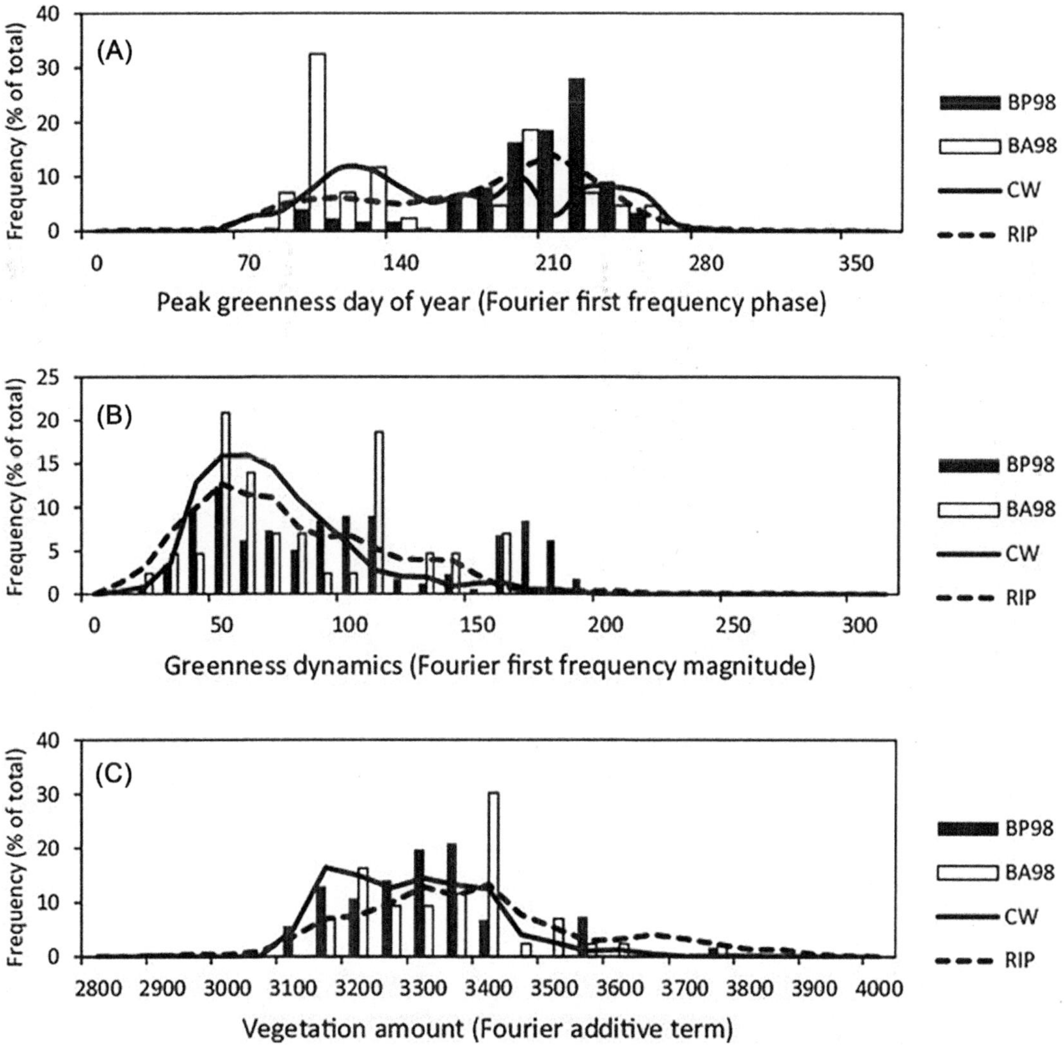

FIGURE 4.71 Histograms of the phenometrics for the landscapes used to model the habitat for the yellow-billed cuckoo. As defined previously the four landscapes are riparian vegetation (RIP), cottonwood-willow vegetation (CW), surveyed area with birds present (BP98), and surveyed areas with birds absent (BA98). Panel (A) gives the timing of the peak vegetation greenness by day of year, panel (B) shows the dynamics of the vegetation, and panel (C) compares the amount of vegetation present (Wallace et al., 2015).

Wallace et al. (2015) found that the models created from the 1998 phenometric values (derived from the satellite data) were effective at predicting the locations of yellow-billed cuckoos sighted in 1999 (Fig. 4.72).

Of these models the rank-based (Fig. 4.72C) is the most generalized and identifies more areas as highly preferred by the cuckoos while the standard deviation stratified model is the least generalized. They considered the standard deviation stratified model to be slightly superior in that it identified a less generalized mapping. This stratified map (Fig. 4.72B) also shows that large areas of Arizona have phenometrics favorable for the yellow-billed cuckoo. To accurately map habitat, it is necessary to consider both the dynamics (phenology) and composition (vegetation type and structure) of the landscape.

These authors then went on to apply the 1998 model to other later years and found a similar outcome as for 1999. Thus they concluded that phenometrics computed from satellite data can be used to create models that successfully predict yellow-billed cuckoo presence within areas of recognized habitat. The synoptic view and frequent repeat sampling allow the creation of these models and their success. The models further reveal the cuckoo response to changes in vegetation within these habitats. These results demonstrate that the cuckoo prefer areas within their cottonwood-willow that experience peak greenness in response to the American Monsoon rainfall some 29 days later. These areas are 36% more dynamic and slightly more productive than the average CW landscape. Models based on satellite imagery can be used in the management of these habitat areas.

4.15.2 Mapping the phenology preference of the red-billed African Quelea

A similar study was carried out by Wallin et al. (1992) using AVHRR data applied to the breeding habitat of the African weaver bird. This species is considered to be an agricultural pest, affecting cereal grain production throughout sub-Saharan Africa. Once again, the frequent temporal coverage over large areas of the AVHRR is key to the success of this study. In this region, the environment exhibits significant patchiness leading to a nonuniform distribution of vegetative resources in time and space. This patchiness affects many ecological processes and thus becomes a determining factor in the study of related biological organisms (McIntosh, 1985). Frequently logistical constraints, rather than organism specific

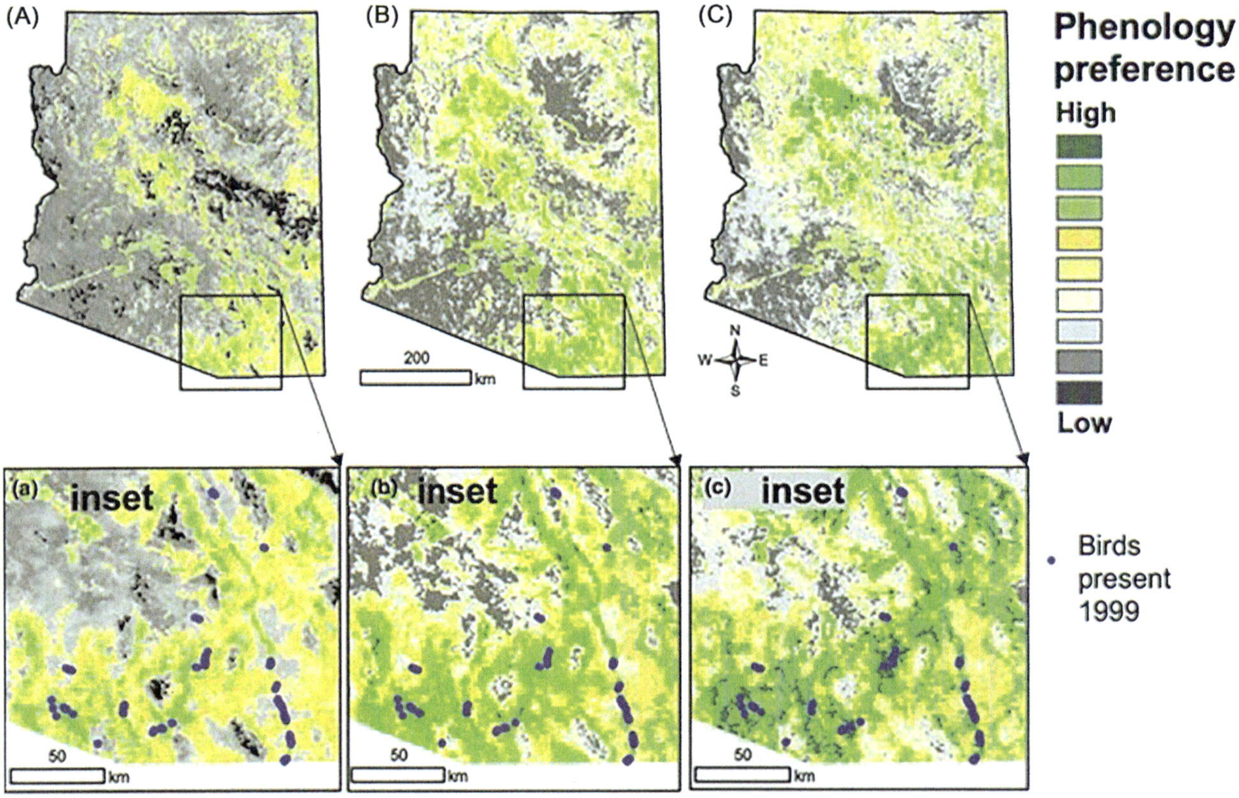

FIGURE 4.72 Phenology for 1999 inferred from the models of 1998 (A) standard deviation distance model, (B) standard deviation stratified model, and (C) rank-based model. Note, the presence of observed birds marked by dots in the inset panels (Wallace et al., 2015). Map lines delineate study areas and do not necessarily depict accepted national boundaries.

characteristics, end up defining the time and space scales of the target organisms (Levin, 1976; Southwood, 1977). Karieva and Anderson (1988) found that 95% of recent papers in the ecological literature were based on study plots of less than 1 ha and that 50% of these studies used plots no bigger than 1 m^2. There is growing interest in the effect of landscape level patch dynamics on ecological systems (Godron, 1986; Turner, 19887). Simply scaling up the small scale studies has many difficulties since small scale dynamics may be unimportant for larger scale processes (Allen and Starr, 1982; Allen et al., 1984; O'Neil et al., 1986; Urban et al., 1987).

For example, while the study of individual feeding patterns on scales of 1–100 m^2, which exist for perhaps minutes to days, may provide insight into foraging behaviors territorially or even mating systems, it is not clear that this information can be scaled up to explain seasonal migration patterns, range changes, or subpopulation formation. All of these latter phenomena occur over much larger spatial and temporal scales, which may not be properly characterized when considering only one small patch. To overcome this problem, satellite imagery has been invoked. Of course, this assumes a close association between the key features of the target organism's life history and the environmental conditions that can be sensed by the satellite imagery. Since the life history is generally closely tied to time/space changes of the vegetation, which can be monitored remotely by aircraft and satellites, as discussed in earlier sections of this book.

The red-billed quelea breeding colonies can often contain hundreds of thousands of breeding pairs that may cover a few tens of hectares (Ward, 1971; Wiens and Dyer, 1977). A few colonies contain well over a million pairs. Field studies (Elliot, 1979; Ward, 1965a) demonstrate that under certain conditions the foraging activities of these birds can substantially reduce the yield of small-grain cereal crops. Individual subsistence farmers can lose their entire crop, while larger farms can lose up to 20% of production (Elliott, 1979). It must be considered, however, that flocks of quelea feeding on insects that are a threat to the crops can significantly reduce the cereal crop loss to insect pests. In spite of this factor, the more visible quelea are often blamed for all of the crop loss. In a geographic region that frequently faces famine, it is often difficult to objectively evaluate the impact of quelea on crop production both on local and national levels. As a result, the red-billed quelea has been described as the most destructive and most numerous pest bird in the world (Ward, 1973; Elliot, 1989). Past efforts to eradicate these birds have evolved into more enlightened and pragmatic approaches to eliminate only those colonies that pose an immediate threat to cereal crops at a critical stage of their development (Bruggers and Elliot, 1989; Ward, 1979). A satellite-based monitoring program able to provide timely information on breeding habitat could play an important role in estimating and predicting the potential impact of the quelea on cereal production. In addition, the satellite data can be employed to assess crop damage, making it possible to objectively design mitigation approaches to protect the crops.

As with the other study of bird breeding, Wallin et al. (1992) used AVHRR data to compute the NDVI, in which they used 4 km GAC data to define and monitor the breeding habitat of the red-billed quelea, and they believed that this spatial resolution was adequate. This is because adults are known to forage out to at least 3 km from the colony site (Brugers et al., 1983). They hypothesized that the 4 km NDVI provides a reasonable index of vegetation condition, and hence food availability for any breeding colony located within an AVHRR pixel.

The ground truth data for this study are based on a compilation of existing data that were collected as part of ongoing crop protection efforts. Ground-based and helicopter surveys to locate quelea breeding colonies in East Africa have been carried out in Tanzania, Kenya, and Somalia. Additional data were compiled from unpublished reports and field notes archived by the national bird control units in each country. A complete listing of colony locations and their histories for all breeding colonies used in this study can be found in Wallin (1990). These surveys were not comprehensive and with the exception of Kenya, most survey efforts were carried out in important agricultural areas. Since the nesting success of each colony is not known and many colonies were destroyed, from aerial surveys these investigators calculated an initiation date (±2 days) by back dating from the age of the young or from the age of the eggs. They found that not all breeding sites surveyed were occupied from year to year. They distinguished between occupancy which meant that breeding birds were present and suitability which is a site that could be used for breeding but was not being used. They also found that for a fixed quelea population, the proportion of suitable breeding sites that remains unoccupied will increase as the as the regional quelea population decreases.

The ultimate goal of the Wallin et al. (1992) study was to determine whether suitable and unsuitable sites have different remotely sensed signatures. Lacking direct ground truth data they focused instead on examining satellite data for occupied and unoccupied breeding sites. Due to the presence of birds, the occupied sites are clearly suitable for breeding, while the unoccupied sites are made up of a mixture of suitable and unsuitable areas. They further assumed that the proportion of unoccupied sites that are actually unsuitable varies from year to year. Thus even small differences in even a single year can determine if the area is suitable or unsuitable for a breeding habitat.

Fewer surveys were conducted in Kenya and Somalia compared with Tanzania. In addition, the unoccupied sites were not recorded in these countries. Therefore the results in this study are based largely on the Tanzanian data supplemented with the Kenya and Somalia data where possible. The NDVI was calculated from daily AVHRR data, and thermal infrared data of channel 5 were used for cloud masking. Any pixel with an early afternoon BT below 10°C is assumed to be cloudy and not included in the analysis. The daily data were used to construct bimonthly composite images. This maximum value composite yields images that are largely cloud free, and it also minimizes the effects of aerosols and atmospheric water vapor. Each bimonthly composite image was corrected for temporal drift in the sensor calibration using values derived by Kaufman and Holben (1995). These corrections account for the slight differences in the three AVHRR instruments used in this study. Such corrections are critical for sensing temporal variations in the NDVI. As mentioned earlier in this text, the NDVI can theoretically range from −1.0 to +1.0. Values below zero are likely due to the presence of subpixel sized clouds. Since the NDVI over vegetated surfaces is essentially a proportion, the arcsine-square root transformation was used (Sokal and Rohlf, 1981). They also examined the untransformed NDVI, which is what they based their conclusions on. The next figure (Fig. 4.73) is a bit

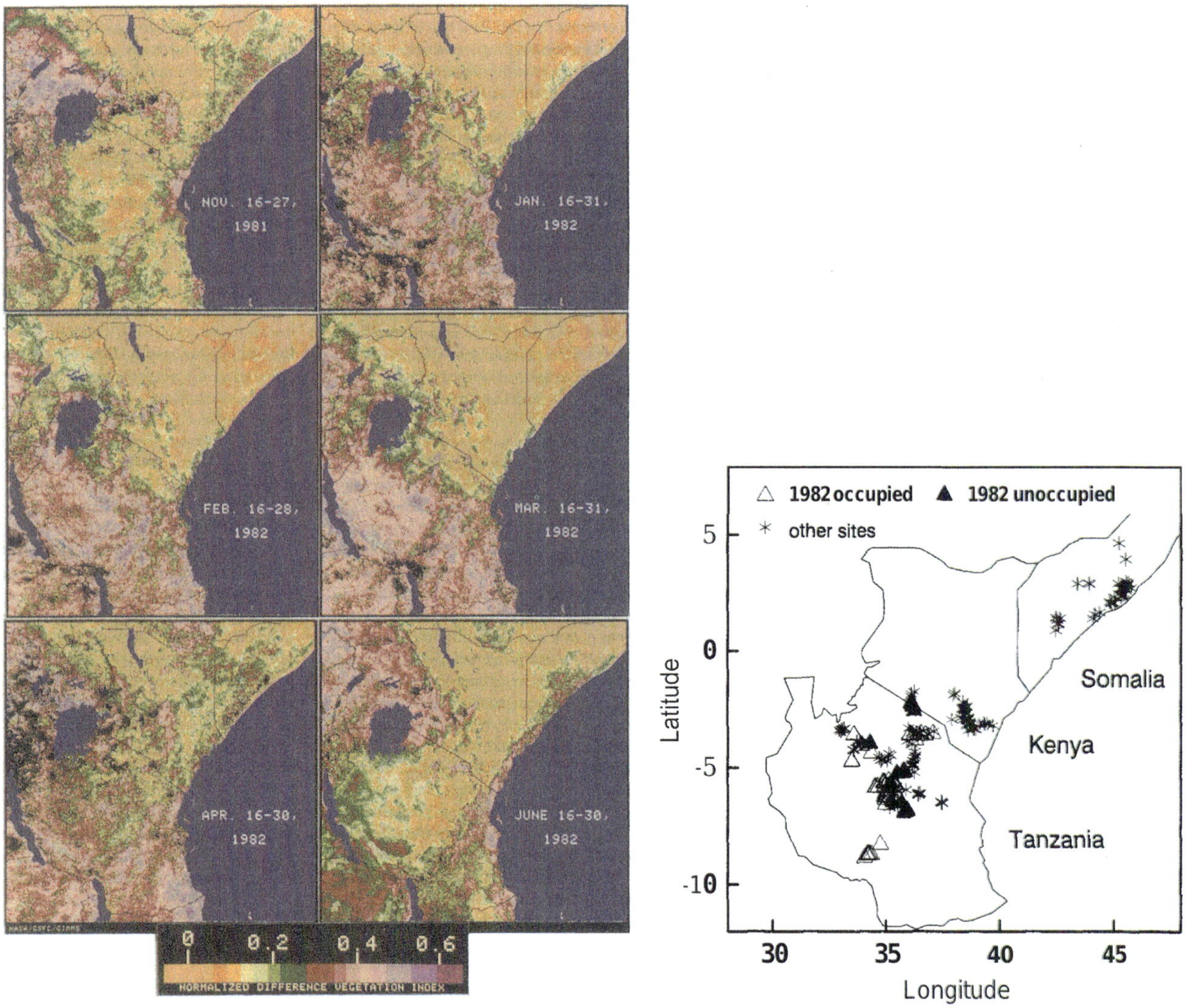

FIGURE 4.73 (left) Biweekly normalized difference vegetation index for the study area by month from Nov. 1981 to June 1982. (right) Quelea breeding sites in East Africa which were occupied at some time during 1978–1988. Locations of occupied and unoccupied sites are known for Tanzania for the 1983 and 1984 breeding seasons. Unoccupied sites were not recorded for Kenya and Somalia (Wallin et al., 1992). Map lines delineate study areas and do not necessarily depict accepted national boundaries.

complex in that it presents the bi-weekly NDVI between November 1981 and June 1982. Separately and on a completely different spatial scale, they include the 1982 sites where breeding pairs were observed.

It should be noted that quelea breeding sites are found in many other places, but these investigators have tried to identify the major colonies. There may be colonies that were missed by the surveys. A careful examination of the NDVI images in the April 16–30 image at the mid-breeding season suggests qualitative differences in the NDVI between area containing bird populations versus unoccupied sites. Temporal changes in mean NDVI for occupied and unoccupied sites are shown as histograms in Fig. 4.74.

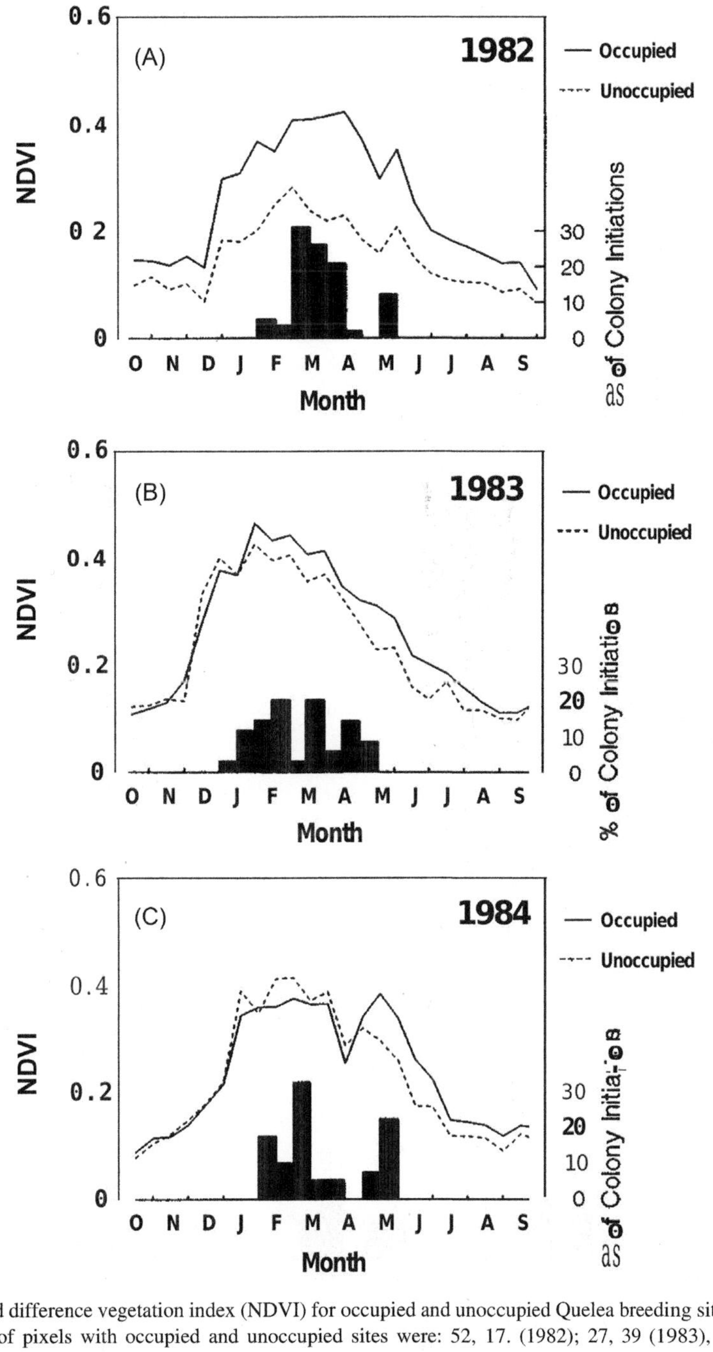

FIGURE 4.74 Mean normalized difference vegetation index (NDVI) for occupied and unoccupied Quelea breeding sites in Tanania. *Black* bars indicate colony initiations. The number of pixels with occupied and unoccupied sites were: 52, 17. (1982); 27, 39 (1983), and 35, 55 (1984). Some pixels contained more than one breeding site (Wallin et al., 1992).

The cumulative frequency distributions of the NDVI for the mean breeding season (January–May) are shown in Fig. 4.74 for 146 pixels in Tanzania with a history of quelea breeding activity. These curves suggest that suitable breeding habitat (sites with a high mean breeding season NDVI) was least abundant in 1982. In 1983 the site abundance was average, and in 1984 it was more abundant. If the observed differences in occupied and unoccupied sites in Figure 1.90 are due to changes in the available sites for breeding, then Fig. 4.75 indicates that this change is the result of interannual differences in the abundance of suitable breeding habitat.

Similar comparisons for Kenya and Somalia support the conclusions based on the Tanzania analysis. One big difference is that Tanzania has a single growth period for its vegetation while Kenya and Somalia have two separate peaks of the NDVI. An example is given here in Somalia for 1987 in Fig. 4.76.

You can clearly see the two growth peaks in the Somalia data of Fig. 4.77. The colony initiation peaks coincide nicely with the NDVI peaks indicating that the quelea are waiting to breed for the second vegetation peak. Similar results are found in Fig. 4.74 which is again the mean NDVI annual curve for Kenya for the year 1987.

Again you can clearly see the two peaks in the NDVI in the 1987 annual cycle which correspond well to the dates of colony initiation.

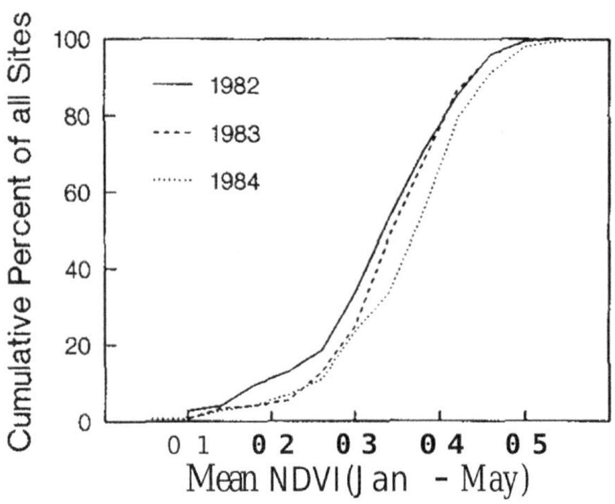

FIGURE 4.75 Mean normalized difference vegetation index (NDVI) during the breeding season (January–May) for 146 pixels in Tanzania with a history of use as breeding sites (Wallin et al., 1992).

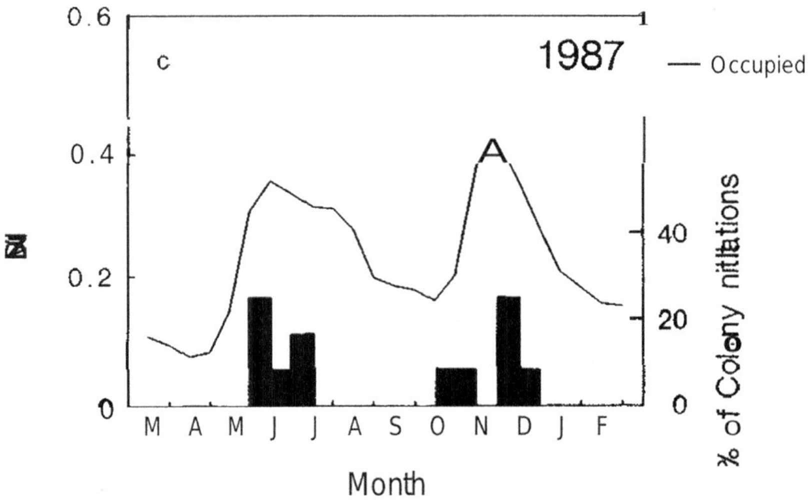

FIGURE 4.76 Mean normalized difference vegetation index curve of the occupied quelea breeding sites in Somalia. The black bars again indicate colony initiation dates (Wallin et al., 1992).

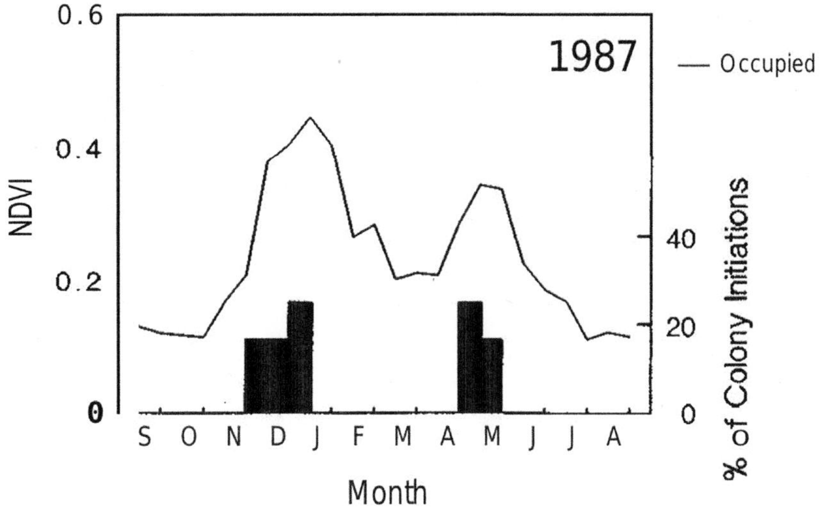

FIGURE 4.77 Mean normalized difference vegetation index (NDVI) curve of the occupied quelea breeding site in Kenya. The *black bars* indicated colony initiation dates (Wallin et al., 992).

Wallin et al. (1992) conclude that coarse resolution satellite data can be used to monitor potential breeding habitat for the red-billed quelea in East Africa. The satellite data offer a source of information to judge the areas of breeding site suitability. The same satellite data can be used to estimate when a site becomes suitable for colony establishment. Thus satellite data can be used to map the spatial and temporal distributions of potential breeding habitat for this species at regional to continental scales. This information can then be used to estimate the possible impact of the bird population and its potential impact on cereal production in Africa.

Chapter 5

Major modern satellite systems, NOAA, Defense Meteorological Satellite Program, and follow-ons

5.1 Introduction

The preceding discussion of the AVHRR is relevant to large satellite programs due to its continued operation by NOAA on its polar orbiting spacecraft. This continued operation of the TIROS-N satellites carrying the AVHRR has produced a significantly long time series of measurements that can be applied to a great many important variables. It is difficult to separate the earlier applications of AVHRR data from later ones, so they are all presented together in the previous chapter. The AVHRR has also been carried by the European polar orbiters known as MetOp, as will be discussed later in this book.

5.2 NOAA's atmospheric profiling instruments

These same TIROS-N satellites carried a suite of instruments to profile the atmosphere. This was the primary source of atmospheric profile data from a system known as TOVS (TIROS Operational Vertical Sounder). The TOVS system consisted of three instruments: the High Resolution Infrared Radiation Sounder (HIRS), the microwave sounding unit (MSU), and the stratospheric sounding unit (SSU). The MSU and SSU were replaced by AMSU-A and AMSU-B on later satellites.

The HIRS-2 had 20 channels with channels 1–5 in the 15 μm carbon dioxide band (15.0, 14.7, 14.5, 14.2, and 14.0 μm), channels 6 and 7 the 13.7 and 13.4 μm CO_2 and H_2O bands, channel 8 and 11.1 μm open window bands, channel 9 the 9.7 μm ozone band, channels 10, 11, and 12 the 6 μm water vapor bands (8.3, 7.3, and 6.7 μm), channels 14 and 15 the 4.57 and 4.52 μm N_2O bands, channels 15 and 16 the 4.46 and 4.40 μm CO_2/N_2O bands, channel 17 the 4.24 μm CO_2 band, channels 18 and 19 the 4.0 and 3.7 μm window bands, and channel 20 the 0.7 μm visible band. HIRS had a field of view (FOV) of 30 km diameter at nadir, while the MSU had an FOV of 110 km diameter. HIRS sampled 56 FOVs in each 2250 km scan line, while the MSU sampled only 11 FOVs along the same scan line.

In addition to profiling the atmospheric temperature and moisture, HIRS was also used to compute surface skin temperature, sea surface temperature, outgoing longwave radiation, cloud fraction, cloud-top height, total integrated column ozone, precipitation, and wind speed. HIRS-3 was introduced with NOAA-15.

The SSU was provided to NOAA by the British Meteorological Office with three 15.0 μm channels using selective absorption by passing the sample air through three pressure-modulated cells containing CO_2. The MSU was similar to an earlier instrument flown on Nimbus and had one channel in the 50.31 GHz window region and three channels in the 55 GHz oxygen band (53.73, 54.96, and 57.95 μm) intended to retrieve moisture profiles independent of cloud cover. With NOAA-15, the MSU was replaced with the 20-channel AMSU-A and AMSU-B.

The AMSU-A is a multichannel microwave radiometer for measuring atmospheric temperature profiles. It is a cross-track (whisk broom) scanning radiometer with 15 channels. It was designed to measure the temperature profile from 3 millibars (~45 km) to the Earth's surface. The FOV was 48 km at nadir, with a swath width of 1,650 km. The frequencies of the 15 channels are given here in Table 5.1.

The AMSU-B was designed to measure the water vapor profiles with an FOV of 15 km. There are approximately nine AMSU-B FOVs in a 3X3 array for each AMSU-A FOV. This is due to the higher spatial variability of the water vapor compared with temperature. The five AMSU-B channels are given here in Table 5.2.

Together, AMSUA,B were used to observe temperature profiles (3 mb [45 km] to the Earth's surface). Water vapor profiles, snow and ice coverage, cloud liquid water, rain rate, ozone, cloud properties, and cloud-cleared IR radiances.

Satellite Remote Sensing. DOI: https://doi.org/10.1016/B978-0-443-22086-9.00008-3

TABLE 5.1 Channel frequencies of AMSU-A.

Channel number	Frequency (GHz)	Polarization at nadir	Number of bands	Instrument sensitivity (K)	Primary function
1	23.8	Vertical	1	0.30	Water vapor
2	31.4	Vertical	1	0.30	Water vapor
3	50.3	Vertical	1	0.40	Water vapor
4	52.8	Vertical	1	0.25	Water vapor
5	53.596±0.115	Horizontal	2	0.25	Tropo temp
6	54.4	Horizontal	1	0.25	Tropo temp
7	54.94	Vertical	1	0.25	Tropo temp
8	55.5	Horizontal	1	0.25	Tropo temp
9	57.290	Horizontal	1	0.25	Strato temp
10	57.290±0.217	Horizontal	2	0.40	Strato temp
11	57.290±0.3222±0.048	Horizontal	4	0.40	Strato temp
12	57.290±0.3222±0.022	Horizontal	4	0.60	Strato temp
13	57.290±0.3222±0.010	Horizontal	4	0.80	Strato temp
14	57.290±0.3222±0.0045	Horizontal	4	1.20	Strato temp
15	89.0	Vertical	1	0.50	Cloud top/snow

TABLE 5.2 Channel frequencies of AMSU-B.

Channel number	Frequency (GHz)	Polarization (at nadir)	Number of bands	Instrument sensitivity NEΔT(K)
16	89±0.9	Vertical	2	0.37
17	150±0.8	Vertical	2	0.84
18	183.31±1.0	Vertical	2	1.06
19	183.31±3.0	Vertical	2	0.70
20	183.31±7.0	Vertical	2	0.60

A lot of the AVHRR and atmospheric profiling heritage of the NOAA satellite programs was carried forward by the European MetOp program, which flew AVHRR and AMSU instruments in their early satellites, thus continuing the time series of data established by the NOAA polar orbiting satellite program.

5.3 The Defense Meteorological Satellite Program

Parallel to the NOAA polar orbiter satellite program, the US military started its own polar orbiting meteorological satellite program called the Defense Meteorological Satellite Program (DMSP) (Cargill Hall, 2001). This was part of the National Reconnaissance Program (NRP) and was initially a classified program. In the 1950's the Rand Corp. had warned the Air Force that successful overhead reconnaissance depended on accurate and timely meteorological forecasts over the Sino-Soviet landmass. Such forecasts were needed to ensure cloud-free photography over areas of interest. At this time, NASA and NOAA were given the responsibility to develop meteorological satellites for both the Department of Commerce and the Department of Defense.

At the Pentagon in 1961, under Secretary of the Air Force Joseph V. Charyk, who also headed the NRP, remained unconvinced that this arrangement would suit their needs. Knowing then that the early TIROS satellites could only point at a limited latitude band whil the NRP had a need for measurements over other earth locations. NOAA would soon overcome this problem with their outward looking wheel satellites that could collect meteorological data over the entire globe. Charyk then decided that what the NRP needed was an interim project for the National Reconnaissance Office (NRO). On June 21, 1961, Charyk directed Major General Robert El Greer, Director of the Office of the Secretary of the Air Force, to prepare a minimum proposal for four Earth-referenced wheel-mode weather satellites to be launched on NASA Scout boosters.

Greer responded with a plan for a 22-month program with a small, fixed budget, and a first launch in 10 months. Greer's deputy, Lt. Colonel Thomas O. Haig, got the job and was made the first director of the DMSP. Haig, a meteorologist and an electrical engineer, accepted the job on the condition that he could select his own staff and could use fixed price, fixed delivery contracts during the project. If the launch could not be made on schedule or if costs exceeded certain limits, Haig was to terminate the program. In the months that followed, the development was carried on outside of the normal NRO offices. In addition, Haig was able to negotiate a very favorable contract with RCA for the weather satellites. Part of Haig's team worked with the Scout launch vehicle. They decided to substitute a high acceleration Lockheed Propulsion Co. MG-18 solid propellant motor for the standard Scout fourth stage Altair motor. The RCA fixed price came in handy when a major flaw of the spacecraft was found in the base plan. RCA requested a 3-month delay to fix the problem, but Haig advised RCA that they had 10 days to fix the problem or the contract would be canceled at no cost to the government. The RCA program manager came back with a revised schedule that met the original launch date.

Neither the Scout rocket nor the RCA satellite had any backup equipment, and a failure anywhere in the system would mean a loss of the mission. Smaller and lighter than the original, the 100-pound TIROS-derived RCA satellite was shaped like a 10-sided polyhedron, 23-in (~58 cm) across and 21 in (~53 cm) high. A spinning motion, introduced on injection into orbit, was maintained on the early NRO weather satellites at about 12 rpm by small spin rockets. Adopting a concept advanced by Haig and Lt. Ralph Hoffman, the spin axis was maintained perpendicular to the orbit plane by torquing the satellite against the Earth's magnetic field; the torquing forces were supplied through a direct-current loop around the satellite's perimeter.

When operated correctly, the RCA shuttered television camera (a photosensitive vidicon tube) would be pointed directly at the Earth once each time the satellite rotated, and the camera would take a picture of an 800-mile (~1,287 km) square area of the surface with the image recorded on tape for a later transmission to the ground. The satellite operated in a 450 nautical mile (~833 km; note the shift to standard units preferred by the US military) near circular orbit, and the camera would provide 100% daily coverage of the Northern Hemisphere at latitudes above the equator and 55% coverage at the equator. Readout of the encrypted tape-recorded images was planned to occur at ground stations and relayed to the Air Weather Service's Air Force Global Weather Center at Offutt AFB near Omaha, Nebraska.

Haig's 10-month development schedule was met, but when it came time for launch, it took second priority to an NRO GRAB satellite launch at Vandenberg AFB, located near Lompoc, CA. This launch was to be the same four-stage rocket to be used for the DMSP satellite. This launch took place on April 25, 1962, and ended in a complete failure of the launch vehicle. The same launch vehicle, this time outfitted with the MG-18 fourth stage, failed during launch of the first NRO weather satellite. The second DMSP launch of August 23, 1962, was a success, but the Lockheed ground-control team failed at first to track the satellite. When it began to operate properly, it took pictures daily as it transited the Soviet Union. Weather pictures of the Caribbean collected by this satellite in October proved to be crucial during the "Cuban Missile Crisis."

The first satellite finally ceased transmissions on March 23, 1963, but there were already plans for the next series of satellites. NASA had announced in September of 1962 that their Nimbus weather satellite would be delayed, and so Charyk advised Haig that he should plan an additional year for the interim NRO meteorological program. Haig anticipated this turn of events and had his budgets ready. Lockheed wanted a major part of the allocated budget for ground support operations, but Haig insisted he could build two ground stations and operate them with "blue-suiters" for about one-eighth the cost of the contractor. Charyk approved this cost-saving plan immediately. Charyk then arranged to meet with Curtis E. LeMay, the Air Force Chief of Staff General, to seek the allocation of Air Force personnel for this project. LeMay quickly agreed and said, "if anybody gets in your way call me."

This began an intensive period where program staff worked day and night. They found surplus Nike antiaircraft rocket sites in Maine (near Loring AFB) and Washington (near Fairchild AFB), procured six large van bodies from Norton AFB in San Bernadino, located two abandoned antenna mounts on Antigua Island in the Caribbean, and wrote a fixed-price contract with Radiation Inc. for two 40 ft (~12 m) radar dishes and associated electronics. At the same time, they trained SAC military personnel who would work at these stations, and by July 1963, 10 months after their initial go-ahead, the DMSP satellite tracking and readout moved from Lockheed to their own stations in Maine and Washington. At the same time, a command and control center for the DMSP manned by SAC personnel began operation one floor below the Air Force Global Weather Central building D, in the old Martin bomber plant, next to SAC headquarters at Offutt AFB, Nebraska.

The first DMSP satellite to be controlled by Air Force personnel was flight number 3, launched on February 19, 1963, from Vandenberg AFB. In this case, the Scout booster upper stages again malfunctioned, resulting in a satellite orbit not well suited to strategic weather reconnaissance. The satellite only operated for a few months. The primary tape recorder circuit was first to fail, so only direct readout of the vidicon signal was possible. A new experiment, however, continued to function well for many months. It was an infrared radiometer that measured the Earth's background radiation and indicated the extent of nighttime cloud cover that Global Weather Central used to produce crude operational maps of cloud cover at night up until January 1964. This unit proved to be so successful that it was included on all successive DMSP satellites up to Block 4, eventually providing measurements of cloud height and the Earth's heat budget.

The fourth and fifth DMSP Block 4 launches on April 26 and September 27, 1963 both resulted in Scout booster failures, creating a weather observation gap until January 1964. Despite appeals for changes in design and testing of the Scout boosters, NASA refused to make them. Haig, with the approval of Charyk's successor Air Force Undersecretary Brockway McMillan, canceled the last two Scout vehicles on the present contract and all six of them on the follow-on contract. Haig terminated all NASA Scout-related activities on 25 October. Five launches in 2 years had resulted in three Scout booster failures and increasing NASA intransigence.

Even before the fourth Scout launch failure, Haig had started looking for a replacement booster that could provide the reliability he needed for future DMSP launches. They were aware that a number of liquid-propellant SM-75 Thor intermediate-range ballistic missiles had recently returned from England after the Cuban Missile Crisis and were stored in San Bernadino. They also knew about an FW-4S solid-propellant rocket motor produced by the United Technology Corp. This rocket was cast in the same motor casing as the Scout fourth stage and would become the second stage of what became known as the "Burner I" launch vehicle. For this launch vehicle, Douglas Aircraft replaced the IRBM inertial guidance system with a Bell Telephone Laboratory guidance package and added a cold gas coast attitude control section on top of the Thor to keep the axis of the FW-4S injection stage properly aligned. A few months later, in March, they received approval to develop a second stage known as "Burner II" to add to the Thor-based launch vehicle. The source selection committee chose Boeing to produce the all new self-guided solid-propellant upper stage. This combination of the Thor/Burner II continued to be used by the program until the early 1980s.

Before the program could use the Thor/Burner system, Brockway McMillan had approved the acquisition of two Thor-Agena launch vehicles. This system was larger than was needed for the DMSP program, but it could carry two RCA weather satellites into orbit. It did just that on January 19 and June 17, 1964 when the two Thor-Agena rockets successfully placed four DMSP satellites into orbit. Now, the DMSP program had all the meteorological data they wanted. These signals were encrypted and not available to anyone but the DMSP program. The first launch of the Thor/Burner took place on January 18, 1965. The failure this time was not in the rocket, but the fairing refused to separate on orbit, rendering the satellite useless. Nevertheless, the Thor/Burner launch vehicle was subsequentially able to achieve an enviable 86% success ratio before being replaced by the Atlas booster in the 1980s.

At first, the "interim" DMSP program was extended from year to year, awaiting the arrival of NASA's weather satellites. By mid 1965 the NRO's interim weather satellite program was looking and acting more like a formal military space program. As its primary mission, DMSP provided the NRP with daily strategic (morning coverage during the first few years) mesoscale observation of cloud distribution over the Eurasian landmass. Beginning in 1966 DMSP operated two polar orbiting satellites in near circular orbits with different equator crossing times. The morning bird passed over the Soviet Union about 07:00 local time while a later "noon" bird covered the same area at about 11:00 a.m. local time, which revealed the change in cloud cover during the atmospheric heating of the day.

Looking back, Haig identified four early DMSP program contributions to astronautics. First, they demonstrated the successful operation of such a program with a small group rather than the "assistance" of a system's engineering contractor. The office used fixed price, fixed delivery contracts, and the offices were staffed with military personnel. This successful operation was carried out on an annual basis at about half the cost of the equivalent NASA/NOAA weather satellite program. Second, because the spin axis of the RCA wheel satellite was maintained perpendicular to the orbit plane by electronically torquing it against the Earth's magnetic field, Haig reasoned one could control and maintain a constant spin rate electronically, eliminating the need for solid rockets to maintain the spin. Third, when the DMSP ground sites were set up in 1963, they eliminated the costly "boresight tower" used to determine a tracking/readout antenna's

FIGURE 5.1 Defense Meteorological Satellite Program Block 1 satellite which became the NOAA TIROS Operational System satellite (Hall, 2001).

point vector coupled with a transmitter to check the receiving system sensitivity. They instead substituted a technique of scanning the sun to establish the pointing vector with a hermetically sealed low-energy transmitter in the center of the antenna to monitor its sensitivity. The DMSP tracking procedure was very successful, operating at a far lower cost that became a standard practice for nearly all satellite tracking/readout systems.

When Colonel Haig finally stepped down as the program director, DMSP had set a new standard for weather satellite operations. DMSP had pioneered the space technology so well, so quickly and so inexpensively that NASA, prodded by NOAA, adopted copies of the DMSP Block 1 wheel satellites (Fig. 5.1) as the TIROS Operational System. In addition to their success with the two ground stations, DMSP developed portable ground stations that could be operated in overseas locations. This was necessary to fulfill the primary mission of the DMSP to deliver weather satellite imagery to military field operations.

By 1963 it was clear that NASA's three-axis stabilized polar orbiter would again be delayed and not able to fill the DoD needs. The political climate at the time did not support an independent military weather satellite program. In early 1964 the DoD wanted to assess the role of weather satellite data in military operations, and the NRO was tasked with performing a test of satellite data in tactical operations as part of the 1964 Strike Command Goldfire exercise at Fort Leonard Wood in southwest Missouri.

Air Force Global Weather Central at Offutt AFB relayed weather reconnaissance images directly to the Army and Air Force supporting ground and paratroop exercises at the fort and for the deployment of fighter aircraft. Between 24 and 26 November of that year, Global Weather Central furnished images over central Africa to the Military Airlift Command, which proved to be crucial in the successful airlift of Belgian paratroopers from Europe to Stanleyville in the Congo, where hostages seized in an uprising were freed. The tactical applications proved the utility of the weather satellite data but also established the need for local readout stations so that the imagery could be received at the operational site.

In September 1964 Radio Hanoi quit broadcasting local weather observation, and the Air Weather Service in Saigon faced forecasting with limited and unreliable data. When the United States began air strikes against North Vietnam, weather information became critical, so much so that on March 18, 1965, DMSP launched a noontime satellite that could record and playback weather data to support tactical operations in Southeast Asia. As one of his last duties, Haig supervised the DMSP ground station at Tan Son Nhut Air Base in Saigon. This station was up and running in time to

FIGURE 5.2 Defense Meteorological Satellite Program Block 4 satellite (Hall, 2001).

support the new satellite and could supply, within 30 minutes of receipt, complete cloud-cover data for North Vietnam, South Vietnam, and parts of Laos, China, and the Gulf of Tonkin.

In 1965 all three military services put to immediate use the DMSP tactical meteorological data and could scrub, delay, or recall aerial sorties or divert them to secondary targets based on accurate weather information. The Naval Advisory Group used DMSP-generated forecasts to schedule the operation of their fleets of small boats operating along the coast of the Indochina Sea and the Gulf of Tonkin. Soon, mobile, air-transportable DMSP ground stations were installed at Udorn AB, Thailand, and Osan AB, South Korea. Another fixed site was located at Hickam AFB, Hawaii.

On May 20, 1965 from Vandenberg AFB, DMSP launched a special meteorological satellite reserved exclusively for tactical applications. Weather data from this satellite so improved the timeliness and accuracy of forecasts in Southeast Asia that the military canceled all daily routine aerial weather reconnaissance flights. These impressive results prompted DoD to break the NASA/NOAA lock on meteorological satellites and propose to run in parallel their own meteorological satellite program. DMSP was then transferred on July 1, 1965 from the NRO to the USAF, who would receive funding to continue the program. The Strategic Air Command would continue to launch the satellites and operate the DMSP control centers and ground stations in the continental US, while the US Air Weather Service would man the direct readout stations overseas and continue to operate Air Force Global Weather Central at Offutt AFB. Operational requirements now flowed from the NRO through the Air Weather Service, and technical guidance was provided by the Deputy Chief of Staff for Research and Development of the Air Force Systems Command. The conversion to a permanent program occurred on September 28, 1965. As a consequence, NASA converted the Nimbus program into a research and development testbed. In March 1973, the Under Secretary of the Air Force, also the NRO director, publicly announced the existence of the DMSP program in a Pentagon press conference.

The DMSP Blocks 1–3 were the same wheel satellites that NOAA flew. The Block 4 satellite had a different design (Fig. 5.2) and was 30 in in diameter, 29 in high and weighed 175 pounds.

Still spin stabilized (note the solar panels on the outside surfaces), this satellite included two one-inch focal length vidicon cameras canted at 26 degrees from the vertical, which provided global coverage of the Earth (contiguous coverage at the equator) with a swath width of 1,500 nautical miles (nm). The spatial resolution varied from 0.8 nm (~1.5 km at nadir) to 3 nm (~5.6 km) at the edge of the scan. In addition to the visible cameras, the satellite carried a multisensor infrared system and a high-resolution radiometer that provided cloud height profiles. A tape recorder with increased storage capacity could store images of the entire northern hemisphere each day, while the satellite furnished real-time direct local tactical weather information to mobile ground stations and shipboard terminals. Seven Block 4 satellites were successfully launched between

FIGURE 5.3 Defense Meteorological Satellite Program Block 5 A satellite (Hall, 2001).

1966 and 1969. During this time, the DMSP director was Major John E. "Jack" Kulpa, Jr. Shortly after assuming command, Kulpa started work on the next series of weather satellites, DMSP Block 5. Promoted to Lt. Colonel, Kulpa delegated a lot of the requirements and design of the spacecraft to two able subordinates, Captain Richard Geer and Major James Blankenship, who had just finished a PhD in atmospheric chemistry at the University of Stockholm.

The revolutionary Block 5 spacecraft (Fig. 5.3) that resulted from the efforts of Geer and Blankenship departed completely from the TIROS-derived technology of its predecessors. They visited meteorologists at work and then examined what industry could provide. They decided that they could have the satellites along their orbit increment the image collected by a line scanner. This would require that the satellite not spin and always look down at the surface. Thus, the satellite had to be three-axis stabilized. A total of 200 hours of testing on a T-39 aircraft demonstrated the concept, and the newly promoted Lt. Colonel Haig was able to secure funding for the Block 5 series of satellites. Westinghouse won the contract to build the constant resolution oscillating telescope known as the "Operational Line Scanner" (OLS). This instrument would continue to be the DMSP primary payload until the death of the program. The OLS produced images of the Earth and its cloud cover in both the visible and thermal infrared spectral regions. This system had a visible resolution of 0.3 nm (~5.6 km) during the daytime and 2 nm (3.7 km) at night when only the infrared imagery was available. This 2 nm resolution of the infrared channel was also available during the day.

One of the unique aspects to the DMSP during this period was the military requirement for low-light imaging. This capability made it possible to view the lights on the Earth's surface at night and was even used to image the aurora (Fig. 5.4). This night light sensing capability will be discussed further in a later section of this book.

To achieve the pointing accuracy required for the Block 5 OLS, the spacecraft it used a novel momentum-bias attitude-control system. It was made up of a momentum wheel, a horizon scanner, and magnetic coils. The wheel and the scanner controlled the pitch axis while the magnetic coils dictated the roll and yaw axes. This slab-sided satellite (Fig. 5.3) remained 30 in (~76 cm) in diameter but increased in height to 48 in (122 cm) and its weight to 230 pounds (~194 kg). On orbit, it closely resembled an overturned garbage can. Three of these Block 5 A satellites were built before the evolving DoD needs dictated further changes.

In 1969 all three military services developed a "joint-service utilization plan" for DMSP. The result was Block 5B and 5C, which were longer at 84 in (~213 cm) in height and heavier at 425 pounds (~193 kg). For a launch vehicle, these satellites required the use of an upgraded booster, the Thor/Burner IIA. In addition, the Block 5B spacecraft added a large sunshade on the morning satellites and a more powerful 20-watt traveling-wave-tube amplifier (TWTA) transmitter needed for receipt of the signal on board ships. The satellites also carried a second primary data recorder

FIGURE 5.4 Defense Meteorological Satellite Program Nighttime image of the Aurora Borealis taken by the first Block 5 A in 1971. Moreover, clearly apparent at the city lights of North America (Hall, 2001).

and a gamma-radiation detector. Block 5C added a vertical temperature/moisture profile sensor and an improved IR sensor that was now able to achieve a spatial resolution of 0.3 nm (~0.5 km). A total of three Block 5 A, five Block 5B, and three Block 5C satellites were built and launched between February 1970 and February 1976. Their operational life expectancy on orbit averaged about 10 months.

By the early 1970s the Block 5 series of satellites had reached the end of its growth potential. The tubular shape once useful for a spin stabilized satellite was an odd fit for Block 5 and Block 6 which needed a completely new design that would better meet the needs of an Earth pointed satellite during its operation. Moreover, the short lifetimes of the Block 5 series recommended a larger satellite with greater endurance on orbit (16 months minimum). But the DMSP Block 6 was not to be, and pressure from Congress to combine the military and civilian satellite programs led the Air Force to designate the new satellite as Block 5D. Congress appropriated funds for five of the Block 5D satellites.

The DMSP program had decided that Block 5D would be a completely new satellite bus, which could satisfy the Earth-oriented pointing accuracy that was much greater than that of Block 5C. After a competition between Boeing and RCA, a contract was signed with RCA in the fall of 1972 for five Block 5D satellites. The greater pointing accuracy, however, increased the cost of these future satellites, which introduced the risk of delays to the program. Once again, the Office of Management and Budget (OMB) asked the Departments of Commerce and Defense to reexamine the possibility of consolidating the civil and military weather satellite programs. A steering group composed of people from NOAA, the DoD, and NASA was formed to discuss the situation and come up with a recommendation. Since the Block 5C technical capabilities already exceeded those of the NOAA's planned successor TIROS-N, the group concluded in their report, issued in mid-1973, that the greatest savings would be realized in a single national meteorological satellite program managed by the Air Force using the planned Block 5D satellites. This recommendation was quickly rejected by Henry Kissinger, President Nixon's National Security Advisor, who argued that this would violate the separation of military and civil spacefaring. Instead, interagency deliberations led to a decision to achieve major cost savings by using the same satellite bus for the Block 5D satellite series as for the NOAA TIROS-N series.

The Block 5D satellite design developed in the early 1970s had to fit in the Burner IIA upper stage of the launch vehicle. This satellite, built by RCA, consisted of three sections: a square precision mounting platform on the forward end supporting the Earth sensors and those needed for precise pointing accuracy; in the center, a five-sided equipment-support module that contained more of the electronics and featured one or two pinwheel louvers on four sides for thermal control; and at the aft end, a circular reaction and control equipment support structure that housed the spent third-stage solid propellant rocket motor along with reaction control equipment and a 16 by 6 ft (~4.9 m) sun-tracking solar array (Fig. 5.5). This larger satellite weighed 1,150 pounds (522 kg), almost twice the size of its predecessor. To launch this additional mass, DMSP contracted with Boeing for a new, larger, and solid-propellant second stage. The original Burner IIA second stage was now adapted as a third stage and fixed to the satellite for injection into its sun-synchronous 450 nautical mile (~833 km) circular orbit.

On orbit, the spacecraft had to control the attitude to within 0.01 degree, making the satellite a spaceborne optical bench. This was achieved by automatic momentum exchange between three momentum wheels, one each for yaw, roll, and pitch, in addition to magnetic coils that interacted with the Earth's magnetic field to prevent the accumulation of wheel secular momentum. There was a backup system that consisted of an Earth horizon sensor that furnished pitch and roll information

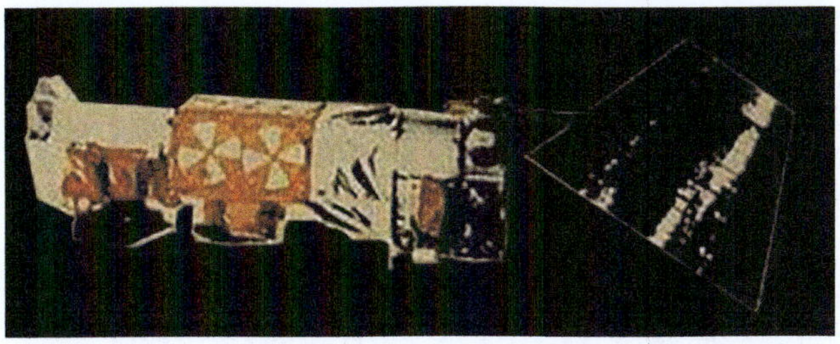

FIGURE 5.5 Defense Meteorological Satellite Program Block 5D-1 satellite (note, the similarity with TIROS-N (Hall, 2001)).

and a sun sensor that added the yaw information. This satellite was equipped with an improved OLS that had a scanning optical telescope that oscillated in a side-to-side motion by counter-reacting springs and a pulsed motor. The swath of this new OLS was 1,600 nm (~2963 km) with a nadir resolution of 0.3 nm (~0.6 km) and a resolution of 0.5 (~0.9 km) nm at the edge of scan. The visual and thermal image data could be stored on one of three tape recorders for transmission to Earth in an encrypted, digital format. Direct readout continued to be available to field terminals. Within the US data transfer volumes were too great for the landlines used earlier, and data transfer was now done with communication satellites.

Unfortunately, the complexity of the new Block 5D series delayed the launch of Block 5D-1 from 1974 to 1976. At the same time, the short lifetime of the last Block 5C satellite resulted in a significant gap in meteorological coverage at the end of the decade. The value of an autonomous attitude control system was clearly demonstrated upon the launch of Block 5D-1, which, after orbit injection, began tumbling end over end in space. The mission was thought to be lost, but in the next few months, intermittent communications with the satellite were established, and the ground controllers were able to reprogram the computers, slow the rate of tumbling, and eventually bring the satellite back to its intended position operating as planned. Nevertheless, this late launch caused a gap, and the weather coverage was compromised between 1975 and 1977. This was made worse when the second Block 5D-1 satellite launched on June 5, 1977 strayed into a drifting orbit, and the OLS data was useless. The third and fourth 5D-1 launches on April 30, 1978 and June 6, 1979 were fortunately more successful, and the DMSP program was back on track. The last 5D-1 satellite was held in reserve for when it might be needed.

During this time, development proceeded slowly with the Block 5D-2 satellites under contract to RCA since 1975. Technical changes dictated by both the civilian and military users and studies of the best booster for 5D-2 both caused delays and increased cost. The DMSP office in El Segundo found it necessary to slip the first 5D-2 launch from 1980 to 1982. At the same time, the leadership of the DMSP program went through a number of personnel changes none of whom stayed long enough to take ownership of the program. The 5D-2 satellites increased in length from 20 to 22.5 feet (~7 m), which increase was absorbed in the middle section of the satellite. This module now contained a second 22.5 amp-hour battery with three temperature control pinwheels on four of its five sides. Two very important sensors were added to those available in Block 5D-1: a topside ionospheric sounder that provided global measurements of the electron distribution in the Earth's ionosphere, and a microwave imager (flown only on the last few 5D-2 satellites). Called the special sensor microwave imager (SSM/I) that will be discussed at length in a later section. One unique aspect of this imager was that through a program called "shared processing," the SSM/I data were decrypted and made available through NOAA and NASA data centers to the large number of interested scientists and researchers in a nearly real-time basis.

With all of these changes, the weight of the 5D-2 satellite (Fig. 5.6) increased to 1,792 pounds (~813 kg), which was too heavy for the Thor/Burner booster combination.

After considerable discussion, it was decided that the General Dynamics Atlas E booster was selected as the launch vehicle. This was an improved version of a liquid-propellant intercontinental ballistic missile that had been deployed in the early 1960s. The Burner IIA upper stage was retained and again used to drive the satellite into its 450 nm circular polar orbit.

Meanwhile, the Block 5D-1 satellites continued to fail and the last one that had been held in reserve was pressed into service. On July 15, 1980 at Vandenberg AFB, a Thor/Burner launch vehicle carrying the last Block 5D-1 satellite and the trusted launch vehicle failed. The second and third stage rockets did not separate, and the satellite fell into the South Pacific. Four weeks later, in August, the last 5D-1 satellite on orbit ceased to function. Now, there was going to be a substantial gap in DMSP weather satellite coverage. A detailed investigation of the DMSP by Air Force Systems Command identified funding cutbacks and a weak program management plagued by a rapid turnover of program directors to be the primary causes. During this gap, the DoD would have to rely on NOAA weather satellites for its meteorological information. This was a hard thing for the Air Force who had argued strongly for separate programs. During this period, it

FIGURE 5.6 Defense Meteorological Satellite Program Block 5D-2 satellite (Note, the special sensor microwave imager and its rotating dish mounted on the top of the spacecraft. Part of its scan will be blocked by the satellite itself.) (Hall, 2001).

FIGURE 5.7 Defense Meteorological Satellite Program Block 5D-3 satellite (Hall, 2001).

became evident that while the NOAA satellites were not specifically designed to meet DoD requirements, their data went a long way to fulfilling the needs of the three services.

In December of 1982 the first of the Block 5D-2 satellites was launched on an Atlas booster and inserted into a morning polar orbit. The second and third followed in November 1983 and June 1987. The OLS on the 1982 bird operated until August 1987, while the second ceased in November of that year, and the third continued until August 1991. During this time, the DoD was making plans for an even larger Block 5D-3 satellite (Fig. 5.7), which increased its length to 24 feet (~7 m) and the weight to 2278 pounds (~1033 kg).

The satellite had a larger solar array, three 50 amp-hour batteries, and a redesigned sunshade. The center section now has four pinwheel thermal control panels on four of its five sides. The first of six 5D-3 satellites was due to be delivered to the Air Force in June 1990. They were originally designed to be launched on the Space Shuttle but after the Challenger disaster in January 1986 the satellites were redesigned for launch atop modified Titan-II intercontinental ballistic missiles.

Early in 1993 OMB and Congress prompted a study of the possibility of merging the DMSP and NOAA weather satellite programs. Representative of the DoD, NOAA, and NASA convened to study the issue. It is most likely that this group was unaware of the similar efforts carried out in 1961 that had produced plans for a National Operational Meteorological Satellite System. Because of security restrictions, they were basically unfamiliar of the history of the DMSP. Not to anyone's surprise, the 1993 study reached conclusions similar to those of the 1961 panel. That is, a single satellite program would eliminate the need for duplication in satellites and ground stations and thereby greatly reduce cost. This led to a "tri-agency" plan to combine the two weather satellite programs into one, but the word combine would not be used. The plan created an Integrated Program Office that would develop, acquire, and operate the National Polar-orbiting Operational Environmental Satellite System (NPOESS). NOAA held overall responsibility for the converged system, including operating the satellites and representing the NPOESS program to the general public. The NPOESS program office, located in Silver Springs, Maryland, opened its doors on October 3, 1994. Four years later, the program director took over responsibility for controlling the DMSP satellites along with the NOAA polar orbiting satellites. There will be a lot more on NPOESS later in this book.

5.3.1 The Defense Meteorological Satellite Program low light capability

As mentioned earlier, the DMSP satellites OLS sensors were required to have a low-light sensing capability, and this capability has been explored extensively by Chris Elvidge, formerly at the National Geophysical Data Center (NGDC) of NOAA in Boulder, Colorado, and now at the School of Mines in Golden, Colorado. His 2009 paper (Elvidge et al., 2009) reviews the use of DMSP nighttime imagery for mapping light sources on the Earth. The OLS has collected this type of imagery starting in 1972, but the archive of useful nighttime data only extends back to 1992. NGDC has produced a set of global cloud-free nighttime lights products, specifically to be used for the detection of changes in light emissions for human settlements over the period from 1992 to 2008.

Elvidge et al. (2009) state that the remote sensing of nocturnal lighting provides an accurate and economical way to map the global distribution and density of developed areas. Since the early days of the electric light bulb, nocturnal lighting has emerged as one of the indicators of modern urban development. While there are some technological differences in geographic regions, the main discriminator is wealth. Areas with high per capita income have lots of lights and those with less monetary assets have less. Nighttime lights have been used as a proxy for many different socioeconomic variables, such as population density, gross domestic product, and poverty. There is an increasing amount of research that suggests that nighttime lights interfere with the human circadian cycle, leading to sleep disorders.

Historically, the only satellite capable of sensing global nighttime lights is the DMSP satellite with its OLS instrument. The NOAA NGDC has been producing nighttime light data since 1994 and has worked with the science community on various applications of these data. The OLS has a specific channel to collect imagery under very low-light conditions. The satellites in sun-synchronous orbits view the Earth's surface with overpasses at 7 to 9 p.m. local time. For a swath width of 3,000 km and 14 orbits per day, each OLS instrument is capable of collecting a complete global coverage in the visible (0.5–0.9 μm)and thermal infrared (10.5–12.5 μm). At night, the OLS visible band is intensified with a photo-multiplier tube (PMT) to make it possible to view clouds illuminated by moonlight. This light intensification makes it possible to view light emitted at the Earth's surface under cloud-free conditions. Using this intensified light, one can map cities, towns, villages, gas flares, heavily lit fishing boats, and fires. This low light capability extends down to radiances of 10^{-9} watts/cm^2/sr.

Under the guidance of Chris Elvidge, the NGDC of NOAA has been producing global cloud-free composites of DMSP nighttime like since 1994. The first step to create such a product was to extract the data from the archive and geolocate it into 30 arc-second grids. To improve the accuracy of the geolocation, the center half of each swath was used in these composites. Clouds were removed from the imagery by comparing the OLS thermal band brightness temperatures with the surface temperature grids from the NOAA National Center for Environmental Prediction. The cloud-free images are then examined for outliers that are usually due to fires. An example of a global nighttime lights composite is shown here in Fig. 5.8.

Elvidge et al. (2009) then compare this map of nighttime lights with a representation of economic strata that will have different levels of light illumination. They discuss the problems of accurately defining and mapping global poverty and turn to a formulation of international poverty data which has been defined as the number of individuals living on either $1 or $2 per day (Fig. 5.9). The source data for these maps are surveys conducted by the World Bank since 1990.

There are a number of inadequacies in the World Bank poverty line data. Some countries do not conduct recent surveys, and the available data are extracted from the surveys collected between 1988 and 2004. Intercomparisons are compromised by variations in how consumption and income are counted and the degree to which individual governments can influence the surveys. Elvidge et al. (2009) then decided to develop a poverty index by dividing the population count by the brightness of the DMSP lights (Fig. 5.10).

To calibrate this estimate of the poverty index, the World Bank estimate of Fig. 5.9 was used for comparison. Merging the DMSP-based poverty estimate in Fig. 5.10 with data from Fig. 5.9, they derived poverty levels for administrative units. The result is presented here in Fig. 5.11.

FIGURE 5.8 Global cloud-free composite of Defense Meteorological Satellite Program (DMSP) Operational Line Scanner nighttime lights. The image was derived from observations made from the DMSP satellite F-15 during 2003 (Elvidge et al., 2009).

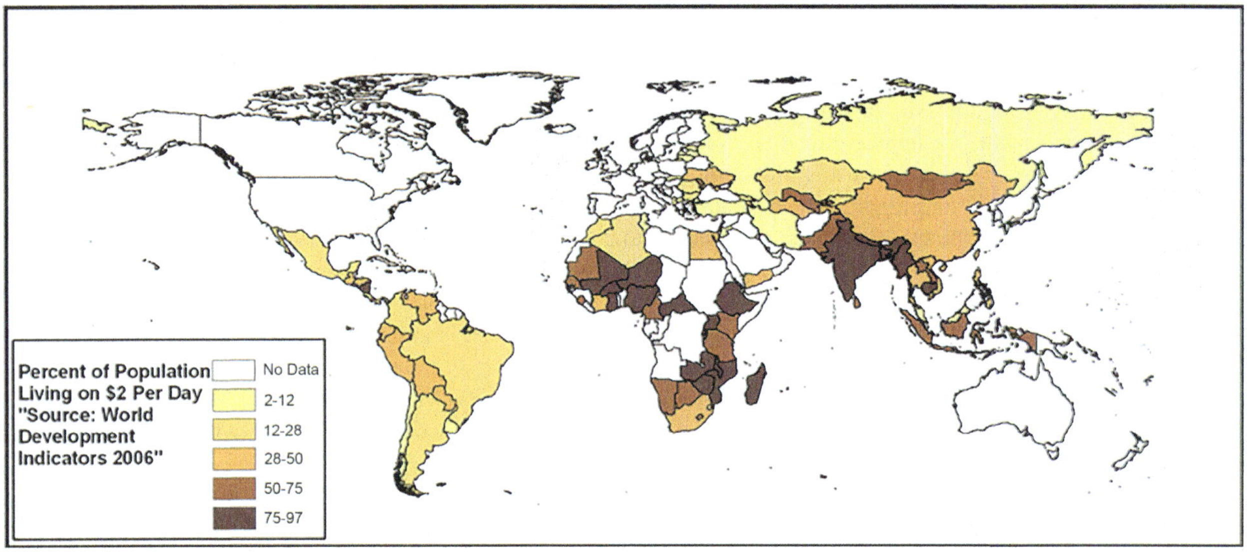

FIGURE 5.9 Global poverty map from the World Development Indicators 2006 database (Elvidge et al., 2009). Map lines delineate study areas and do not necessarily depict accepted national boundaries.

FIGURE 5.10 Poverty index is defined as the population count divided by the brightness of the Defense Meteorological Satellite Program lights (Elvidge et al., 2009). Map lines delineate study areas and do not necessarily depict accepted national boundaries.

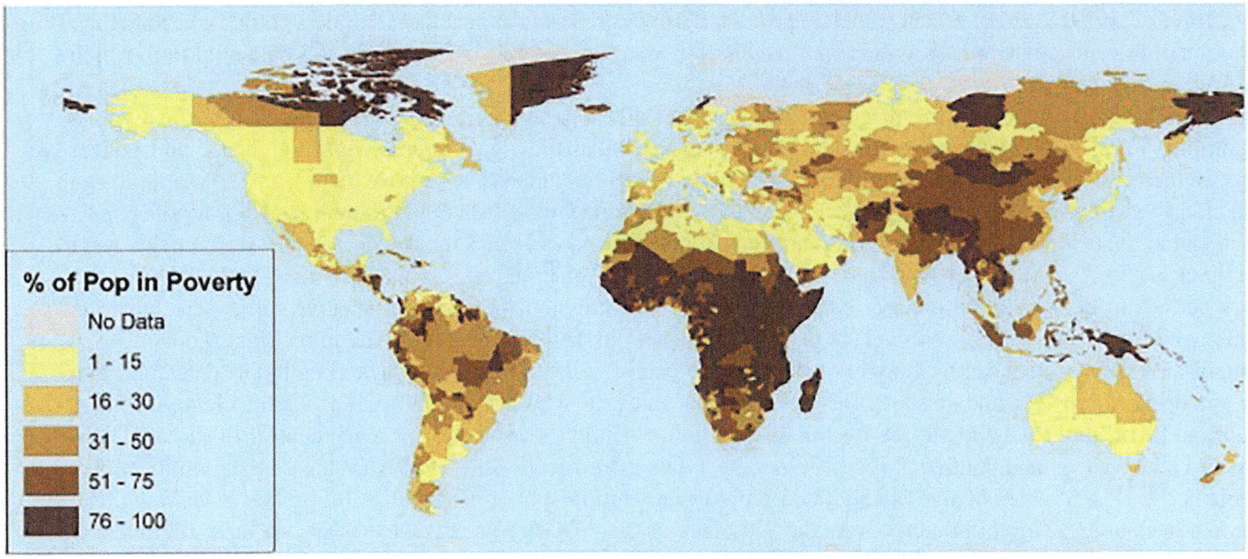

FIGURE 5.11 Poverty levels computed by merging the Defense Meteorological Satellite Program lights derived poverty levels of Fig. 5.10 with those of Fig. 5.9 (Elvidge et al., 2009). Map lines delineate study areas and do not necessarily depict accepted national boundaries.

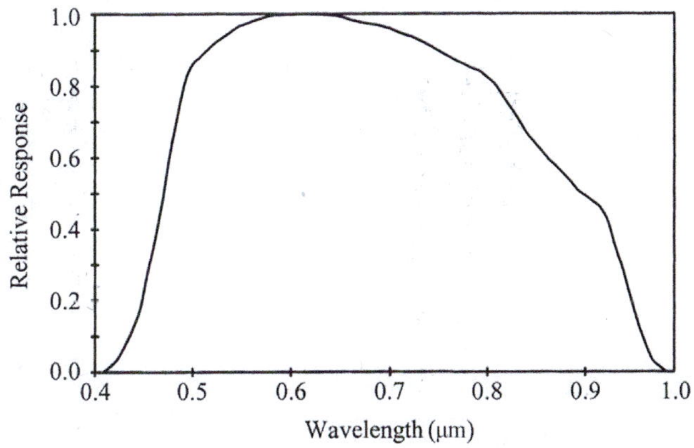

FIGURE 5.12 Visible channel for the Operational Line Scanner sensor onboard Defense Meteorological Satellite Program F-12 (Elvidge et al., 1999).

A study by Hsu et al. (2015) outlines a method that expands the dynamic range of the OLS using observations to make different fixed-gain settings, and by adopting the areas not affected by saturation from stable light sources. They then compare their radiance estimates with preflight OLS calibrations. The result is a calibrated OLS nighttime lights product. The photo-multiplier tube (PMT) responsible for the ability to collect images under low solar illumination has a gain that can be changed over a wide range but is set to automatically adjust day and night conditions. The gain automatically responds to the brightness in the scene. Unfortunately, the OLS only has a 6-bit quantization with digital numbers (DNs) ranging from 0 to 63 coupled with a limited dynamic range. As a consequence, pixels in bright areas under low light sensing (such as city centers) often reach the digital value of 63 and are, therefore, saturated with no ability to resolve further detail. The nature of low-light sensing is at night, when the gain of the OLS visible channel is adjusted to that the instrument can view moonlit clouds.

To overcome this problem, Elvidge et al. (1999) proposed a "radiance calibration" method that examined the collection of DMSP OLS VIS data at three different fixed-gain levels: low, medium, and high. The OLS has a response function given here in Fig. 5.12.

This visible channel covers both the visible reflectance (VNIR) and the near-infrared reflectance. The thermal channel ranges from 10 to 13 μm with an 8-bit quantization, while, as stated earlier, the visible has only a 6-bit quantization. The potential for using nighttime OLS data for observing city lights and other near-infrared emissions was first noted in the

1970's by Croft (1973, 1978, 1979). Later, Welch (1980) and Foster (1983) speculated that OLS data could be used to map the distribution of human activities such as energy consumption. Many of these early studies of OLS data used film, which greatly restricted their analyzes.

Starting in 1992, the US Air Force, together with NOAA, established a digital archive of DMSP data at the NGDC. As a result, algorithms have been developed to use DMSP OLS data to map fires, gas flares, fishing lightboats, and human settlements. In these studies, the thermal infrared channel is used to identify clouds and remove them from the visible image to produce cloud-free images. The remaining data are then threshold limited to exclude noise, lightning, and fires. The results are maps of stable lights within the limits of the cloud removal. This product shows the frequency with which lights were detected within a set of cloud-free observations, with no consideration of brightness of these lights. The high gain used during dark periods made it possible to detect faint sources of VNIR emissions but had the negative consequence that the urban lights would saturate the OLS sensor. This made it impossible to calibrate the earlier nighttime light products.

The gain of an optical system can be thought of as an amplification of the signal that is picked up by the detectors. The OLS has analog preamplifiers and postamplifiers with fixed gains as well and the Variable Digital Gain Amplifier (VDGA). In addition, the PMT adds to the overall system gain for night scenes. The primary control of the low-light imaging gain is through ground control of the VDGA, which has values from 0 to 63 dB (the log of the amplification). Before launch, the OLS visible channel is calibrated with an integrating sphere which aims the sensor at light sources having known irradiances. The calibration covers the full range of the VDGA converting the telescope input illuminance to DNs at specific VDGA gain settings. The relationship between OLS VDGA gain settings and the observed radiances is shown here in Fig. 5.13.

Using the information in Fig. 5.13, we can relate the OLS DNs to radiance in terms of $W/cm^2/sr/\mu m$. Saturation radiance (dn=63) is indicated by the solid line while the dashed diagonal line corresponds to a dn of 1, the lowest detectable radiance. For a VDGA gain setting of 63, the final system gain is 136 dB. This results in a light intensification of 10^{136} times the original input signal. This allows detection of radiances as weak as 10^{-10} $W/cm^2/sr/\mu m$. This is a radiance approximately six orders of magnitude lower than the standard OLS daytime visible band. The goal of the DMSP program for the utilization of this capability was to be able to map moonlit clouds for a variety of viewing angles. Thus the lowest gain settings occur under full moon conditions. Gain settings gradually increase as lunar illumination decreases. Gains are modified every 0.4 ms by an onboard along scan gain control algorithm. Moreover, a bidirectional reflectance distribution function (BRDF) further adjusts the gain control in the scan segment where the illumination angle equals the observation angle. In the period of greatest darkness of 10 nights in the lunar cycle the OLS is unable to detect clouds even

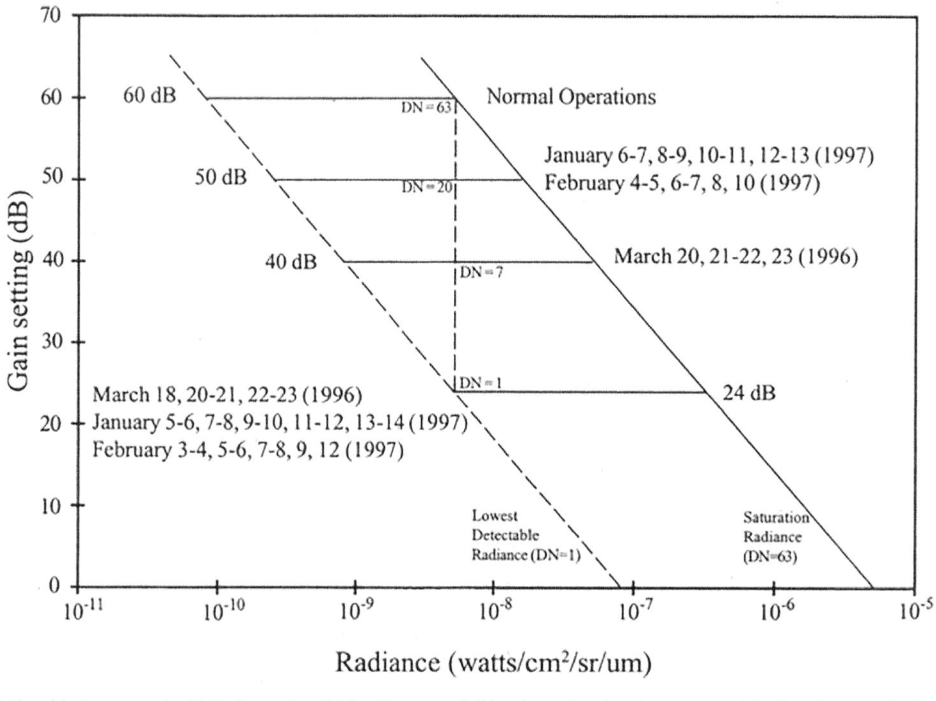

FIGURE 5.13 Relationship between the F-12 Operational Line Scanner visible channel gain when operated in the photo-multiplier tube mode, digital numbers and observed radiances derived from preflight calibrations (Elvidge et al., 1999).

with the highest gain settings. During these dark periods, both the along scan gain and the BRDF algorithm are minimized, and the gain rises to its maximum level.

While the DMSP downlink in encrypted, the NGDC made an arrangement with the DMSP program office to acquire OLS data at reduced gain settings and SSGC and BRDF algorithms turned off. The data was requested for an 8-night period in March 1996. To avoid interference with the nighttime lights observation by the leading edge of the aurora at high northern latitudes, data were restricted to a range between −55 and +55 degrees latitude, but the northern limit was further restricted to 45° N due to the presence of solar glare. Since at this point there was no experience with data from a variety of VDGA settings, a series was requested that stepped the gain function for the first 24 hours of the 8-day period, as shown, for example, in Fig. 5.14.

In this figure, the changes in gain are shown on the right, and one can clearly see the changes in background and sensitivity to light emissions from the Earth. This exercise led to the selection of a VDGA gain setting of 26 dB in the next day of sampling. It was then realized that this gain resulted in saturated data in the major cities, and the gain was reduced again to 24 dB. This then only responded to the strongest light sources and excluded any of the fainter lights. To enable the detection of these weaker signals, it was decided to collect data with a gain of 40 dB and then to alternate every 24 hours period between gains of 24 and 40 dB (Fig. 5.14). Study of this first sample of 1996 low-gain data revealed that it was possible to detect brightness variations within urban centers. It was also realized that a gain setting greater than 40 dB would be needed to detect the myriad small towns that had been detected in the 1994–95 stable lights products. NGDC requested and received additional DMSP data from 5–14 January and 3–12 February, which were the darkest nights of January and February 1997.

The area of interest was first extracted from the global data files and then processed with the following steps: (1) establishment of a reference grid, (2) identification and geolocation of lights, clouds in the coverage area, (3) defining the final DN to radiance relationship for the target area, (4) cloud-free compositing for two different gain values (high and low), (5) applying a threshold to eliminate erroneous isolated detections, and (6) combine the calibrated images from the two gain ranges by averaging based on the number of light detections. The reference grid had a spatial resolution finer than the input imagery. They used the 1 km equal area Interrupted Goode Homolosine Projection (Goode, 1925; Steinwand, 1993). This grid was optimized to yield a uniform grid cell size for all latitudes. Data gaps in the maps of lights and clouds were filled with zeros. Lights were detected using an automatic algorithm that established a local threshold based on background brightness levels (Elvidge et al., 1997). This local thresholding algorithm runs very quickly to detect major cities and towns under a variety of conditions. It does not, however, generally detect "diffuse lighting," which is dim and scattered across the landscape. They found that noise greatly reduced the ability of this algorithm to detect light.

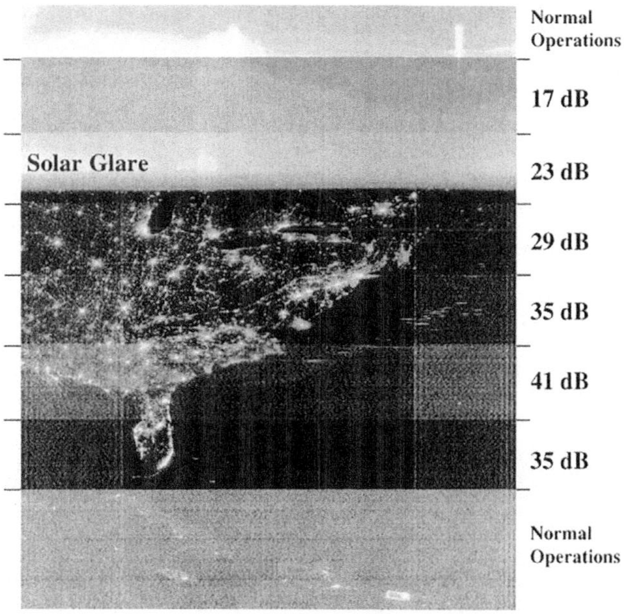

FIGURE 5.14 Defense Meteorological Satellite Program F-12 satellite OLS visible band acquired with a range of gain settings on 16 March 1996 (Elvidge et al., 1999).

In an attempt to detect and map area of diffuse lighting, Elvidge et al. (1999) developed a software tool that an image analyst could use to select a threshold single DN about the noise level encountered in the sample. The analyst then reviewed each suborbit image section, selecting thresholds, which were used to extract pixels with detectable levels of VNIR emissions. The analyst then identifies defective scan lines and scan lines with lighting. The image analyst also created image masks for areas of excessive noise or solar glare. These data were then discarded for the compositing procedure. Clouds could not be identified in the visible image due to the low level of lunar illumination. Cloud identification used the thermal infrared band using a simple thresholding procedure. This cloud filtering is fairly easy at low latitudes where there is a large difference between Earth's surface temperature and cloud top temperature. At other latitudes, it became necessary to have different cloud thresholds every 200-pixel-wide latitude band. These thresholds were selected subjectively by the analyst for cloud removal.

Their geolocation algorithm finds the latitude and longitude of the pixel centers using the geodetics subtrack of the satellite orbit, the satellite orbit, the satellite altitude, OLS scan angle equations, an Earth and sea level model, and digital terrain information. Naval Space Command provided the geodetic subtrack of each orbit based on daily radar bevel vector sightings of the satellite. This information was then fed into an Air Force orbital mechanics model that calculates the satellite position every 0.4208 seconds. The satellite heading is found by computing the tangent to the orbital subtrack. Elvidge et al. (1999) at first divided the suborbit data into two roughly equal data volume groups: (1) data acquired at gains less than 30 dB and (2) data collected at gains greater than 30 dB. They then found the number of "cloud-free" observations by taking the data coverage polygons and subtracting the cloud detection grid values. The result is the number of cloud-free observations in the time series. The second step in the preliminary compositing state is to sum up the number of light detections and to produce an average radiance for each reference grid pixel with light detection.

Further, thresholding filtering was applied to remove any large signals that only occurred once. This step removes noise and other transient signals such as fire or lightning.

The result is the color-coded image in Fig. 5.15 that represents the radiance-calibrated light of the central United States.

This image shows the cities and many of the smaller towns in the central US and provides better spatial detail than the previous nighttime lights map (Elvidge et al., 1997) related to brightness difference within urban centers (Fig. 5.15) that

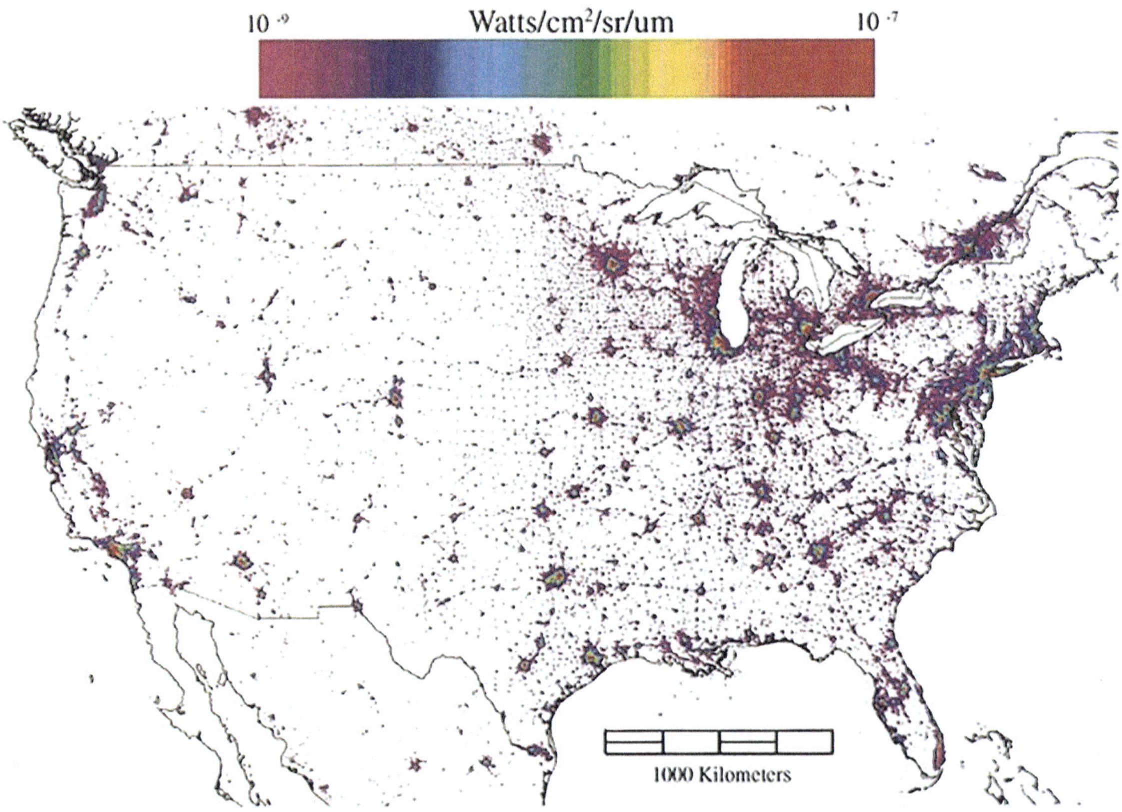

FIGURE 5.15 Radiance calibrated nighttime lights of the central United States (Elvidge et al., 1999). Map lines delineate study areas and do not necessarily depict accepted national boundaries.

were not present in the earlier study. Here, it is clear that the radiance calibrated lights show levels of diffuse lighting in densely populated areas, particularly in the Washington DC to Boston corridor, in western Pennsylvania, Indiana, Michigan, Wisconsin, and Minneapolis-St. Paul, Minnesota.

A typical false color presentation of the St. Louis area is given in Fig. 5.16, which does not emphasize the urban community but rather is best at distinguishing land cover types. In Fig. 5.16C, the radiance calibrated nighttime lights are a good approximation of the color presentation of the population density in Fig. 5.16B. It should be noted that outside of the urban areas Fig. 5.16B shows population densities in rather large homogenous blocks. This is due to the fact that the census uses postal ZIP codes which can cover rather large areas in the rural environment. Comparing Fig. 5.16B and C indicates that the St. Louis airport and surrounding commercial and industrial areas are quite bright in the radiance

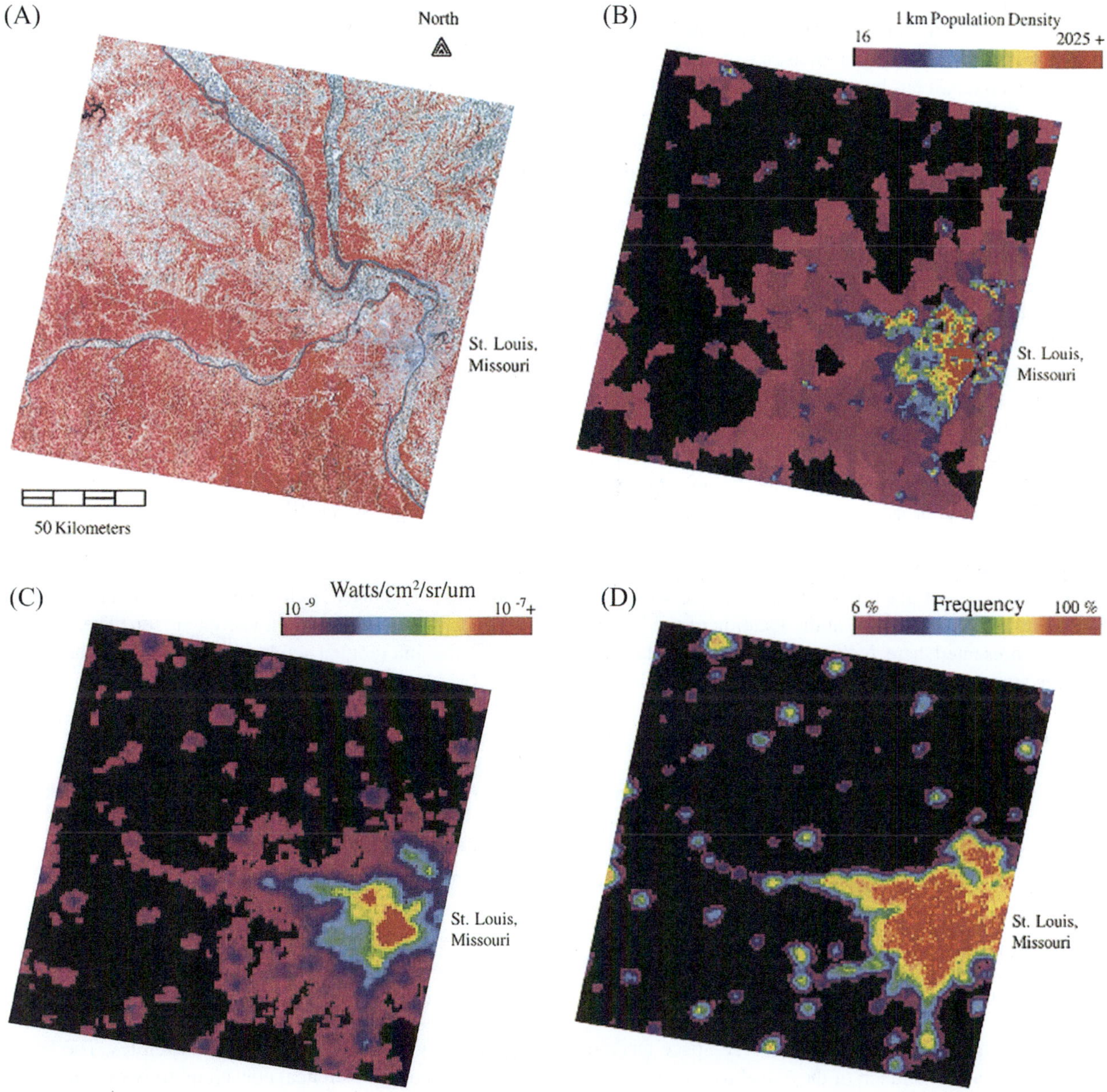

FIGURE 5.16 (A) Landsat MSS image of the St. Louis, Missouri area, bands 4, 2, and 1 as *red*, *green*, and *blue* on October 3, 1992. (B) Population density of the St. Louis, Missouri area in a 1 km grid (CIESIN, 1996). Derived from the 1990 US Census Bureau data. (C) Radiance calibrated nighttime lights of the St. Louis, Missouri area derived from Defense Meteorological Satellite Program (DMSP)-OLS data acquired during March 19,966 and January-February 1997. (D) Stable lights of the St. Louis, Missouri area derived from a 1994–95 DMSP-Operational Line Scanner time series (Elvidge et al., 1997). Map lines delineate study areas and do not necessarily depict accepted national boundaries.

calibrated nighttime lights even though they have a relatively low population density. This demonstrates the ability of the nighttime lights to report human activities that are not associated with residences. Fig. 5.16D shows that the earlier stable light product is not able to show the detail in the radiance image of Fig. 5.16C.

Elvidge et al. (1999) conclude that the radiance-adjusted values are highly correlated with electric power consumption. They suggest that the nighttime lights may prove useful in mapping human activities such as greenhouse gas emissions. Population is typically based on residence location rather than workplace. There are many locations, such as airports, industrial zones, and commercial centers that have relatively low population density but emit high levels of nighttime lights. They suggest that for many applications, the radiance-calibrated nighttime lights may be a better indicator of human activity or human impacts on the environment than of population density.

This type of research has been continued, and a 2015 paper by Hsu et al. (2015) addresses radiance-calibrated nighttime lights, adding the intercalibration between DMSP satellites. They provide an interesting comparison of the Elvidge et al. (1977) stable lights with the radiance calibrated lights presented here in Fig. 5.17.

Tokyo was selected for this intercomparison because it is one of the world's brightest areas. By comparing the transects, we can see how much detail has been lost in the stable light product when compared with the radiance calibrated product. The comparisons in Fig. 5.17 show how the stable lights not only lose detail, but there is a marked difference in overall amplitude, with the stable light product severely underestimating the level of the radiance calibrated lights. Hsu et al. (2015) extend the radiance product by carrying out an intercalibration between DMSP satellites. This intersatellite calibration adjusts the DN in the radiance calibrated product to account for the differences caused by sensitivity discrepancies between the OLS instruments on different satellites. These intercalibrations were made based on the highest fixed gain setting available, which is called the base gain in their study. Hence the DN values in the Radiance Calibrated products can be calculated as the reading of sensors at the base gain, having a very large range. Using the multipliers given in Table 5.3, applied to the appropriate radiance calibrated products using their satellite and base gain, the adjusted DN values are now theoretically equivalent to the OLS onboard DMSP F016 at a gain setting of 55 dB.

Biases other than instrument differences can be addressed by carrying out interannual calibration of the Radiance Calibrated product time series. The Radiance Calibrated products require a reference location that is not only stable but also provides brightness values ranging from very low light to the world's highest light levels. The city of Los Angeles, California, was selected as the reference site for two reasons. First, Los Angeles is a mature metropolis, and the light change over time is negligible. Second, as a metropolis, it contains a wide range of DN values from the intense city center to the lower DNs in the suburban areas. In addition, the year of the previous stable light study (F12 1999) was selected as the reference year. The numerical values of regressions for the calibrated products are given in Table 5.4.

These results verify the strong relationship between the analyzed products.

To evaluate the success of this intercalibration they chose 12 cities to compare the adjusted with the nonadjusted values. An 11X11 pixel area with the center at the location of the brightest pixel in each city was used in this comparison. These comparisons are presented here in Fig. 5.18. These plots demonstrate that the interannual calibration suppresses the amplitude of the annual cycle, and it is clear that before 2006, the nonadjusted values overestimated the lights, while in the year 2010, they underestimated.

Hsu et al. (2015) conclude that their methods provide a way to intercalibrate data from the DMSP satellites to produce time series of nighttime lights. Since the DMSP satellites do not carry visible calibration references onboard, it is necessary to develop intercalibration procedures that can yield useful time series information. The DMSP nighttime lights ended in 2010 since the future satellites will not collect fixed-gain imagery anymore. The sensing of nighttime lights, however, has been reborn in the launch and operation of the Visible Infrared Radiometer Suite flying on present day NOAA polar orbiter weather satellites, but that discussion is reserved for a later chapter.

5.3.2 The special sensor microwave imager (SSM/I)

The SSM/I was a joint US Navy and Air Force project designed to meet Navy needs for all weather atmospheric and ocean remote sensing. The sensor was designed and built by the Hughes Aircraft Company under the direction of the Navy Space Systems Activity (NSSA) and the Air Force Space Division, and it was designed to fly on the DMSP satellites. The Space Sensing Applications Branch of the Naval Research Laboratory (NRL) served as a technical consultant to NSSA on the SSM/I beginning in FY 1982 (Hollinger and Lo, 1983). NRL carried out a number of studies and analyzes requested by the NSSA. They also prepared a draft SSM/I calibration/validation plan and conducted a calibration/validation workshop in 1982.

The SSM/I was to be a departure from the Dicke type of microwave radiometer that had been flown previously. It was instead a full power microwave radiometer taking full advantage of the new low-noise amplifiers that reduced the noise in

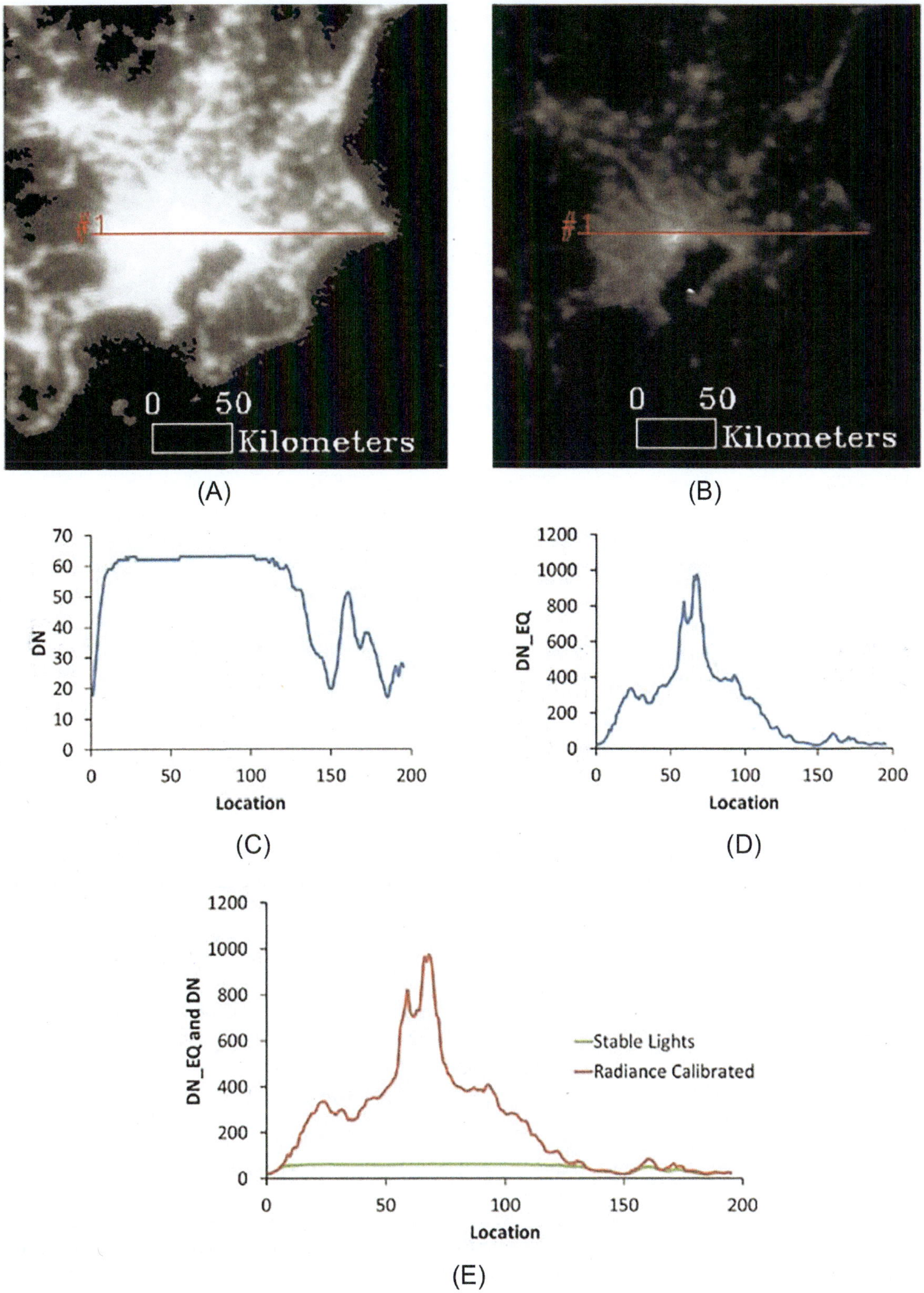

FIGURE 5.17 Comparison of the stable light product (F16, 2006) and radiance calibrated product (F16, 2005–2006) over Tokyo Metropolitan area (A) stable light product of Tokyo Metropolitan area, (B) radiance calibrated product over Tokyo Metropolitan area, (C) transect of stable light project over the center of Tokyo, (D) transect of radiance calibrated product over the center of Tokyo, and (E) overlay of transects from stable light (*green*) and radiance calibrated product (*red*) (Hsu et al., 2015).

TABLE 5.3 Intersatellite calibration multiplier.

Satellite	Base gain (dB)	Multiplier	Radiance @ DN1 ($W/cm^2/sr$)
F12	55	0.96	1.44E−10
F14	55	0.82	1.23E−10
F15	55	0.90	1.35E−10
F16	55	1.00	1.50E−10

TABLE 5.4 Interannual calibration regression results.

Satellite_Year	Slope (a)	Intercept (b)	R^2	Number of pixels[a]
F12_19960316-19970212	0.915	4.336	0.971	20,540
F12_19990119-19991211	0.780	1.423	0.980	20,846
F12–F15_20000103-20001229	0.710	3.658	0.980	20,866
F14–F15_20021230-20031127	0.797	3.736	0.980	20,733
F14_20040118-20041216	0.761	1.062	0.984	20,848
F16_20051128-20061224	1.000	0.000	1.000	21,044
F16_20100111-20101209	1.195	2.196	0.981	20,848
F16_20100111-20110731	1.246	-1.987	0.981	20,848

[a] *Number of pixels included in the regression analysis.*

the system electronics to the point where you could receive and process the full passive microwave signal. Previously, the only way to acquire the weak microwave emissions of the Earth was to switch between the target measurement and a known noise source (called the Dicke switch after R.H. Dicke, who first introduced the method). This made it possible to remove the noise in the system that, in the past, overwhelmed the passive microwave signal from the Earth. Thus the SSM/I was unique in that it no longer retained this switching mechanism to sense the full passive microwave emissions from the Earth.

The SSM/I is a seven channel, four frequency, linearly polarized, passive microwave radiometer. The instrument measures atmosphere/ocean surface brightness temperatures at 19.3, 22.2, 37.0, and 85.5 GHz. Its predecessor was NASA's scanning multichannel microwave radiometer (SMMR) that briefly flew on Seasat and was then deployed on Nimbus 7 in the same year that Seasat was launched. To achieve this rapid turnaround, the engineering model of the SMMR was quickly space qualified and added to the Nimbus 7 instrument complement. This continued the measurements that started with Seasat.

As a late add on to the DMSP satellite series, some compromises had to be made, and the SSM/I was located on the top of the spacecraft (Fig. 5.19).

This meant that every scan of the SSM/I was blocked by the satellite on a portion of the instrument's rotation (Fig. 5.19). The instrument consists of an offset parabolic reflector 61 cm × 66 cm, which feeds a corrugated, broad-band, seven-port horn antenna feed (Fig. 5.20). The SSM/I rotates at a uniform rate, making one complete revolution in 1.9 seconds, during which time the satellite advances 12.5 km. The reflector and the feed are mounted on a drum which contains the radiometer, digital data subsystem, mechanical scanning subsystem, and power subsystem. The reflector-feed-drum assembly is rotated about the central axis by a coaxially mounted bearing and power transfer assembly (BAPTA). All data, commands, timing, and telemetry signals, as well as power, pass through the BAPTA on slip ring connectors to the rotating assembly.

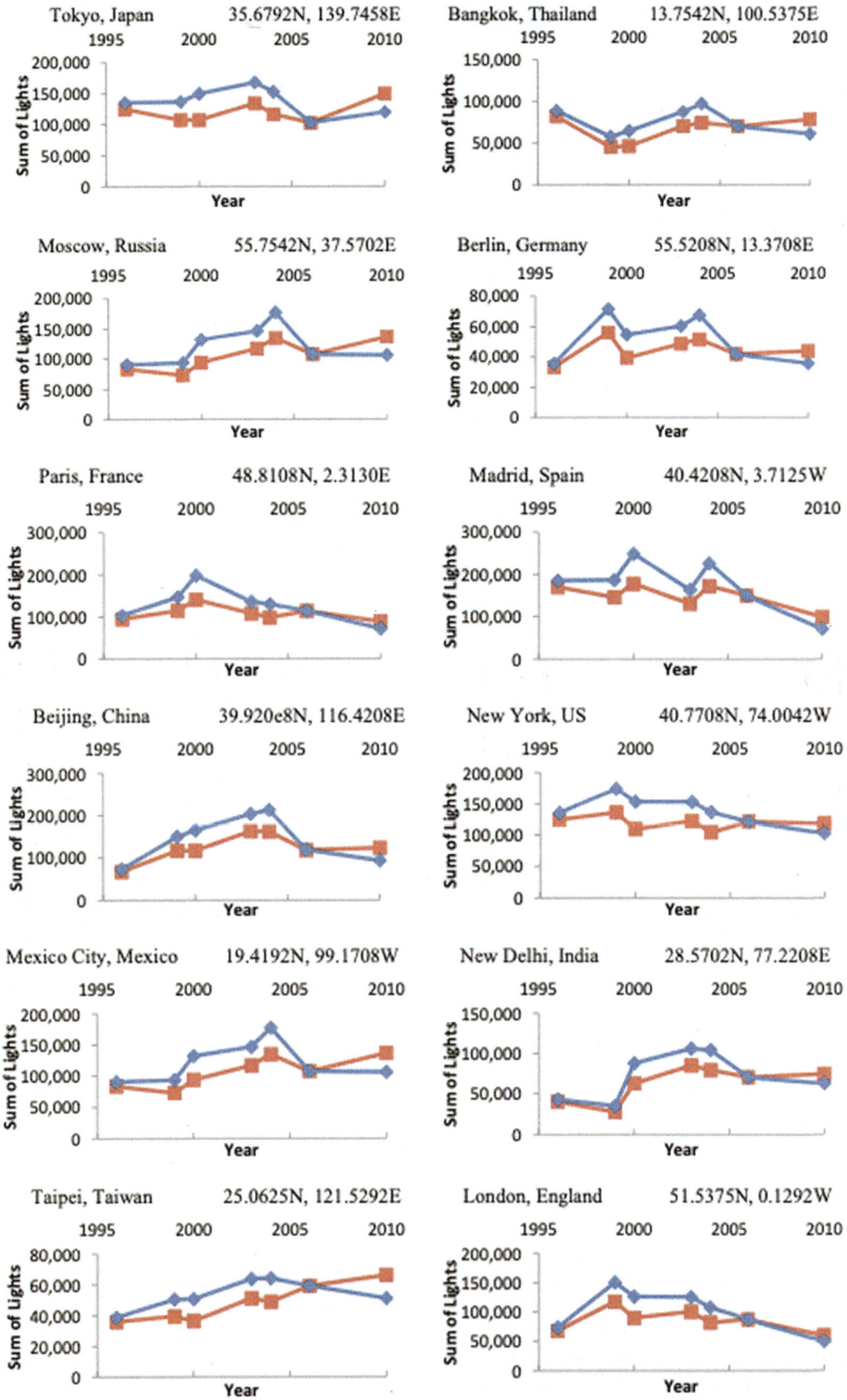

FIGURE 5.18 Time series of adjusted radiance light and nonadjusted. The red line is the interannual calibrated and the blue line is not interannual calibrated lights (Hsu et al., 2015).

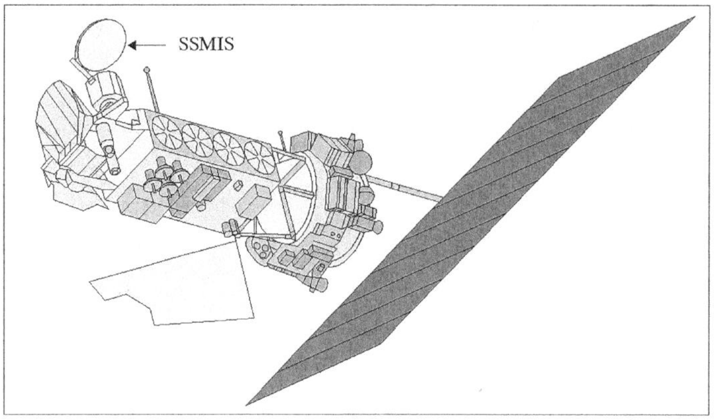

FIGURE 5.19 The Defense Meteorological Satellite Program satellite with the special sensor microwave imager mounted on the top front (Hollinger and Lo, 1983).

A small mirror and a hot reference load absorber are mounted on the BAPTA and do not rotate with the drum assembly. They are positioned off-axis such that they pass between the feed horn and the parabolic reflector. They block the feed once each scan. The mirror reflects cold sky radiation into the feed serving along with the hot load as calibration references for the SSM/I each scan. Corrections for spillover and antenna pattern effects from the reflector are incorporated into the data processing algorithm.

The feed is a unique aspect of the SSM/I (Fig. 5.20).

Here, the dimensions are in inches in keeping with the DoD requirements to have things in standard American units. It should be noted here that the feed horn is designed to cover all of the frequencies of the system. Later, microwave radiometers preferred to have different feeds to represent the different frequencies. In Fig. 5.21 we show the expanded view of the SSM/I.

The performance characteristics of the SSM/I are given here in Table 5.5.

Note that all but the 22.235 GHz channel are both horizontally and vertically polarized. These polarizations would prove to be very useful in connecting SSM/I data to geophysical parameters. The radiometric accuracy was estimated to be 1.5 K. Note, as with all microwave radiometers, the ground resolution improves as you go to higher frequencies. The reason that the 22.235 GHz channel was not dual polarized is that this is the channel that measures atmospheric water vapor, and there was no requirement for it to be dual polarized. The radiometers are all total power, balanced mixer, and superheterodyne receivers. The hot load reference element is shown in Fig. 5.22.

The hot load is designed to be an excellent absorber and emitter with a known temperature measured by three thermistors embedded in the hot load. Errors in the thermistor readings will translate into errors in the SSM/I antenna temperatures. Moreover, gradients across the hot load will degrade SSM/I performance. In addition, if the mirror reflecting cold space (2.73 K) is not a perfect reflector or if it is receiving radiation from other than cold space, this will compromise the calibration of SSM/I (Wentz, 2010).

The SSM/I scan geometry is shown here in Fig. 5.23.

The SSM/I drum and reflector rotate at a uniform rate of one revolution in 1.9 seconds (114 rpm) during which time the satellite advances its position by 12.5 km. The antenna beams are 45 degrees off of the BAPTA rotational axis, which is normal to the Earth's surface. As the antenna rotates, the beams sample the surface of a cone, which at the surface has an incidence angle of 53.1 degrees. The scene is viewed over a scan angle of 102.4 degrees centered on the ground track following the spacecraft motion, which then yields a scene swath width of 1,394 km. The outputs of the radiometers are sampled on two alternate scans. During scan A, the five lower frequency channels are each sampled over 64 equal 1.6 degrees intervals. For the alternate B scan, only the two 85.5 GHz channels are sampled at 128 equal 0.8 degree intervals or approximately each 11 km along the scan. The 12 bit precision was achieved using the integrate, hold, and dump method with an integration period of 7.95 ms for the five lower frequency channels and 3.89 ms for the two 85.5 GHz channels. Alternate 0.8 degree intervals are centered on the midpoints of the 1.6 degrees intervals so that the samples of seven channels are collocated with additional 85.5 GHz samples equally spaced between them. As a result, the five lower channels are all sampled on an approximately 25 km grid across the scan and along the track. The two 85.5 GHz channels have one half this spacing, both cross and along track.

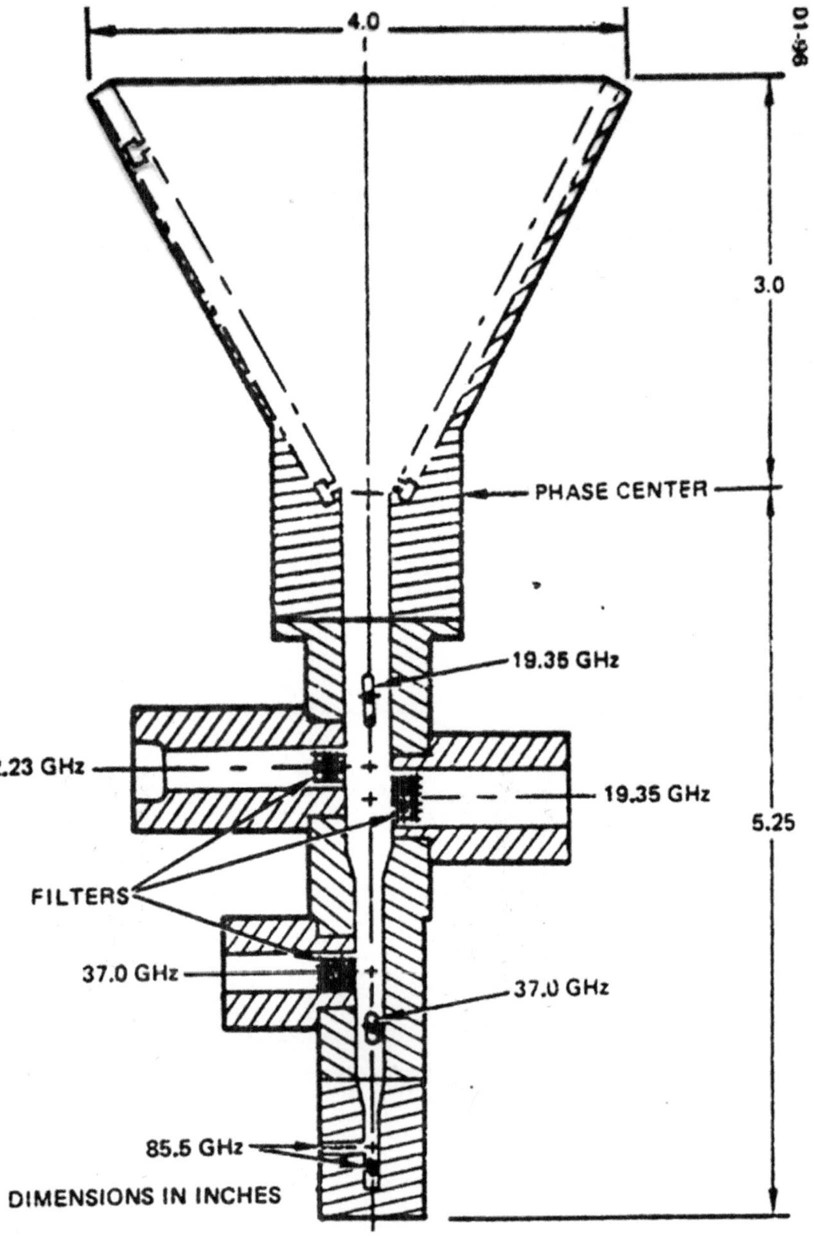

FIGURE 5.20 Special sensor microwave imager feed assembly (Hollinger and Lo, 1983).

The reference hot load and cold sky reflector are sampled in a similar manner as the scan, but over total angles of 8.0 degrees, rather than 102.4 degrees as is true for the scene data, as they obscure the feed horn during the remainder of the scan. The scene data, along with the hot load and space reflector measurements, along with the temperature of the hot load from three embedded thermistors, added to various instrument parameters, result in 3,636 bit blocks of data each second. The radiometer outputs are converted to absolute antenna temperatures using the hot load and cold sky reflectance measurements and are then further corrected for antenna pattern effects, resulting in absolute brightness temperatures.

The brightness temperatures are then used with a "D" matrix retrieval algorithm to estimate various environmental parameters. This D-matrix was developed by the Environmental Aircraft Company under a subcontract to Hughes Aircraft Company (Burke et al., 1980a; Burke et al, 1980b). A summary of their results is given in Table 5.6.

Following Hollinger and Lo (1983), we will describe some of these parameters retrieved using the D-matrix solution.

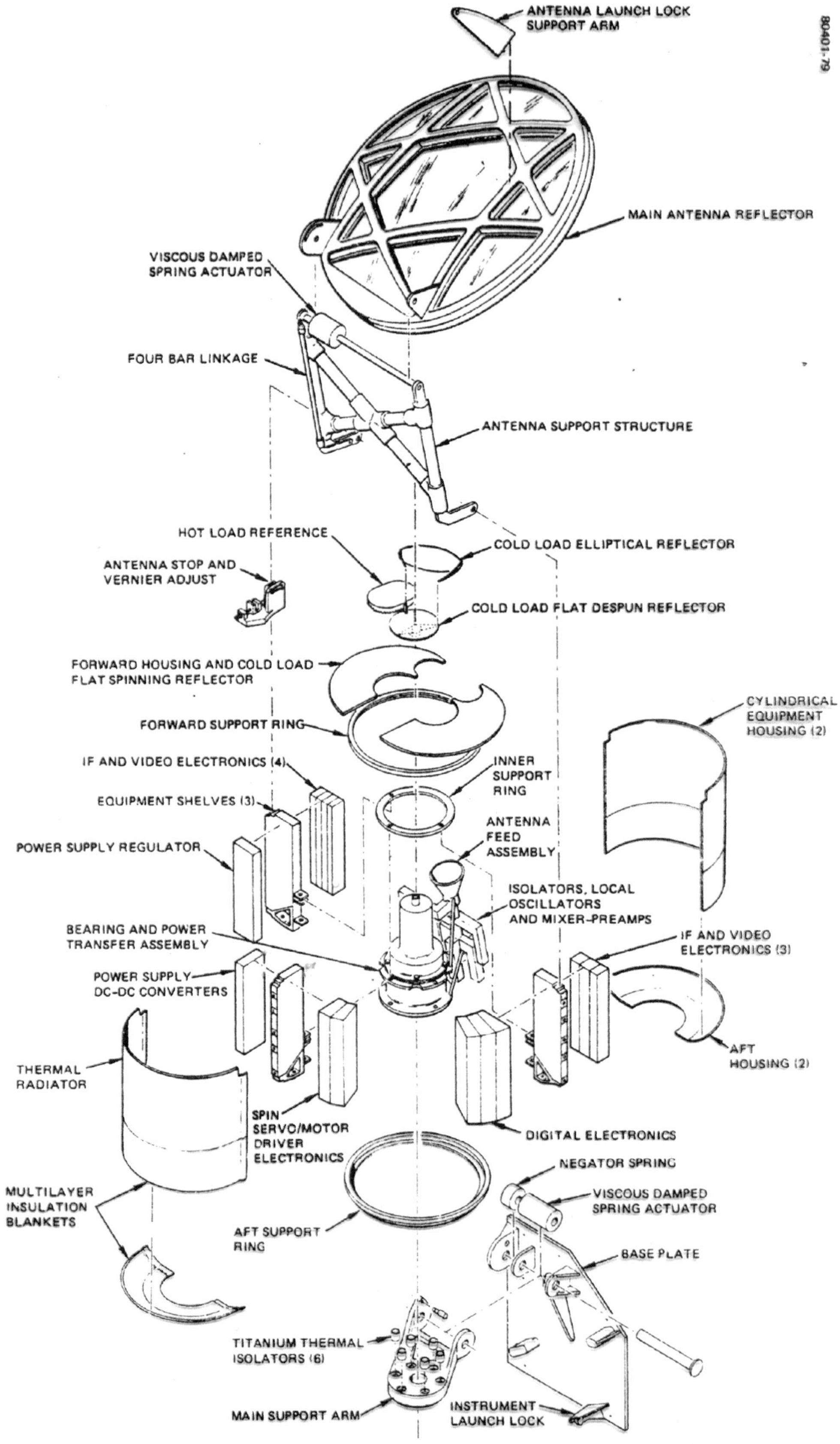

FIGURE 5.21 The special sensor microwave imager (SSM/I) (Hollinger and Lo, 1983).

TABLE 5.5 Special sensor microwave imager (SSM/I) performance characteristics.

Parameter	Radiometric performance			
Center FREQ (GHz)	19.35±0.05	22.235±0.05	37.0±0.1	85.5±0.3
Polarization	VERT and HORIZ	VERT	VERT and HORIZ	VERT and HORIZ
IF bandpass (MHz)	10 to 250	10 to 250	100 to 1000	100 to 1500
Radiometric accuracy (K)	1.5	1.5	1.5	1.5
ΔT (K)	0.8	0.8	0.6	1.1
Dynamic Temp Range (K)	375	375	375	375
Effective FOV (3 dB antenna beamwidth; includes INTEG time) reference (KM)	70x45	60x40	38x30	16x14
Integration time (mS)	7.95	7.95	7.95	3.39

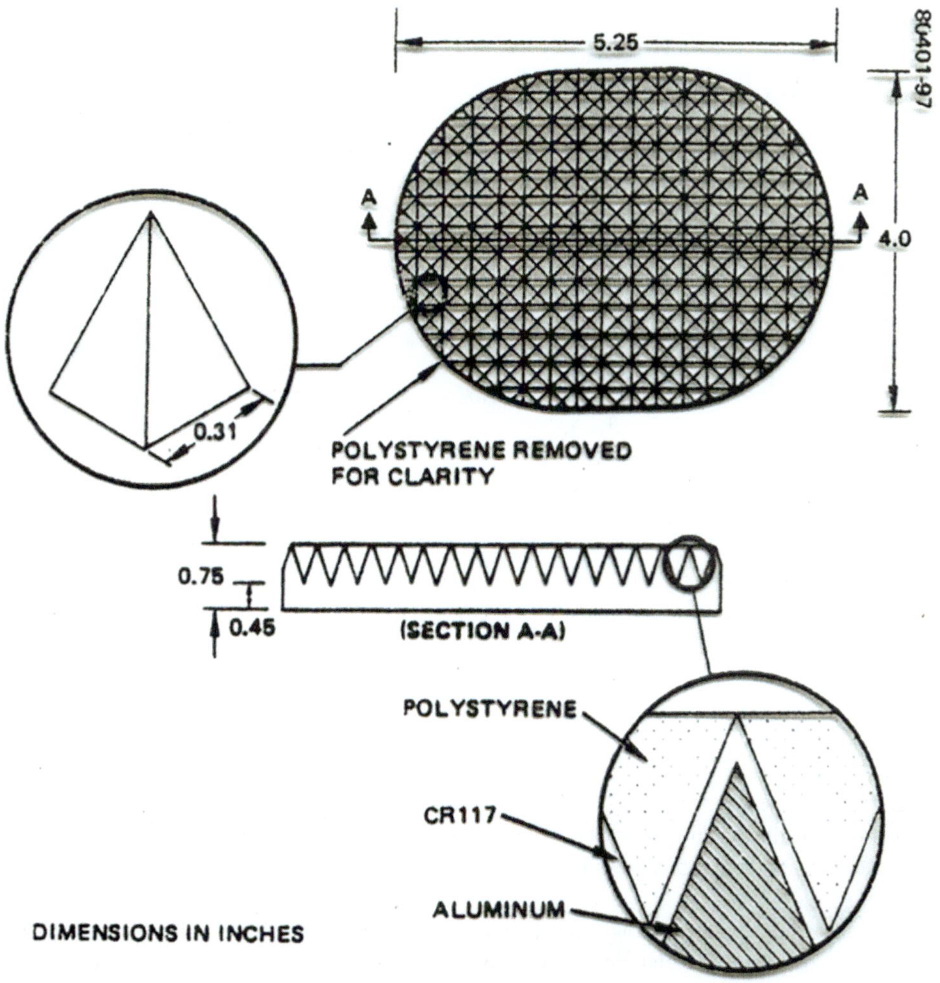

FIGURE 5.22 Special sensor microwave imager hot load reference (Hollinger and Lo, 1983).

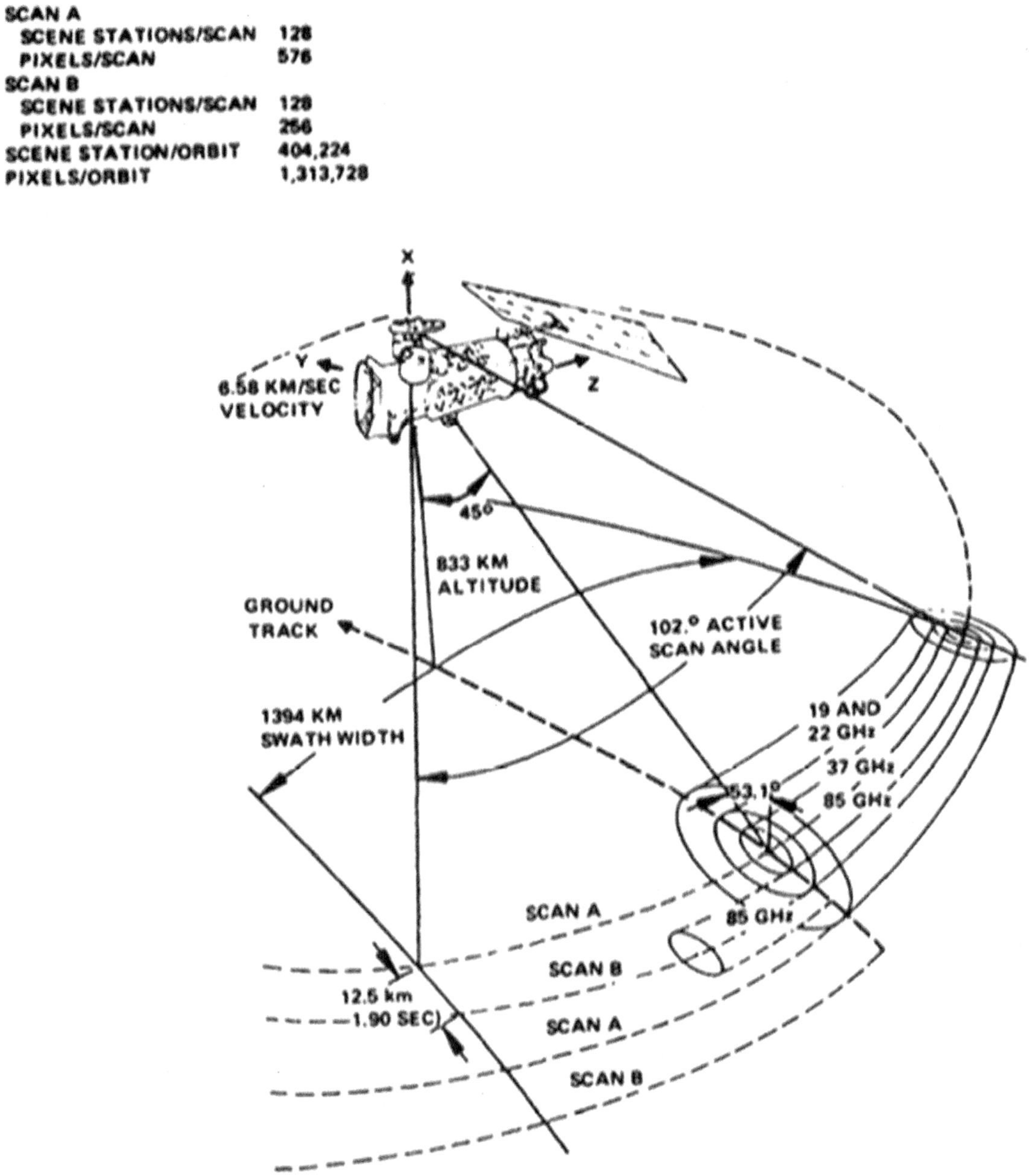

FIGURE 5.23 The Special sensor microwave imager scan geometry (Hollinger and Lo, 1983).

5.3.2.1 Ocean surface wind speed

The basic concept of the SSM/I wind speed algorithm is to measure the surface roughness and foam on the ocean. This surface roughness is related to the friction velocity at the air-sea interface. The emissivity of a smooth water surface is well known (Chang and Wilheit, 1979). The Fresnel equations for a plane dielectric interface can be used to calculate the emissivity of a smooth surface for a given view angle and polarization. Since ocean wind speeds have traditionally been measured at 20 m height, the relationship developed by Cardone (1969) is used to relate the SSM/I inferred wind to the wind at a 20 m height.

The roughness effect is modeled using Cox and Munk (1954), and treating the ocean surface as a collection of plane facets which are large compared with the observed wavelengths. Sea surface slope variance is set as a function of the wind speed

TABLE 5.6 Special sensor microwave imager (SSM/I) D-matrix retrieval characteristics.

Parameter	Desired geometric[a] resolution, KM	Range of values	Quantization levels	Absolute accuracy
Ocean surface wind speed	25	3 to 25	1	±2 M/S
ICE				
• Area covered	25	0 to 100	5	±12%
• AGE	50	1ST Year, multiyear	1 Year, > 2 Years	None
• Edge location	25	NA	NA	±12.5 KM
Precipitation over land areas	25	0 to 25	0, 5, 10, 15, 20, >25	±5 MM/HR
Cloud water[b]	25	0 to 1	0.05	±0.1 KG/M^2
Liquid water[c]	25	0 to 6	0.10	±2.0 KG/M^2
Precipitation over water	25	0 to 25	0, 5, 10, 15, 20, >25	±5MM/HR
Soil moisture	50	Dry—Saturated	Dry—Moist Wet—Saturated	None

[a] *Goals for system.*
[b] *<100 μm diameter.*
[c] *>100 μm diameter.*

and is then used to calculate emissivity, assuming a Gaussian distribution for the slope facet surfaces and once again employing the Fresnel equations (the ERT model). The wind speed dependence on the surface slope variance σ^2 is given by

$$\begin{aligned} \sigma^2(f) &= (0.3 + 0.02\ f)(0.003 + 0.48\ W) \text{ for } f < \mathbf{35} \text{ GHz, and} \\ &= 0.003 + 0.48\ w \text{ for } f > 35 \text{ GHz} \end{aligned} \tag{5.1}$$

where f is frequency in GHz and W is the wind speed at a 20 m height in m/s. In the ERT model, foam is treated as partially obscuring the surface in a way that is independent of polarization but dependent upon frequency. The foam fraction, K, is defined as

$$\begin{aligned} K &= a(l{-}e\ \ o)(W{-}7) \text{ for } W > 7 \text{ m/ s, and} \\ &= 0 \text{ for } W < 7 \text{ m/ s.} \end{aligned} \tag{5.2}$$

where a=0.006 second/m and f_0 is 7.5 GHz. This equation has its basis in Norberg's (1971) observation that the brightness temperature increases linearly with wind speed when it exceeds 7 m/s and that no foam forms below that wind speed. Fig. 5.24 shows the relationship between wind speed and brightness temperature for the 19 V and H, 22 V, and 37 V and H, assuming a tropical, clear atmosphere. Note that the strong dependence of wind speed on brightness temperature for 19 and 37 H and the relatively weaker dependence of all three vertical channels.

5.3.2.2 Precipitation over ocean and land

One of the most difficult environmental parameters to measure in the ocean is rainfall. Rain gauges mounted on ships are plagued by sea spray, breaking waves, and the motion of the ship, all of which make it extremely difficult to make a conventional rain gauge measurement at sea. Precipitation size droplets in the atmosphere are the primary contributors to atmospheric attenuation of the microwave signal from the Earth's surface.

The degree of attenuation depends upon wavelength, drop size distribution, and precipitation layer thickness. If we look at this attenuation as the signal, we have a way to use microwave observations for the detection of precipitation and a method to estimate rain rate as well. Over the ocean, this provides a capability not available by other conventional methods. At lower frequencies, such as 19.35 GHz, rain is highly absorptive and results in an apparent warm brightness temperature over a cold background such as the ocean. At frequencies greater than 80 GHz, scattering effects become dominant, resulting in an apparent cool brightness temperature relative to a warm background, such as land. These effects increase with increasing rain rate.

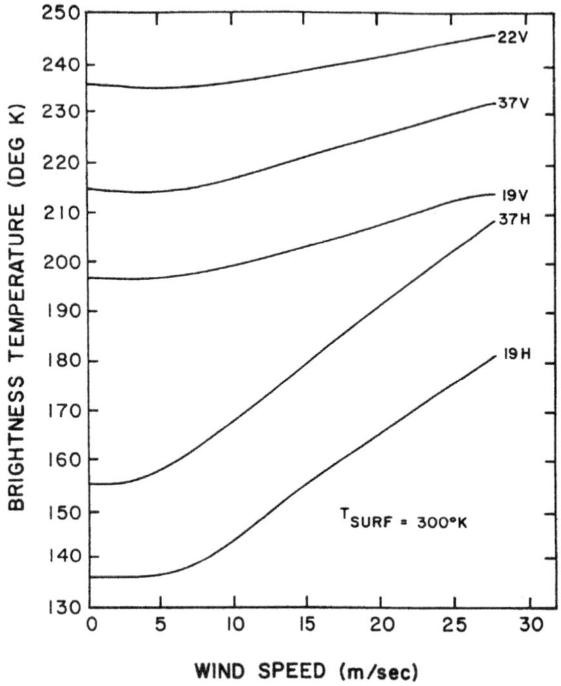

FIGURE 5.24 Ocean surface wind speed as a function of special sensor microwave imager brightness temperature for 19 V and H, 22 V, and 37 V and H channels (Hollinger and Lo, 1983).

For the SSM/I, the drop size distribution was modeled by fitting the Deirmendjian distribution to the empirically observed spectra of Laws and Parson (1943). The Deirmendjian distribution is defined as follows:

$$n(r) = \mathrm{A}r^{C_1}\exp(-\mathrm{B}r^{C_2}) \tag{5.3}$$

where $n(r)$ is the drop size distribution in $\mathrm{cm}^{-3}/\mu\mathrm{m}$, r is the radius of raindrops in μm, and A and B are scale parameters. Here A is

$$A = \left(\frac{MC_2}{\frac{4\pi}{3}\rho_w 10^6}\right)\left(\frac{B}{r}\right)^{[(C_1} + 4)/C_2] \tag{5.4}$$

where M is the total liquid water content in gm/m^3 and

$$M = \rho_w \int_0^{\infty} \frac{4\pi}{3} r^3 n(r)\,dr \tag{5.5}$$

And

$$B = \left(\frac{C_1}{C_2}\right) r_0^{-C_2} \tag{5.6}$$

Here, r_0 is the mode radius in μm. The shape parameters are C_1 and C_2, where the former affects the distribution of the smaller radii while the latter affects the larger radii. The results of fitting the Laws–Parson (1943) empirical spectrum are as follows:

$$M = 0.0636(R^{0.881})\ \mathrm{gm/cm^3}, \quad \text{where } R \text{ is the rain rate in mm/h.} \tag{5.7}$$

$r_0 = 225 + 9.16\ (R^{0.881})$; and $(r_0)_{max} = 375$ μm, $C_1 = 4.0$, $C_2 = 0.70–0.00458\ (R^{0.881})$ and $(C_2)_{min} = 0.625$.

Thus, we can relate the rain rate to the mode radius and liquid water density. The parameterization is in terms of rain rate only and simplifies the solution to the Deirmendjian distribution. At the same time, it retains the sensitivity of scattering to drop size. In Fig. 5.25 we can see the good agreement between the Deirmendjian model and the Laws and Parson distribution.

For weak storms with low rain rates ($R \leq 8$ mm/h), the ceiling for the rain layer does not exceed the 0°C isotherm. For a more intensive convective storm, the ceiling is higher with a supercooled water layer on top. The thickness of this layer depends upon the intensity of the convection. Climatological radiosonde data stratified by tropics, mid-latitudes, and polar

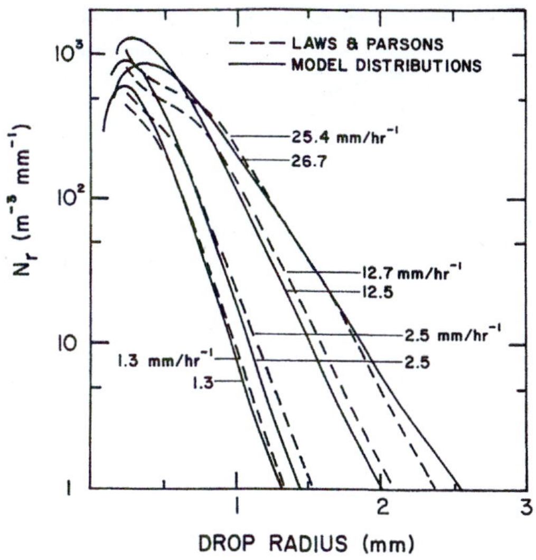

FIGURE 5.25 Comparison of the Deirmendjian raindrop size distribution used in the model computations and the Laws–Parsons empirical distributions (Hollinger and Lo, 1983).

regions are also separated by surface type (land and water) and are used to provide estimates of the supercooled water layer as a function of observed rain rate.

At lower frequencies, the absorption effect of ice particles is insignificant, but large ice or snow crystals, which may occur at the top of a storm, can cause significant scattering at 85.5 GHz. Simulation studies have shown that the presence of layers of ice particles of 400 μm would result in the loss of all sensitivity to rain rate. For a 200 μm particle size, the effect could also be significant depending on the layer thickness. In general, for typical rain conditions, the typical ice particle size is 100 μm, which does not cause significant attenuation at 85.5 GHz. Brightness temperature calculations are based on radiative transfer models (RTMs). The brightness temperature versus rain rate over the mid-latitude ocean is presented in Fig. 5.26 and over mid-latitude land in Fig. 5.27. These calculations include the presence of supercooled water layers for rain rates above 8 mm/h. The 19.35 and 37.0 GHz frequencies are used for retrieval over the ocean, while the 37.0 and 85.5 GHz channels are added to the 19.35 GHz for use over land.

Over the ocean, the rain rate depends very strongly on brightness temperature up to about 10 mm/h. Above that, the dependency is much weaker, making the retrievals noisier. For the land, by using all of the available channels, it should be possible to get an accurate estimate of rain rate up to about 20 mm/h.

5.3.2.3 Cloud liquid water over ocean and land

Cloud liquid water is defined to consist of droplets smaller than 0.01 mm in diameter. Droplets of this size produce negligible scattering at the SSM/I frequencies and only absorption by the water molecules needs to be considered. The absorption coefficient, γ is given by Staelin (1976) for droplets with diameters small compared to the wavelength as follows:

$$\gamma = \frac{(1.0016X10^4)(10^{0.0122(291-T)})}{\lambda^2} \tag{5.8}$$

where m is liquid water content in gm/m^3, T is atmospheric temperature in K, and λ is the wavelength in cm.

In general, the retrieval of cloud liquid water over the ocean can be achieved with a high degree of accuracy due to the fact that the ocean surface is fairly homogeneous and cold in the microwave range. This retrieval involves the 19 V and H, the 22 V, and the 37 V and H channels on the SSM/I. The dependency of brightness temperature relative to cloud liquid water is represented graphically in Fig. 5.28.

Here, one can see the value of the 19 H GHz channel which then hands off the sensing to the 37 H channel. Then the 19 V and the 22 V come into play with the last sensing by the 37 V. It is quite easy to see how an algorithm could be composed from these dependencies. The retrieval of cloud liquid water over land is complicated by the fact that the ground is much warmer than the ocean resulting in a poorer contrast between cloud and the surface making it difficult to sense the thermal variations over the land surface.

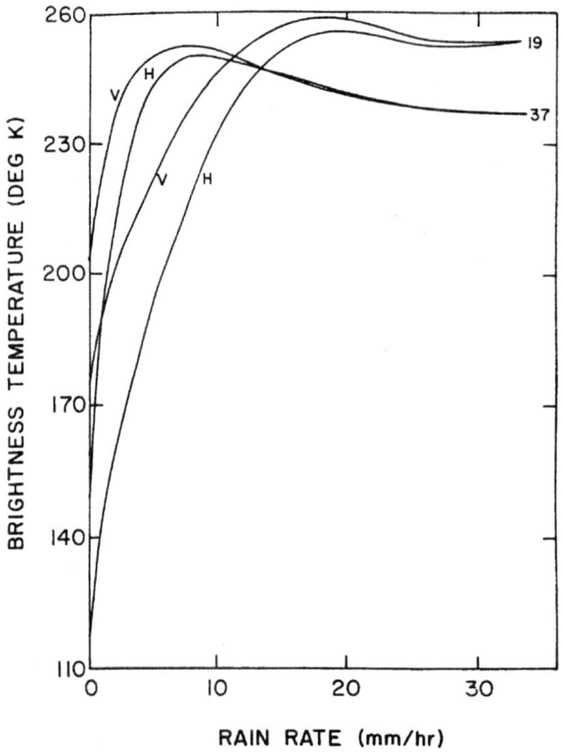

FIGURE 5.26 Brightness temperature versus rain rate over a mid-latitude ocean background at 19.35 and 37.0 GHz for both vertical and horizontal polarizations (look angle=50 degrees) (Hollinger and Lo, 1983).

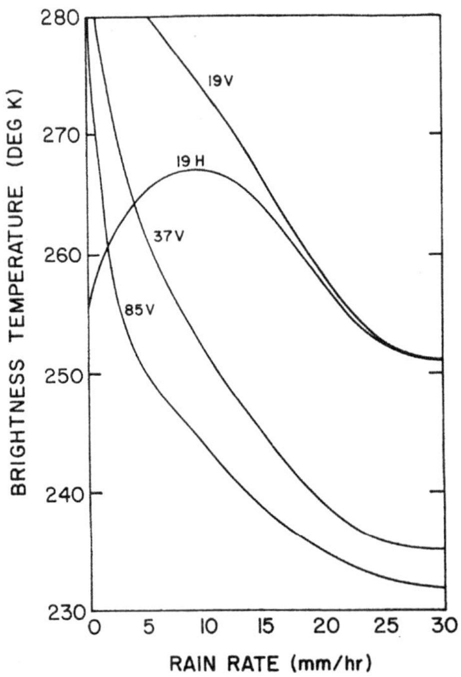

FIGURE 5.27 Brightness temperature versus rain rate over a mid-latitude land surface at 85.5, 37.0, and 19.35 GHz. The look angle is 50 degrees (Hollinger and Lo, 1983).

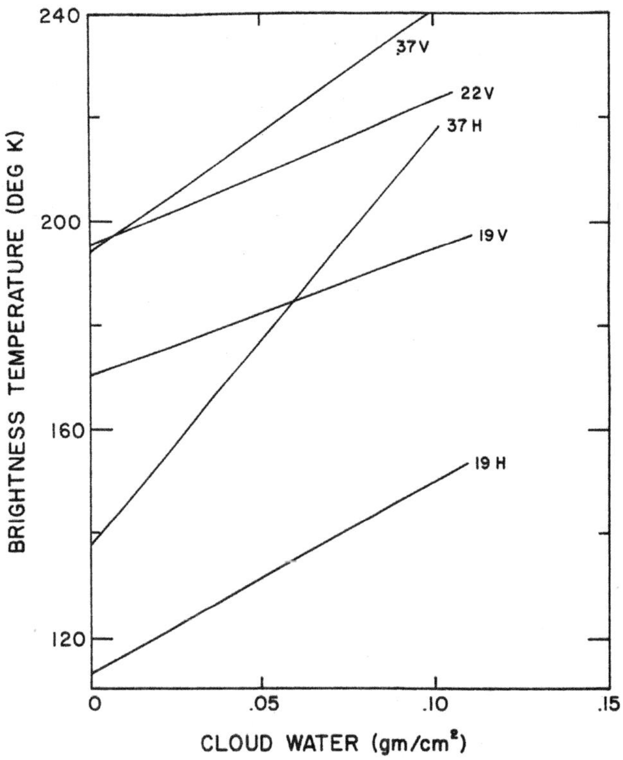

FIGURE 5.28 Brightness temperature versus cloud liquid water over a mid-latitude ocean (Hollinger and Lo, 1983).

When rain is actually falling, the ability to retrieve cloud liquid water is lost. Differences between the H and V polarization of the 85.5 GHz channel have been used to determine if rain is present over land. Similar criteria have been established for the 37.0 V and H channels over the ocean. These brightness temperature criteria are different for different latitude bands and seasons.

5.3.2.4 Soil moisture

The soil moisture algorithm developed for SSM/I was based on the idea that thermal microwave radiation in soils results from the random, microscopic current loops within the soil volume (Stogryn, 1970). The intensity of radiation at a given point depends on the local dielectric constant and the physical temperature of the soil. Moisture produces a marked increase in both the real and imaginary parts of the dielectric constant of the soil, resulting in a decrease in the soil emissivity. Both experimental and theoretical studies have shown that the emissivity of soil at microwave frequencies can range from >0.9 for dry soils to ≤0.6 for very wet soils.

A general incoherent layered surface model was used to simulate the expected SSM/I soil moisture signatures. In this model, layers of soil with different moisture contents are modeled as multilayer media, with the reflection coefficients between layers and opacities within the layer calculated from the dielectric properties of the medium. The soil moisture model uses existing measurements of the dielectric properties of the soil. The layered surfaces are combined with a radiative transfer model (RTM) of the atmosphere to obtain brightness temperatures for the inversion algorithm. The predicted emissivity at 1.43, 18, and 37 GHz is plotted in Fig. 5.29.

Choudhary and Schumugge (1979) developed a simple correction factor to the reflectivity to take into account the surface roughness effect.

5.3.2.5 Sea ice

One of the great benefits of sensing in the passive microwave is that for many frequencies and polarizations the atmosphere is transparent, and the Earth's surface can be viewed in any weather condition. This is particularly important in the polar regions, where persistent cloud cover obscures the Earth's surface most of the time. Thus, SSM/I offered an altogether new way to view sea ice. In this case, a geophysical model was not used, but the sea ice parameters are obtained

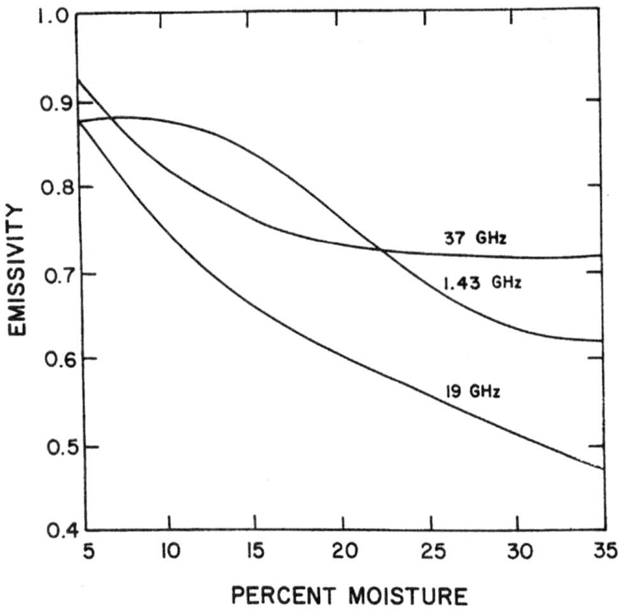

FIGURE 5.29 Predicted emissivity at 37, 18, and 1.43 GHz (Hollinger and Lo, 1983).

from deterministic equations with empirically determined constants. The upwelling brightness temperature sensed by the SSM/I radiometer for a scene containing sea ice can be written as follows:

$$T_{Bp=e^{-\gamma}}[C \cdot F(\varepsilon_{fp}T_f + (1 - \varepsilon_{fp})T_{\text{sky}}) + C \cdot (1 - F)(\varepsilon_{mp}T_m + (1 - \varepsilon_{mp})T_{\text{sky}}) + (1 - C)(\varepsilon_{wp}Tw + (1 - \varepsilon_{wp})T_{\text{sky}})] + T_{\text{atm}} \tag{5.9}$$

where γ is the total atmospheric opacity, p indicates the polarization (H or V), C is the fraction of sea ice including all ice types, and F and M are the fractions of first year ice and multiyear ice relative to the total ice present. The emissivity is given by ε_{fp}, and T is the ice surface temperature in Celsius. Subscripts f, m, and w indicate first year ice, multiyear ice, and seawater. T_{sky} is the incident sky temperature at the surface, and T_{atm} is the atmospheric contribution to the upwelling radiation.

The difference in V and H polarizations for 37 GHz shows less dependence on ice age than does the V and H difference (Table 5.7) in the 19.35 GHz channel or the differences in the vertical and horizontal polarizations at either frequency. This characteristic is used to find the sea ice concentration using only the differences in the horizontal and vertical polarizations of the 37 GHz brightness temperatures using an inversion model.

5.3.2.6 Shared processing

All of the algorithms given above were developed prior to the launch of the first SSM/I in June 1987. One of the very unique aspects of the SSM/I is that while the DMSP data downlink was encrypted, NOAA and the DMSP program office developed what became known as shared processing, whereby the SSM/I portion of the DMSP downlinked data was extracted on the ground decrypted and made readily available to the research community via NASA and NOAA data systems that made satellite data available online. Associated with shared processing was the effort by the Hughes Aircraft Company to familiarize people with the SSM/I and what it could do. Dr. Richard (Doc) Savage of Hughes traveled the country giving talks on SSM/I and encouraging researchers to use this unique source of passive microwave satellite imagery.

5.3.2.7 Modern special sensor microwave imager parameter retrievals

One of the people who took advantage of these data was Frank Wentz, who formed a company Remote Sensing System (RSS) of Santa Rosa, California, to process the SSM/I data into geophysical parameters using a RTM he had developed in his studies. This model served as the basis for many advances in the model to invert the data from an increasing number of microwave satellite instruments.

We will start with Wentz's (1997) well-calibrated ocean algorithm for the SSM/I. As pointed out earlier, retrieving parameters over the ocean is a lot easier than over the land since the ocean is cold and relatively homogeneous in the

TABLE 5.7 Emissivity of sea ice.

Ice type	My		Fy ice		Fy thin ice		Water calm sea	
Polarization[a]	V	H	V	H	V	H	V	H
FREQ (GHz)								
19.35	0.86	0.73	0.97	0.84	0.96	0.78	0.60	0.33
37	0.69	0.64	0.97	0.95	0.96	0.87	0.70	040
37 (V H)	0.05		0.02		0.09		0.30	

[a] *V—Vertical; H—Horizontal; 50 degrees Earth incidence angle.*

microwave. Wentz (1997) observed that a microwave radiometer over the ocean can measure air temperature and the three phases of atmospheric water (vapor, liquid, and ice). In addition, they can measure sea surface temperature along with sea ice type and concentration. These data can be used to study a wide variety of hydrologic and radiative processes, such as air–sea heat and momentum flux, and the effect of clouds on radiative fluxes. This makes the microwave radiometer a unique and valuable tool for studying the Earth.

Focusing now on the SSM/I, Wentz (1997) relates a method to retrieve geophysical variables from SSM/I brightness temperatures over the ocean. Using the SSM/I channels (Table 5.5), it is possible to retrieve three important geophysical variables over the ocean: near-surface wind speed W (m/s), total column water vapor V (mm), and columnar cloud liquid water L (mm). In addition, rainfall can be estimated, but Wentz restricts his study to ocean areas without rainfall. This is done to avoid the complexity to the first three variables introduced by the presence of rainfall. There is a unique relationship between the SSM/I brightness temperature (T_B) and the parameters W, V, and L. This relationship is found using a radiative transfer equation (RTE) called the T_B model function (Wentz, 1983). The retrieval of W, V, and L is done by varying these parameters until the T_B fits the SSM/I observations. Thus, the accuracy of the W, V, and L retrievals depends on the accuracy of the T_B model. The objective must be to achieve the highest possible accuracy, which means that the model must contain all of the relevant parameters and the model must be well calibrated. This model can have the three focus parameters of W, V, and L as primary parameters in addition to four secondary parameters: sea surface temperature T_s, (K), effective atmospheric temperature T_E (K), effective atmospheric pressure P (millibars) of the water vapor column, and wind direction ϕ. In this retrieval, Wentz uses climatological values to provide the secondary variables that are needed for the retrieval. It is noted that the RTE model is more sensitive to the W, V, and L primary variables that we seek to retrieve than to the secondary variables.

Wind direction is too variable to be specified from climatology and hence Wentz included it as a fourth retrieval parameter in addition to W, V, and L. The wind direction enters the T_B model in terms of the line-of-sight wind component W_{LS} which is the component of the wind vector that is along the SSM/I observation direction. Including wind direction in the model has two benefits: first, it improves the accuracy of W the wind speed retrieval and second, it provides new information on the wind direction over the ocean. For each SSM/I pixel, the values of W, V, L, and W_{LS} are found by the algorithm, and when substituted into the T_B model, they are varied in the model until the T_B values match the SSM/I observations at 19 GHz V_{pol}, 22 GHz V_{pol}, 37 GHz $V_{pol,}$and 37 GHz H_{pol}. The T_B model is quasilinear in terms of the four primary parameters, and hence the retrieval amounts to solving four equations with four unknowns.

The RTM for the retrieval was first published by Wentz in 1983. The brightness temperature at the top of the atmosphere, as seen by a satellite, is expressed as the sum of the upwelling atmospheric radiation, the downwelling atmospheric radiation that is reflected back upward by the sea surface and the direct emission of the sea surface attenuated by the intervening atmosphere. Thus we can write this as follows:

$$F(W, V, L) = T_{BU} + \iota[ET_s + (1 - E)(\Omega T_{BD} + \iota T_{BC})] \quad (5.10)$$

where T_{BU} and T_{BD} are the upwelling and downwelling atmospheric brightness temperatures, and ι is the transmittance through the atmosphere. E is the sea surface emissivity, and T_{BC} is the cosmic background radiation temperature which equals 2.7 K. The Ω accounts for how the rough sea surface reflects radiation other than specular reflection.

The upwelling and downwelling atmospheric brightness temperatures are defined in terms of effective air temperature as follows:

$$T_U = T_{BU}/(1 - \iota) \quad (5.11a)$$

$$T_D = T_{BD}/(1 - \iota) \quad (5.11b)$$

It should be noted that T_U and T_D are very close in value, with T_U being slightly colder. For the SSM/I model, both T_U and T_D are calculated from a set of 42,195 radiosonde profiles using the complete integral form of the RTE. Wentz found that they are both highly correlated with radiosonde columnar water vapor V (millimeters) and sea surface temperatures T_S (K) at the radiosonde site. The following least squares regressions were used:

$$T_D = c_0 + c_1 V + c_2 V^2 + c_3 V^3 + c_4 V^4 + c_5(T_S - T_V) \quad (5.12a)$$

$$T_U = T_D + c_6 + c_7 V \quad (5.12b)$$

$$T_V = 273.16 + 0.8337V - 3.029EV^{3.33} \ \mathrm{V} \leq 48 \quad (5.13a)$$

$$T_V = 301.16. \ \mathrm{V} > 48 \quad (5.13b)$$

In Equation 5.12a, if V is greater than 58 mm, the expression is linearly extrapolated. The regression coefficients are given here in Table 5.8.

Eq. 5.12a and b are solved by regressing the climatological sea surface temperature at the radiosonde site on V. Hence T_V represents a sea surface temperature that corresponds to the water vapor V. The term T_S–T_V in Eq. 5.12a accounts for the fact that the effective air temperature is typically higher (lower) for unusually warm (cold) sea surfaces.

In the absence of rainfall, the total absorption along the SSM/I viewing path is the sum of the following: oxygen, water vapor, and cloud liquid water. For convenience, the absorption is normalized in terms of a vertically integrated quantity rather than along a viewing path. This separates the dependence of the absorption on viewing angle (θ). For incidence angles less than 60°, the ratio of the viewing path length through the troposphere to the height of the troposphere is sec θ. If we let A_O, A_V, and A_ι be the vertically integrated absorption components due to oxygen, water vapor, and liquid water, respectively, we can write the atmospheric transmittance along the SSM/I viewing path as

$$\iota = \exp[-\sec\theta A_O + A_V + A_\iota) \quad (5.14)$$

As stated earlier, A_O and A_V were computed from 42,195 radiosonde profiles using the complete RTE formulation. As a result, it was found that the A_O values were nearly constant over the globe with a small dependence on air temperature. The following equation was used for a least squares fit of the radiosonde-derived A_O and the downwelling atmospheric brightness temperature T_D.

$$A_O = (\alpha_O/T_D)1/4 \quad (5.15)$$

TABLE 5.8 Radiative transfer equation (RTE) model coefficients for the atmosphere.

Parameter	19 GHz	22 GHz	37 GHz
C_0 (K)	240.58E+0	242.04E+0	239.55E+0
C_1 (K/mm)	305.96E−2	297.16E−2	248.15E−2
C_2 (K/mm$^{2)}$	−764.41E−4	−769.38E−4	−438.59E−4
C_3 (K/mm$^{3)}$	885.95E−6	931.80E−6	278.71E−6
C_4 (K/mm$^{4)}$	−40.80E−7	−44.85E−7	−3.23E−7
C_5	0.60E+0	0.20E+0	0.60E+0
C_6 (K)	−0.16E+0	−0.15E+0	−0.57E+0
C_7 (K/mm$^{4)}$	−2.13E−2	−7.51E−2	−2.61E−2
a_0 (K)	11.80E+0	13.01E+0	28.10E+0
$a_{V1,\ \mathrm{LIEBE}}$ (mm$^{-1)}$	2.28E−3	6.16E−3	2.06E−3
$a_{V1,\ \mathrm{SSM/I}}$ (mm$^{-1)}$	2.23E−3	6.16E−3	1.85E−3
$a_{V2,\ \mathrm{LIEBE}}$ (mm$^{-2)}$	0.06E−5	1.05E−5	0.49E−5
a_{V2}, LIEBE (mm$^{-1)}$	0.00E−5	0.67E−5	0.17E−5

Subscript LIEBE refers to the use of the vapor absorption expression by Liebe (1985). See text for details. All numbers are reported in exponential format. For example, read 0.06E-5 as 0.06×10^{-5}, equal to 0.0000006.

The water vapor absorption A_V is a function of V as follows:

$$A_V = \alpha_{V1} V + \alpha_{V2} V^2 \tag{5.16}$$

The α_0 and α_v coefficients are given here in Table 5.9 for three different SSM/I frequencies.

Two different sets of coefficients are listed. Those denoted by a subscript Liebe are derived from the radiosonde data using the vapor absorption expression given by Liebe (1985). Wentz found that the Liebe coefficients produced an erroneous correlation between the cloud liquid water retrieval and water vapor. He, therefore, rederived the coefficients using the collocated SSM/I and RAOB. These coefficients are denoted by an SSM/I subscript in Table 2.8a, and the Liebe coefficients are given for comparison purposes.

In the no rain condition, the radiative transfer through the cloud droplets, which are much smaller than the radiative wavelengths, is governed by Rayleigh scattering, and the absorption is proportional to the columnar cloud liquid water (Goldstein, 1951). The absorption also depends on the temperature of the water droplets. At 37 GHz, the Rayleigh absorption A_{L37} is given by

$$A_{l37} = 0.208[1 - 0.026(T_L - 283)]L \tag{5.17}$$

where L is in millimeters. T_L is approximated by $(T_3+273)/2$, which is the mean temperature between the surface and the freezing level. The temperature dependence is nearly the same at both 19 and 22 GHz, where the Rayleigh scattering gives as follows:

$$A_{L19} = 0.2858\ A_{L37} \tag{5.18a}$$

$$A_{L22} = 0.3751\ A_{L37} \tag{5.18b}$$

The wind dependence on TS is implicit in the emissivity E and the scattering term Ω. The emissivity E is given by

$$E = E_0 + E_W \tag{5.19}$$

where E_0 is the specular emissivity, and E_W is the wind-induced emissivity. The specular emissivity can be found using the Fresnel equation, which depends on polarization, incidence angle, and the dielectric constant of seawater. The dielectric constant is itself a function of frequency, water temperature, and salinity. The salinity dependence is so small that it is

TABLE 5.9 RTE model coefficients for the atmosphere.

Parameter	19 GHz	22 GHz	37 GHz
c_0 (K)	240.58E+0	242.04E+0	239.55E+0
c_1 (K/mm)	305.96E−2	297.16E−2	248.15E−2
c_2 (K/mm$^{2)}$	−764.41E−4	−769.38E−4	−438.59E−4
c_3 (K/mm$^{3)}$	885.95E−6	931.80E−6	278.71E−6
c_4 (K/mm$^{4)}$	−40.80E−7	−44.85E−7	−3.23E−7
c_5	0.60E+0	0.20E+0	0.60E+0
c_6 (K)	−0.16E+0	−0.15E+0	−0.57E+0
c_7 (K/mm)	−2.13E−2	−7.51E−2	−2.61E−2
a_0 (K)	11.80E+0	13.01E+0	28.10E+0
$a_{V1,\ LIEBE}$ (mm$^{-1)}$	2.28E−3	6.16E−3	2.06E−3
$a_{V1,\ SSM/I}$ (mm$^{-1)}$	2.23E−3	6.16E−3	1.85E−3
$a_{V2,\ LIEBE}$ (mm$^{-2)}$	0.06E−5	1.05E−5	0.49E−5
$a_{V2,\ SSM/I}$ (mm$^{-1)}$	0.00E−5	0.67E−5	0.17E−5

Subscript LIEBE refers to the use of the vapor absorption expression by Liebe (1985). See text for details. All numbers are reported in exponential format. For example, read 0.06E-5 as 0.06×10^{-5}, equal to 0.0000006.

acceptable to use a constant value of 35 parts per thousand for the salinity. In an earlier paper, Wentz (1992) derived the following equation for the Fresnel emissivity:

$$E_0 = (\varepsilon_0 + \epsilon_1 \iota + \epsilon_2 \iota^2 + \varepsilon_3 \iota^3 + \varepsilon_4 q + \epsilon_5 \iota q + \epsilon_6 q^2 + \varepsilon_7 \iota^2 q)/T_s \tag{5.20}$$

where $q = \theta - 51$ degrees and $\iota = T_s - 273.16$. *This equation is valid for* θ between 48 and 55 degrees. The ε coefficients are given here in Table 5.10.

The sea surface roughness is accounted for by the wind-induced emissivity. This surface roughness changes the local incidence angle, rotates the polarization states, and diffracts the radiation. Added to this, sea foam acts as an impedance match between the air and water. Wentz (1975) characterizes this effect with a two-scale scattering model, which results in the wind-induced emissivity being a monotonic function of wind speed. We can write E_W as

$$E_W = M_1 W \quad W \le W_1 \tag{5.21a}$$

$$E_W = M_1 W + 0.5(M_2 = M_1)(W - W_1)^2/(W_2 - W_1) W_1 < W < W_2 \tag{5.21b}$$

$$E_W = M_2 W - 0.5(M_2 - M_1)(W_2 + W_1). \quad W \ge W_2 \tag{5.21c}$$

These equations represent two linear segments connected by a quadratic spline such that its first derivative in W is continuous. The spline points are W_1=7 m/s and W_2=12 m/s. The slopes of the linear segments are M_1 and M_2, respectively. The two-scale scattering model indicates that M_1 and M_2 have an incidence angle dependence and a slight dependence on sea surface temperature.

$$M_j = m_j + \beta(\theta - 53) + \mu(T_s - 288) \tag{5.22}$$

where the subscript j equals 1 or 2. The m_1, m_2, β, and μ coefficients are also in Table 5.9 for three SSM/I frequencies and two polarizations. The values of β and μ are theoretically derived from the two-scale scattering model, where the sea surface is represented by a collection of tilted facets with each facet acting as an independent specular surface (Wentz, 1975). The values for m_1 and m_2 are found using collocated SSM/I and buoy wind observations.

The main downwelling radiation component is the specular reflection from the zenith angle corresponding to the incidence angle θ. The specular reflection is $(1 - E)\,T_{BD}$ except for a rough sea where there will be additional sky radiation due to the tilted surface facets reflecting radiation from other parts of the sky into the SSM/I. Since the downwelling radiation increases as the secant of the zenith angle, the total radiation scattered from the sea surface is greater than that

TABLE 5.10 Model coefficients for the sea surface.

Parameter	19V	19H	22V	22H	37V	37H
ε_0 (K)	162.53E+0	83.88E+0	166.99E+0	86.98E+0	186.31E+0	101.42E+0
ε_1	−25.70E−2	−52.22E−2	−34.08E−2	−59.52E−2	−56.37E−2	−85.88E−2
ε_2 (K^{-1})	17.29E−3	18.76E−3	17.35E−3	19.38E−3	14.81E−3	20.76E−3
ε_3 (K^{-2})	−11.77E−5	−9.25E−5	−10.36E−5	−8.99E−5	−2.96E−5	−7.07E−5
ε_4 (K/deg)	21.62E−1	−14.72E−1	21.64E−I	−15.15E−1	21.23E−1	−17.01E−1
ε_5 (deg^{-1})	0.70E−2	0.21E−2	0.75E−2	0.30E−2	1.17E−2	0.55E−2
ε_6 (K/deg^2)	0.45E−1	−0.16E−1	0.45E−1	−0.16E−1	0.41E−I	−0.19E−I
ε_7 (K^{-1}/deg)	0.14E−4	−1.10E−4	0.02E−4	−1.17E−4	−0.71E−4	−1.27E−4
β (s/m/deg)	−0.81E−4	0.81E−A	−0.87E−4	0.87E−4	−1.19E−4	1.05E−4
μ (s/m/K)	0 41E−5	−0.13E−5	0.54E−5	−0.16E−5	1.25E−5	−0.29E−5
m_1 (s/m)	0.46E−3	3.01E−3	0.34E−3	3.20E−3	−0.09E−3	3.91E−3
m_2 (s/m)	3.78E−3	7.50E−3	3.48E−3	7.39E−3	2.38E−3	7.00E−3
T_B offsets (K)	0.78E+0	2.10E+0	0.78E+0	–	−1.68E+0	0.13E+0

V and H refer to vertical and horizontal, respectively. All numbers are reported in exponential format. For example, read 0.06E-5 as 0.06×10^{-5}, equal to 0.0000006.

given by a simple specular reflection. The two-scale models indicate that the total scattered radiation can be approximated by multiplying the specular reflection component by the following factors:

$$\text{for } V_{pol} \ \Omega = 1 + 2.5(\sigma^2 - 68\sigma^6)\tau^3 \tag{5.23a}$$

$$\text{for } H_{pol} \ \Omega = 1 + 6.1(\sigma^2 - 68\sigma^6)\tau^2 \tag{5.23b}$$

where σ^2 is the sea surface slope variance. The term $\sigma^2 - 68\sigma^6$ reaches a maximum at σ^2=0.07, and the term is held at its maximum value of 0.0467 for any σ^2>0.07. For moderately high winds (12 m/s) and a moist atmosphere (high vapor or heavy clouds), the scattering process increases the reflected radiation by 1 K for V_{pol} and 5 K for H_{pol}. There is an accuracy of about 0.2 K associated with these assumptions relative to using the theoretical two-scale model. It should be noted that when the atmospheric absorption gets large, Ω tends to unity because the sky radiation for a completely opaque atmosphere is isotropic.

In the two-scale scattering model, the slope variance σ^2 depends on the observation frequency *f*, as well as the wind speed. Ocean waves have wavelengths longer than the radiation wavelengths and do not contribute to the variance σ^2. As it increases with *f*, it reaches an optical limit (Wilheit and Chang, 1980; Wentz, 1983). For 37 GHz, Wentz (1997) used the Cox and Munk (1954) formulation for σ^2. For 19 and 22 GHz, he applied a reduction factor to the variance estimate:

$$\sigma^2 = 5.22\text{E}{-3}\xi W \tag{5.24}$$

where ξ represents the reduction factors 0.688, 0.739, and 1.0 for 19, 22, and 37 GHz, respectively. Since the Cox and Munk (1954) wind speed was measured at 12.5 m elevation, it was necessary to increase their coefficient of 5.12E-3 in their σ^2 expression by 2% to account for the wind being referenced to a 10 m elevation. The last row in Table 5.10 is a set of five offsets that are subtracted from the T_s observations. These offsets remove the overall bias between the model T_s and the measurements. The offset for the 22 V channel is subtracted from the antenna temperature rather than the brightness temperature. All of the offsets represent a combination of instrument and modeling absolute errors. They are all quite small, which indicates that SSM/I is a well-calibrated sensor and that the T_s model is an accurate representation. To summarize equations 1.39 to 1.57, together with the coefficients in Tables 5.8 and 5.9, completely specify the isotropic T_s model as a function of T_s, *W*, *V*, and *L*.

Thus, a detailed description of the Wentz model was provided to give the reader a view into the many detailed steps required to accurately retrieve geophysical parameters from passive microwave satellite data. Once developed, however, this model can be easily modified for additional sensors and used for retrievals. For this reason, RSS, the company founded by Frank Wentz, has become one of the primary processing centers for a wide variety of NASA and NOAA passive microwave satellite data. RSS makes these processed data and parameters readily available through NOAA and NASA online archive systems.

In the development of the model functions and retrievals, Wentz (1997) used a lot of observations. Wind measurements were used from buoys managed by (1) the US National Data Buoy Center (NDBC), (2) NOAA's Pacific Marine Environmental Laboratory (PMEL), and (3) the Japanese Meteorological Association (JMA). All of these data were collected for the 4-year period between 1987 and 1990. NDBC manages 75 moored buoys and 50 manned stations in the northeast Pacific, in the Gulf of Mexico, in the northwest Atlantic, near Hawaii, and one off the coast of Peru. Of these, Wentz (1997) used only the 42 stations that were at least 30 km from the coast. PMEL manages the TOGA-TAO array of 20 moored buoys in the equatorial Pacific. JMA operates four buoys off the Pacific coast of Japan. All together, this was a total of 66 sites measuring wind speed and direction (Fig. 5.30).

The sampling times are different for the various buoy types (Wentz, 1997). NDBC buoys sample for 8.5 minutes at 1-hour intervals, TOGA-TAO buoys take continuous measurements and report averaged wind at various intervals (1, 2, 6, and 24 hours) depending on the buoy's electronics (which are updated as the buoys are revisited and refreshed.), JMA buoys sample for 10 minutes at 3-hour intervals. For each buoy location, it was necessary to find the SSM/I overpasses for which any portion of the swath was within 30 km of the buoy. A temporal interpolation was used to specify the buoy wind speed at the time of the SSM/I overpass. In an effort to accommodate the different sampling times and intervals, a triangular weighting function was used as follows:

$$W_B = \frac{\Sigma_{j=1}^{N}(P - |\, t_{SSM/I} - t_{Bi} \,|)W_{Bi}}{\Sigma_{j=1}^{N}(P - |\, t_{SSM/I} - t_{Bi} \,|)} \text{ only for } |t|_{SSM/I} - t_{B\tau}| \le P \tag{5.25}$$

where W_{Bi} is the ith buoy wind taken at time t_{Bi}, $t_{SSM/I}$ is the SSM/I overpass time, and *P* is the time interval between the buoy observations. The summation is over all the buoy observations that are within *P* hours of the SSM/I time. This results

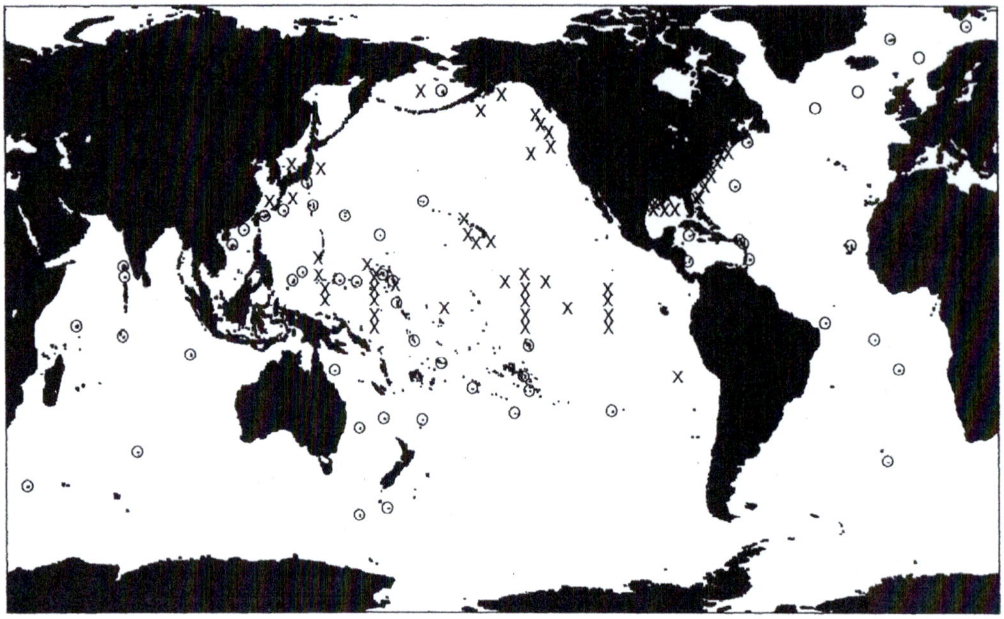

FIGURE 5.30 Locations of the buoys (x) and radiosonde site (open circle) used for calibrating the special sensor microwave imager brightness temperatures model algorithm (Wentz, 1997). Map lines delineate study areas and do not necessarily depict accepted national boundaries.

in the time window for the SSM/I-buoy collocations increasing with the sampling time interval. For the NDBC buoys, when the sampling interval is less than 3 hours P is set to 3. As a result, the minimum time window for collocation is ±3 hours, and the maximum time window is ±24 hours for the PMEL daily averages. For sampling intervals ≥3 hours, 30 is used, and it is equivalent to a linear interpolation in time using the two buoy winds that bracket the SSM/I overpass time. For shorter (<3 hours) periods, 30 is the average of the buoy winds within a ±3.0 hour window, weighted by the SSM/I-buoy time difference. This triangular weighting leads to rms values of $t|_{SSM/I}-t_{B\tau}$ having a typical value of one hour, which is in agreement with the 30 km spatial window.

All buoy observations were subjected to the usual set of quality control (QC) procedures, including checks for missing data, blank fields, and out-of-bounds measurements. Moreover, it needed to be within the specified time window, as when the winds vary by more than 10 m/s, then the SSM/I overpass is discarded, as this higher level of wind variability compromises the satellite-to-buoy comparisons. The anemometer heights H for the buoys varies. The NDBC buoys have H=5 or 10 m but some of the coastal manned stations have anemometers as high as 60 m. The PMEL buoys all have anemometers at 3.8 m above the sea surface, and the JMA anemometers are all at 7.5 m. To account for this variability, all buoy winds were normalized to 10 m assuming a logarithmic wind profile as given by the following:

$$W_{B,10\mathrm{m}} = \left[\ln\left(\frac{10}{z_0}\right)/\ln\left(\frac{H}{z_0}\right)\right] W_{B,H} \tag{5.26}$$

where z_0 is the surface roughness length, which equals 1.52E–4 m assuming a drag coefficient of 1.3E–3 (Peixoto and Oort, 1992).

The radiosonde observations (RAOB) for 1987–90 were obtained from the National Center for Atmospheric Research (NCAR). Since the accuracy of the SSM/I retrievals degrades when land is nearby, Wentz (1997) selected only those 55 radiosonde sites that are on weather ships or small islands (Fig. 5.30). For each RAOB site, software locates all the SSM/I overpasses for which any portion of the pass is within 60 km of the RAOB site. Most RAOBs are collected at 0 UTC and 1200 UTC. The imposition of too restrictive a time window would eliminate a lot of RAOB sites. For example, using a 1-hour time window would select only those sites with longitudes near 90°E and 90°W. Thus Wentz (1997) decided to use a 6-hour time window so that all the sites in Fig. 5.30 are included. It is more important to have a good global distribution of RAOB sites than to restrict things to near-simultaneous observations from a few sites. Using the ±6 hour window, Wentz (1997) simply took the RAOB measurement closest in time rather than averaging or interpolating observations.

An objective QC procedure is used to detect and discard incomplete and anomalous soundings. Each RAOB profile consists of a number of vertical levels, each level containing observations of pressure, temperature, and dew point depression. If any of these three parameters is missing, the corresponding level is discarded. Wentz (1997) defined the tropospheric RAOB levels as those for which the pressure is greater than 180 mbar. This cutoff of 180 mbar guarantees that the 200 mbar level is included. This first QC step is to identify and remove soundings for which the measurements are outside reasonable physical bounds. Only 3% of the RAOB soundings were rejected for this restriction.

The next QC step is to make sure that the sounding has a valid surface level reading, which is important as most of the water vapor is close to the surface. The RAOB data set contains an individual flag that identifies the surface level and indicates if it agrees with the in situ surface report. Any soundings with poor quality surface levels are eliminated from consideration. This removed about 8% of the soundings. In addition, Wentz (1997) discarded 1% that have surface pressures that are 30 mbar below the average for each site. There are also QC procedures that address the water vapor profile. Wentz (1997) required that there are at least seven levels in the troposphere and that the vertical gap between values is always less than 3 km. Moreover, he required that the highest tropospheric level has an air pressure less than 520 mbar and a water vapor pressure less than 0.5 mbar. Another 10% of the soundings were lost due to these restrictions. The QC procedure also discarded any large temperature spikes (>6 K) and vapor pressure spikes (>7 mbar). In some cases, these spikes may be real but it is better to remove them since the water vapor profile is probably not adequately sampled to get an accurate columnar vapor content. About 2% of the RAOB profiles were lost to this restriction.

The remaining soundings were then extrapolated from the elevation of the radiosonde station down to the sea surface. The sea surface air pressure and vapor pressure are found by assuming they vary exponentially with height. The assumed exponential decay rate for vapor pressure is $-0.63\ \text{km}^{-1}$, and the assumed air temperature lapse rate is $-5.8\ \text{Kkm}^{-1}$. These are both global averages for all of the soundings. Since all but 3 of the RAOB sites are at elevations less than 100 m the extrapolation to the sea surface is a small correction that adds about 3% to the total column water vapor. Another small correction is needed to account for the water vapor above the troposphere. Wentz (1997) did not use the RAOB measurements above the troposphere as he considered them as unreliable. Instead, he assumed an exponential decay rate of $-0.63\ \text{km}^{-1}$ and extrapolates up from the highest tropospheric level. This upward extrapolation up to 50 km typically adds about 0.2% to the columnar water vapor content.

The columnar water vapor V (mm) is computed by vertically integrating the water vapor profile as:

$$V = 10^{-3} \sum_{i=0}^{N-1} (h_{i+1} - h_i)\left(\frac{1}{4}\rho_{V,i} + \frac{1}{4}\rho_{V,i+1} + \frac{1}{2}\sqrt{\rho_{V,i}\rho_{V,i+1}}\right) \tag{5.27}$$

in which $\rho_{V,I}$ is the water vapor density (g/m^3) for the ith level, and h_i is the altitude (m) of the ith level. Level 0 is the sea level and level N is the last radiosonde level at 50 km. To specify the vapor density ρ_V as a function of air and dew point temperatures, Wentz (1997) used the formulation given by Liebe (1985). The altitude is found from the standard hydrostatic equations that give potential height as the integral of pressure and temperature (Peixoto and Oort, 1992). The factor of 10^{-3} is a conversion factor from units of grams per square meter to millimeters. This equation is a compromise between assuming that ρ_V varies linearly with h and that ρ_V varies exponentially with h between levels. In the case where relative humidity varies linearly with h, Eq. 5.27 is accurate to about 0.1%.

The RAOB data set is used to compute simulated brightness temperatures which has two advantages. First, it allows the computation of the effective air temperatures for oxygen and water vapor absorption. Second, it makes possible an error analysis where the simulated brightness temperatures are inputs to the retrieval algorithm. The upwelling and downwelling atmospheric brightness temperatures and the atmospheric transmittance are computed from the standard RTEs (Wentz, 1983):

$$T_{BU} = \int_0^H [\alpha_O(h) + \alpha_V(h)]\tau(h, H)\sec\theta dh. \tag{5.28}$$

$$T_{BD} = \int_0^H [\alpha_O(h) + \alpha_V(h)]\tau(0, h)\sec\theta dh. \tag{5.29}$$

$$\tau = \tau(0, H) \tag{5.30}$$

$$\tau(h_1, h_2) = \exp[-\int_{h_i}^{h} [\alpha_O(h) + \alpha_V(h)]\sec\theta dh]. \tag{5.31}$$

The integrals are from the sea surface (h=0) to h=50 km above which atmospheric absorption is negligible. Here, $T(h)$ is the air temperature, and $\alpha_0(h)$ and $\alpha_V(h)$ are the absorption coefficients for oxygen and water vapor. The function $\tau(h_1, h_2)$ gives the transmittance between altitudes h_1 and h_2 at the SSM/I incidence angle θ. The slant path through the atmosphere is given by sec θdh. The absorption coefficients are computed using the RAOB observations of pressure, temperature, and dew point depression using the formulations from Liebe (1985).

After applying all the Q/C procedures, 42,195 soundings remained. They were used with equations 5.28 through 5.30 to compute T_{BU}, T_{BD} and t for this set of radiosonde profiles. The effective upwelling and downwelling air temperature, T_U and T_D were then computed from (5.11a). Least squares regression (Eq. 5.12a and b) is then used to relate the air temperatures (T_U and T_D) to the RAOB columnar water vapor V and the sea surface temperature T_s. Plots of the downwelling air temperature resulting are shown here in Fig. 5.31. Corresponding plots for T_U are very similar and are not presented.

Sea surface temperature (T_s) is handled in much the same way except that T_{BU}, $T_{BD,}$ and τ are calculated from the RAOB profiles using Eq. 12a,12b–14. For this calculation, a clear sky and a wind speed of 7 m/s were assumed. Wentz (1997) used the Liebe regression for A_V (i.e., Liebe values of a_{V1} and a_{V2} in Table 5.10). The retrieved values are then compared with their true values to evaluate the algorithm's accuracy. This simulation establishes the retrieval error due to variations in the shape of the atmospheric profiles and due to approximations made in the regressions for T_U, T_D, $A_{0,}$ and A_V. The rms values of the retrieval errors in W, V, L, ΔT_{BV}, and ΔT_{BH} are given in the first column of Table 5.11 for the atmospheric model error. Moreover, in this table are the other error sources that contribute to the overall observed rms variability of W, V, L, ΔT_{BV}, and $\Delta T_{BH.}$

Wentz (1997) found a relationship between effective air pressure P of the water vapor column and the water vapor volume. An increase in P broadens the water line and thereby increases the absorption at 19 and 37 GHz and decreases it at 22 GHz. As an example, a pressure increase of 10 mbar changes the 19, 22, and 37 GHz water vapor absorptions by +0.3%, −0.9%, and +0.8%, respectively. Globally, the water vapor column height tends to increase with increasing V, and therefore P tends to decrease with increasing V. We can relate P (mbars) to V (mm) as follows:

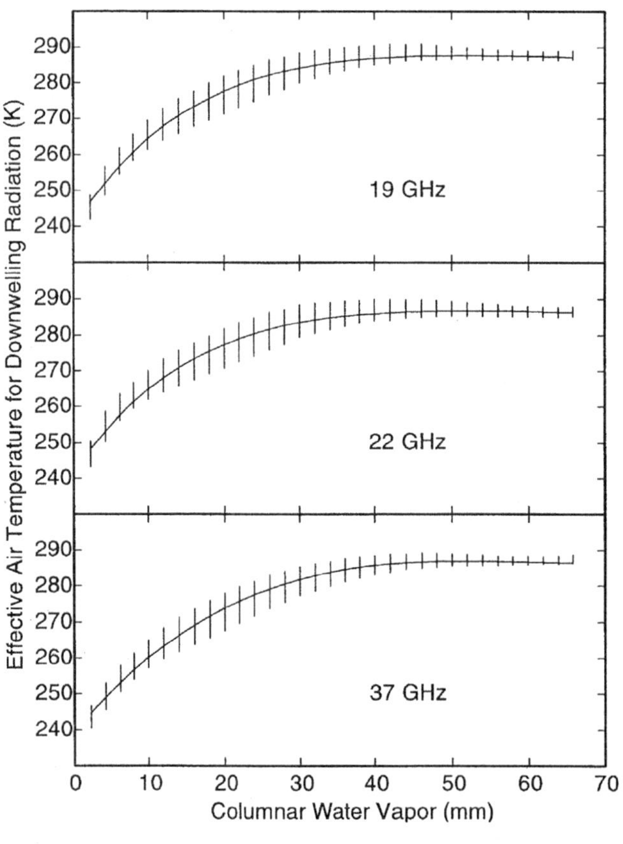

FIGURE 5.31 The effective air temperature T_D for downwelling radiation plotted against the radiosonde observations columnar water vapor. The solid lines are the model values, and the vertical bars represent ±1.0 standard deviation of T_D derived from the radiosonde profiles (Wentz, 1997).

TABLE 5.11 Root-mean-square error budget for the retrieved parameters.

Retrieval	Atmospheric model	Wind direction	Radiometer noise	Sampling mismatch	Other	Total observed
W (ms$^{-1)}$	0.51	0.35	0.53	0.94	0.41	1.31
V (mm)	0.81	0.21	0.43	3.68	0.74	3.87
L (mm)	0.019	0.004	0.007	0	0.014	0.025
ΔT_{BV} (K)	0.50	0.27	0.42	0	0.56	0.90
ΔT_{BH} (K)	0.54	0.20	0.60	0	0.59	1.02

W is near-surface wind speed; V is columnar water vapor; L is columnar cloud liquid water; and T_{BV} and T_{BH} are vertical and horizontal brightness temperatures, respectively.

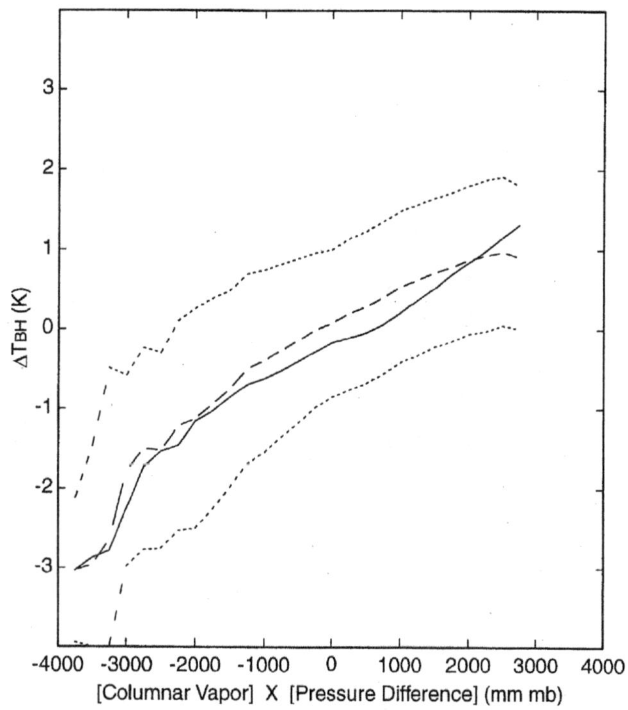

FIGURE 5.32 The T_{BH} modeling error produced by variations in air pressure P. The model error in T_{BH} is plotted versus the columnar water vapor V time the air pressure variation. The solid line In Fig. 5.32 the solid line represents theoretical simulations while the long-dashed line is the mean of the special sensor microwave imager retrievals, and the short-dashed lines are the ±1 standard deviation envelope (Wentz, 1997).

$$P = 860{-}1.15V \qquad (5.32)$$

This global correlation of P and V is given by the relation, but variations from the typical values given by (5.32) are not accounted for in the RTE model, and these P variations will show up as modeling errors in ΔT_{BV} and ΔT_{BH}. The relationship between ΔT_{BH} and the product of ΔP times V (ΔP=the difference between the effective pressure for a given RAOB profile minus the typical pressure given by (2.30)) is presented here in Fig. 5.32.

At very low pressures the true water vapor line is narrower than that assumed by the model, resulting in overcorrections at 19 and 37 GHz. These overcorrections show up as negative values of T_{BH} at these low pressures. The curve for T_{BV} is very similar. The dependence of ΔT_{BH} on the effective pressure is also seen in the actual SSM/I measurements. The long dashes in Fig. 5.32 are the retrieved values from the SSM/I observations, and the short-dashed lines are the ±1 standard deviation of the retrieved ΔT_{BH}.

Variations in P lead to errors in the SSM/I retrieval of liquid water L due to overcorrecting or undercorrecting for the water vapor absorption at 37 GHz. This error can be as large as 0.1 mm for the extreme case where $V\Delta P_V$=− 4000 mm mbar. To minimize this error, Wentz (1997) used the following correction for the 37 GHz retrieval:

$$A_{L37,\mathrm{corr}} = A_{L37,\mathrm{ret}} - 0.003\Delta T_{\mathrm{BH}} \tag{5.33}$$

Using this correction, the appropriate liquid water content can be found from Eq. (5.27). This correction is applied after the retrieval algorithm is run, and it reduces the error in *L* due to *P* variations by a factor of 2. The errors in *W* and *L* due to *P* variations are relatively small (0.2 m/s and 0.4 mm), so no additional correction is made. The information on *P* as contained in ΔT_{BH} might reveal anomalies in the shape of the water vapor profile. Schulz et al. (1993) suggest that SSM/I observations contain useful information on the shape of the water vapor profile.

Returning to the T_s model, some of the coefficients are derived from the collocated SSM/I-buoy and SSM/I-RAOB data. This set of coefficients is listed here in Table 5.12.

Now, we let the difference between the SSM/I wind and the buoy wind be ΔW and the difference between the SSM/I water vapor and the RAOB water vapor be ΔV. Then the model coefficients are derived such that: (1) The mean value of ΔW is 0 over the full range of T_s, *W*, *V*, and *L*, (2) the mean value of ΔV is 0 over the same full range, (3) the mean v_{pol} residual ΔT_{BH} is 0 over the same full range of T_s, *W*, *V*, and *L*, (4) the mean h_{pol} residual ΔT_{BH} is 0 over the same full range, and (5) the *L*=0 point of liquid histograms is centered on the steep lefthand side of the histogram and does not vary over the full range of T_s, *W*, and *V*.

Together, this makes for 19 conditions. Sets 1 through 4 each have four conditions, while set 5 has three conditions. Deriving the model coefficients in this way is known as inverse modeling since the model derivation is based on the outputs of the model's inverse (the retrievals).

In steps 1 through 4 the value of T_s comes from climatology (Shea et al., 1990). Moreover, the values for *W*, *V*, and *L* come from the SSM/I retrieval algorithm except for the following two conditions. For set 1, *W* is the average of the SSM/I wind and the corresponding buoy wind and for set 2, *V* is the average of the SSM/I vapor and the RAOB vapor. This allows the model retrievals and in situ values to have similar weights in the ΔW versus *W* and the ΔV versus *V* curves. Condition 5 requires that the lefthand side half-power point of the *L* probability density function (pdf) be at *L*=0 for all T_s, *W*, and *V*.

The principal conditions to derive each component are given in Table 5.12 in functional form. For example, the condition that ΔW is 0 over the range of T_s is denoted in the expression $\Delta W(T_s)=0$, and the condition that the liquid water histograms are aligned for all *V* is given by $L_0(V)=0$. Some coefficients are not strictly given by the expressions in Table 5.12 but by other conditions as well. In addition, there are more conditions to satisfy than there are coefficients. Thus the coefficient derivation problem is overdetermined and a priori there is no guarantee that the model will be able to accurately satisfy all conditions.

TABLE 5.12 Conditions for deriving coefficients in T_s mode.

Coefficient	Channel	Condition
α_{V1}, α_{V2}	19	$\Delta T_{\mathrm{BV}}(V)=0$
α_{V1}, α_{V2}	22	$\Delta V(V)=0$
α_{V1}, α_{V2}	37	$L_0(V)=0$
m_1, m_2	19V	$\Delta T_{\mathrm{BV}}(W)=0$
m_1, m_2	19H	$\Delta T_{\mathrm{BH}}(W)=0$
m_1, m_2	22V	Interpolation between 19V and 37V
m_1, m_2	22H	Interpolation between 19H and 37H
m_1, m_2	37V	$L_0(W)=0$
m_1, m_2	37H	$\Delta W(W)=0$
$\varepsilon_0, \varepsilon_1, \varepsilon_2$	19H.22H.37H	$\Delta W(T_s)=0$, $\Delta T_{\mathrm{BH}}(T_s)=0$

Before using the model to retrieve the coefficients, the SSM/I-buoy and SSM/I-RAOB data sets are subjected to one final QC procedure. Observations near land that are affected by rain are discarded. If the land contamination is greater than 0.2 K, then the observation is dropped. Most of the observations are completely free of land contamination. For the buoys, which are primarily in the open ocean, Wentz (1997) used a very tight restriction of 30 km radius around the buoy. For the RAOBs, which originate from mostly small islands, Wentz (1997) used a larger window of 60 km in an effort to avoid the island. Moreover, when $L>0.18$ mm drizzle or light rain is likely; then Wentz used the requirement that $L \geq 0.18$ to indicate that rain is present, and the observation is discarded. The wind speed retrieval is very sensitive to rain. This excluded about 22% of the buoy observations. The water vapor retrieval is much more robust and is not seriously affected by light rain. Thus, for the SSM/I-RAOB data set, an observation is excluded only when $L \geq 0.5$ mm, which excludes about 4% of the observations.

After all of the QC adjustments, Wentz was left with 37,650 SSM/I overpasses of buoy sites and 35,108 overpasses of RAOB sites. For each overpass there are 4–5 and 6–7 high quality SSM/I data sets within a 30 km radius of the buoys and 60 km of the RAOB sites. This resulted in 167,264 SSM/I-buoy matchups and 238,627 SSM/I-RAOB matchups. The coefficients are then found by varying their values until the 19 conditions are satisfied. Earlier studies provided good starting points, and coefficients are adjusted by trial and error until the conditions are met. In addition to determining the model coefficients, a set of five TB offsets are found so that the global mean values for ΔW, ΔV, ΔT_{BV}, and ΔT_{BH} are 0 and the global cloud histogram is positioned at $L=0$.

The levels to which the model and algorithm fulfill the 19 conditions are produced by first binning the data and then computing the mean and rms statistics for each bin. An ideal algorithm would produce flat curves along the zero axis over the entire range. Deviations from a flat curve are a measure of systematic errors or crosstalk among the retrievals. The total rms variations are as the bracketing curves, and the values are in the rightmost column of Table 5.12. It should be noted that the systematic errors (deviations of the solid curves in Fig. 5.33 from the zero baseline) in ΔW, ΔV, ΔT_{BV}, and ΔT_{BH} are smaller than the rms errors. The typical systematic errors in ΔW, ΔV, ΔT_{BV}, and ΔT_{BH} are −0.3 m/s, 0.6 mm, 0.3 K, and 0.3 K, respectively. There are a few cases for which the systematic error is much larger than these values, such as the 1.7 mm drop in ΔV that occurs at $V=62$ mm. These larger errors occur for bins with fewer observations, such as the ΔV dip, which has only 1271 observations, whereas the typical bin has about 7,000 observations.

The remaining three conditions are shown here in Fig. 5.34 where the liquid water pdfs are separated into T_s, W, and$V$. The top panel of Fig. 5.34 shows six different histograms for six different ranges of sea surface temperature (SST; i.e., 0°C–5°C, 5°C–10°C, 10°C–15°C, 15°C–20°C, 20°C–25°C, and 25°C–30°C). The middle and bottom panels of Fig. 5.34 show analogous PDFs for levels of wind and water vapor. All the PDF peaks are at about $L=0.025$ mm and at $L=0$ THE PDFs are about half the peak value. Wentz (1997) used this half power point (i.e., 0.025 mm) as an indicator of the rms error in L as can be found in Table 5.11. To specify the systematic error in L, Wentz used the alignment of the left side of the histograms, which is at about 0.005 mm. Regarding the cloud liquid water accuracy, Wentz assumes that the average temperature T_L of the cloud is precisely known. In the model T_L equals the mean temperature between the surface and the freezing level. When the actual cloud temperature differs from T_L, there will be an additional error in L which can be computed from Equation 1.50. As an example, an error of 4 K in T_L leads to a 10% error in L.

Wentz (1997) compared the water vapor absorption coefficients a_{v1} and a_{v2} derived from the model with those derived from Liebe's (1985) expression. Both sets of coefficients are entered into Table 5.11. At 19 and 22 GHz, the Wentz model coefficients agree well, with the Liebe coefficients with the Liebe values being slightly higher. At 37 GHz, however, the Liebe absorption is about 20% higher than those from the model. Wentz tested this case by putting the Liebe values into the retrieval algorithm, which significantly displaced the water vapor PDFs given in Fig. 5.34, suggesting an overcorrection for water vapor at 37 GHz. Wentz suspects the water vapor continuum used by Liebe as the cause of this problem.

To test the sensitivity of the inverse modeling to variations in the in situ data, Wentz carried out numerical experiments where he used only half of the in situ data to determine the model coefficients leaving the other half of the data to test the algorithm's performance. As an example, he randomly selected data for every other day and compared them. Because the data set is so large, the random division into two halves had essentially no effect on the values of the model parameters nor the performance statistics. This confirms that the calibration data set is sufficiently large enough to ensure a stable derivation of the coefficients.

While the SSM/I was not designed to retrieve wind direction, Wentz (1997) accounts for wind direction in his retrieval model. Having buoy wind measurements that include wind speed and direction it is possible to take the 37,650 SSM/I-buoy matchups to evaluate the influence of wind direction on the retrieval accuracy. The angle ϕ is the SSM/I look angle minus the buoy wind direction, with $\phi=0$ corresponding to an upwind observation. We present in Fig. 5.35 the line-of-sight wind retrieval W_{LS} plotted again the relative wind direction ϕ for all of the SSM/I-buoy matchups.

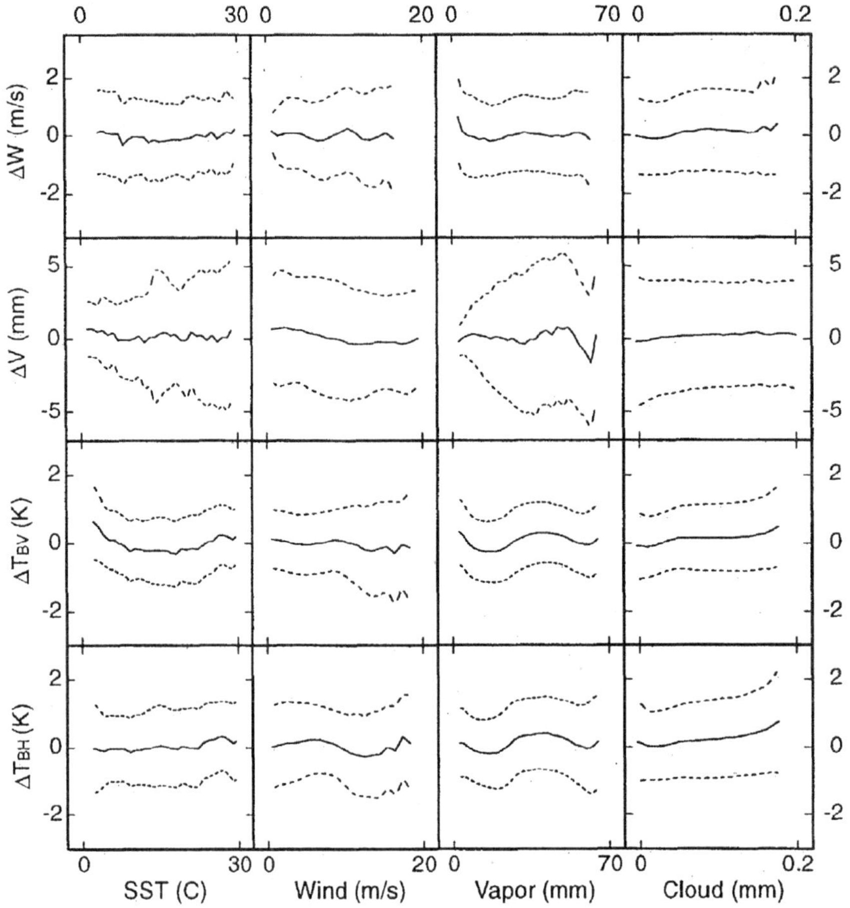

FIGURE 5.33 Special sensor microwave imager (SSM/I) retrievals for 37,650 overpasses of buoy sites and 35,108 overpasses of RAOB sites. The SSM/I—buoy wind difference ΔW, the SSM/I—RAOB vapor difference ΔV, and the V_{pol}, H_{pol} residuals ΔT_{BV} and ΔT_{BH} are plotted again sea surface temperature, wind speed, columnar water vapor, and columnar cloud water. The solid lines are the mean values, and the dashed lines represent ±1 standard deviation (Wentz, 1997).

In Fig. 5.35, the data have been averaged into bins that are 15 degrees in width. The solid line is the buoy reported wind speed (W_{BUOY} cos ø) for the directions, while the long-dashed line is the mean of the W_{LS} retrievals. As before, the short-dashed lines represent the ±1 standard deviation of the W_{LS} retrievals. The mean W_{LS} agrees fairly well with the buoy values except that the downwind minimum of the retrieved W_{LS} is flatter than that for a cos ø function. This deviation is due to the influence of the h_{pol} directional signal on the retrieval. The signal-to-noise ratio (SNR) for the W_{LS} retrieval is very weak. If, however, W_{LS} is averaged over sufficiently large spatial and temporal scales then the SNR increases, and useful information can be retrieved such as monthly SSM/I wind vector maps (Wentz, 1992). In addition, SSM/I swath images show that at high wind speeds the retrieved W_{LS} shows the general direction of the wind relative to the SSM/I look direction.

An advantage to include W_{LS} in the retrieval algorithm is that it reduces the retrieval error in the wind speed due to variations in the wind direction. This is clearly demonstrated in Fig. 5.35 that plots retrieval wind speed versus wind direction for retrievals done with and without the inclusion of W_{LS} in the retrieval. To show the difference more clearly, Fig. 5.36 is plotted only for wind speeds greater than 7.5 m/s. It should be noted that by including W_{LS} in the retrieval the wind error is reduced by a factor of 3. When all observations (both low and high winds) are included in the statistics the wind error due to wind direction is about 0.5 m/s with a rms effort of 0.35 m/s. Variations in wind direction also produce errors in the other retrieval parameters. Computer simulations indicate that the rms errors in V, L, ΔT_{BV}, ΔT_{BH} due to wind direction are 0.21 mm, 0.004 mm, 0.27 K, and 0.20 K, respectively. These values are listed in Table 5.12.

The two remaining errors given in Table 5.12 are the error due to radiometer noise and the error due to the spatial–temporal sampling mismatch between the SSM/I footprint and in situ point observations. The radiometer noise when averaged for the F08 SSM/I is given as the error in T_A measurement noise of 0.4 K for both 19 and 22 GHz and

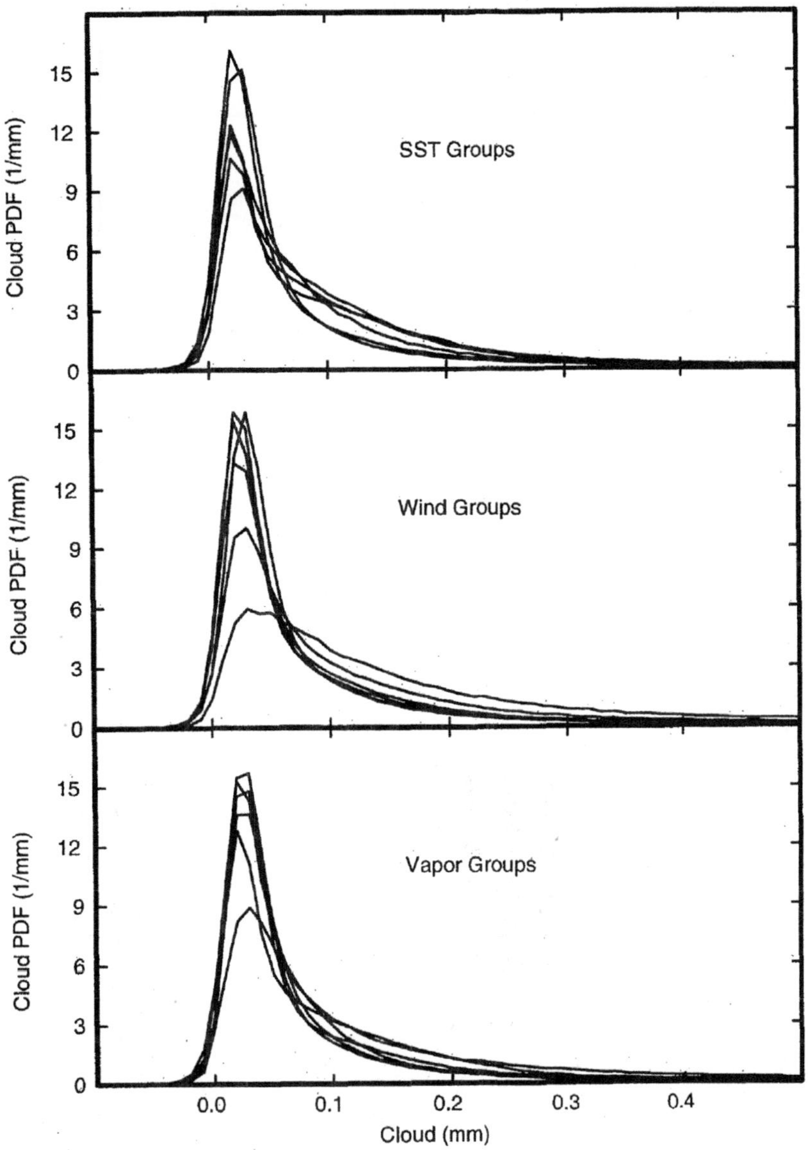

FIGURE 5.34 Probability density functions (PDFs) for liquid cloud water. The PDFs are separated by sea surface temperature, wind speed, and water vapor. Each curve expresses the PDF for a particular stratification (Wentz, 1997).

0.2 K for 37 GHz. The values in Table 5.12 are the rms retrieval errors resulting from the T_A measurement noise assuming clear skies and a 7 m/s wind speed.

The more difficult error to estimate is the spatial–temporal sampling error. To estimate the SSM/I versus buoy wind sampling error Wentz assumed the wind field has a linear gradient in space that is advecting in time. He let the x-axis be in the direction of the spatial gradient and v be the velocity of the advection. Then, the wind at a position x and a time t is

$$W(x, t) = W_0 + g(x - vt\cos\phi) + \varepsilon \tag{5.34}$$

where g is the gradient (m/s/km), and ϕ is the angle of the advection relative to the x-axis. The ε represents a random wind component. Wentz used an average advection speed of v=28.8 km/hour (8 m/s). When averaged over the 50 km scales of the SSM/I footprint the mean of ε is 0, and its variance $\langle \varepsilon^2 \rangle$ is independent of position and time. Thus the second term in 5.34 accounts for variations in the wind field on spatial scales equal to the SSM/I footprint, and the third term in this equation accounts for the wind variations on spatial scales smaller than the footprint. Wentz let the buoy observation at x=0, t=0 be

$$W_B = W_0 + \varepsilon \tag{5.35}$$

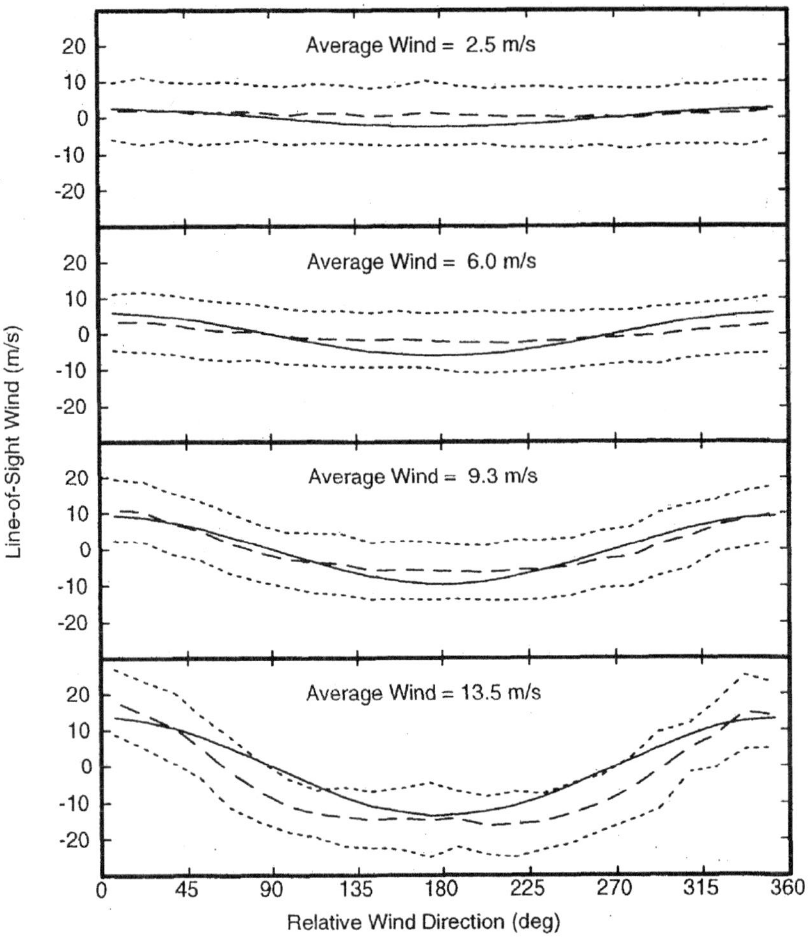

FIGURE 5.35 The line-of-sight wind W_{LS} from special sensor microwave imager versus the relative buoy wind direction (deg). The solid lines show the buoy W_{LS}. The long-dashed curves are the mean value of the W_{LS} retrievals, and the short-dashed lines represent the ± standard deviations for the W_{LS} retrievals. Results are presented for four different average wind speeds (Wentz, 1997).

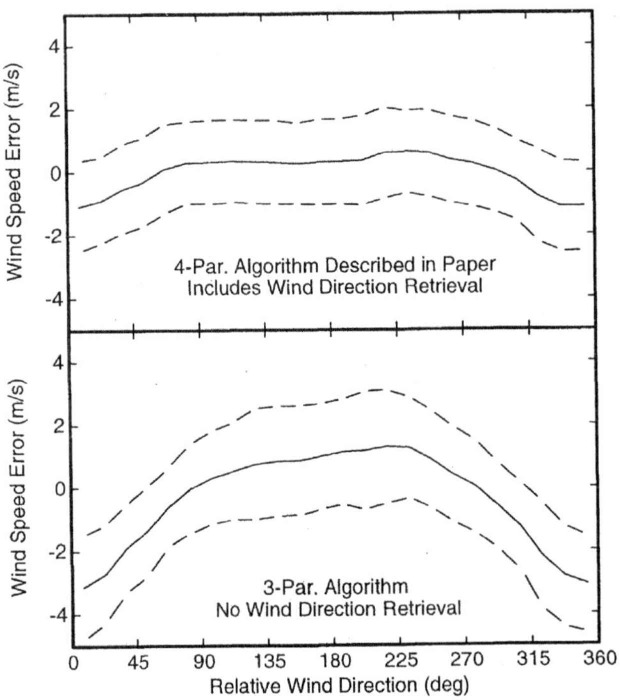

FIGURE 5.36 The error in wind speed due to wind direction variations. (top) Wind speed error versus wind direction for a retrieval, including W_{LS}. (bottom) Results when the retrieval algorithm does not include W_{LS}. The solid lines represent the mean, and the dashed lines are ±1 standard deviation of the error (Wentz, 1997).

Here, the SSM/I observation time is t, and the footprint center is at a position $x=r\cos\psi$ and $y=r\sin\psi$ where r is the radial distance from the buoy to the footprint center. Averaging $W(x,t)$ over the footprint gives the SSM/I wind speed:

$$W_S = W_0 + g(r\cos\psi - v\ r\cos\phi) \tag{5.36}$$

Now considering an ensemble of SSM/I observations over a buoy site, the distribution of angles ψ and ϕ will be fairly uniform over 0 to 2π. Let $\Delta W=W_S-W_B$ and find the ensemble average of ΔW^2 to be

$$\langle \Delta W^2 \rangle = \frac{1}{2} g^2 R^2 + \langle \varepsilon^2 \rangle \tag{5.37}$$

$$R^2 = \langle r^2 + v^2 t^2 \rangle \tag{5.38}$$

For the buoy comparisons, the spatial window of 30 km and the triangular time weighting method results in R=36 km. Thus the spatial–temporal sampling error can be estimated using Eq. 5.37 once we have the gradient g and the interfootprint spatial variability $\langle \varepsilon^2 \rangle$. To estimate g, we use the wind speed spatial variability observed by the SSM/I. Using Eq. 5.37, the wind speed difference between neighboring SSM/I footprints is simply $\Delta W_S=g\ r\cos\psi$, where r is now the radial distance between footprints. Taking the ensemble average of ΔW_S^2 over many observations separated by a distance r gives

$$\sqrt{<\Delta W_S^2>} = gr/\sqrt{2} \tag{5.39}$$

The rms values of ΔW_S derived from the SSM/I are plotted against r for three different wind speeds in Fig. 5.37.

As expected, the spatial variability increases with wind speed and for the larger separations in r. The curves flatten out as they approach the decorrelation length for winds on the ocean. The radiometer noise will contribute to the SSM/I error, so Wentz subtracted in a root-mean-square-sense, the radiometer noise component. Wentz used a 60 km separation to specify g. The curve for W=8 m/s at r=60 km gives a value of 0.95 m/s for the rms wind speed difference. Substituting this value into Equation 1.52 gives the value of 0.022 m/s km for g.

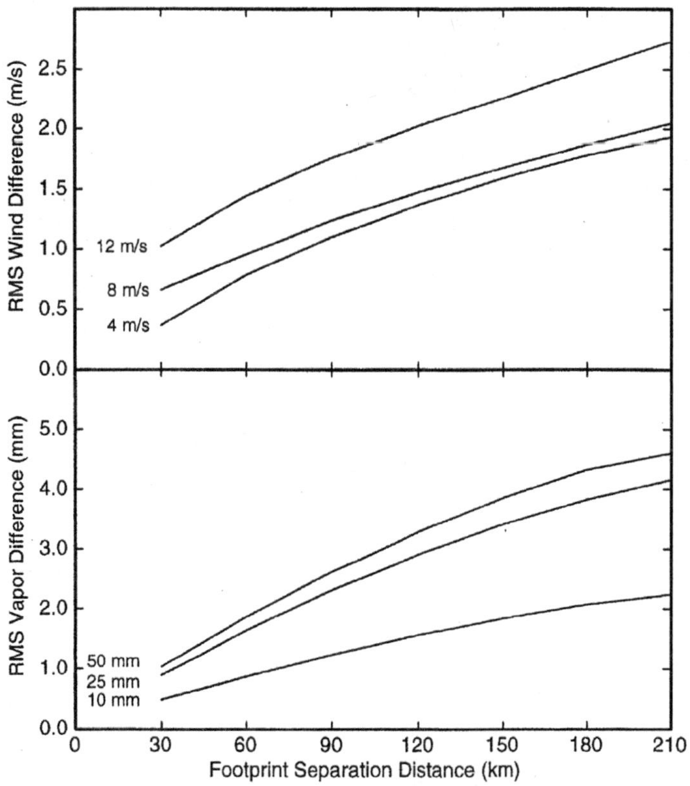

FIGURE 5.37 The rms difference of special sensor microwave imager retrievals of wind speed and water vapor for neighboring footprints. The three curves in each panel indicates the results for low, medium, and high values of wind speed and water vapor (Wentz, 1997).

To get an estimate for $\langle \varepsilon^2 \rangle$ Wentz examined the wind speed temporal variability measured by the buoys. Using Eq. 5.37, we can write the difference in wind speed at the same buoy but for two different times as

$$\Delta W_S = g \;\; v \;\; t \;\; \cos \phi + \varepsilon_t - \varepsilon_0 \tag{5.40}$$

where t is the time separation between the buoy observations and ε_0 and ε_t are the random wind components at times 0 and t. Taking the ensemble average of $\Delta W_S{}^2$ over many observations separated by time t gives us

$$\langle \Delta W_B{}^2 \rangle = \frac{1}{2} g^2 \langle v^2 t^2 \rangle n + 2\langle \varepsilon^2 \rangle \tag{5.41}$$

The rms value of ΔW_S from the buoy data set was 1.17 m/s for a t=1 hour separation. Using Eq. 5.41, we can now estimate the rms value of the SSM/I minus buoy wind speed difference due to the spatial–temporal sampling error, which is $<\Delta W^2>$=0.94 m/s. These results are very similar to the wind speed spatial variability reported by Monaldo (1988).

The sampling error for the RAOB comparisons is calculated in the same way. For the RAOB comparisons, the space-time window is 60 km and 6 hours, which then results in an R=122 km. The rms vapor differences are shown in the bottom panel of Fig. 5.37. For the RAOBs, the shortest time interval is 6 hours and the rms difference for this period is 4.8 mm. To be consistent with this time separation of 6 hours, Wentz used the spatial gradient g that corresponds to an 1,880 km separation distance when computing $\langle \varepsilon^2 \rangle$. When computing the RAOB-SSM/I sampling-error, he used the spatial gradient that corresponds to a separation distance of 122 km. For a V=25 mm, which is the average of the RAOB data set, he finds a g=0.343 mm/km for a separation of 120 km and 0.03 mm/km for a separation of 180 km. He then finds $\langle \varepsilon^2 \rangle$ to be 2.18 mm and the RAOB-SSM/I sampling-error to be 3.68 mm.

At this point, you are probably wondering why we went through such a detailed discussion of this retrieval method. As I mentioned earlier, I wanted the readers to realize what goes into the retrieval of parameters from passive microwave measurements. Fortunately, Frank Wentz and his company RSSs have been contracted to process the majority of microwave data collected from satellite radiometers. Once retrieved using the RSS model the parameters as well as the original data are made freely available through a variety of national data systems such as the Physical Oceanography Distributed Active Archive Center (PO-DAAC) operated by NASA's Jet Propulsion Lab in California.

5.3.3 Follow on sensors to special sensor microwave imager

5.3.3.1 The special sensor microwave-temperature

The SSM/I spawned a number of follow on instruments all of which carried the special sensor microwave (SSM) part of their title. This was because they were intended to fly on future DMSP satellites. The first of these was the SSM-temperature (SSM/T) intended for atmospheric temperature sounding in nearly all-weather conditions. It was a 7-channel microwave radiometer covering the 54.0–59.5 GHz band (Table 5.13). The instrument is a cross-track, nadir scanning radiometer with a FOV of 14.4 degrees. At the orbital altitude of 833 km, the subtrack spatial resolution is an approximate circle of 174 km diameter at nadir. There are seven cross-track scan positions separated by 12 degrees with a maximum

TABLE 5.13 SSMT channels.

Channel #	Central frequency (GHz)	Bandwidth (MHz)	Polarization	NEΔT (K)
1	50.50	400	H	0.60
2	53.20	400	H	0.40
3	54.35	400	H	0.40
4	54.990	400	H	0.40
5	58.40	350	H	0.50
6	58.825	300	H	0.40
7	59.40	250	H	0.40

cross-track scan angle of 36 degrees. At the far end of each scan, the resolution degrades to an ellipse of 213 X 304 km. The SSMT swath is about 1.500 km creating a data gap between successive orbits.

An example of an SSM/T, atmospheric profile is presented here in Figure 1.40, which is a series of atmospheric temperature layers taken when Typhoon Zelda was present on November 1, 1994 (Fig. 5.38).

Goldberg et al. (1998) used pattern recognition to derive SSM/T2 moisture profiles. The algorithm uses pattern recognition to select the first guess to be used in a regression based moisture profile retrieval. It also uses an adjustment procedure to remove cloud liquid water contamination from the SSM/T2 pixels. The SSM/T2 channels are given in Table 5.14.

The SSM/T2 measurements are influenced by the surface emissivity, the temperature profile, the atmospheric transmittance and the moisture signal. In the SSM/T2 measurements arrive from the strong moisture dependence on the atmospheric transmittance in the SSM/T2 spectral regions.

The approximate maximum water vapor burden in each channel is given in Table 5.14 along with the atmospheric pressure. Measurements from the 91 and 150 GHz channels are highly correlated with total precipitable water (TPW) over the ocean due to the particularly strong signal from the surface. The relationships between the TPW and simulated 91 and 150 GHz measurements are presented here in Fig. 5.39 for an assumed ocean emissivity of 0.65 and zero cloud liquid water. As TPW decreases, the surface contribution increases. Since the surface emissivity is relatively low, the signal in the measurement due to a change in water vapor is very high.

Moisture retrieval from the SSM/T2 involves the following steps:

1. The measurements are limb adjusted to nadir to remove angular dependence.

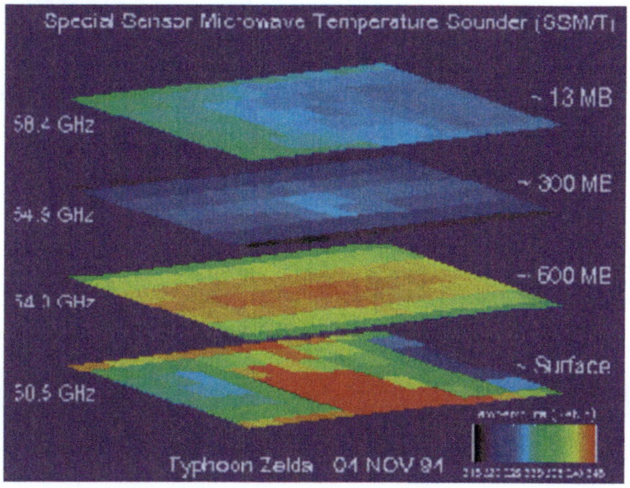

FIGURE 5.38 Special sensor microwave-temperature temperatures for Typhoon Zelda.

TABLE 5.14 Contains SSM/T2 channel frequencies, maximum water vapor burden sensitivity, and the pressure of maximum atmospheric contribution for typical tropical and midlatitude atmospheres.

Channel	Frequency (GHz)	H_2O burden (cm)	Atm max contribut.
1	183±3	0.2	400 hPa tropics/650 hPa Mid latitude
2	183±1	0.05	300 hPac tropics/500 hPa mid latitudes
3	183±7	0.8	500 hPa tropics/800 mid latitudes
4	91.65	8.0	Surface
5	150	3.0	700 hPa tropics/900 hPa mid latitudes

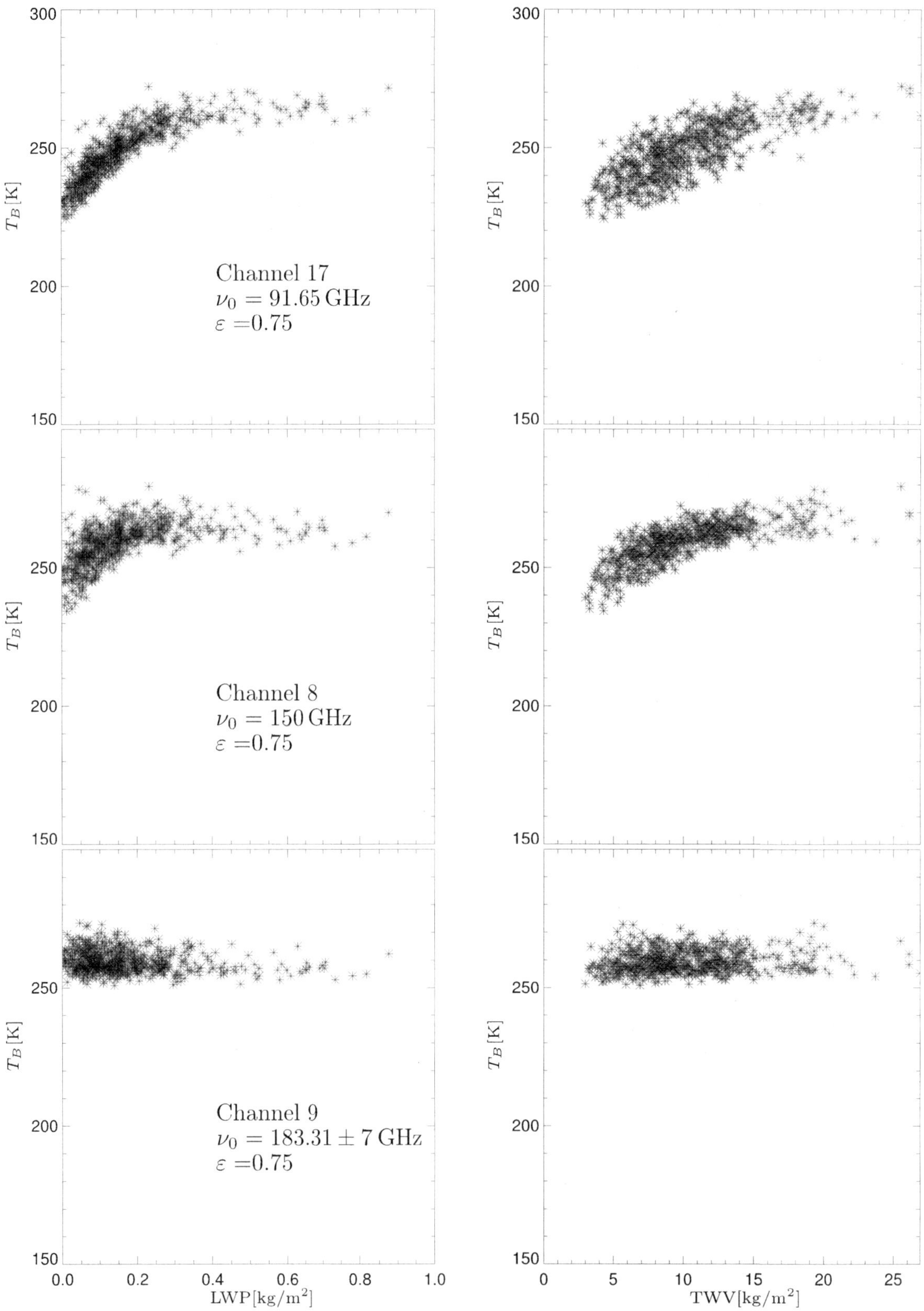

FIGURE 5.39 Scatter between simulate T2 brightness temperature and total precipitable water for the 91 (left) and 150 (rights) GHz channels assuming zero cloud liquid water and surface emissivity of 0.65.

TABLE 5.15 The final retrieval errors each type of retrieval.

	Library (%)	Library+Regression (%)	Regression only (%)	Mean water burden (cm)
300–500 hPa	58	58	67	0.102
500–700 hPa	43	42	51	0.385
700–1,000 hPa	21	20	25	1.810

2. Cloud liquid water is computed for precipitation screening.
3. The measured brightness temperatures are adjusted to "look like" computed brightness temperatures for zero cloud liquid water and a fixed emissivity.
4. A pattern recognition library search moisture retrieval is derived and used as the initial estimate in a regression algorithm.

Goldberg et al. (1998) focused on steps 3 and 4. Details of the limb adjustment and cloud liquid water computation can be found in Wark (1993) and Grody (1991), respectively. Goldberg et al. (1998) notes that the regression retrieval was only temporary, while an accurate and rapid RTM is to be developed as found in (McMillin et al., 1995).

A very important step before the library search and regression retrievals was removing the effects of cloud liquid water and variable surface emissivity. A suite of collocated radiosonde profiles and satellite observations was used to calculate the brightness temperatures that were computed for zero cloud liquid water and a fixed emissivity of 0.65 over the ocean. Regression was then used to calculate the brightness temperature from the SSM/T2 observed brightness temperatures. Moreover, the surface temperature was derived from AVHRR and HIRS infrared data. Rather than using ordinary regression, which may give a particular channel more weight than physically expected, a constrained regression was used so that the channel being adjusted was the most significant predictor. More information on the regression can be found in Crone et al. (1996). The retrieval errors are given in Table 5.15.

These retrieval errors appear large, but most of the error is due to the collocation problem. For example, retrievals based on the same ensemble but using a simulation result in the library search results in errors of 22%, 23%, and 5% for the 300–500, 500–700, and 700–1000 hPa layers, respectively. Normally, an estimate of the atmospheric temperature is needed for water vapor retrievals, which should have been provided by the DMSP SSM/T1 temperature sounding that unfortunately malfunctioned. The temperature data required by the library approach generally has a much smaller effect that the moisture signal. If they had had perfect temperature information, their results would have improved by about 2%–5%.

5.3.3.2 *The special sensor microwave imager/sounder*

The SSMIS combined the imaging and sounding capabilities of the earlier instruments. It covered 21 frequencies with 24 channels covering the 54–183 GHz bands (Table 5.16).

The scanning mechanism was conical with a 53.1 degrees zenith angle over a 1700 km swath and a scan rate of 31.0 scans/min for 12.5 km/scan. It provided global coverage once per day. It flew on DMSP satellites since 2003.

Kunkee et al. (2008) discuss the design and evaluation of the very first special sensor microwave imager/sounder (SSMIS) instrument (Fig. 5.40). They state that as originally conceived, the SSMIS integrates the imaging capabilities of the SSM/I and the profiling capabilities of the SSM/T and SSM/T-2 into a single conically scanning 24-channel instrument with extended sounding capability to also profile the mesosphere. The first SSMIS was onboard DMSP satellite F-16 launched by the last Titan II vehicle from Vandenberg, AFB, CA. The SSMIS was built by the Northrop-Grumman Electronics System (NGES) Company under the direction of the Air Force DMSP Office. The SSMIS represents a collaboration between the Air Force and the Navy to provide synoptic maps of critically important atmospheric, oceanic, and land parameters to support numerical weather prediction (NWP) for both military and civilian users.

The scan geometry of the SSMIS is shown in Fig. 5.41.

The bandwidths and polarizations of SSMIS were selected based on the earlier SSM/sensors. The lower air temperature sounding channels 1–7 are similar to those in SSM/T, while the frequencies, passbands, and stabilities of channels 8–11 are nearly the same as those from SSM/T-2. Based on the success of the SSM/I instrument, the same frequencies and

TABLE 5.16 Special sensor microwave imagery sounder channels.

Number	Central frequency (GHz)	Bandwidth (MHz)	Polarization	NEΔ*T* (K)	IFOV (km)	Pixels (km)
12	19.35	355	H	0.34	42.4X70.1	25.0X12.5
13	19.35	356.7	V	0.34	42.4X70.1	25.0X12.5
14	22.235	407.5	V	0.45	42.4X70.1	25.0X12.5
15	37.0	1545	H	0.24	27.5X44.2	25.0X12.5
16	37.0	1615	V	0.24	27.5X44.2	25.0X12.5
1	50.3	380	H (V F16)	0.21	17.5X25.8	37.5X12.5
2	52.8	388.8	H (V F16)	0.20	17.5X25.8	37.5X12.5
3	53.596	380	H (V F16)	0.21	17.5X25.8	37.5X12.5
4	54.4	382.5	H (V F16)	0.20	17.5X25.8	37.5X12.5
5	55.5	391.3	H (V F16)	0.22	17.5X25.8	37.5X12.5
6	57.29	330	RC	0.26	17.5X25.8	37.5X12.5
7	59.4	238.8	RC	0.25	17.5X25.8	37.5X12.5
24	60.79±0.36	26.4816	RC	0.38	17.5X25.8	75X12.5
23	60.79±0.36	7.312	RC	0.37	17.5X25.8	75X12.5
22	60.79±0.36	2.6239	RC	0.58	17.5X25.8	75X12.5
21	60.79±0.36	1.2883	RC	0.86	17.5X25.8	75X12.5
20	60.79±0.36	1.3556	RC	1.18	17.5X25.8	75X12.5
19	63.28±0.28	1.3584	RC	1.23	17.5X25.8	75X12.5
17	91.655	1411	V	0.19	13.1X14.4	12.5X12.5
18	91.655	1418	H	0.19	13.1X14.4	12.5X12.5
8	150±1.25	1642	H	0.53	13.1X14.4	12.5X12.5
9	183.31±6.6	1526	H	0.56	13.1X14.4	12.5X12.5
10	183.31±3.0	1019	H	0.39	13.1X14.4	12.5X12.5
11	183.31±1.0	512.5	H	0.38	13.1X14.4	12.5X12.5

polarizations were selected with the exception of changing the 85.5→91.655 GH to reduce the hardware complexity. In addition, the same conical scan of the SSM/I was adopted along with the external calibration approach. The upper air channel frequencies and passbands were selected from studies conducted by the contractor NGES.

The weighting functions for the atmospheric soundings are presented here as Fig. 5.42 for a standard atmosphere along with nominal polar and equatorial weighting functions for the upper air channels (Poe et al., 2001). Spacecraft interface requirements dictated the limits of the instrument FOV, size, power, and weight, which in turn restricted the antenna size, the number of channels, the data rate, and the swath width. The antenna reflector is the same as the SSM/I, but to accommodate the wide range of frequencies, a multiple-horn frequency multiplexing approach was selected to obtain the desired spatial resolutions and high mean beam efficiencies with low feedhorn spillover loss and cross polarization coupling. The task of integrating a large number of highly sensitive receiver electronics into a single instrument presented some unique challenges and a state-of-the-art component development and electronic compatibility issues to be able to achieve the requisite calibration accuracies and radiometer sensitivities. Many lessons were learned along the way as the sensor was developed.

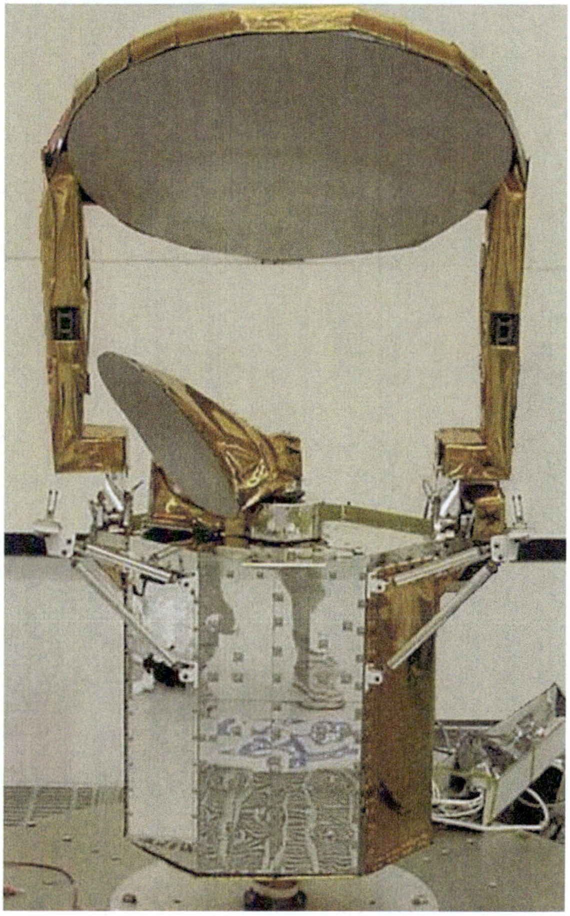

FIGURE 5.40 The special sensor microwave imager/sounder instrument showing the canister, the main reflector antenna, and calibration load assembly (Kunkee et al., 2008).

The main reflector is a lightweight graphite/epoxy offset parabolic laminate shell with a diameter of 0.6096 m and D=0.3826 m. It is illuminated by six broadband corrugated feedhorns, resulting in multiplexing in channels 1–5, 6, 7, 19–24, channels 12–14, channels 15–16, channels 8–11, and channels 17–18. The main reflector and cold space reflector have vacuum deposited coatings of aluminum and silicon oxide to obtain high reflectivity at microwave frequencies and prevent high on-orbit temperatures due to solar illumination. The antenna feeds are positioned in the reflector focal plane to produce centroids of the secondary antenna beams. The instrument spins in the counterclockwise direction with a spin rate of 31.6 rpm.

All channels are sampled every scan with a fixed receiver integration time of 4.1 ms and a sampling period of 4.22 ms (which corresponds to an 0.8 degree azimuth scan increment). All samples are then subjected to onboard averaging in the along-scan direction to improve the NEΔT, meet the instrument data rate of 14.22 kb/s, and produce 60 uniformly spaced samples per scan for channels 1–7; 180 for channels 8–11 and 17–18; 90 for channels 12–16; and 30 for channels 19–24. Antenna beam registration of all channels along with additional beam averaging is carried out in the ground data processing software to yield nearly circular concentric antenna beams in the along-track and the along-scan directions.

On each scan, the feedhorns pass beneath the stationary warm-load reference body and the cold-space reflector calibration targets, interrupting the FOV of the main reflector to provide the two point calibrations for each scan. Four samples of each calibration target are each scanned for all channels and are averaged by the onboard software before being transmitted to the ground. The warm load consists of tapered pyramids coated with a magnetically loaded microwave absorbing material. Its temperature is maintained at 302 K, which is monitored periodically with three high-precision platinum resistive temperature probes.

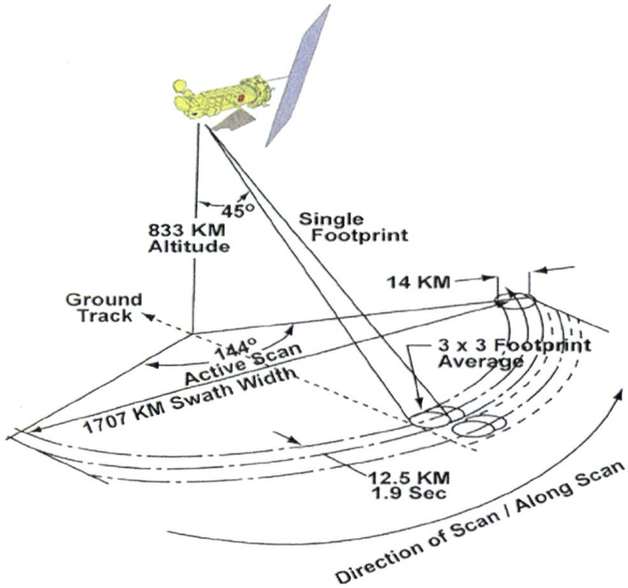

FIGURE 5.41 The special sensor microwave imager/sounder scan geometry showing the direction of scan, the scan width, the ground track, and the footprint averages (Kunkee et al., 2008).

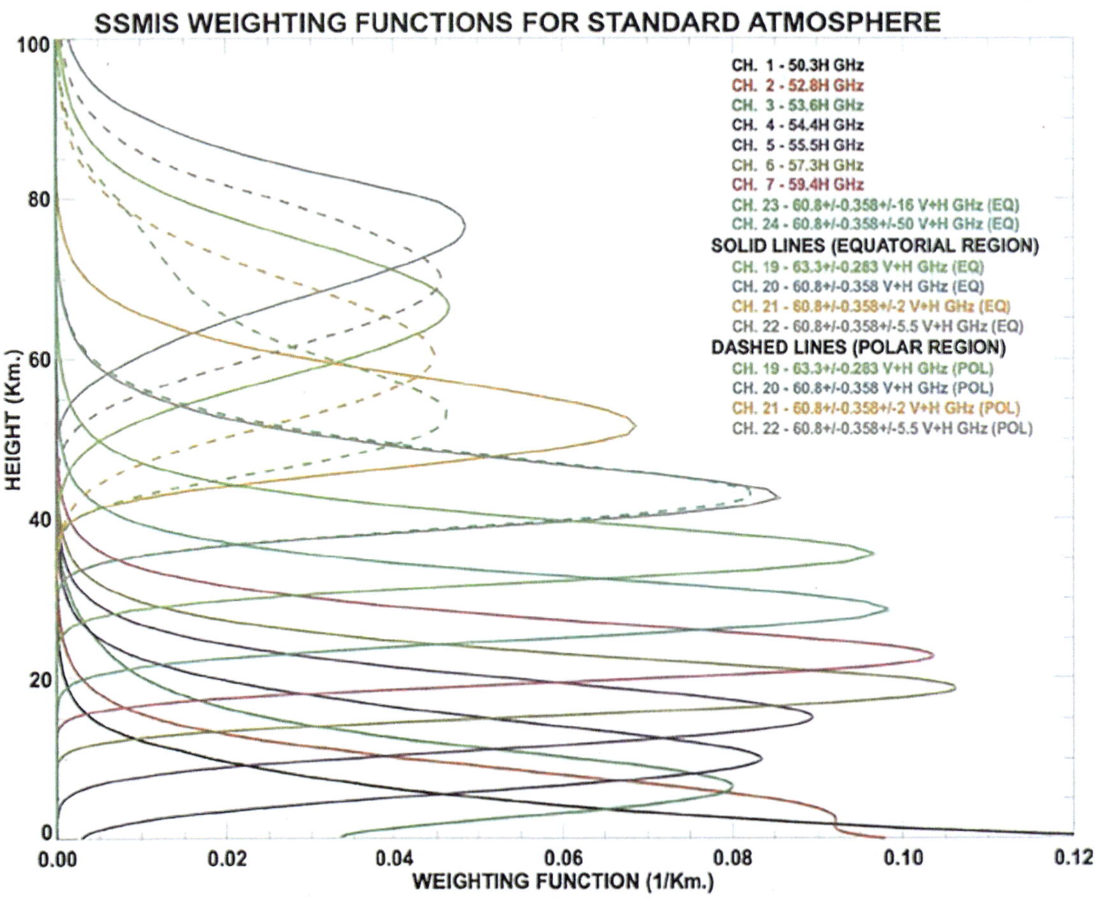

FIGURE 5.42 Special sensor microwave imager/sounder weighting function for a standard atmosphere (Kunkee et al., 2008).

The receiver subsystem accepts the energy from the six antenna feeds directly and provides amplification, filtering, and additional frequency multiplexing to output 24 discrete frequency bands to a signal processing system. The 19, 22, and 37 GHz channel radiation is directly input by the direct detection receivers and then amplified by low-noise amplifiers and directly detected at the microwave frequency without any frequency down conversion as is required in conventional superheterodyne receivers. This approach eliminates the need for a local oscillator, mixer, and intermediate amplifier, thus, saving a lot of weight and improving the NEΔT. The other channels are implemented in the more conventional way.

The SSMIS cal/val program was designed to (1) verify the end-to-end instrument radiometric performance, including long- and short-term stability, NEΔT, and absolute calibration accuracy and (2) validate the imaging and sounding environmental data records (geophysical parameters) for the needs of the DMSP mission and other data users. This cal/val effort is well described in Fig. 5.43.

Matchup data at specific locations revealed that a systematic bias was occurring during certain times of the year. It was not until comparisons were made between global Numerical Weather Prediction (NWP) fields and a graphic simulation of the SSMIS sensor on-orbit that they learned that thermal emission coupled with large temperature swings of the main reflector was the likely explanation of the systematic biases observed earlier in the matchup data. They concluded that careful multiperspective analyzes enabled by the availability of many sources of data are essential to the cal/val of precision state-of-the-art on-orbit microwave sensors.

The SSMIS cal/val program also found solar intrusions into the SSMIS hot load, which led to an investigation that a similar effect had been present on earlier sensor missions but had not been detected or corrected. In the SSMIS cal/val program there was a limited analyzes of coincident ocean buoy measurements with LIDAR and radiosonde (RAOB)

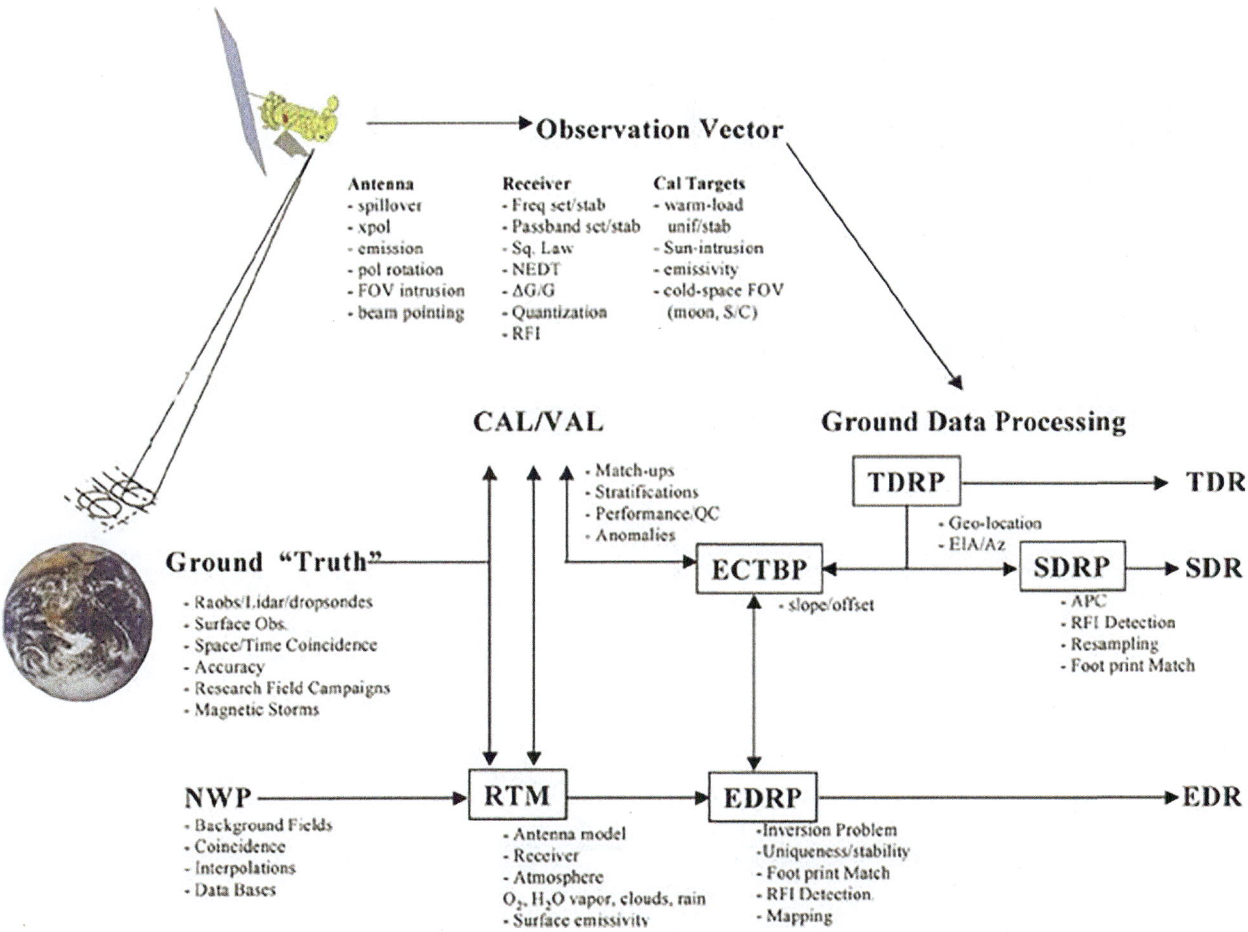

FIGURE 5.43 The uncertainties and errors in the cal/val process of the special sensor microwave imager/sounder (SSMIS). Error sources are in the on-orbit brightness temperatures, ground truth measurements, aspects of the numerical weather prediction (NWP) models, radiative transfer model (RTM), the temperature data record processor (TDRP), the sensor data processor (SDR), the Environmental Data Processor (EDRP), and the Empirically Corrected Brightness Temperature Processor (ECTBP) (Kunkee et al., 2008).

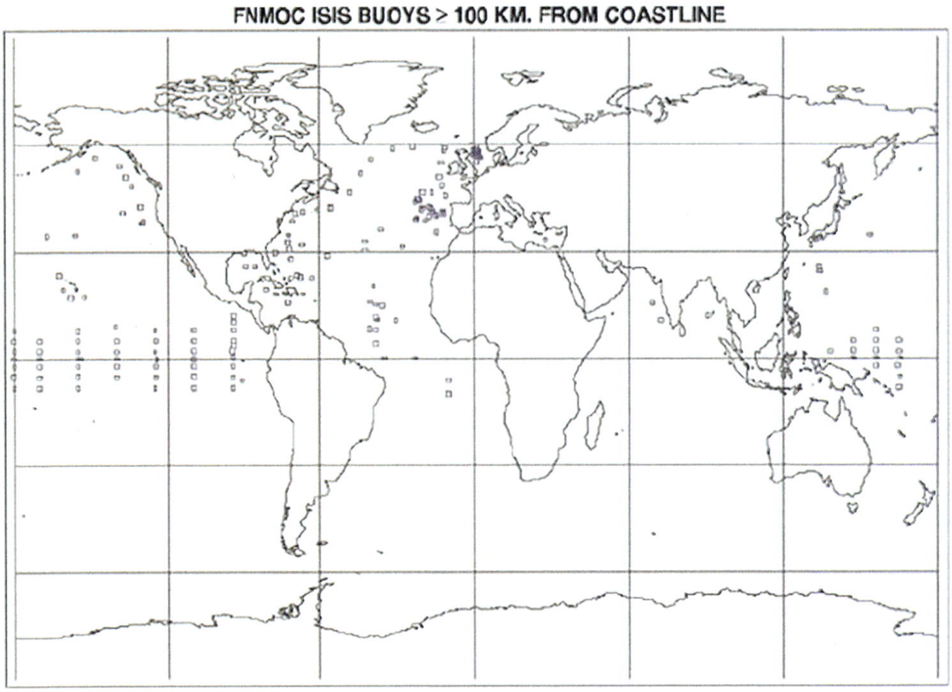

(A)

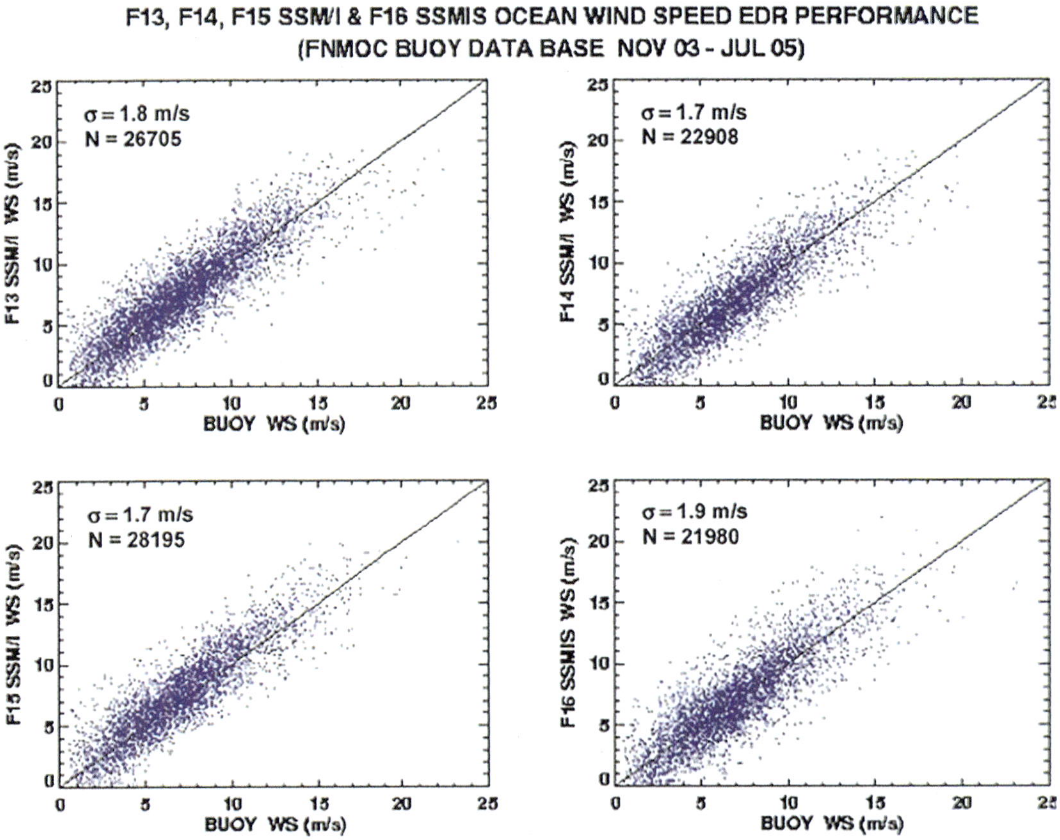

(B)

FIGURE 5.44 (A) Comparison ocean buoy locations and (B) Defense Meteorological Satellite Program F-13, F14, and F-15 special sensor microwave imager,Underflights and F-16 SSMIS ocean wind speed compared with nearly coincident buoy wind speeds (November 2003–July 2005). Map lines delineate study areas and do not necessarily depict accepted national boundaries (Kunkee et al., 2008).

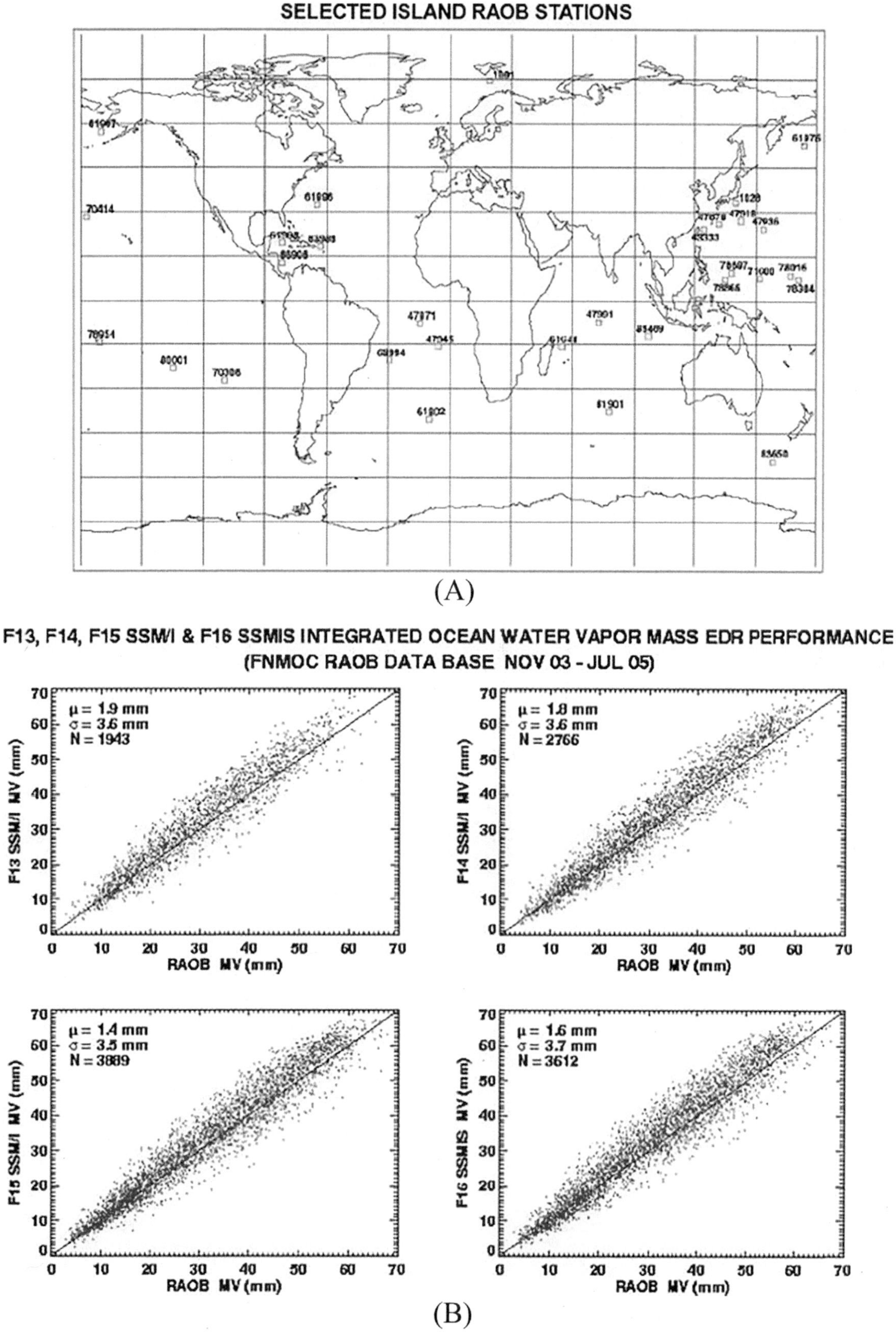

FIGURE 5.45 (A) Map of the radiosonde observatiSons (RAOB) database, (B) Defense Meteorological Satellite Program F-13, 14, 15 special sensor microwave imager and F-16 special sensor microwave imager/sounders integrated water vapor, November 2003–July 2005. Map lines delineate study areas and do not necessarily depict accepted national boundaries (Kunkee et al., 2008).

observations. SSMIS Environmental Data Records (EDRs) were also compared with NWP fields. Underflights were carried out with the Airborne Polarimetric Microwave Imaging Radiometer and the Conical Scanning Microwave Imaging Radiometer (CoSMIR). During the cal/val flight, the CoSMIR instrument helped to identify a design flaw in the F-16 SSMIS, where channels 1–5 were measuring *V*-pol instead of *H*-pol as designed (Table 5.14).

Another way to evaluate the SSMIS sensor data record (SDR) accuracy is to compare them with coincident global analysis fields generated by the European Center Medium Range Weather Forecasting (ECMWF) NWP model applied to the RTTOVS-7 forward RTM. Channels for the SSMIS lower atmosphere temperature sounding suite that do not contain significant contributions from the surface show a smooth, slowly varying background over which there are readily apparent sensor anomalies. Residual biases for channels 3, 4, and 5 were observed to be about 2–3 K and larger biases were seen in channels 9–11. The cause of these residual biases turned out to be emissions from the SSMIS main reflector after it was confirmed that sun-illumination of the SSMIS main reflector was highly correlated with changes in the bias. Other independent measurements confirmed this conclusion.

As with the SSM/I the, primary ocean source of calibration measurement was the moored buoys for wind and surface temperature and small island RAOB measurements. The comparisons with the buoy measurements are presented in Fig. 5.44. The number of observations was >20,000 with a mean difference <0.2 m/s with a standard deviation between 1.7 and 1.9 m/s. The integrated water vapor retrieval using the island RAOB data is shown in Fig. 5.45. Here the number of comparison points was ~2,000 with a mean difference between 1.4 and 1.9 mm with a standard deviation between 3.5 and 3.7 mm. Overall the SSMIS performed up to the expectations set by the success of the SSM/I.

The SSMIS had a unique capability to monitor the atmosphere at altitudes well into the mesosphere (~0.01 mb). This was achieved using radiometers with analog passbands with multiple ~0.002% bandwidths, frequency stability to within ~1 ppm, and a Doppler compensation for the conical scan geometry to a frequency scale well below that. Thus the SSMIS is an engineering achievement for space-based radiometry.

Chapter 6

Individual satellite programs, Seasat, Nimbus, SeaWIFS and Tropical Rainfall Mapping Mission

6.1 The Seasat satellite (also referred to as Seasat-A and Seasat-1) (https://www.eoportal.org/satellite-missions/seasat#space-and-hardware-components)

A very ambitious NASA satellite called Seasat was first launched on June 26, 1978, and unfortunately lasted only to October 19, 1978. This short-lived mission demonstrated the capabilities of four totally new instruments: (1) radar altimeter,(2) synthetic aperture radar (SAR), (3) satellite scatterometer (SASS), and the scanning multichannel microwave radiometer (SMMR). Seasat also carried the visible and infrared radiometer (VHRR) that was the forerunner of the core of later NOAA satellite missions in the form of the advanced very high resolution radiometer (AVHRR) discussed at length in previous sections of this book. Of the four new instruments on Seasat, NASA was in the position to space qualify the SMMR instrument's engineering model and get it flown on Nimbus 7, the last of the Nimbus series, in the same year that Seasat failed.

Seasat was launched into an 800 km nonsun-synchronous, near circular polar orbit. It had a repeat cycle of 17 days with some areas repeating every 3 days. The satellite and its instruments were a project at NASA's Jet Propulsion Lab in California (Fig. 6.1).

It had three main objectives: (1) to demonstrate new techniques to monitor the Earth's oceanographic phenomena and features from space on a global scale, (2) to provide timely oceanographic data to scientists studying marine phenomena, and to users of the oceans as a resource (ocean shipping, fishermen, marine geologists, etc.), and (3) to determine the key features of an operational full-time ocean-monitoring system. The short-lived success of the four original instruments clearly fulfilled the first objective. Seasat's SAR demonstrated for the first time from space the value of high spatial resolution images independent of atmospheric interference. The scanning multichannel microwave radiometer confirmed many of the observations that had just been established by the first flight of the special sensor microwave/imager (SSM/I). In fact, it was so successful that NASA decided to space qualify the engineering model and fly it on Nimbus 7. The radar altimeter followed on some very preliminary measurements made from Skylab but demonstrated the value of this instrument, which later became a primary sensor for physical oceanography. The SASS scatterometer demonstrated for the first time that it was possible to observe the wind vector over the ocean from space.

The spacecraft was designed and developed by the Lockheed Missiles and Space Company (LMSC, now Lockheed-Martin) as the prime contractor, supported by Ball Aerospace Systems in Boulder, CO. The satellite used the Agena upper stage to provide satellite bus functions, including power, telemetry (S-band), attitude control, and command and control function. You can see the Agena bus in the orange section of Fig. 6.2.

Comparing Figs. 6.1 and 6.2, you can see how another section was added that carried the sensors. The satellite was three-axis stabilized using a momentum wheel/horizon sensing system to accurately point the sensors at the Earth's surface. The attitude control system further used hydrazine reaction control thrusters and a gyro reference unit as an attitude reference.

NASA began planning for the Seasat satellite mission in 1972, which would be the first multisensor spacecraft dedicated to studying the ocean. Specific objectives were to collect data on sea surface winds, sea surface temperature, wave heights, ocean surface topography, internal waves, atmospheric water (vapor+cloud liquid), and sea ice parameters (extent and concentration). The requirements for this mission were developed by a User Working Group consisting of the Office of the Oceanographer of the Navy, Fleet Numerical Weather Center in Monterey, CA, Navy Surface Weapons Center in Dahlgren, VA, Naval Research Laboratory, the Johns Hopkins University Applied Physics Lab (APL), the

Satellite Remote Sensing. DOI: https://doi.org/10.1016/B978-0-443-22086-9.00018-6

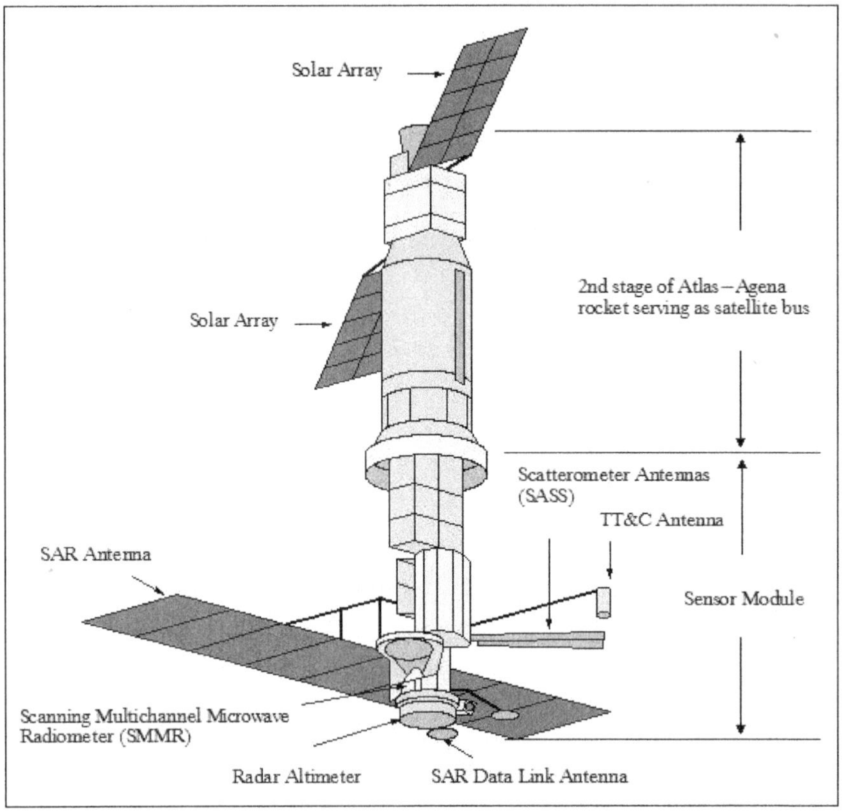

FIGURE 6.1 The Seasat satellite (https://www.eoportal.org/satellite-missions/seasat#space-and-hardware-components). *SAR*, Synthetic aperture radar.

Office of Naval Research and the Navy/NOAA Joint Ice Center. NOAA was further represented by members of the various NOAA labs in Suitland, MD, the Pacific Marine Environmental Lab in Seattle, WA, NOAA's Marine Fisheries Office in Bay St Louis, MS, the Defense Mapping Agency, the United States Geological Survey (USGS), the US Coast Guard, the Department of the Interior, and the Department of Agriculture. It is amazing that a group this large could come up with a coherent list of recommendations for Seasat.

As stated earlier, the project was managed by JPL for NASA with significant participation from NASA's Goddard Space Flight Center in Greenbelt, MD, NASA Wallops Flight Facility, Wallops Island, VA, NASA's Langley Research Center, Hampton, VA, NASA's Glenn Research Center, Cleveland, OH, the Johns Hopkins University Applied Physics Lab (APL), Laurel, MD, Lockheed Missles and Space Systems, Sunnyvale, CA, and the National Oceanic and Atmospheric Administration, Washington, DC.

The specific mission objectives are as follows:

- Provide an evaluation of sensor capabilities to measure the following geophysical parameters:
  - Wave heights
  - Wavelength and direction
  - Surface wind speed and direction
  - Ocean surface temperature
  - Atmospheric water content (liquid and vapor)
  - Sea ice morphology and dynamics
  - Icebergs.
  - Provide oceanographic data for participating users and, following geophysical evaluation, for distribution to the general user community:
  - Predictions of wave height, directional spectra, and wind fields for ship routing, ship design, storm damage avoidance, coastal disaster warning, coastal protection and development, and deep water port development.
  - Maps of current patterns and temperatures for ship routing, fishing, pollution dispersion, and iceberg hazard avoidance.
  - Charts of ice fields and leads for navigation and weather prediction.

FIGURE 6.2 Seasat satellite (https://www.eoportal.org/satellite-missions/seasat#space-and-hardware-components).

- Charts of the ocean geoid fine structure.
- Determine key features of an operational ocean dynamics monitoring system including the following:
  - Sensor operation
  - Global sampling
  - Production of geophysical data records
  - Near real-time data handling
  - User operations interaction
  - Precision orbit determination
  - Demonstrate the economic and social benefits of user agency products.

The program was designed as a US interagency program with NASA as the lead agency. All subsequent data analysis would be funded by agencies other than NASA. Commercial users were an important part of the program from the very beginning. In this collaboration. Naval Fleet Numerical Weather Central would process and distribute near-time weather information to support weather forecasting and maritime operations. NOAA Environmental Data Services would archive and distribute geophysical data processed by NASA for nonreal-time studies. NOAA, Navy, Coast Guard, Geological Survey, and the National Science Foundation (NSF) would fund scientific experiments based on Seasat data. Experimental teams composed of scientists supported by these agencies would evaluate the performance of all the new Seasat sensors.

Unfortunately, this cooperation was mostly unfunded and never happened. NOAA never received congressional approval and funding for its part of the mission. After the early failure of the satellite, funds programmed for future operations of the Seasat satellite were reprogrammed by Dr. W. Stanley Wilson to fund a Seasat Data Utilization Project. Seasat managers, advised by their science communities, released data to all with a demonstrated interest in evaluating the data. This led to a very rapid understanding of the instrument measurement accuracies, the usefulness of the data, and science/applied results.

In the commercial user area:

1. 16 users completed 18 2-year studies at no cost to NASA and found:

   a. Seasat data can significantly improve the forecast of winds, waves, and the location of storms.
   b. Improved forecasts lead to better routing of ships with savings in operating costs.
   c. Seasat data are useful for establishing the climatology of offshore areas, resulting in the better selection of proper offshore equipment.
   d. Seasat data can lead to a more efficient fishery operation.
2. The usefulness of the data led to continuation of many aspects of the program.

The original development of the sensors was divided among the NASA partners. JPL was the project center to develop the SAR, NASA Langley was to develop the wind scatterometer, NASA Wallops for the Altimeter in conjunction with APL, and NASA GSFC was responsible for the Passive Microwave Radiometer. Within these groups, subgroups were organized to develop experiments and validation procedures for each instrument relative to the guidelines set down by the User Working Group. Systems assembly, integration, and test took place at Lockheed facilities in Sunnyvale, CA. The program met cost and schedule milestones by adhering to careful design goals and plans and was ready to ship 28 months from the project go-ahead.

Seasat was launched on June 26, 1978, from Vandenberg AFB near Lompoc, California, using an Atlas-Agena launch vehicle into a nonsun-synchronous orbit of inclination 108 degrees, altitude of 799 km, period of 101 min, and a repeat cycle of 17 days. All systems were correctly deployed on schedule, and initial power on and checkout sequences were successful. From late June to October 1978, Seasat continued to collect global data from all sensors, performing special validation experiments at selected locations. International participation added validation sites and global experiments, expanding Seasat observations and applications. An unexpected spacecraft power failure terminated the mission prematurely. There were rumors that the SAR had been able to detect the presence and movements of submarines and that the US Department of Defense shut the mission down, but there has been no evidence from subsequent SAR missions to validate this claim, and the power failure continues to be the proper justification for the end of the Seasat mission. The legacy of the Seasat mission is realized in the many subsequent missions flying instruments first flown on Seasat.

6.1.1 The Seasat synthetic aperture radar

While SAR instruments had been flown on aircraft, this would be the first spaceborne SAR. It was designed and built by JPL, the prime contractor for the entire satellite. The SAR instrument had a HH polarization with a look angle of 20 degrees, a pixel size of 25 m×25 m (spatial resolution after the collection of four looks), and a radiometric resolution of five-bit raw data. The sensor transmission frequency was 1.275 GHz (L band) with a wavelength of 23.5 cm and a swath width of 100 km. The antenna was a 1024-element phased array of 10.74 m×2.16 m, the PRF was 1464–1640 Hz, the pulse duration was 33.4 μs, the bandwidth (linear FM) was 19.077 MHz, and the transmitted peak power was 1 kW (nominal).

The antenna planar array was made up of eight 1.3 m × 2.16 m rigid and structurally identical fiberglass honeycomb panels. The panels were hinged together in series but were individually supported by a deployable tripod structure that governed the deployment of the truss and provided the interface of the antenna structure with the spacecraft bus. As the first SAR in Earth orbit, the antenna was mounted with its boresight oriented at 20 degrees from the vertical direction (look angle) pointing to the right of the satellite orbital path. The antenna beamwidth was 6.2 degrees in elevation and 1 degree in azimuth, resulting in a footprint of 100 km×15 km (3 dB contour). The swath extends from 290 to 390 km to the right of the spacecraft ground track (Fig. 6.3).

An image produced by the Seasat SAR is presented here in Fig. 6.4.

It should be noted that the demands of SAR processing were so great in the late 1970's that the early processing of Seasat SAR imagery was done optically. This led to stripes of different exposure that detracted from the interpretation of the SAR image features. MacDonald Dettweiler Inc of Vancouver, BC did the first completely digital reconstruction of Seasat SAR images, which eliminated the exposure stripes in the images. Digital processing became the standard method for SAR processing as digital computers increased in speed and capability. The image in Fig. 6.4 was processed digitally by JPL.

6.1.2 The scanning multichannel microwave radiometer (SMMR)

The SSMR is a five-frequency radiometer that was flown later on Nimbus-7. The instrument was designed and built at JPL (not GSFC as stated earlier). The sensor measurement objectives were monitoring sea surface temperature (SST), wind speed, rain rate, atmospheric water content both vapor and cloud liquid water and sea ice conditions. SMMR was dual polarized with frequencies at: 6.6 (45.4 mm), 10.7 (28 mm), 18.0 (16.6 mm), 21.0 (14.2 mm), and 37.0 GHz (8.1 mm). SMMR utilizes six Dicke-type radiometers and those operating at the four longest wavelengths measured alternate polarizations during successive scans of the antenna, while the others operated continuously for both polarizations. (A reminder here that the Dicke radiometer was first introduced in 1946 by Robert Dicke at the Radiation Laboratory of the

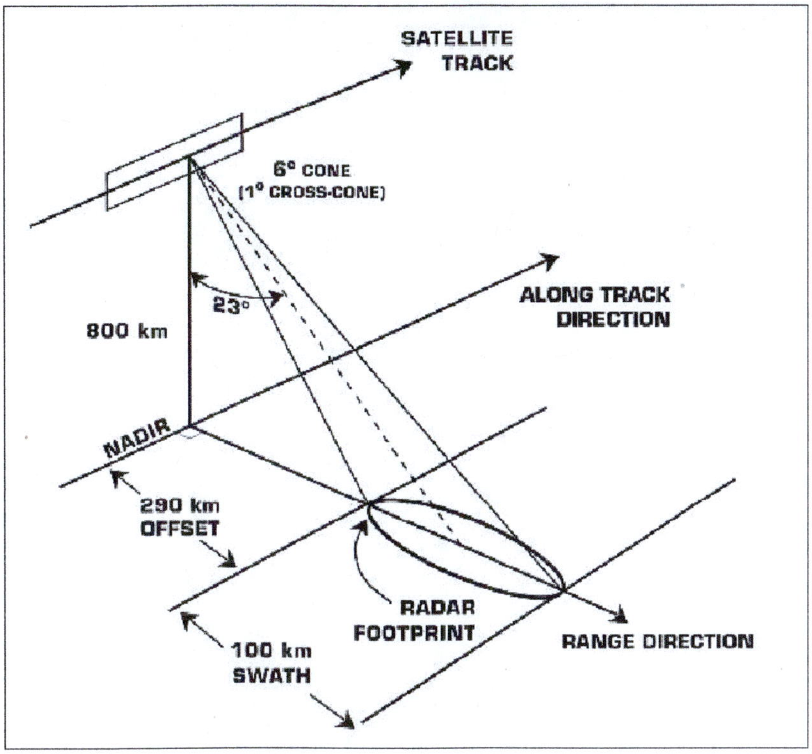

FIGURE 6.3 Seasat synthetic aperture radar viewing geometry.
Image courtesy JPL.

Massachusetts Institute of Technology [MIT] to better determine the temperature of the microwave background radiation. He was frustrated by the noise entered into this measurement by instrument noise and reasoned that he could switch the measurement system between the target and an onboard cryogenic load at a known temperature. The instrument noise level could then be subtracted, yielding the correct target brightness temperature. This became known as the Dicke Switch. The SSM/I discussed earlier was one of the first "full power" microwave radiometers that took advantage of the dramatic noise reduction in the low-noise amplifiers that had been developed.)

The SMMR consisted of five hardware elements:

1. The antenna assembly consisting of the reflector made of graphite epoxy and the feed horn.
2. The scan mechanism, including momentum compensation devices.
3. An RF module containing the input and reference switching networks, the IF-mixer preamplifiers, and the Gunn local oscillators.
4. An electronics module containing the main IF amplifiers, all the post-detection electronics, and the power supplies for the scan and data subsystems.
5. A power supply module containing the ac-to-dc converters and regulators for the rest of the instrument.

The antenna was a parabolic reflector offset from nadir by 42 degrees. The motion of the reflector provided observations from within a conical volume along the ground track of the satellite (Fig. 6.5).

SMMR had a swath width of 600 km, and the spatial resolution ranged from about 22 km at 37 GHz to about 100 km at 6.6 GHz. The absolute accuracy of the SST was estimated to be 2 K with a rms of 0.5 K. The wind speed accuracy was 2 m/s for winds ranging from 7 to 50 m/s. The SMMR had a mass of 53.39 kg and power consumption of about 60 W. An identical instrument was flown on Nimbus-7, launched October 24, 1978.

Unlike SSM/I, the SMMR had channels specifically designed to measure SST. Njoku and Swanson (1983) discussed that most of the initial analysis effort with SMMR data was on the SST measurement. Hofer et al. (1981) produced a 2-week average map of SST in the Pacific Ocean from SMMR data using an early version of the processing algorithms. This study revealed occasional contamination of the data by sunglint and residual crosstalk calibration errors. In spite of these problems, the data correctly portrayed the major SST features and demonstrated the microwave insensitivity to cloud cover. In addition, visual inspection of the SST contour distortions near land boundaries made it possible to make a

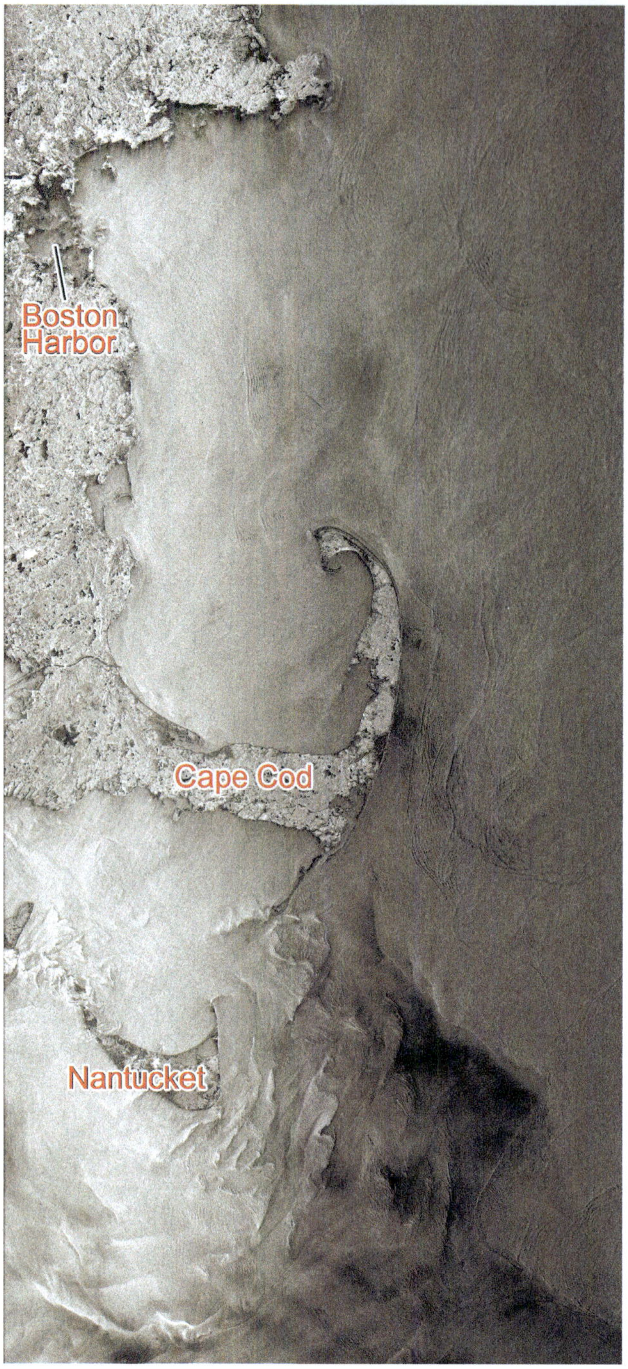

FIGURE 6.4 Internal waves and shallow subsea features imaged by the Seasat synthetic aperture radar on August 27, 1978, near Cape Cod, Massachusetts. Both the internal waves and shallow feature signatures were generated by strong tidal currents in the region.
Image courtesy JPL.

qualitative estimate of land contamination in the antenna sidelobes. A safe distance from coastlines to be able to ensure an accurate SST retrieval was estimated to be about 600 km. This distance was far less for the other parameters (wind speed, water vapor, and liquid water) due mainly to the fact that spatial resolution of the SMMR antenna improves at higher frequencies and the SST retrieval was largely dependent upon the lowest 6.6 GHz frequency. In other studies of SMMR data, Bernstein (1982) and Bernstein and Marris (1983) analyzed SMMR SST measurements in the northern and tropical eastern Pacific by producing 10-day and monthly averaged SST maps. After corrections for some systematic errors, the SMMR SSTs agree with similarly averaged ship SSTs to within 0.5°C rms.

6.1.3 The radar altimeter on Seasat

The altimeter is built on heritage gained in flights of the S-193 instrument flown on Skylab and the Geos-3 missions. The goals of the Seasat ALT were to measure the height of the satellite above the ocean surface with an accuracy of ±10 cm

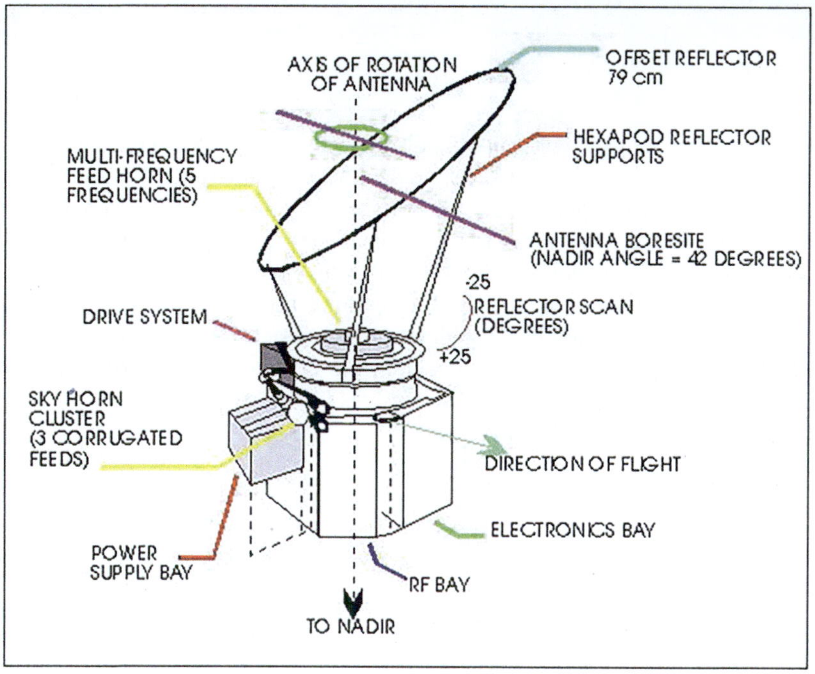

FIGURE 6.5 The SMMR instrument.
Image courtesy JPL.

FIGURE 6.6 The Radar altimeter flown on Seasat.
Image courtesy JPL.

once per second, to measure wave height at nadir with an accuracy of ±10% or 0.5 m, whichever is greater, once per second for wave heights from 1 to 20 m, to measure the radar backscatter coefficients with an accuracy of ±1.0 dB and to combine the altimeter height measurements with an accurate ephemeris to determine the marine geoid, sea surface currents, tides, storm surges, etc.

The ALT instrument was a Ku-band compressed pulse radar altimeter (first use of the full-deramp technique). With the new full-deramp technique, no compression filter is needed in the receiver. From Seasat onwards, all radar altimeters have been built using this technique, achieving significant improvement in the resolution. The ALT instrument was designed and developed by JHU/APL. Two of its most unique features were a linear FM transmitter with a 320 MHz bandwidth, which yielded a 3.125 ns time-delay resolution, microprocessor-implemented closed-loop range tracking, automatic gain control, and real-time estimation of significant wave height. This instrument flew the first microprocessor (8080-based controller/tracker) in space. The altimeter operated at 13.56 GHz (chirp signal at 2 kW peak power) using a 1 m parabolic antenna pointed off nadir and had a swath width which varied from 2.4 to 12 km depending on sea state (Fig. 6.6).

ALT operated in chirp pulse mode with a 3.2 μs uncompressed pulse width and a 3.125 ns compressed pulse width. The ALT had a mass of 93.8 kg and a power consumption of 177 W. In the Seasat design, the ALT number of echo samples is increased compared to Geos-3. The samples are spaced 3.125 ns apart to cover the anticipated spread in ocean return for wave heights up to 20 m. In this case, waveform sampling is implemented by a bank of filters with a 312.5 kHz bandwidth and spacing. In contrast to previous designs, the samples are an integral part of the altitude tracking process and are used in such a way that the system adapts as a function of wave height to optimize tracker performance. The altitude tracking loop is closed in two parts: a coarse adjustment of the local oscillator pulse timing in 12.5 ns steps and a fine adjustment. Calibration of the ALT for height bias was carried out using four overflights of Bermuda supported by the Bermuda laser. Seasat carried a laser reflector for this purpose. The estimated height bias was 0.0±0.07 m.

6.1.4 The Seasat scatterometer

The heritage of SASS was again the S-193 flown on Skylab. It is a fan-beam (Fig. 6.7), dual-polarized (VV, HH) Doppler scatterometer with the objective of radar backscatter measurements (sigma naught) over ocean surfaces for the estimation of the wind vector.

Pulse transmit frequency is 14.599 GHz (Ku-band). SASS illuminated the surface with four fan-shaped beams (two orthogonal beams, each 500 km wide, on each side of the ground track). Doppler filters were used to discriminate resolution cells in the long dimension of the fan beam, resulting in 500 km swaths on either side of the satellite. The high wind swaths added an additional 250 km to each side. The spatial resolution was 50 km over the region from 200 to 700 km on either side of the spacecraft. This experimental instrument first demonstrated the ability to accurately infer vector winds over the ocean surface from space. It should be noted here that SASS measurements had a 180 degree ambiguity that could be easily resolved with large-scale geostrophic wind maps. The SASS instrument has a total mass of 103 kg with an electronics assembly of 59 kg. Each antenna has a mass of 11 kg, and the SASS had a power consumption of 100 W (peak).

SASS was a proof-of-concept experiment to measure the ocean surface wind vector under day/night and near-all-weather conditions. The physical basis for this remote sensing technique is the generation of a capillary wave on the ocean surface by the friction velocity of the overlying wind. The amplitude of the cm-long wavelength capillary ocean waves is in equilibrium with the local wind, and the two-dimensional wave spectrum is highly anisotropic with the wind direction.

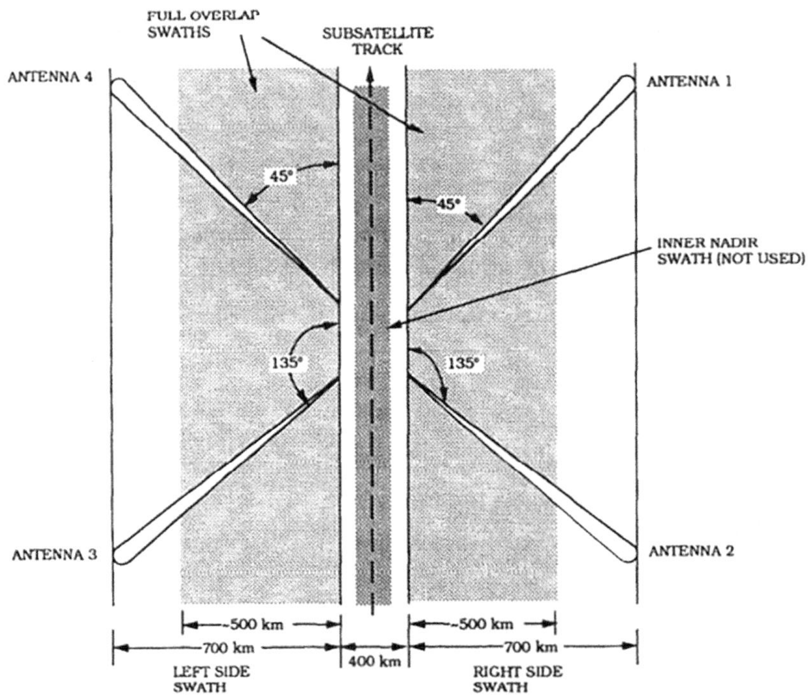

FIGURE 6.7 Viewing geometry of the satellite scatterometer instrument.
Image courtesy JPL.

TABLE 6.1 Seasat VHRR specifications.

Parameter	Visible channel	Infrared channel
Spectral region	0.49–0.94 μm	10.5–12.5 μm
Resolution at ground	2.3 km	4.4 km
Sensitivity (NEΔT)	N/A	4 K with the scene at 185 K, 1k with the scene at 300 K
Dynamic range	65–10,000 fl (scene brightness)	180–330 K (scene temperature)
Detector	Silicon photovoltaic	Thermistor bolometer
Scan rate	48 rpm	48 rpm

The ocean radar backscatter results from Bragg scattering from these capillary waves, and the normalized radar cross-section (σ°) grows approximately as a power series of the wind speed.

6.1.5 The visible and infrared radiometer

The VHRR is a supporting instrument on Seasat with heritage of the VHRR on NOAA satellites. It had the objective to provide images of visible reflections and thermal infrared emissions from the oceanic, coastal, and atmospheric features that may be combined with the other Seasat instruments. It was intended to measure infrared SST and cloud top height. Scanning was accomplished by a rotating mirror mounted at 45 degrees to the optical axis of the collecting telescope. VHRR used a 12.7 cm diameter Cassegrain-type telescope, focusing the radiation on a field stop. The optical system transmitted the radiation to a dichroic beamsplitter, separating it into visible and infrared wavelengths. The swath width of the VHRR is about 2,280 km wide centered at nadir. The VHRR instrument had a mass of 8.1 kg and a power consumption of 7.3 W. The other parameters of the VHRR are given here in Table 6.1.

6.1.6 Seasat conclusions

Our experience over the past 40+ years has clearly shown that Seasat was a dramatically visionary satellite system. It provided the precursors to many subsequent spaceborne instruments. The Seasat SAR alone revealed that SAR could image a lot more than what was originally anticipated. A multitude of dedicated SARs have been and are flying to provide a great deal of imagery over land and oceans. The scatterometer ushered in a new world of ocean wind vector measurements that led to important follow-on missions such as Quickscat. The SMMR engineering model was quickly upgraded to a spaceflight instrument and flown on Nimbus-7, providing a long time series of microwave radiometer measurements. The world owes a debt of gratitude to the dedicated engineers and scientists who made Seasat possible. It is unfortunate that it did not continue to operate longer, but then maybe the science community was not quite ready for that plethora of new data. At least the measurement principles had been established, giving the global science community the motivation to make follow- on measurements of the global ocean.

6.2 Nimbus (rain cloud) satellite series

The Nimbus satellite program was initiated by NASA in 1960 to provide a stable Earth-oriented space platform to test newly developed instrumentation, most of which would then be transferred to other agencies for on-orbit operations. The meaning of Nimbus is "rain cloud" or "luminous cloud" and was perhaps intended to suggest that this satellite program covered a lot of different areas. They were designed to be three-axis stabilized (pitch, roll, and yaw). It took advantage of gravity gradient stabilization with mass concentrations at the top and bottom of the satellite (Fig. 6.8). Considered as the first in a second-generation of meteorological research and development satellites, Nimbus 1 (Fig. 6.8) shows the main components of the satellite as well as the payload carried by it. The main focus was to developed instruments for meteorology and weather forecasting. The satellite orbits were sun-synchronous, nearly circular, with an altitude of 634 km. Launched on August 28, 1964, with a Thor-Agena rocket from Vandenberg, AFB, CA.

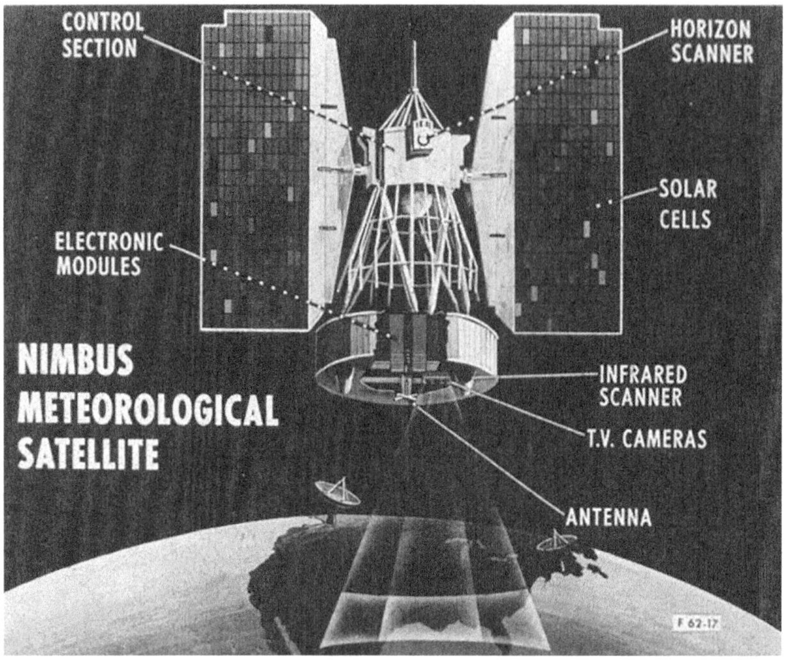

FIGURE 6.8 Nimbus 1 satellite (Press, 1970). *AVCS*, Advanced vidicon camera system.

The automatic picture transmission (APT) instrument was to provide real-time images of cloud cover to weather field forecast offices and mobile platforms such as ships at sea. This instrument was first installed on TIROS-8 and then later on Nimbus 1 and 2, followed by installation on the operational ESSA series of satellites discussed earlier in this text. These images were broadcast on a VHF radio frequency that could be collected by a simple antenna with no need to track the satellite. Images were taken in 8 sec and then scanned by 800 lines × 800 pixels/line and transmitted in real time at the rate of 240 lines/min. Swatch ranged from 1200 to 2200 km, depending on the satellite orbital height, yielding a ground resolution of 1.8–3 km.

The Advanced Vidicon Camera System (AVCS) was first installed on Nimbus 1 and 2 and then on the ESSA series of satellites and on the first ITOS satellite, that represent the changeover from TV cameras to scanning radiometers. Resolution ranged from 0.9 km at nadir to 3.7 at the edge of the image. This instrument provided global coverage night and day. The high resolution infrared radiometer (HRIR) was an instrument to test collecting Earth imagery by a radiometer rather than as a camera. It was a single-channel infrared radiometer in the 3.4–4.2 μm band covering a temperature range from 210 to 300 K with an NEΔT of 1 K @ 260 K. Radiometers later became standard for all meteorological satellites as discussed earlier in this book. Nimbus 2 added to this same payload the medium resolution infrared radiometer (MRIR), which was a five-channel radiometer, with two broad band (0.2–4.0 and 5–30 μm) and three narrow band (I6.4–6.9, 10–11, and 14–16 μm).

Nimbus 3 launched on April 14, 1969, dropped the APT and the AVCS while keeping the HRIR and the MRIR, and added the Image dissector camera system, intended to blend the characteristics of the APT and AVCS flown earlier. This instrument would not be continued beyond Nimbus 4 (launched April 8, 1970), and the discontinued use of APT suggests that it was not needed. The IRIS-B (InfraRed Interferometer Spectrometer B) was an experiment to further explore the infrared spectrum. The interrogation, recording, and location system collected data from Earth platforms and, using two contacts, could determine position. This evolved into the data collection system satellite mission that demonstrated the ability of the satellite to collect from many platforms simultaneously. The MUSE (monitor of ultraviolet solar energy) was added to begin testing instruments for space weather monitoring. The Satellite InfraRed Spectrometer was a first effort to profile the atmospheric temperature from space. It was a grating spectrometer with 8 spectral bands and sampled only at nadir directly beneath the spacecraft.

Nimbus-5 (launched December 11, 1972) flew a completely new suite of sensors. An advanced attitude-control system made it possible to control the spacecraft orientation to within 1 degree in all three axes. Primary experiments included (1) a temperature–humidity infrared radiometer (THIR) for measuring day and night surface and cloud-top temperatures, as well as the water vapor content of the upper atmosphere, (2) an electrically scanning microwave radiometer (ESMR) for mapping the microwave radiation from the earth's surface and atmosphere, (3) an infrared temperature profile radiometer for obtaining vertical profiles of temperature and moisture, (4) a Nimbus E microwave spectrometer for determining tropospheric temperature profiles, atmospheric water vapor abundances, and cloud liquid water contents, (5) a selective chopper radiometer for observing the global temperature structure of the atmosphere, and (6) a surface composition mapping radiometer for measuring the differences in the thermal emission characteristics of the earth's surface. Nimbus 6 repeated some of these but carried many new sensors. The nine experiments selected for Nimbus 6 (launched June 12, 1975, Fig. 6.9) were (1) earth radiation budget (ERB), (2) ESMR, (3) high-resolution infrared radiation sounder, (4) limb radiance inversion radiometer, (5) pressure modulated radiometer, (6) scanning microwave spectrometer, (7) THIR, (8) tracking and data relay experiment, and (9) tropical wind energy conversion and reference level experiment. This complement of advanced sensors was capable of (1) mapping tropospheric temperature, water vapor abundance, and cloud water content; (2) providing vertical profiles of temperature, ozone, and water vapor; (3) transmitting real-time data to a geostationary spacecraft (ATS 6); and (4) yielding data on the earth's radiation budget.

Nimbus-7 (Fig. 6.10) was launched on April 14, 1969, and finally ceased operation on October 1, 1994, producing a relatively long time series of observations. It was the final satellite in this series of successful proof-of-operation satellites. The author is not going to speculate on the reasons for the termination of this very successful program, but he was told that NASA could no longer afford to run a parallel weather satellite series to NOAA, even if it was primarily to test new instruments. At least Nimbus-7 was a great way to end this series, as it brought in new sensors that have become a central focus for the wider research community. Nimbus-7 carried nine instruments, some repeated and some for the first time.

It should be noted that the Earth Resources Technology Satellite (ERTS), which later became the Landsat program, grew out of the Nimbus program, and the first few ERTS satellites used the Nimbus satellite bus.

The nine instruments on Nimbus-7 included: a new coastal zone color scanner (CZCS) that was a multichannel scanning radiometer to make ocean color measurements for the first time from space. The scanning multispectral microwave radiometer (SMMR), as has been mentioned earlier, was a copy of the SMMR on Seasat and was the result of a rapid effort to space qualify the SMMR engineering model and get it deployed in space. The ERB experiment uses a 22-channel radiometer to determine the Earth's energy balance by studying the radiation budget on both synoptic and planetary scales by simultaneously measuring the incoming solar radiation and the outgoing Earth reflected and emitted radiation. The solar backscatter ultraviolet spectrometer (SBUV) and total ozone mapping spectrometer (TOMS) that mapped the total ozone in the stratosphere and in the troposphere. The Stratospheric Aerosol Measurement (SAM)was a sun photometer that viewed a small portion of the sun's rays through the Earth's atmosphere during sunrise and sunset, which also mapped global concentrations of aerosols in the stratosphere and troposphere as well as the distribution of particles in the troposphere. The Limb Infrared Monitor of the Stratosphere (LIRMS) was a six-channel infrared radiometer that viewed the Earth's limb to map vertical profiles of temperature and the concentrations of ozone, water vapor, nitrogen dioxide, and nitric acid in the lower to middle stratosphere while the stratospheric and mesospheric

FIGURE 6.9 Nimbus-6 (https://en.wikipedia.org/wiki/Nimbus_program).

sounder (SAMS) experiment was an infrared spectrometer the global vertical gas concentrations and temperatures in the mesosphere and stratosphere by viewing radiation from the limn of the atmosphere. The thermal infrared radiometer (THIR) was an infrared radiometer designed to measure surface and cloud top temperatures as well as water vapor in the upper atmosphere. It is important to recognize how many of these instruments led to future satellite missions. The CZCS spawned the sea-viewing wide field-of-view sensor (SeaWIFS) mission, SMMR defined all future microwave radiometers, ERB gave rise to the ERBE, and later the cloud and earth radiation energy sensor (CERES) instruments to study the Earth's radiation budget, the SBUV was then carried on many future satellites, and TOMS was the precursor to many future ozone mapping/monitoring instruments.

6.2.1 The coastal zone color scanner

First of its kind, the CZCS was a multispectral whiskbroom radiometer with six spectral bands fed by a rotating plane mirror canted at 45 degrees to the optical axis of a Cassegrain telescope. The rotating mirror scanned 360 degrees but only data within 40 degrees of the spacecraft nadir were retained for the ocean color measurements. The instrument field of view (IFOV) of each detector was 0.865 mrad, resulting in a spatial resolution of 1km at the satellite subpoint (nadir view). The swath extended over 1,566 km, and the data were transmitted to Earth at a rate of 800 kbps, with a digital resolution of 8 bits. The CZCS was designed to sense important biological parameters in the coastal ocean: chlorophyll

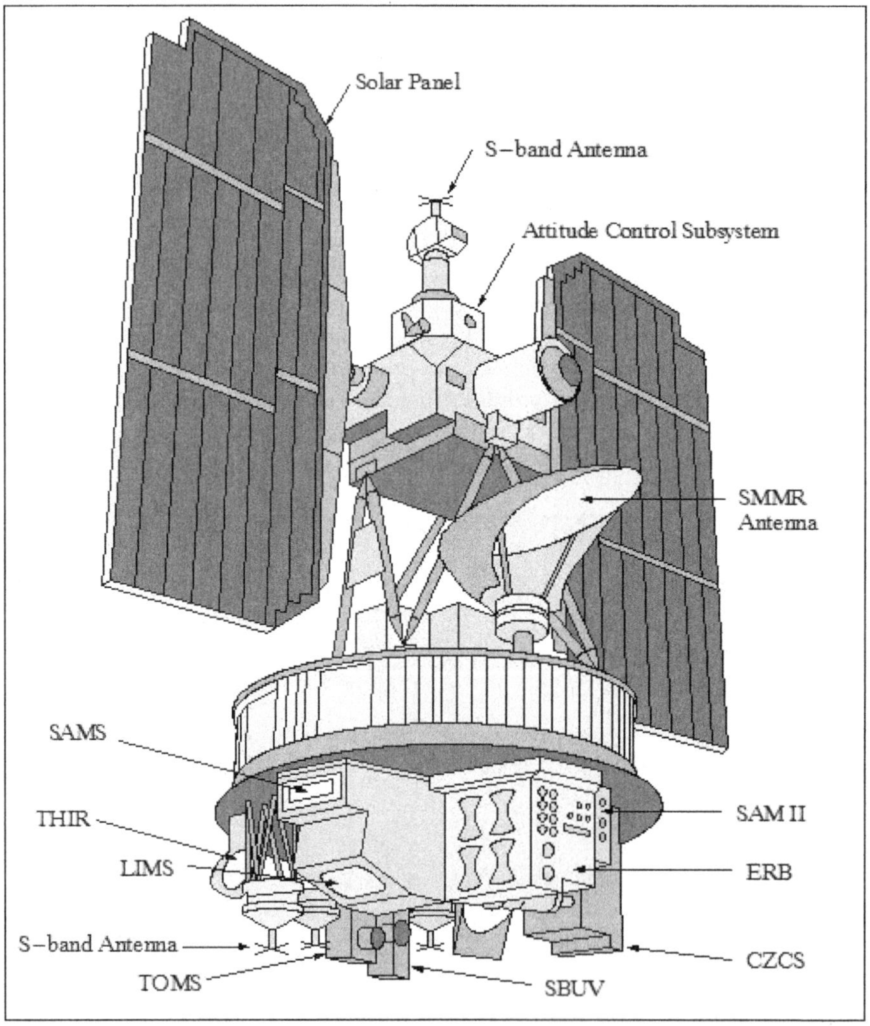

FIGURE 6.10 Nimbus-7 satellite. *CZCS*, Coastal zone color scanner; *ERB*, earth radiation budget; *LIMS*, limb infrared monitor of the stratosphere; *SAM-II*, Stratospheric Aerosol Measurement II; *SAMS*, stratospheric and mesospheric sounder; *SBUV*, solar backscatter ultraviolet spectrometer; *THIR*, temperature–humidity infrared radiometer; *TOMS*, total ozone mapping spectrometer.
Image credit courtesy NASA.

absorption was measured in the 0.433–0.453 μm band and the band between 0.51 and 0.53 μm. Gelbstoff (yellow substance) concentration was measured in the 0.54–0.56 band, aerosol absorption used the 0.66–0.68 μm band, while land and cloud detection used the 0.7–0.8 μm band, and total internal reflection and surface temperature were measured with the thermal infrared band between 10.5 and 12.5 μm.

The performance of these channels is shown here in Table 6.2.

The CZCS ushered in a whole new area of ocean remote sensing. Ocean biologists could at last see large-scale maps of primary productivity and its changes over time. The fact that Nimbus-7 lived quite a long time provided data to work with in studying ocean biological productivity.

Yoder et al. (2000) used CZCS imagery off the US East Coast to study variability in biological productivity. They used the 7.5-year time series (October 1978–July 1986) to study general patterns in near-surface phytoplankton chlorophyll concentrations in the coastal waters. They defined 21 relatively large study areas (>100 km^2, Fig. 6.11) within the Mid-Atlantic Bight (MAB) and the South Atlantic Bight (SAB), previous to setting boundaries for averaging the CZCS data. Their objective was to separate the CZCS chlorophyll concentration variability into three general categories based on time scale: daily, seasonal, and interannual. In addition, they wanted to learn about the relationships between the temporal patterns in the 21 study areas.

Previous in situ and satellite studies had identified many different processes influencing the spatial and temporal variability in phytoplankton biomass and productivity in the SAB and MAB. Upwelling and intrusion of nutrient-rich water associated with frontal eddies and meanders of the Gulf Stream fronts stimulate surface and near-surface phytoplankton blooms (Verity et al., 1993; Pribble et al., 1994). The frontal region itself often has enhanced phytoplankton chlorophyll concentrations caused by mixing due to wind and current interactions (Ryan and Yoder, 1996). Seasonal phytoplankton blooms, including the spring bloom, are important in shelf and slope water of the MAB (Riley, 1946, 1947; Ryther and Yentsch, 1958) and have a strong signal in ocean color imagery (Brown et al., 1985). Gulf Stream rings also influence phytoplankton patterns in slope waters (Garcia-Molinear and Yoder, 1994). Mixing associated with frontal dynamics leads to enhanced phytoplankton biomass concentrations and productivity in the shelf-slope frontal region compared to adjacent waters (Marra et al., 1990; Ryan et al., 1999a, b). MAB wind events throughout the year alter the distribution of pigment in the water column, resulting in higher phytoplankton growth and productivity due to increased nutrient flux (Walsh et al., 1978).

More than 3,500 scenes of Level 1 satellite images(i.e., top-of-the atmosphere radiance in satellite swath coordinates) were used that covered their (Yoder et al., 2000) area of interest (Fig. 6.11). These images had the full (1×1 km) nominal pixel resolution acquired from the NASA Goddard Space Flight Center. Using the University of Miami's DSP software, the Level 1 imagery was processed to Level 2 data (water-leaving radiance, chlorophyll concentration, and other parameters, all in satellite swath coordinates). They then used Gordon's multiple scattering atmospheric correction algorithm with revised calibration coefficients (Gordon et al., 1983; Evans and Gordon, 1994). Based on their evaluation and analysis of East Coast imagery, they used mean East Coast epsilon values (Bisagni et al., 1997) to remove the aerosol contribution from the top-of-the-atmosphere radiance.

At this point, it should be noted that contamination of the CZCS signal by atmospheric aerosols could easily overwhelm the signal in the imagery. Some form of atmospheric correction was needed to work with any ocean color imagery. Later ocean

TABLE 6.2 Performance of the coastal zone color scanner (CZCS).

Band	Wavelength nanom	Signal/Noise	Radiance (mW/cm^2/nanom/ster)
1	443	150	5.41
2	520	140	3.50
3	550	125	2.86
4	670	100	1.34
5	750	100	10.8
Band	Wavelength	NEΔT	Brightness temperature
6	1150	0.220	270

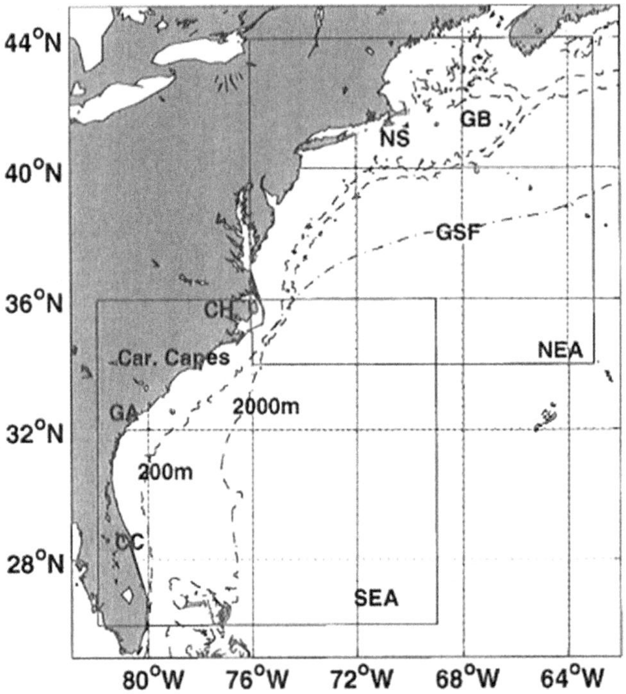

FIGURE 6.11 The study is off the US East Coast. The two large boxes show the NEA (Northeast Atlantic) and SEA (Southeast Atlantic) regions used for mapping coastal zone color scanner imagery. Letters refer to place names: CC=Cape Canaveral, GA=State of Georgia, Car. Capes=Carolina Capes, CH=Cape Hatteras, NS=Nantucket Shoals and GB=Georges Bank. The 200 and 2000 m isobaths are also shown as is the mean position of the Gulf Stream Front (GSF) north of Cape Hatteras. (NASA). Map lines delineate study areas and do not necessarily depict accepted national boundaries. (Yoder et ar., 2000).

color programs would lead to the establishment of vicarious sites to calibrate the ocean color imagery and subtract out the aerosol influence on the signal. Much more will be said about this in a later section of this book.

Yoder et al. (2000) used a three-band biooptical algorithm (Clark, 1981) to convert the water-leaving radiance at 443, 520, and 550 nm to estimates of chlorophyll concentration, which is generally interpreted as the sum of chlorophyll a +pheopigments (Clark, 1981), which they referred to as CSAT (mgm^{-3}). The three-band algorithm yields a log-normal CSAT distribution in continental shelf and slope waters, rather than a bimodal distribution (Denman and Abbott, 1988), which is typical of high-chlorophyll imagery processed with the NASA standard CZCS biooptical algorithm, which switches between two two-band ratio algorithms depending upon the chlorophyll concentration (Feldman et al., 1989).

The CSAT data were mapped to a 512×512 pixel equi-rectangular projection off the US East Coast (Southeast Atlantic [SEA] and Northeast Atlantic [NEA]). Daily composites were formed, resulting in 1,121 daily composite images for SEA and 1,114 for NEA. After mapping, the images had a nominal resolution of 2×2 km, and the final geolocation of the imagery was done by hand using the coastline for reference. The automatic cloud filter allowed some clouds to filter through; additional filtering was carried out to be sure that the images being analyzed were "cloud free." The 21 study areas are shown here in Fig. 6.12 from which the CSAT concentrations were extracted on a pixel-by-pixel basis from the CZCS daily composites.

To generate spatially averaged (geometric mean due to the log-normal distribution) time series, the CSAT values were extracted on a pixel basis from each of these 21 study areas. Theoretically, the CZCS should have covered a given location of the East Coast every day or two. Cloud cover and the fact that CZCS did not operate continuously dictated that the number of daily images covering some or all of each area was significantly less than the total number of days during the 7.5 years of the time series. Frequently, study areas would be covered over with clouds that obscured the surface. Yoder et al. (2000) found that the time series patterns and overall daily mean CSAT concentrations were very similar for the SEA and MEA series based on at least 25% and 50% unmasked pixels.

The monthly mean composite CSAT images for the SEA and NEA regions, consisting of all the available data from the 7.5 year record, are presented here in Figs. 6.13 and 6.14. The images for the SEA show very persistent spatial patterns dominated by the across-shelf gradient with little along-shelf variability (Fig. 6.13). The coastal band of relatively high CSAT extends farther offshore in winter than during the summer. The monthly mean composites for the NEA (Fig. 6.14) exhibit a lot more spatial variability than for the SEA. The Nantucket Shoals (immediately off Cape Cod) and Georges Bank (offshore and east of Cape Cod) have a much greater degree of variability in time and space. The effects of Gulf Stream rings are clearly visible in the April composite of Fig. 6.14.

These monthly composites conceal a great deal of the temporal and spatial variability in CSAT. Statistics from the 21 study areas demonstrate that the standard deviations are often as large as the arithmetic mean (Table 6.3). Most of the study areas have N values that are over 100, indicating the relatively good coverage of the CSAT data. The small standard deviations in the open ocean area reflect the uniformity of the low levels of productivity in the open ocean. This is confirmed by the relatively low values of both the arithmetic mean and the geometric mean.

In an effort to compare satellite-derived products with in situ measurements, CSAT concentrations were compared with MARMAP in situ near-surface chlorophyll a+pheopigments for the NEA study areas (Fig. 6.11). MARMAP was an extensive, NOAA–National Marine Fisheries Service multiyear study of the MAB ecosystem. MARMAP investigators measured chlorophyll a concentrations from 78 oceanographic cruises between 1977 and 1988 (O'Reily and Zetlin, 1988), which overlapped with the CZCS period.

Table 6.3 contains statistics for the 21 time series of daily average CSAT concentrations (>25% pixel coverage) from the study areas shown in Fig. 6.11, as well as statistics for the NEA areas 11–21 and SEA areas 1–8. N is the number of days having a valid (>25% pixel coverage) during the 7.5 year time series.

Previous studies have shown that CSAT values based on the Clark (1981) algorithm underestimate the in situ chlorophyll a+pheopigment concentrations in the MAB. This can also be seen in Fig. 6.13 as the geometric mean of CSAT for areas 13, 14, 17, 18, 20, and 21 was 59% of the geometric mean of the corresponding in situ measurements. Fig. 6.15 shows no systematic difference between the mean seasonal distribution of CSAT and MARMAP data. Both have maximum concentrations occurring in fall-winter, particularly in the middle and outer shelf waters. Slope waters (20

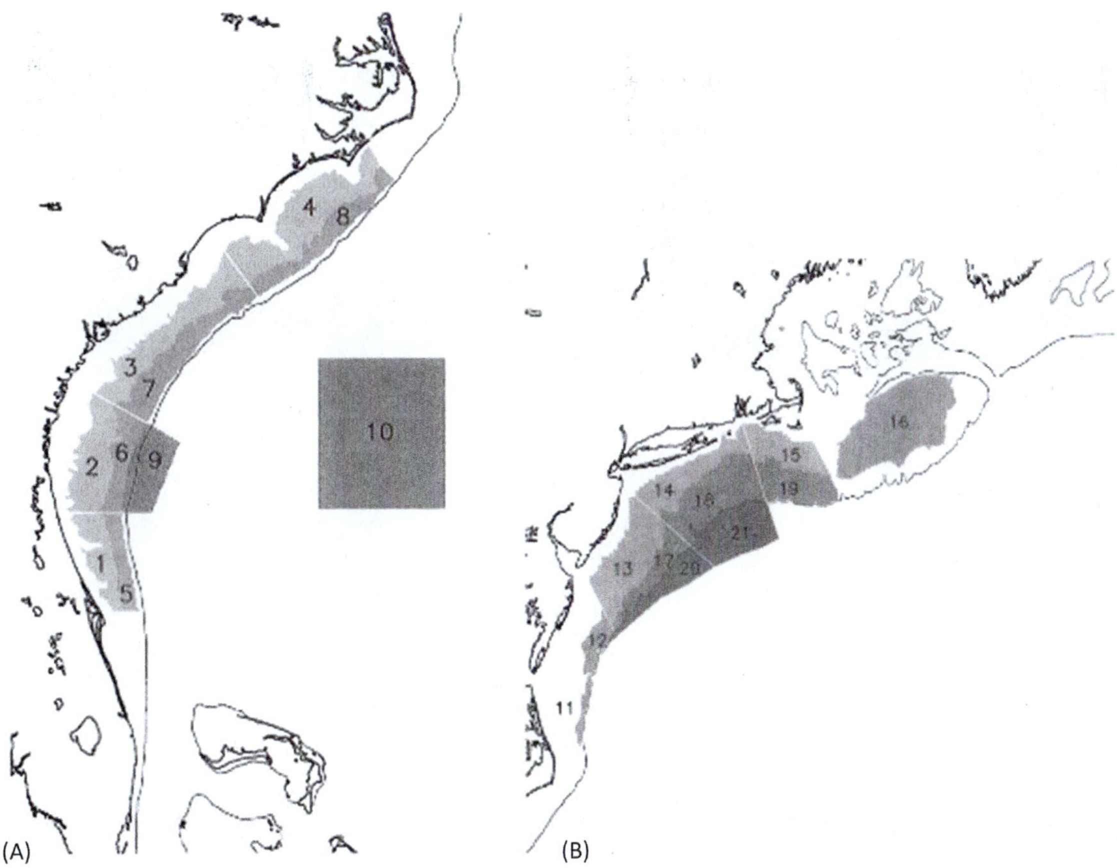

FIGURE 6.12 Study areas off the US East Coast. (A) Middle shelf areas 1–4 are bounded by the 20 and 40 m isobaths, outer shelf areas 5–8 by the 40 and 200 m isobaths, area 9 is the Gulf Stream, and area 10 is in open ocean waters. (B) Middle shelf areas 11 and 12 are bounded by the 20 and 40 m isobaths, and the middle shelf areas 13, 14, and 15 by the 25 and 60 m isobaths. Area 16 is enclosed by the 80 m isobath, outer shelf areas 17, 18, and 19 are bounded by the 60 and 200 m isobaths, and the slope water areas 20 and 21 are bounded by the 200 m isobath and the mean position of the Gulf Stream (Yoder et al., 2000). Map lines delineate study areas and do not necessarily depict accepted national boundaries. Yoder et al. (2000)

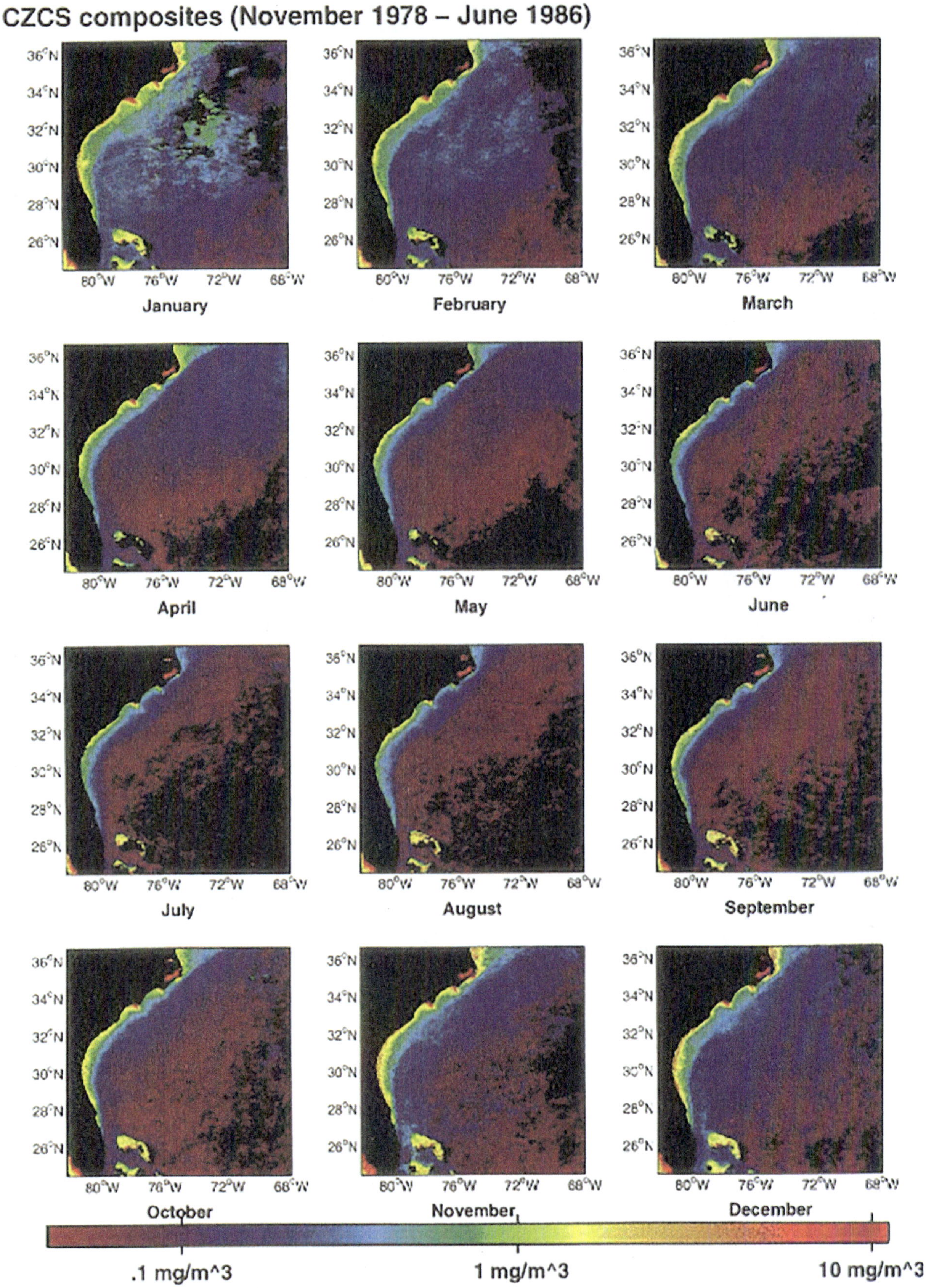

FIGURE 6.13 Monthly mean composite CSAT images for the Southeast Atlantic study area. The color bar represents the CSAT concentration (mgm^{-3}), and black pixels indicate no valid data. The high concentrations (>0.50) surrounding the Bahama Banks (off Florida) are an artifact due to bottom reflection in shallow, clear water (Yoder et al., 2000). *CZCS*, Coastal zone color scanner. Map lines delineate study areas and do not necessarily depict accepted national boundaries.

TABLE 6.3 Statistics for the daily averages of CSAT concentrations.

Area	*N*	Arith. mean	S.D.	Geom. mean
1	231	1.17	1.45	0.80
2	252	0.94	0.98	0.70
3	446	0.75	0.69	0.59
4	467	0.65	0.50	0.53
5	205	0.39	0.30	0.32
6	244	0.42	0.45	0.31
7	89	0.41	0.41	0.32
8	370	0.31	0.25	0.25
9	194	0.13	0.11	0.11
10	189	0.10	0.07	0.09
SEA (1–8)	2604	0.62	0.73	0.44
11	251	0.88	0.95	0.62
12	281	0.96	1.22	0.66
13	421	1.36	1.46	0.94
14	460	1.77	1.86	1.17
15	414	2.16	2.13	1.48
16	309	1.60	1.40	1.21
17	422	0.85	1.01	0.59
18	409	1.02	1.20	0.69
19	384	0.97	1.19	0.65
20	369	0.57	0.60	0.41
21	344	0.58	0.72	0.39
NEA	4064	1.19	1.44	0.75

NEA, Northeast Atlantic; *SEA*, Southeast Atlantic.

and 21) also have a fall-winter maximum after year-day 300 but then have a distinct spring maximum near year-day 100 as previously reported (Brown et al., 1985). It is also clear from Fig. 6.14 that CSAT has a much wider range than do the MARMAP observations during fall-winter-spring when concentrations are generally quite low as compared to the productive summer season. Mean trends in the CSAT data are consistent with those of the MARMAP averages. In area 20, the CSAT values during the summer are systematically lower than the in situ values due to a pronounced subsurface chlorophyll maximum layer, 25–35 m below the surface. Further comparisons with the in situ data suggested that CSAT data may overemphasize the importance of the early winter period to the annual mean chlorophyll concentration compared with the in situ values. Both CSAT and the MARMAP data show that chlorophyll concentrations begin to increase near year-day 290 and continue up until year-day 150 of the following year.

Using 7.5 years of CZCS imagery, Yoder et al. (2000) were able to quantify major time/space CSAT distribution patterns in the ocean margin of the US East Coast. They focused on patterns in mean CSAT concentrations averaged for 21 study areas representing rather large (>100 km^2) shelf/slope areas. They did not examine spatial variability within the 21 study sites, but earlier studies (Yoder et al., 1987; Smith et al., 1988; Denman and Abbott, 1988) indicate that such variability is significant. Their nominal sampling interval was one day, but often consecutive-day sampling was not realized due to cloud cover or the fact that CZCS was not operating. They compensated by averaging CSAT over 45-day bins to yield a continuous time series. Their results can be summarized to show:

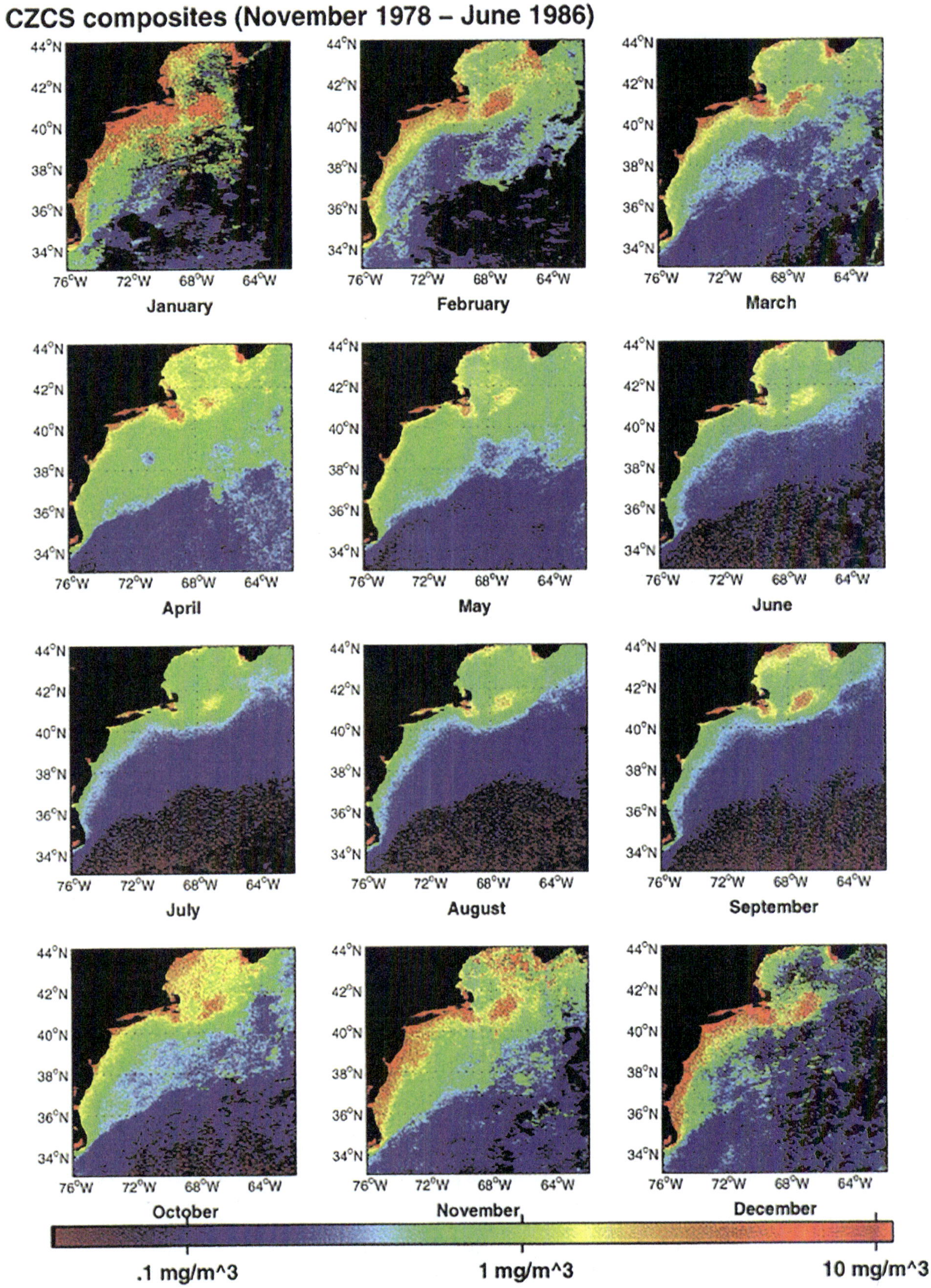

FIGURE 6.14 Monthly mean composite CSAT images for the Northeast Atlantic study area (Yoder et al., 2000). *CZCS*, Coastal zone color scanner. Map lines delineate study areas and do not necessarily depict accepted national boundaries

1. A simple annual CSAT cycle common to all ocean margin waters along the entire East Coast, with a broad peak during winter and a minimum during the summer.
2. Relatively subtle cross-and along-shelf changes in the timing and relative magnitude of the winter CSAT maxima and summer minima, as well as the presence of secondary seasonal peaks in some areas (e.g., spring peak in the outer shelf and slope waters of the NOEA).
3. High variability at time scales of days to weeks superimposed on the seasonal pattern.
4. High spatial coherence of the seasonal component between all 21 study areas.
5. High coherence of the days-to-weeks components between study areas near each other but generally low coherence for study areas far apart.
6. Detectable but weak interannual variability.

Ocean color sensors such as the CZCS require some System of Vicarious Calibration (SVC) after launch and during operation (Barnard et al., 2004). This is because the prelaunch and onboard calibration uncertainties are insufficient to meet the ocean color product requirements (Gordon, 1987). The SVC process has been used to evaluate and adjust the calibration factors of visible bands of the ocean color sensors. This aspect will be discussed more completely when the NASA SeaWIFs mission is discussed.

6.2.2 The scanning multichannel microwave radiometer

6.2.2.1 *Retrieval of sea surface temperature*

As discussed earlier, the SMMR first flew on the Seasat satellite, and its continuation on Nimbus-7 was an effort to deploy the same instrument on another platform. A great effort was made to integrate the SMMR onto the Nimbus satellite, which launched only a few months after the demise of Seasat. The SMMR was a 5-frequency and 10-channel microwave radiometer operating in the range of 6.6–37 GHz. It was the prototype of many multichannel conical scanning microwave radiometers of the future. It was a conical scanner with an 80-cm antenna, rotating at 15 rpm with an incidence angle of 50.2 degrees, resulting in a swath width of 780 km. The resolution changes with frequency, ranging from 22 km at 37 GHz to 120 km at 6.6 GHz. It had a mass of 52.3 kg, a power requirement of 60 W and a data rate of 3 kbps. It flew on Nimbus-7 from late 1978 to 1994. The SMMR channels are described in Table 6.4.

As was mentioned in the discussion of the SMMR on Seasat, one of the primary objectives of the SMMR was to measure SST in cloudy conditions. While it was true that the spatial resolution of the SMMR SST would be significantly worse than that of the AVHRR, the ability to view the SST in the presence of clouds was a pretty strong motivation.

Gloersen (1983) computed global SST from Nimbus SMMR observations and used January 1979 as an example. The ability of SMMR to provide accurate SST measurements is limited by the noise in the radiometer, which for the Nimbus SMMR was ±0.4 K in the 6.6 GHz band. This noise level was observed in prelaunch laboratory test viewing a 150 K target and was later confirmed on orbit by analyzing the radiometric signal relative to the cold space calibration target. As pointed out by Gloersen (1983), this noise level can be reduced by averaging over several footprints since the footprints are considered independent. Other sources of noise in the SMMR SST measurements are variations in humidity and the near-surface winds. In addition, the roll and pitch of the spacecraft can contribute to this variability.

The Numbus-7 SMMR experiment team for SST used spatial averaging to compute a global map of SST for the first 12 months of SMMR data (Gloersen et al., 1984; Wilheit et al., 1984). In summary, this method consists of averaging (with weights that take into account the antenna pattern) 14 6.6 GHz footprints into 5 cells across the swath and about 6 footprints along track, with each cell being about 156 km on a side. Other SMMR channels required for this calculation (wind and water vapor corrections) are also averaged into cells of this size. Comparison of these SSTs with colocated ocean buoys has yielded SST rms variability of 1.5°C for individual observations and less than 1.0°C for 2-week averages for 150 km cells.

In his study, Gloersen (1983) mapped the SSTs onto a 1 degree × 1 degree latitude–longitude grid after applying a polarization mixing correction and calibration adjustments, which are improvements over those discussed by Gloersen (1981). The SST retrieval algorithm used in Fig. 6.16 was relatively simple and involved corrections only for sea surface wind effects and precipitable water effects.

The choice of 135 K in the 18 H channel data as the cutoff point to be included in Fig. 6.16 was based on a histogram of all of the 18 H data in January of 1979 over the oceans. This cutoff deleted less than 10% of the total January 1979 data. Additional filtering was carried out with the SST relevant channels:

140K<10V<180K

70K<10H<130K

135K<6V<175K.

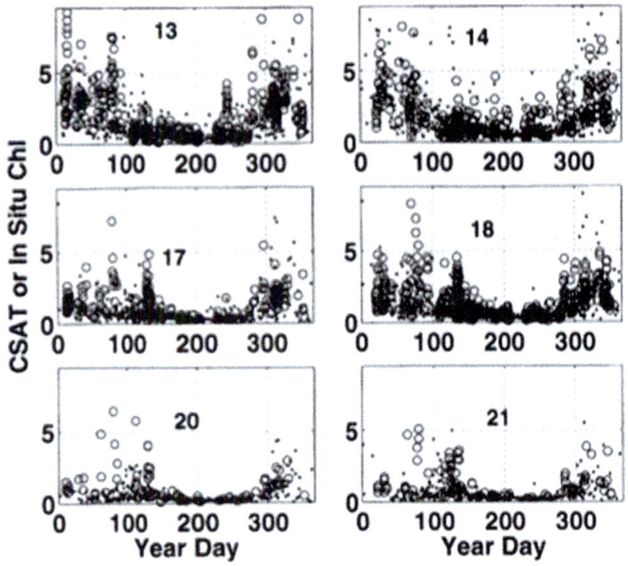

FIGURE 6.15 MARMAP in situ near-surface chlorophyll a+pheopigments (open circles) and CSAT concentrations (dots, mgm-3) versus year day for the representative Northeast Atlantic study areas (13 and 14 middle shelf; 17 and 18 outer shelf; and 20, 21 slope). In situ concentrations are single station values, while the CSAT concentrations are the daily means for that portion of the image covering the station location (Yoder et al., 2000).

TABLE 6.4 SMMR characteristics.

Central frequency (GHz)	Bandwidth (MHz)	Polarization	NEΔT (K)	IFOV (km)	Pixel size
6.6	250	V, H	0.9	95×160	40×26
10.7	250	V, H	0.9	60×100	20×26
18.0	250	V, H	1.2	35×60	20×26
21.0	250	V, H	1.5	30×50	20×26
37.0	250	V, H	1.5	17×29	10×26

IFOV, Instrument field of view.

Fig. 6.16 suggests that land effects extend into ocean areas as a result of the antenna patterns. Radiation from land can be observed in Fig. 6.16 as far out as 600 km from shore when the satellite proceeds from land to the ocean. Only descending (nighttime) orbits were used to produce Fig. 6.16 in an effort to avoid any possible sunglint contamination. The limits on the 6.6 V channel correspond to an SST range of about 80°C, which greatly exceeds any temperature of the ocean.

The map of SST in Fig. 6.16 is consistent with the accepted global distribution of SST. The poleward gradients agree with the SST from the Navy climatological reports. Coastal upwelling signatures are influenced by land contamination, making it difficult to discern the coastal upwelling regions. One of the most significant features of Fig. 6.16 is the tongue of cold water that extends northwestward from further south in the eastern South Pacific. This tongue reaches the equator at longitudes of 105 degrees–130 degrees W and the temperature increases by about 3°C as one progresses from east to west. This tongue was also observed in the analysis of ship intake SSTs (Liu, 1982). There are many other features of Fig. 6.16 that agree with climatological SST maps, showing a unique ability of the SMMR to map and monitor SST due to the lower frequency channels.

6.2.2.2 *Retrieval of precipitation*

The bulk of all tropical precipitation occurs in tropical cloud clusters (Petty and Katsaros, 1989), which can easily be identified in satellite imagery by their mesoscale cirriform cloud shields. Within these cloud clusters, most of the rain falls in mesoscale precipitation features that are made up of convective and stratiform components. While tropical cloud clusters are formed and driven by convective activity, studies have shown that a third to a half of the total rainfall in tropical clusters may be due to the stratiform component (Houze and Betts, 1981; Houze and Hobbs, 1982). Petty and Katsaros (1998) compare SMMR observations of three tropical cloud clusters with operational visible and infrared satellite imagery and with radar estimates of precipitation parameters. The goal of their study was to demonstrate the utility of passive microwave satellite data for the global monitoring of rainfall and to begin creating a quantitative method to use microwave observations to study tropical cloud clusters. In support of this goal, these authors introduced a normalized form of the observed 37 GHz brightness temperature polarization difference, which they interpret as microwave attenuation by rain, different from earlier emission-based rain algorithms (Wilheit et al, 1977) and scattering-based algorithms (Spencer, 1986).

The ability of a microwave radiometer to detect rainfall over the ocean is made possible by the strong contrast between the thermal emission of rain and that of the sea surface (Wilheit et al., 1977). At 37 GHz, the microwave emissivity of the ocean is low (~0.5), resulting in the ocean appearing as a relatively cold radiometric background. The emissivity of an optically thick layer of rain is closer to 1.0. Rain over the ocean will give a much higher brightness temperature for the SMMR than does the ocean surface. Raindrops can also introduce scattering effects by precipitation-sized particles. This scattering effect increases with increasing particle size (Savage and Weinman, 1975) and acts to depress brightness temperatures below that which would occur in a layer of similar opacity but with no scattering. So as rain starts, and the rain rate increases, the microwave brightness temperature first increases rapidly and then saturates for some critical value of rain rate (typically near 4 mm/h for a 4 km think layer). After this, there is a reduction in brightness temperature as the rain rate increases further(Savage, 1976; Wu and Weinman, 1984; Spencer, 1986).

Other geophysical factors influence the 37 GHz brightness temperatures to a lesser degree than rain. These are atmospheric water vapor content, cloud-liquid water, SST, and wind-induced sea surface roughness. (Wilheit and Change,

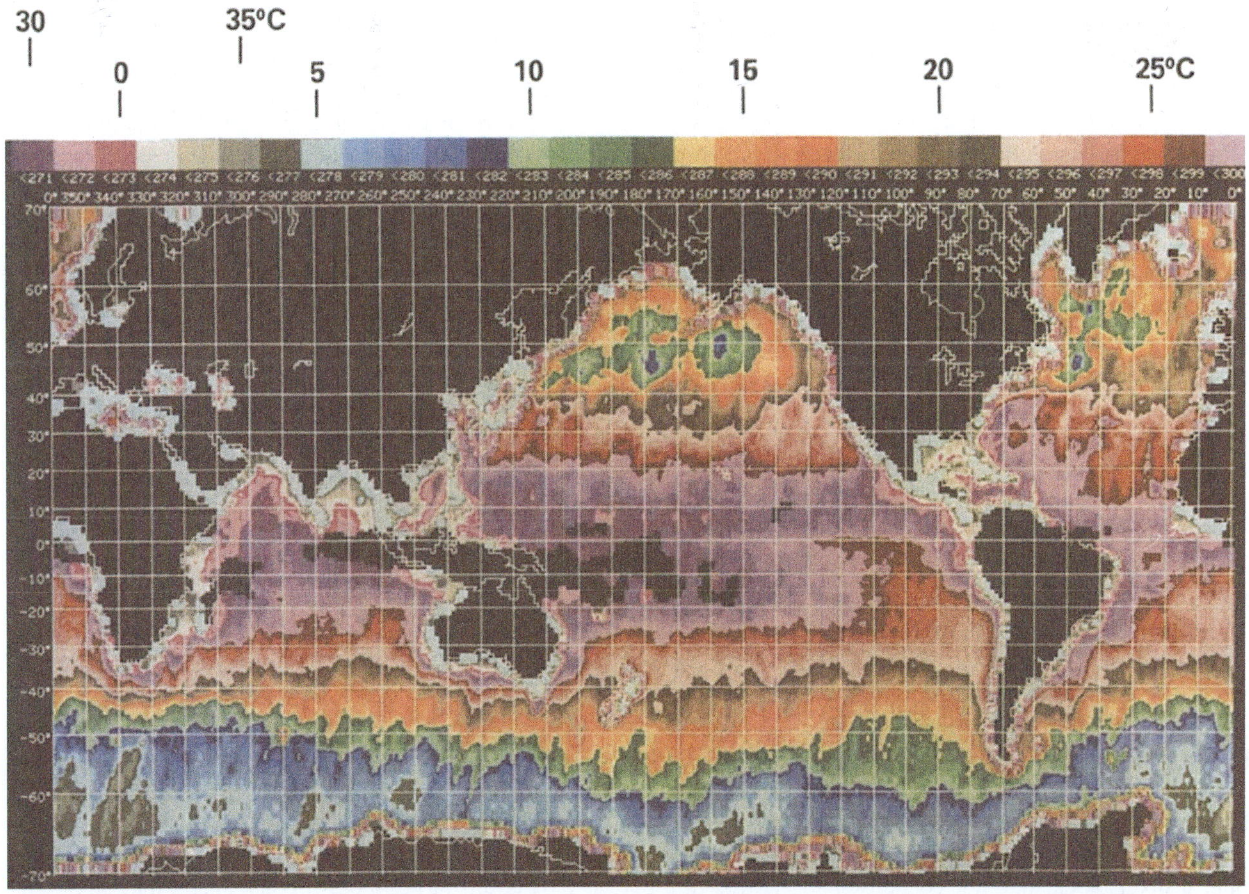

FIGURE 6.16 Global sea surface temperature for January 1979, as calculated from Nimbus-7 radiance. Color bar ranges from −2.0°C to 29.0°C (Gloersen, 1983). Map lines delineate study areas and do not necessarily depict accepted national boundaries.

Wentz, 1983). If an accurate rain rate retrieval is desired, particularly for low rain rates, the variability of these other parameters should be considered. Most passive microwave retrieval methods have attempted to exploit the emission of rain clouds by relating changes in brightness temperature in one polarization to variations in rain rate. Spencer (1986) proposed a semiempirical method that relates the observed scattering properties of deep convective clouds at 37 GHz to rain intensity over the ocean. This latter method is primarily intended to be used with large rain rates associated with strong convective clouds that contain a significant amount of frozen precipitation particles (i.e., graupel and hail) above the freezing level and may be a useful complementary method to emission-based algorithms for light rain. This method was developed for conditions over North America, which may not apply elsewhere.

Petty and Katsaros (1998) proposed a new approach to interpret passive microwave observations as rain. The method makes use of the attenuation of rain on the polarization of the observed microwave brightness temperatures. For the oblique viewing angle (50.3 degrees) of the Nimbus-7 SMMR, the microwave sea surface emissivity is greater for vertical than for horizontal polarization, so that over the unobscured ocean, there is typically a 40–70 K difference in the measured 37 GHz brightness temperatures in the two different polarizations. Unfortunately, the atmosphere first attenuates and also emits at this frequency, nearly independent of polarization, thus reducing the polarization difference associated with rain over the ocean. This difference decreases monotonically to near zero with decreasing visibility of the ocean surface (Spencer et al., 1983; Beer, 1980). Petty and Katsaros (1998) express the polarization difference as a fraction of the value expected in the absence of rain and clouds. The result can be interpreted as a rain cloud transmittance with much of the influence of unrelated geophysical parameters having been factored out.

The primary limitation to any rain rate estimation from SMMR data is the coarse spatial resolution of the sensor and the strong spatial inhomogeneity and complex microphysical structure of rain clouds. This has resulted in it being difficult to obtain unbiased rain rate retrievals simply by inverting simple models of the atmosphere. Improvements in theoretical retrieval models require dramatically improved knowledge of the rain cloud structure and its statistical nature. For their study, Petty and Katsaros (1998) felt justified in using a simple model to quantitatively explain the polarization signature of rain. They also relied on data from surface-based radars to develop useful empirical relationships between SMMR observations and tropical precipitation.

During the WMONEX experiment conducted in December 1978 in the South China Sea, MIT operated a C-band digital weather radar at Bintulu on the northern coast of Borneo. The radar operated during the entire period of the experiment without any significant gaps, sampling a volume of the atmosphere at a sequence of elevation angles out to a nominal range of 256 km from Bintulu. Each volume scan took about 8 min to complete, and a new sequence was started every 10 min. Unfortunately, the site was restricted by partial blockage of the radar beam by trees and buildings at low elevation angles. As a consequence, some care needed to be exercised in interpreting the reflectivity patterns. The volume scans of interest were interpolated to Cartesian coordinates using a method that corrected for Earth curvature and refraction under standard atmospheric conditions. The grid spacing was 3.75 km, which corresponds to one-eighth of the diameter of a gridded 37 GHz SMMR pixel. The vertical grid spacing was 1 km. Due to blockage, the lowest 1 km was unreliable and not considered for their analysis. Instead, the largest reflectivity value above the surface but below 4 km was used. This constraint also avoids contamination by bright band effects near the freezing level, which was typically about 4.5 km above sea level. To convert radar reflectivity to rain rate, they used the Z–R relationship of Austin and Geotis (1979), which was derived from observations of near-equatorial oceanic precipitation during GARP (Global Atmosphere Research Program) Atlantic Tropical Experiment (GATE).

$$Z = 180R^{1.35} \tag{6.1}$$

The Nimbus-7 SMMR data at 37 GHz has a spacing of approximately 30 km both along and across the swath, and the total swath width is 26 cells or approximately 780 km. The navigation of the SMMR data was checked by comparing the 37 GHz horizontally polarized brightness temperatures against the northern and southern coasts of Borneo. Corrections of up to 0.7 cell diameters, or about 20 km, were necessary for some orbits to bring the SMMR images of the Borneo coastline into agreement with the true coastline. The clear-sky polarization difference was determined for each SMMR pixel from a square array of pixels (dimension $M \times M$, where M is an odd integer) centered on the pixel in question. For pixels with a polarization difference of less than 35 K, which indicated clearly that they were influenced by clouds, rain, or land, the Nth percentile value of the remaining polarization differences was chosen as the baseline polarization difference.

The choice of M and N is subjective but needs to consider the horizontal scale of the rain clouds being studied in addition to the fraction of SMMR pixels within the $M \times M$ block, which are expected to satisfy the clear-sky condition. If M is too small, a pixel block may not contain enough cloud-free pixels. If it is too large, the clear-sky pixels may not represent conditions at the center of the block. After some experimentation, Petty and Katsaros (1998) found that M=13 and N=90 led to smooth and physically consistent fields of the clear-sky polarization differences. The clear-sky polarization field agrees to within two or three degrees with those predicted by the model of Wentz (1983) when applied to the water vapor and wind speed fields obtained using the lower frequency SMMR channels.

The observed polarization differences for each pixel were divided by the corresponding clear-sky value to yield the field of P. Comparison of maps of P to GMS-I imagery of the area revealed that nearly cloud-free areas were associated with values of P close to unity, as expected. Some pixels with $P>1$ do appear, especially near the Borneo coast. For these pixels, their method underestimates the clear-sky polarization difference. Using a larger value for N would reduce the number of pixels for which the clear-sky polarization difference is underestimated but would also lead to the clear-sky polarization being determined by a very small number of pixels. Small errors in the clear-sky polarization introduce relatively small errors in P, which are small compared with the variations due to precipitation.

The value of P is plotted against horizontally polarized brightness temperatures in Fig. 6.17 for three cases that we will discuss.

Since $T_{B,H}$ is the more familiar microwave index of rain, it is important to show its very high correlation with P. The scatter in the plot can be attributed to (1) variations in the background, which are factored out in P but not in $T_{B,H}$, and (2) differences in the effective emitting temperature of the rain clouds in each pixel, including those due to volume scattering by ice.

The mesoscale cloud clusters in the WMONEX experiments were slow moving, and clouds persisted over a given area for periods of a day or longer. They were also very large with horizontal scales of several hundred kilometers. Houze et al. (1981) found that the convection in WMONEX exhibited a well-defined diurnal cycle with convective activity offshore peaking in the early morning hours due to the interaction between the large-scale monsoon wind and the local land-breeze circulations. By 0800 LST, the convection had merged into an organized mesoscale system, with many convective cores bound together by an extensive (~200 km) stratiform precipitation area. By midday, the mesoscale systems began dissipating, so that stratiform rainfall from the large mid-to-upper level clouds increasingly shielded convection as the dominant source of precipitation.

Petty and Katsaros (1998) examined three cloud clusters in the South China Sea on December 11, 17, and 23, 1978, where the first and the last dates gave the best overlap of SMMR with the MIT radar. This coincident coverage allowed them to make an assessment of P as an index of tropical precipitation activity. The cloud cluster on December 17 was not detected by the radar at the time of the SMMR overpass. The December 11 and 23 cloud clusters were days of unusually strong northeasterly monsoonal flow over the South China Sea, which is called "cold surges." These events frequently bring heavy rainfall to the North Borneo coast. The December 17 cluster occurred during a declining northeasterly flow following a strong surge on December 15.

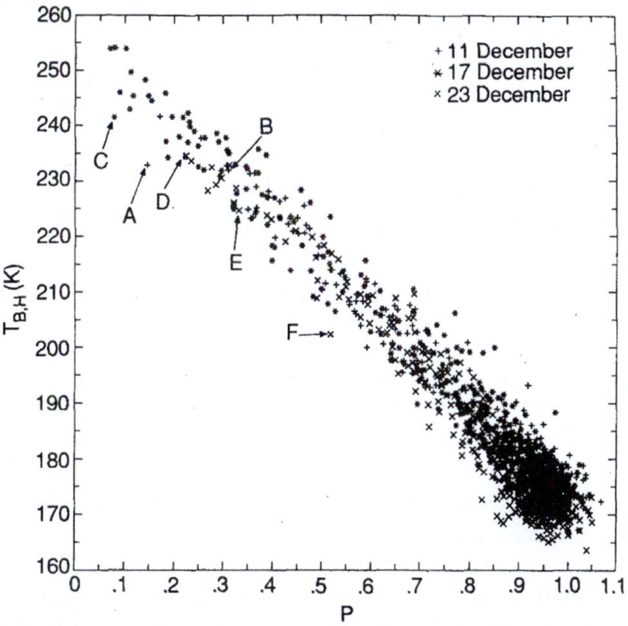

FIGURE 6.17 Scatter plot of SMMR normalized polarization difference(P) versus horizontally polarized brightness temperature ($T_{B,H}$) at 37 GHz for three cloud cluster cases (Petty and Katsaros, 1998).

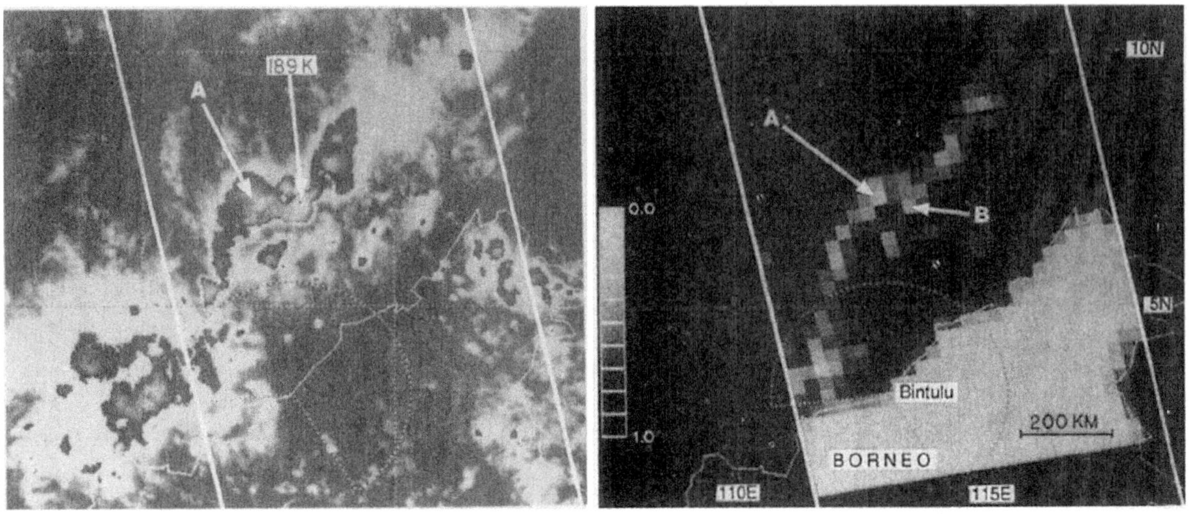

FIGURE 6.18 SMMR and GMS-1 images of tropical cloud cluster over the South China Sea off the Borneo coast, December 11, 1978. Dashed semicircles in the GMS-1 image indicate the radar coverage, while the solid lines are the Borneo coast. Left is the GMS-1 infrared image at 0533 UTC time, and right is the SMMR pass at 0412 UTC. The labels A and B refer to the labels in Fig. 6.17 (Petty and Katsaros, 1998). Map lines delineate study areas and do not necessarily depict accepted national boundaries

In each case study, the cluster was observed during the daytime crossing of Nimbus-7, which occurred around local noon. The observations generally coincide with a weakening phase of the clusters, although new convection is still observed in two of the cases. Visible and infrared (IR) images from the Japanese geostationary meteorological satellite, GMS-1, were obtained roughly 1.5 h before and after each SMMR crossing. Due to the time difference, only qualitative comparisons can be made between the SMMR and GMS-1 images (Fig. 6.18).

Petty and Katsaros (1998) also present the visible GMS-1 images and do so for the time preceding the SMMR pass. They present 3 cases with similar results. The area of low P agrees fairly well with the colder IR brightness temperatures, particularly at 0533 UTC. The point labelled as A in Fig. 6.17 is 233 K, several degrees colder than expected from the corresponding value of P. This suggests that this spot is a location of particularly strong convective activity. They found from all three cases that the ability of the SMMR to delineate the region of precipitation beneath the extensive and rather amorphous cirrus cloud shield to be impressive. It was also clear from both the SMMR and GMS-1 images that the radar sensing of precipitation at this range and azimuth was deficient. When the SMMR pass fell within the radar coverage, they found a very good agreement that was limited by the small number of comparison points. They concluded that the SMMR was capable of mapping precipitation in selected tropical cloud clusters. In every case, the SMMR demonstrates the ability to distinguish between major regions of rain and no rain, even beneath cirrus shields that accompany mesoscale cloud clusters. They found that the SMMR 37 GHz normalized polarization difference is a rather good predictor of fractional coverage by rain. These results have application to the Tropical Rainfall Mapping Mission (TRMM). They also learned that the radar-derived precipitation patterns, often used as ground truth for validating satellite data, are far from infallible.

6.2.3 The solar backscatter ultraviolet spectrometer (SBUV)

The SBUV was a follow-on of the backscatter UV (BUV) flown on Nimbus-4 and had the goal of a vertical profile of ozone and other species in the atmosphere. It was an Ultraviolet spectrometer having 12 discrete 1 nm bandwidth channels in the interval 150–340 nm with a continuous sweep from 160 to 340 nm. It was nadir viewing only at 200 km sampling intervals with a global coverage in 14 days. It was utilized by NASA from 1979 to 1994. It was later carried by the NOAA polar orbiting satellites. The sensor collects radiation reflected off the Earth in wavelengths characteristic of ozone. Ozone is measured as a ratio of sunlight incident on the atmosphere to the amount of sunlight scattered back into space. From this information, the total ozone can be calculated. The SBUV is capable of measuring the global ozone concentration in the atmosphere to an absolute accuracy of 1% and the vertical distribution of atmospheric ozone to 5%.

Bhartia et al. (2013) describe an algorithm that has been used for 42 years to create a record of total ozone and ozone profiles from eight SBUV instruments flying on both NASA and NOAA satellites. While the SBUV first flew on Nimbus-7, it was continued by NOAA on their operational polar orbiting satellites. This history is depicted in Fig. 6.19 which shows the times and the satellites that carried the SBUV and its successor, the SBUV/2.

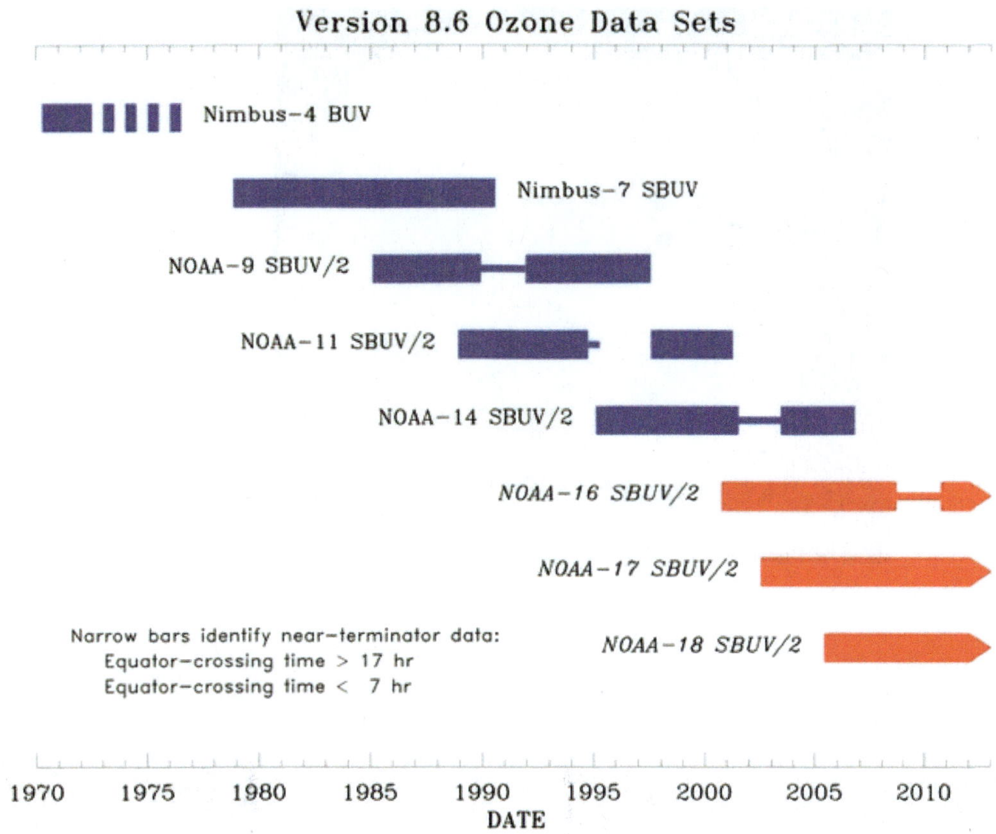

FIGURE 6.19 Timeline of the solar backscatter ultraviolet spectrometer (SBUV) instruments flown on NASA and NOAA satellites (Bhartia et al., 2013).

This algorithm was optimized to derive monthly zonal mean (MZM) anomalies for ozone assessment and model comparisons. This approach minimizes the effects of some errors and noise due to clouds and aerosols. In addition, with this approach, errors due to vertical smoothing of the quasibiennial oscillation and other periodic and aperiodic variations become more important. Using this algorithm, they provide smoothing kernels that can be used to compare anomalies in SBUV profiles and partial ozone columns with models. They show how to use these kernels to compare SBUV data with microwave limb sounder ozone profiles. These comparisons are particularly useful in the lower stratosphere, where SBUV profiles have poor vertical resolution. They also provide their best estimate of smoothing errors associated with SBUV MZM profiles. Smoothing errors are the largest source of uncertainty in these ozone profiles, and they are treated as error bars in deriving interannual variability. They derive total column ozone by integrating the SBUV profiles, rather than using a separate set of wavelengths. This makes it possible to extend the total ozone retrieval to 88 degrees solar zenith angle (SZA). It should be noted that the quality of total column data is affected by a reduced sensitivity to ozone in the lower atmosphere due to cloud and Rayleigh attenuation, which gets worse with increasing SZA.

The Nimbus-7 SBUV was preceded by the BUV that flew on Nimbus-4 in April 1970. Since then, nine additional instruments have flown on NASA and NOAA satellites (Frederick et al., 1986; Heath et al., 1975; Mateer et al., 1971). The SBUV was first launched on Nimbus-7 and was mated with the TOMS, which was designed to produce global maps of total ozone (Heath et al., 1975). An interesting story is that the researchers working with SBUV data since 1978 discounted the low values over Antarctica as errors in the data. It was not until May of 1985 when members of the British Antarctic Survey noted the ozone hole (Fig. 6.20) and published a paper in the journal Nature that NASA researchers reexamined the SBUV data that clearly showed the ozone hole over the Antarctic. Later research by Susan Solomon of NOAA connected this ozone hole to the presence of chlorofluorocarbons in the atmosphere, such as freon, that were widely used for air conditioning. This led to the Montreal Protocol, which banned the use of freon for air conditioning in an effort to maintain the atmospheric ozone layer.

Bhartia et al. (2013) conclude that their algorithm can be used to create a 42-year time series combining data from Nimbus-7 with later data collected from NOAA satellites. Errors in their computations occur mostly in the lower atmosphere,

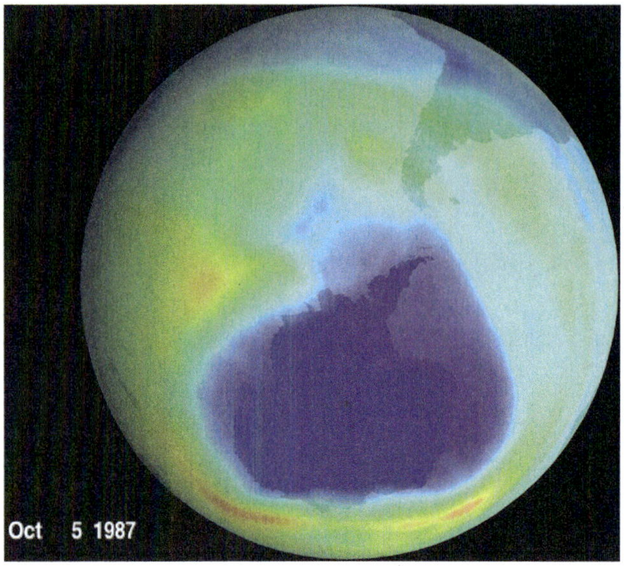

FIGURE 6.20 Total ozone over the Antarctic, October 5, 1987 (Heath et al., 1975).

where the SBUV data are less reliable. They conclude that the vertical resolution of SBUV-retrieved profiles varies from ~6 km near 3 hPa to >15 km in the lower stratosphere. Though the SBUV algorithm does not use the longer wavelengths traditionally used by total ozone measuring systems such as TOMS, the quality of MZM total column ozone derived by integrating the SBUV profiles appears to be quite good. Since the SBUV uses more ozone-sensitive wavelengths than TOMS, it is expected that SBUV total ozone will be less sensitive to instrument drift than TOMS. Since the SBUV derives ozone profiles, while TOMS uses a climatological database of ozone profiles, it is expected that SBUV total ozone will be less affected by profile variations at SZA>80 degrees. They also demonstrate that SBUV partial ozone columns, particularly the stratospheric columns, are also of high quality.

6.2.4 Total Ozone Mapping Spectrometer

The TOMS was a NASA spectrometer specifically built for measuring the global ozone layer. Of the five instruments built, four entered successful orbits. The first flew on Nimbus-7, launched on October 24, 1978. The TOMS on Nimbus-7 provided global measurements of total column ozone on a daily basis (Fig. 6.21).

Bowman and Krueger (1985) produced a global climatology of total ozone from four years of daily Nimbus-7 TOMS observations. Measurements were collected at local noon, and no observations are available in polar regions during polar night. The instrument data have been area averaged onto a 5 degrees × 5 degrees latitude–longitude grid. The global coverage is more than 90% complete during the past 3 years (October 1978–September 1982) and approximately 70% complete in the first year (October 1978–September 1982). A global map of the time-mean global ozone is plotted in Fig. 6.22 along with a time-mean total ozone as interpreted by London and Oltmans (1978/1979) from Dobson and M83 measurements.

The TOMS analysis exhibits a number of features that differ from the ground-based field of total ozone. These differences may have a number of sources: (1) errors in the TOMS instrument or the Dobson/M83 measurements, (2) subjective errors in hand contouring of the ground-based measurements, (3) sampling errors, or (4) a real change in ozone content between the observing period of the ground measurements (1957–1975) and that of the TOMS data (October 1978–September 1983). Since Bhartia et al. (1984) found that small relative errors between TOMS and the Dobson/M83 measurements exist, (1) it is not likely the source of the difference. (2) It is very difficult to quantify. (3) It should not be a problem with the TOMS data, leaving only (4) the source to be the differences in the two ozone fields. These authors conclude that, in view of the known difficulties of intercalibration and spatial coverage with the Dobson and M83 networks, the TOMS total ozone map should be a better estimate.

The TOMS geographical distribution is nearly zonal in the mid-latitudes (Fig. 6.22). The ridges over the ocean and central Asia and the corresponding troughs over the continents are suppressed in the TOMS map. The TOMS analysis does show distinct drops in total ozone over mountains, reflecting the decrease in the size of the atmospheric column. In

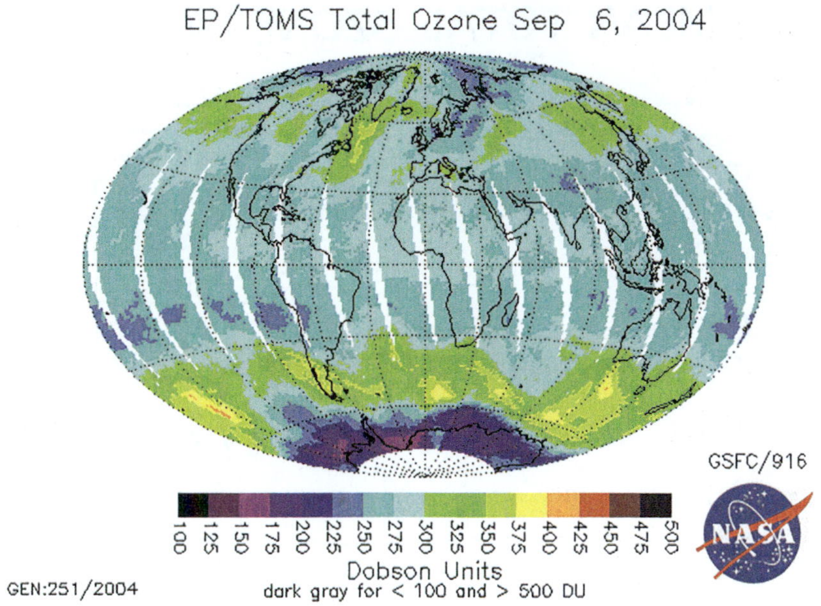

FIGURE 6.21 Total Ozone from total ozone mapping spectrometer (TOMS). Map lines delineate study areas and do not necessarily depict accepted national boundaries.

the southern hemisphere, the large ridges over the oceans seen in the bottom panel of Fig. 6.22 do not appear in the corresponding TOMS map.

The rms deviations of the daily ozone values from the long-term mean are presented here in Fig. 6.23.

The variance is minimal around and just south of the equator and increases towards the poles in both hemispheres. The total variance is larger in the middle and high latitudes of the northern hemisphere. The confidence intervals for the total ozone are proportional to the variance; they will be largely zonal and are reflected in the variance of Fig. 6.23. The TOMS data were then used to plot the annual harmonic and the semiannual harmonic in amplitude and phase. The annual harmonic in the northern hemisphere reaches a maximum in late winter to early spring. The annual harmonic explains a large fraction of the variance over much of the Earth, especially in the tropics.

In another study, Krueger et al. (1995) used TOMS data to study volcanic sulfur dioxide. They used radiative transfer simulations to evaluate the accuracy of retrieving sulfur dioxide measurements from TOMS at low latitudes (to specify the atmosphere for the model). The retrieval algorithm is a simultaneous solution of the absorption optical depth equations, including ozone and sulfur dioxide at the four shortest TOMS wavelengths and an empirical correction based on background condition residuals. Their retrieval algorithm yields model stratospheric sulfur dioxide plume amounts that are within ±10% over most central scan angles and moderate SZAs if no aerosols or ash are present. Aerosols in the plume in a moderate optical depth (0.3) cause an overestimation of the sulfur dioxide of 15%–25%, depending on particle size and composition. Retrievals over land are affected by reflections from the underlying surface or clouds. The precision of individual TOMS sulfur dioxide measurements is limited by quantization to ±6 Dobson units. This accuracy is independent of most instrument calibration errors but depends linearly on the relative sulfur dioxide absorption cross-section errors at the TOMS wavelengths. Volcanic plume mass estimates depend on corrections for background offsets integrated over the plume area. The errors vary with plume mass and area and are highly individual. Overall, they are least for moderate-sized, compact plumes. Errors of 15%–20% are not unusual. Effusive eruption total mass estimates are more uncertain due to difficulties in separating new from old sulfur dioxide in daily observations.

6.3 SeaWIFS (sea-viewing wide field-of-view sensor)

After the termination of the CZCS carried on Nimbus-7 (August 1994), there was an effort to get a new stand-alone satellite for sensing ocean biological activity. The follow-on was SeaWIFS, which was launched on the Pegasus rocket (Fig. 6.24).

This launch vehicle was carried aloft by an L1011 aircraft that served as its first stage. The rocket would then deploy, and its wings would orient the rocket vertically, and the next two stages would carry the payload into space. In the case

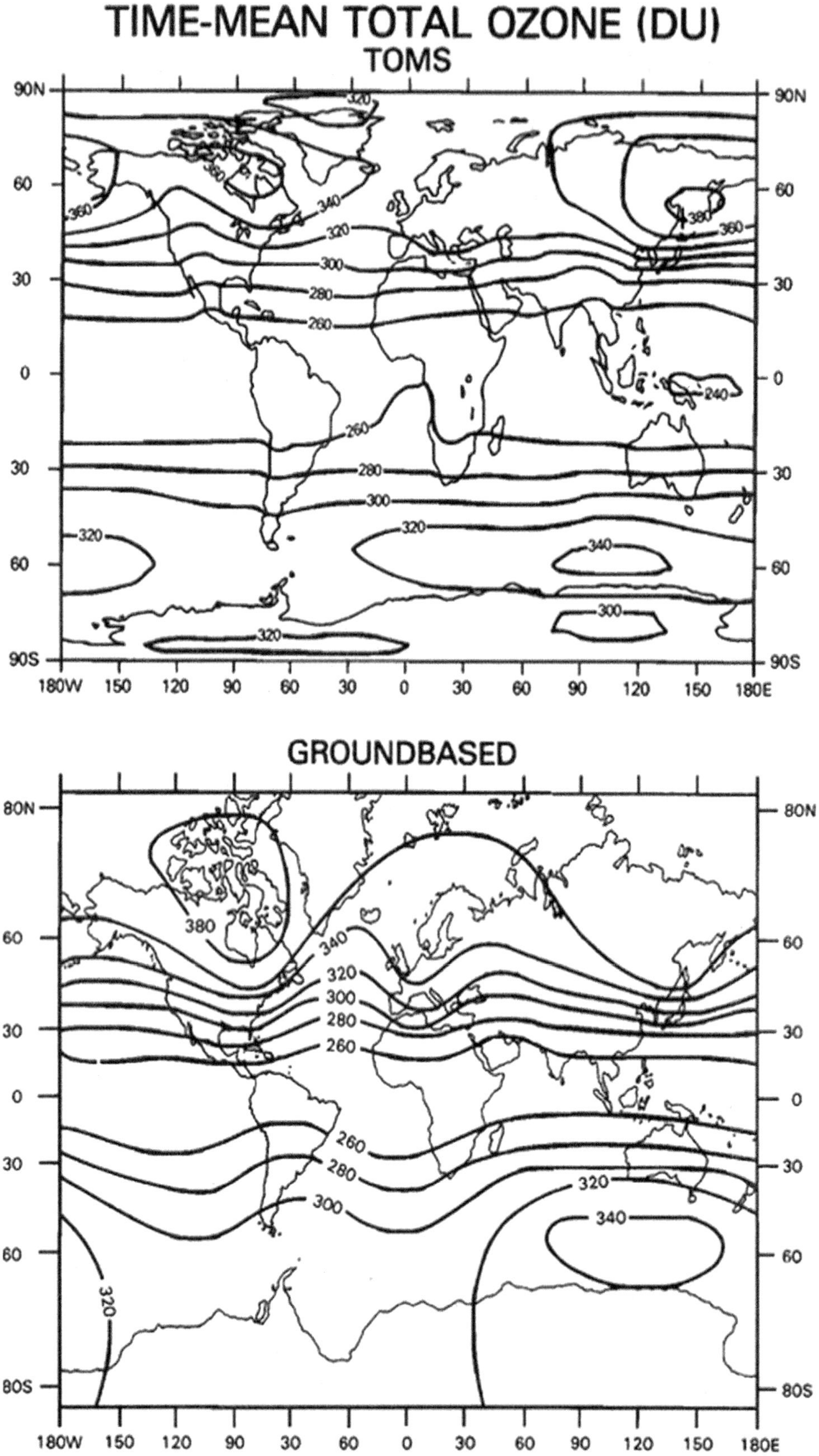

FIGURE 6.22 Total ozone mapping spectrometer (TOMS) time-mean ozone for the 4-year observing period (top) and the time-mean total ozone derived by London and Oltmans (1978/1979) from ground-based Dobson and M83 measurements for the period July 1957–June 1975. Map lines delineate study areas and do not necessarily depict accepted national boundaries.

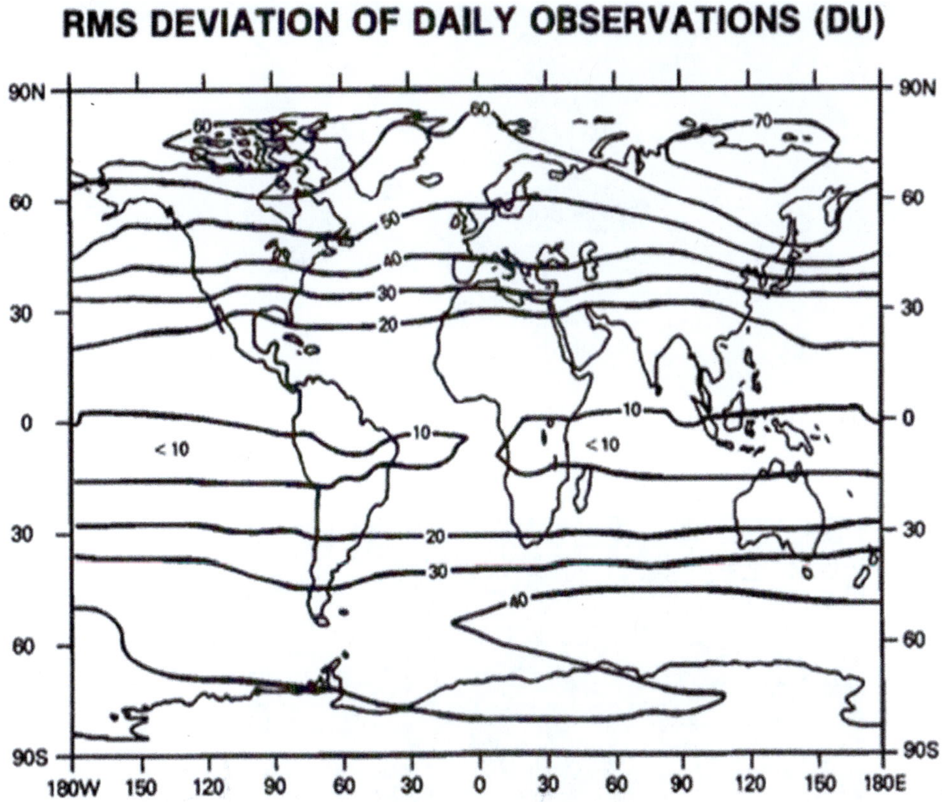

FIGURE 6.23 Root mean square (RMS) deviations of the daily values of total ozone from the long-term mean (Dobson units) (London and Oltmans, 1978/1979). Map lines delineate study areas and do not necessarily depict accepted national boundaries.

of SeaWIFS, the third stage of the Pegasus rocket became the satellite bus known as SeaStar (also known as OrbView), as shown here in Fig. 6.25. SeaWIFS operated from September 1997 to December 2010. Its primary mission was to quantify chlorophyll produced by marine phytoplankton.

Initially, SeaWIFS was a commercial project managed by the Orbital Sciences Corp. NASA was supposed to purchase a large amount of SeaWIFS data for the bio science community. In the long run, this arrangement did not work out, and SeaWIFS became a NASA satellite managed by people at NASA's Goddard Space Flight Center. The sensor's spatial resolution is 1.1 km at nadir. Like the NOAA satellites, SeaStar also had a global area coverage where the spatial resolution was reduced to 4.5 km.

The sensor had the following optical bands.

Band	Wavelength (nm)
1	402–422
2	433–453
3	480–500
4	500–520
5	545–565
6	660–680
7	745–785
8	845–885

The instrument was able to tilt up to 20 degrees to avoid sun glint from the sea surface. This capability was critically important at equatorial latitudes, where glint from the sunlight often obscures watercolor. As mentioned earlier, with CZCS, the atmospheric corrections were very important, and SeaWIFS relied heavily on the Marine Optical Buoy (MOBY) for vicarious calibration. In addition, the satellite would be oriented to sense a full moon, getting the full moon reflection values from the monitoring station at Flagstaff, Arizona.

FIGURE 6.24 Pegasus rocket attached to an L1011 aircraft.
Image credit courtesy NASA.

FIGURE 6.25 The SeaStar satellite, which carried sea-viewing wide field-of-view sensor.
Image credit NASA.

One of the primary observables for SeaWIFS was chlorophyll, which is often calculated from the ratio of blue to green light and relating those ratios to known chlorophyll concentrations (Fig. 6.26).

A typical formula for computing chlorophyll from SeaWIFS data divides the reflectance of the maxima of several wavelengths (443, 490, or 510 nm) by the reflectance at 550 nm. This reflectance (R) is then plugged into a cubic polynomial that relates the band ratio to chlorophyll.

$$\text{Chl} = \text{antilog}\ (0.366\text{–}3.06\ R + 1.93R^2 + 0.64R^3\text{–}1.53\ R^4) \tag{6.2}$$

This formula and others were derived empirically from observed chlorophyll concentrations. NASA maintains a system of related oceanographic and atmospheric data called SeaBASS (SeaWIFS biooptical archive and storage system). A major element of SeaBASS was the MOBY (Fig. 6.27), which began operational use in July 1997 off the coast of Lanai,

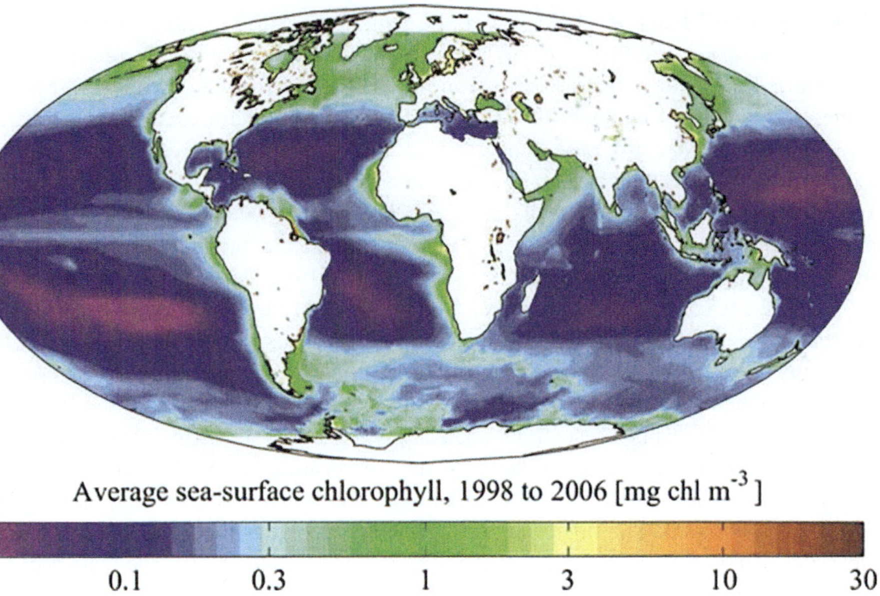

FIGURE 6.26 Sea-viewing wide field-of-view sensor–derived average sea surface chlorophyll for the period 1998–2006. Map lines delineate study areas and do not necessarily depict accepted national boundaries.
Image credit courtesy NASA/GSFC.

Hawaii. This buoy was developed under the direction of Dennis Clark to provide the necessary calibration information required for accurate ocean color measurements. The project was funded jointly by NASA and NOAA. It has continued on under regular NOAA funding. Its primary purpose is to measure visible and near-infrared radiation entering and leaving the ocean. Variations of the visible region-reflected radiation are referred to as ocean color, from which other quantities can be derived, such as the abundance of phytoplankton. The buoy collects data three–five times per day at the same site, allowing oceanographers to monitor the daily fluctuations in biomass concentrations at the site to fine-tune their satellite algorithms. MOBY also serves as a calibration reference for a variety of satellite instruments. This includes nonocean low-end radiance satellite sensors.

MOBY began as a project that Clark initiated in 1986. The buoy consists of six major components, which, because the project was not fully funded at the outset, had to be purchased and integrated over time. These components are a dual spectrograph, fiber optic multiplexer, and a top-end buoy with solar panels, the control electronics, a GOES (Geostationary Operational Environmental Satellite) data relay system, fiber optic radiance and irradiance collectors, and a dichroic water mirror for the dual spectrograph. The optical system uses two spectrographs with a dichroic (water) mirror to measure radiometric properties with high spectral resolution and stray light rejection. The dichroic mirrors allowed the red (630–900 nm) to transmit and the blue (389–600 nm) to be reflected, making the transition from reflectance to transmittance between 590 and 650 nm. Stray light is greatly reduced due to splitting of the visible spectrum at the beginning of the water absorption region.

At 50 feet in length, MOBY is the world's largest marine optical device. Only its antenna, solar panels, strobe light, and surface buoy are above the ocean surface. The mast extending into the ocean is more than 40 feet long, directly beneath the surface buoy with instruments at 6, 16, and 28 feet mounted on 9-foot log booms, or "arms," extending out perpendicular to the mast. Optical collectors (radiance and irradiance) are mounted at the ends of these arms as well as on top of the buoy to collect light coming into the ocean and then light reflected back out of the ocean. The reflected portion of the light has been modified by particles such as phytoplankton suspended at various depths, which modifies the signal available to the satellite sensors. Lenses focus the light onto fiber optic cables, which then transmits the light to a fiber optic multiplexer housed in the instrument bay. The multiplexer relays the light to the dual spectrograph with detectors that measure the spectral radiant energy. These signals are then digitized and then transmitted up to the main computer housed in the surface buoy. The data is then stored on a disk drive, which may be accessed by cell phone, and transmitted via the GOES link.

MOBY was calibrated at the National Institute of Science and Technology (NIST) with two integrating spheres with calibrations based on NIST standards. For irradiance, MOBY uses standard lamps that are routinely calibrated at NIST. Wavelength calibrations are carried out with a series of emission lines observed from a set of low-pressure lamps. Each MOBY instrument views these standards before and after its ocean deployment to estimate any sensor drift. During each deployment, the stability of the MOBY spectrographs and internal optics is monitored using three internal reference sources. In addition, the collection optics for the instruments are cleaned (Fig. 6.27, right) and checked on a monthly basis while the buoy is deployed. This requires monthly visits and SCUBA divers to clean the optics. Divers place lamps over the optics before and after each cleaning to assess the effectiveness of the cleaning. When a buoy is being replaced, the calibration histories of the two MOBYs are compared. In general, the difference between pre- and post-deployment calibration ranges from +1% to 6% with a definite bias towards a negative difference for the post-deployment values. The estimated radiometric uncertainty for MOBY in-water ocean color measurements ranges from about 4% to 8%.

Eplee et al. (2004) describe the lunar calibration of the SeaWIFS instrument to provide a very long-term calibration and evaluate the stability of the instrument. The lunar measurements require a complete reorientation of the satellite to point the sensor at the moon instead of the ocean. These lunar measurements must be normalized to a common viewing geometry. Correction computed from the time and geometry of the observations includes Sun–Moon and instrument-Moon distances, oversampling of the lunar image, and variation in the lunar phase angles. The SeaWIFS project implemented a correction for lunar libration. The lunar calibration method provides top-of-the-atmosphere radiances for SeaWIFS that are stable to better than 0.07% over the life of the mission. The resulting water-leaving radiances are stable to better than 0.7%.

6.4 Tropical Rainfall Mapping Mission

TRMM is the first satellite dedicated to studying tropical and subtropical rainfall, precipitation that falls within 35°N and 35°S of the equator. This tropical rainfall comprises more than two-thirds of the world's total precipitation. The TRMM

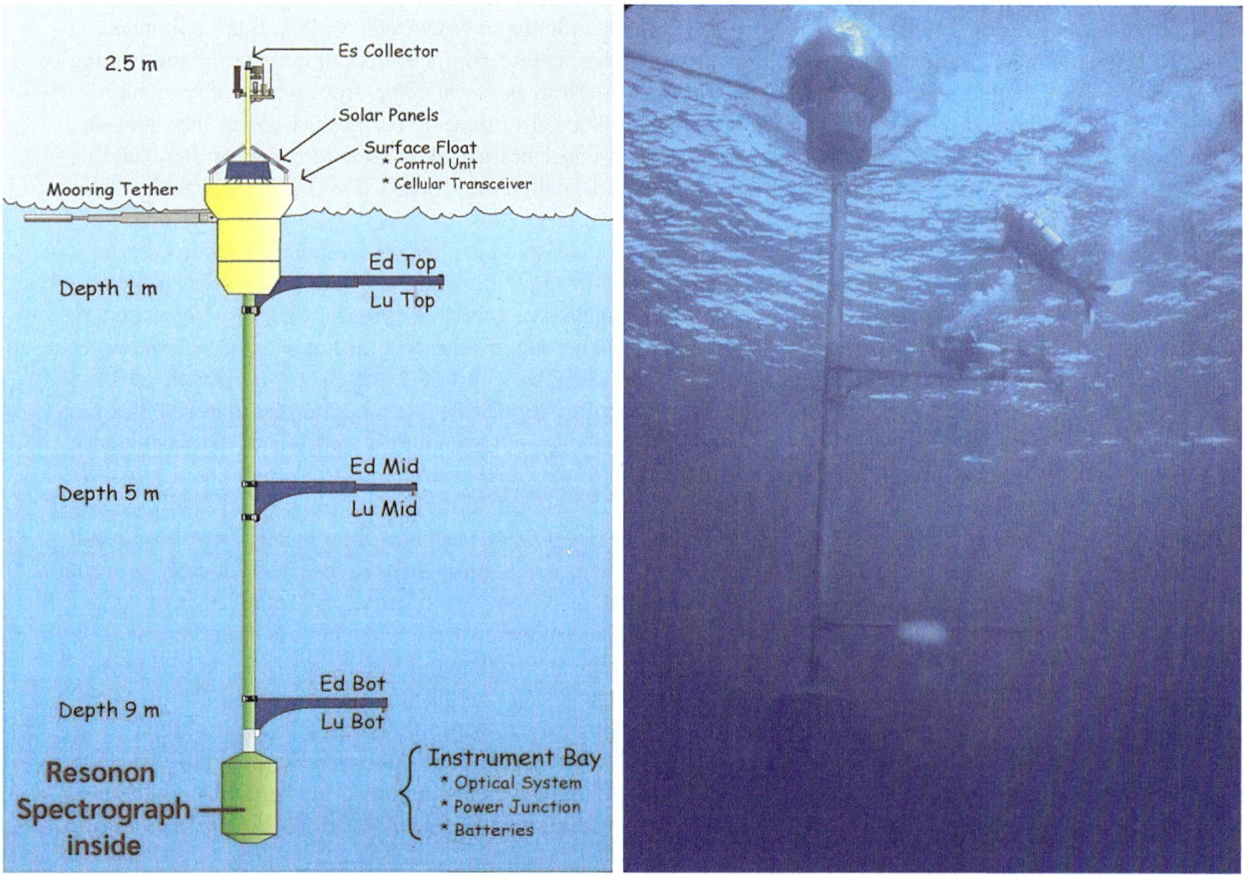

FIGURE 6.27 The marine optical buoy (MOBY) buoy schematic (left) and deployed (right) (Clark et al., 2003).

satellite uses a variety of instruments to detect rainfall and map its effects, including microwave imaging and lightning sensors. The satellite was in a circular, nonsun-synchronous orbit with an inclination of 35 degrees to the equator and a precession period of about 46 days for the orbit nodes with respect to the sun (affecting the time of day of sampling at different latitudes). The orbit altitude varied over the life of the mission, as shown in Fig. 6.28. So from 11/27/1997 to 8/6/2001 it flew at 350±6 km and between 8/24/2001 and 7/15/2024 it was boosted up to 403±6 km. The initial low altitude was selected to ensure sufficient detection capability of the weak radar echoes of the precipitation radar (PR) instrument on TRMM. It was later boosted to a higher average altitude to extend the life of the mission by reducing atmospheric drag.

TRMM carried five instruments to observe tropical rainfall (Fig. 6.29). The visible infrared radiometer is a heritage instrument of the NOAA AVHRR. It is a five-channel, cross-track scanning radiometer sampling at 0.63, 1.6, 3.75, 10.8, and 12 μm to provide high-resolution images of cloud cover, infer cloud type, and measure cloud top temperatures. The TRMM microwave imager (TMI) is also a heritage instrument descended from the SSM/I. It is a multichannel passive microwave radiometer operating at five frequencies: 10.65, 19.35, 37.0, and 85.5 GHz with dual V and H polarizations and at 22.235 GHz with vertical polarization only. The TMI was designed to provide observations of the integrated column precipitation, cloud liquid water, cloud ice content, rain intensity, and rainfall types (e.g., stratiform or convective). While it was not intended to be used for mapping the ocean, it was later very successfully applied to mapping SST. While it is a higher frequency than would be optimal for mapping SST, it was the only passive microwave radiometer flying at the time that could be used to measure SST.

The primary instrument for TRMM was the PR, which was the first of its kind flying in space. It was an electronically scanning radar, operating at 13.8 GHz, intended to measure the 3D rainfall distribution over both land and ocean and to define the layer depth of this precipitation. The CERES was also carried by a variety of other satellites. Its purpose was to

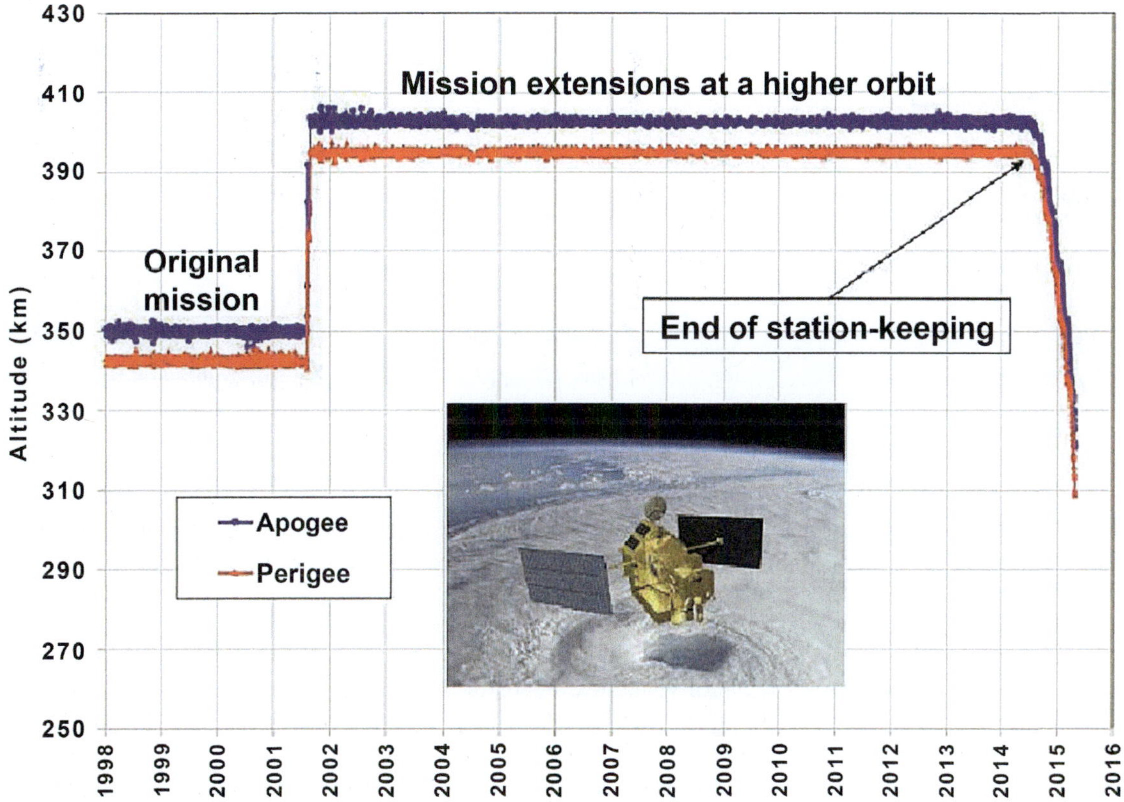

FIGURE 6.28 The maximum (*red*), mean (*black*), and minimum (*blue*) altitude of the Tropical Rainfall Mapping Mission satellite by orbit number of its lifespan (11/27/1997–7/15/2015).
(Image credit Stephen Bilanow, NASA Precipitation Processing System) (image credit NASA).

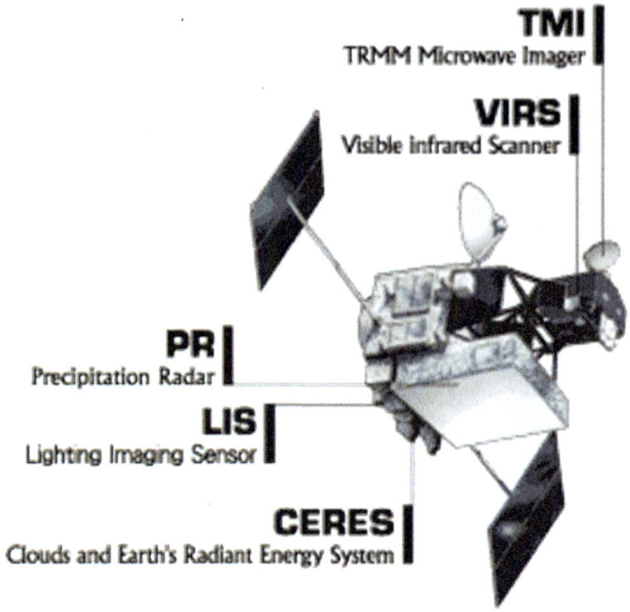

FIGURE 6.29 Tropical Rainfall Mapping Mission instruments.
Image credit NASA.

TABLE 6.5 Instrument specifications.

	Microwave radiometer (TMI)	Radar (PR)	Visible and infrared radiometer
Frequencies	10.7, 19.3, 21.3, 37.0, and 85.5 GHz (dual-polarized except for 21.3: vertical only)	13.8 GHz 13.8	0.63, 1.601.6, 3,715.6, 1,03.8.7, 5, and 12 μm
Resolution	11 km×8 km field of view at 37 GHz	5-km footprint and 250-m vertical resolution	2.5-km resolution
Scanning	Conically scanning (530 inc.)	Cross-track scanning	Cross-track scanning
Swath width	880-km swath	250-km swath	830-km swath

measure the energy at the top of the atmosphere, as well as to estimate energy levels within the atmosphere and at the Earth's surface. The Lightning Imaging Sensor was a small instrument designed to detect and locate lightning over the tropical areas of the Earth.

The instrument specifications are given here in Table 6.5.

The PR was the first spaceborne instrument designed to provide three-dimensional maps of storm structure. These observations provide valuable information on the intensity and distribution of the rain, on rain type, on storm depth, and on the height at which the snow melts into rain. The amount of heat released into the atmosphere at different heights based on these measurements can be used to improve models of the global atmospheric circulation. The radar has a ground resolution of about 5 km and a swath width of 247 km. It can provide vertical profiles of the rain and snow from the Earth's surface up to about 20 km. The PR can detect light rain rates down to about 0.7 mm/h. New methods of data processing have been developed to handle cases of intense rain rates. The PR is able to separate rain echoes for vertical samples of about 250 m, which look straight down. It performs all of these functions while using only 224 watts of electric power. The PR was designed and built by the National Space Development Agency (JAXA) of Japan as part of its contribution to the joint US/Japan TRMM mission.

Unlike the PR, the TMI was not really new, as it was based on the design of the highly successful SSM/I, which had been flying since 1987. The TMI measured the intensity of radiation at five different frequencies: 10.7, 19.4, 21.3, 37, and 85.5 GHz. The 10.7 is a new frequency that was not on the SSM/I, and it was designed to provide a more linear response for the high rainfall rates common in tropical regions. As stated previously, while not intended, this frequency became useful for observing SST and its variations. The other big improvement in the TMI was a better ground resolution. This was not due to any change in the instrument but rather to the fact that TRMM flew at a much lower altitude (402 km) than the DMSP satellites that carried the SSM/I. TMI had an 878 km wide swath, and the ground resolution varied with frequency.

Calculating rainfall rates from TMI involves some rather sophisticated mathematical computations, as we have reviewed in terms of SSM/I processing. The basis of these is Planck's radiation law, which specifies how much energy a body radiates depending on its temperature. Water surfaces such as oceans and lakes emit only about one-half of the microwave energy specified by Planck's law and therefore appear to only have about half of the real temperature of the surface. Therefore water surfaces appear very "cold" to microwave radiometers. Raindrops, however, appear to have an apparent temperature equal to their real temperature. They appear warm against the cold ocean surface; the more raindrops, the warmer the whole scene appears, making it possible to fairly accurately measure rainfall rates over water surfaces based on the microwave-sensed temperature. This is particularly important for the ocean, where it is extremely difficult to directly measure in situ estimates of rain and rain rate.

The land surface is different from the ocean and appears to emit energy equivalent to about 90% of its real temperature. Thus, it is difficult to discriminate the "warm" raindrops from the warm land. Still, there are aspects of rainfall that can be observed. The high-frequency microwaves (85.5 GHz) measured by TMI are strongly scattered by the ice present in many rain clouds. This reduces the microwave signal and provides a contrast with the warm land background.

Taken together, the TRMM instruments were capable of mapping monthly rainfall. In Fig. 6.30 we present the monthly rainfall map wrapped around a sphere. Remember, the TRMM orbit restricts measurements to the tropics and the subtropics. In Fig. 6.30, red indicates the areas of highest rainfall, and blue corresponds to areas of low rainfall. Green represents the median rainfall. One can clearly see the band of tropical rainfall that is broken by areas of lower rainfall. The highest rainfall occurs in pockets embedded in broader rainfall features. These features appear continuous from the ocean to the land, indicating that the data has been properly processed to account for the major differences between the land and ocean surface.

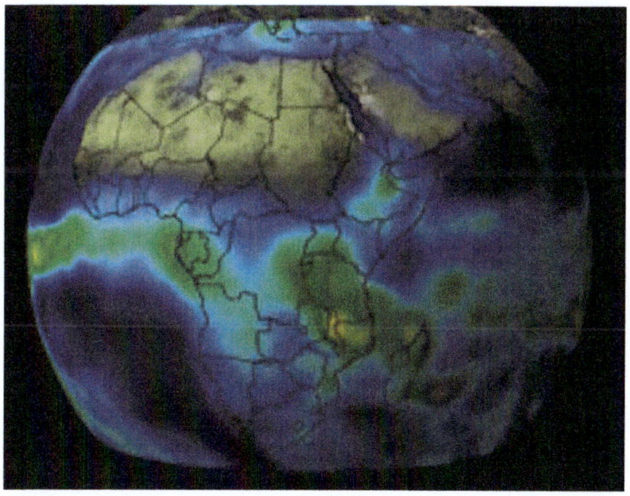

FIGURE 6.30 Tropical Rainfall Mapping Mission monthly rainfall January 1998 through October 2000 (https://svs.gsfc.nasa.gov/2033). Map lines delineate study areas and do not necessarily depict accepted national boundaries.

TRMM data also made it possible to compute the rain map anomalies (Fig. 6.31) using the overall mission mean rainfall as the basis. Again, red indicates high values and blue the low values. One can see the high rainfall anomaly in the southern subtropical Pacific, bordered by low rainfall to the north. This was a La Nina year, which explains this pattern of rainfall.

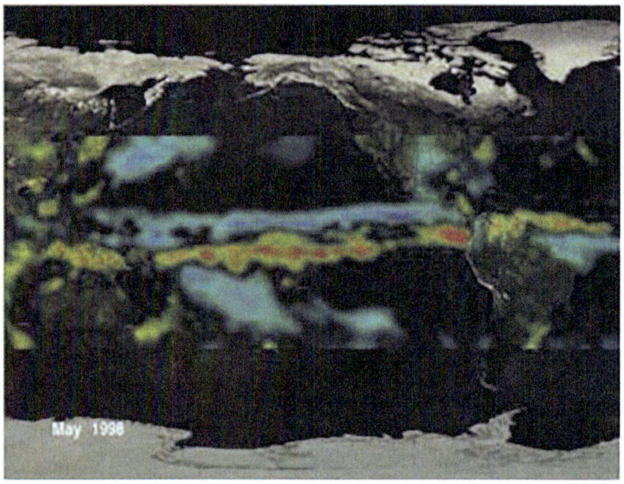

FIGURE 6.31 Tropical Rainfall Mapping Mission monthly average rain map anomaly December 2000 (https://svs.gsfc.nasa.gov/2033). Map lines delineate study areas and do not necessarily depict accepted national boundaries.

Chapter 7

Geostationary weather satellites and land remote sensing

We are accustomed to seeing weather satellite data in our daily forecasts, but most of us don't realize that these data come from satellites flying in geosynchronous orbits that allow them to continually view at least part of the Earth. In the mid 1960's NASA launched a series of six satellites to explore whether it was possible to operate instruments in a geostationary orbit of 35,785 km altitude required to keep the satellite in sync with the rotation of the Earth. They explored spin stabilization, gravity gradient stabilization, and a number of communications experiments. The first five satellites were similar and were built by Hughes Aircraft. The sixth satellite was a new design and was built by Fairchild. Out of the first six, only one, three, and five were able to attain and operate in geostationary orbit.

7.1 The applied technology satellite ATS program

One of the most important instruments tested during the ATS program (https://science.nasa.gov/mission/ats/) was the spin-scan camera first flown on ATS-1 to continuously monitor the global atmospheric circulation. In the mid 1960s, Dr. Verner E. Suomi, working with Robert J. Parent, invented the Spin-Scan Cloud Camera. The idea for this camera was first conceived while Suomi was serving as the first Chief Scientist for the National Weather Service. He wanted to develop a system that could take frequent observations of a single weather phenomenon and provide time-rate-of-change information about the weather. Suomi and Parent solved this problem by designing a camera that could scan the Earth while the geostationary satellite spun, using a small reflective telescope that marched up and down to provide hemispherical scans of the Earth's atmosphere (Fig. 7.1).

It is important to realize that the spinning satellite provided a component of the camera's data collection, as it was important to know both the scan line and position on the scan. This has to be synchronized with the spin of the satellite. The success of this instrument is manifest in the fact that it became the major instrument in all geostationary weather satellites from 1963 to 2013. The Spin-Scan Cloud Camera made it possible to observe weather systems at intervals as small as a few minutes and hence measure the dynamics of many phenomena such as air motion, cloud height, rainfall location, and the extent of atmospheric pollution. This instrument moved satellite remote sensing from qualitative viewing of the Earth to quantitative measurement of the Earth's atmosphere.

One of the first color images of the Earth was collected by the Spin-Scan camera on ATS-3 (Fig. 7.2).

7.2 NOAA's GOES satellites

This image is said to have inspired environmental research around the world. Two operational systems were based on the spin scan design, which led to NASA's Synchronous Meteorological Satellite (SMS) and NOAA's Geosynchronous Operational Environmental Satellite (GOES). The SMS was a pair of satellites to test if two geostationary satellites could be operated at the same time. The spin scan design was adopted by the European Space Agency and by the Japanese Meteorological Agency for their geostationary weather satellites. The reason why it was important to be able to operate two geostationary satellites simultaneously was that the US needed one satellite located over 137°W (GOES West) to detect and monitor the incoming weather to the western US. Another satellite is located at 75.2°W (GOES East) to detect and monitor Atlantic hurricanes. GOES West also serves to provide weather satellite coverage for the Hawaiian Islands and hurricanes spawned in the Eastern North Pacific.

The instrument on the NOAA GOES satellite became known as the Visible Infrared Spin Scan Radiometer or VISSR. This instrument had a visible channel (0.55–0.75 μm) with a resolution of 0.9 km at the subsatellite point (SSP) and a

Satellite Remote Sensing. DOI: https://doi.org/10.1016/B978-0-443-22086-9.00007-1

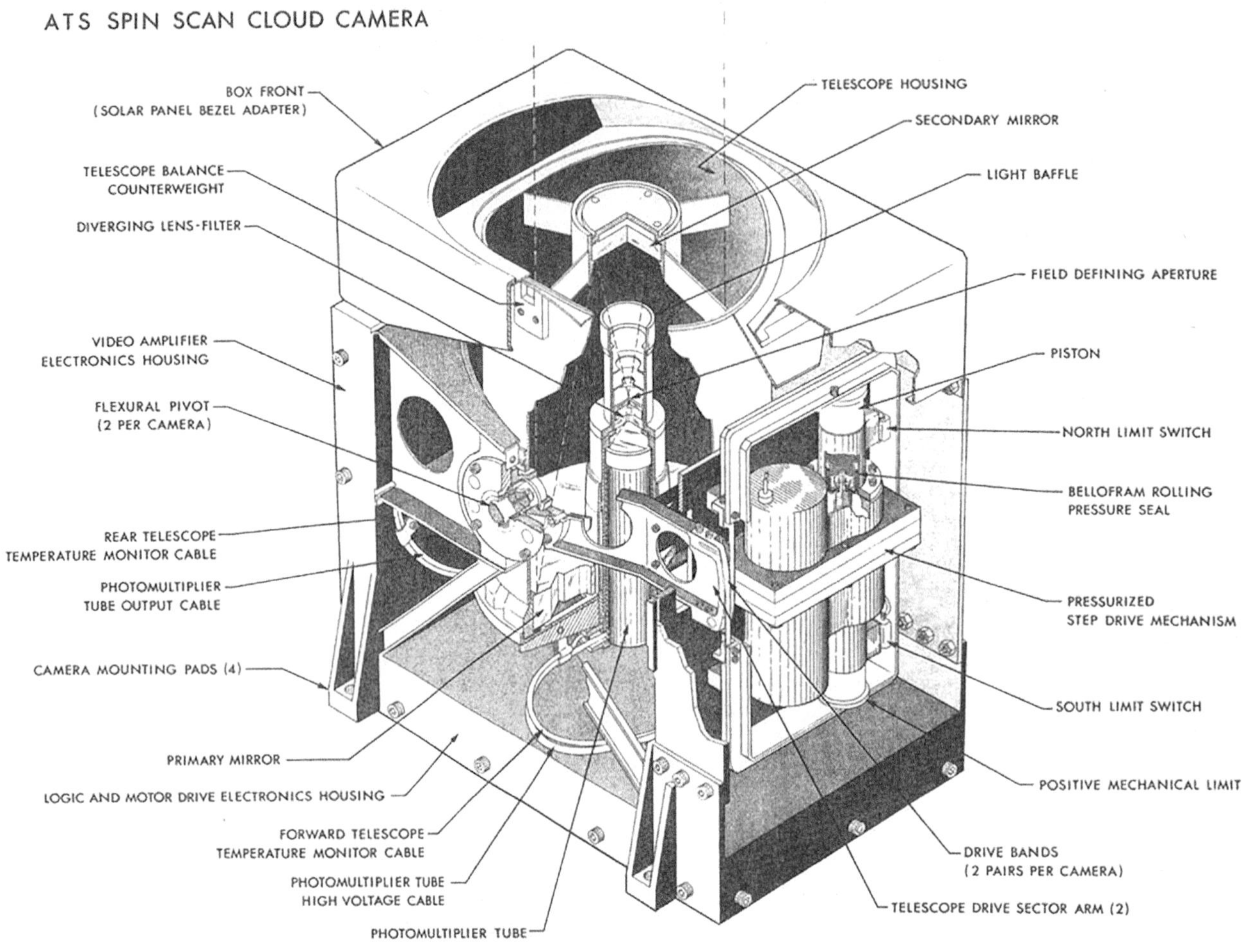

FIGURE 7.1 The Spin-Scan Cloud Camera. (https://science.nasa.gov/mission/ats/)

thermal infrared channel (10.5–12.6 μm) with a resolution of 6.9 km at the same SSP. This instrument was carried by GOES-1, GOES-2, and GOES-3. It was also the instrument on both NASA SMS satellites. Soon meteorologists wanted more information, and Suomi proposed to sound the atmosphere for temperature and water vapor from geostationary orbit. This became known as the VISSR Atmospheric Sounder (VAS). To sound the atmosphere at geostationary orbit was a major challenge as the signals were relatively weak at that altitude. High spatial resolution was needed in spite of this challenge. Considerable new technology was needed to receive sufficient radiation within the instrument's narrow field of view. The VAS was a modification of Suomi's original spin-scan camera, with additional detectors to cover the new spectral bands. The successful performance of the VAS was demonstrated by its first launch on GOES-4 in 1980. Temporal and spatial variations of temperature and water vapor associated with severe convection and weather were observed and retrieved with the accuracy set forth in Suomi's 1971 proposal for the instrument. The VAS continued to be the only instrument with sufficient temporal and spatial resolution to observe the evolution of severe storms over regions of several hundred thousand square miles. Data from VAS was used to predict hurricane tracks much more accurately and to aid in the forecasting of tornadoes. All of this data was used as input to numerical weather forecast models.

GOES-1 (Fig. 7.3) was the first of the operational spin-stabilized geostationary weather satellites.

Due to the spin, it only views the Earth 10% of the time as compared with the modern three-axis stabilized geostationary weather satellites that can constantly view the Earth. While the VISSR viewed the Earth in the visible and thermal infrared, it was not able to provide information on cloud thickness, total moisture content, or the vertical layers of the atmosphere. This satellite bus was used for GOES1, 2, 3 and operated between 1975 and 1993. It should be noted here that it was NOAA's practice to identify satellites with alphabetic nomenclature before launch, so this series was known as GOES A–C. When the satellite is operating on orbit, it is then referred to as GOES 1–3. This method accounts for any launch or other operational failures to keep the naming of the satellites or orbit in a sequential manner. One other

FIGURE 7.2 Earth from ATS-3 on November 10, 1967. (https://www.nasa.gov/image-article/december-1966-applications-technology-satellite-1-ats-1/)

FIGURE 7.3 GOES-1 geostationary weather satellite.
Image credit courtesy NOAA

advantage of a spinning satellite was that the solar panels could be mounted on the outside of the spacecraft and would all have an equal opportunity to receive sunlight for charging the satellite batteries. The same is true for thermal equilibrium since the spinning satellite would continuously pass through the sunlit and dark view experienced by a geostationary satellite. The future three-axis stabilized geostationary satellites had to be designed to account for the fact that one side of the satellite was facing the sun while the other side was not.

FIGURE 7.4 GOES 4–7.

The next series of GOES spacecraft carried Suomi's VAS instrument that provided both imaging and atmospheric profiling. Before launch, these were designated as GOES D–H and later on orbit as GOES 4–7 (Fig. 7.4). The VAS imager and sounder still shared the same optical system, which meant that the VAS alternated between images and soundings. The soundings became very useful for numerical weather prediction as the soundings could be assimilated into the numerical models.

While the satellite was still spin stabilized, the portion of the satellite containing the antennas for both up and down links of data and commands remained fixed, pointing at the ground station with which it communicated. This modification then required "slip rings" to transmit data and information from the spinning portion to the fixed antennas. This modification increased the amount of time the satellite could communicate with the ground due to the lack of spinning.

While the NOAA GOES history page mentions this as a "fun fact," it was a very stressful time when, instead of the two GOES satellites, we were down to only one GOES-7 spinner, which was then placed over the center of the country to be able to "see" both west and east coasts. This condition was the result of two different problems. The first was the premature failure of the other GOES satellites due to the fact that the shaft encoder depended on the light of a small bulb, and these bulbs were failing very early, making the VAS inoperable. GOES-7 did not seem to have the problem. The other issue was that NOAA was developing a three-axis stabilized satellite to replace both GOES West and GOES East. This development ran into serious technical problems that greatly delayed the completion of these new GOES satellites and incurred large cost overruns. GOES-7 served as the sole geostationary weather satellite from 1989 to 1994.

7.3 Three axis stabilized GOES satellites

The GOES I-M (8–12) series ushered in the use of three-axis stabilized spacecraft as geostationary weather satellites (Fig. 7.5). While these satellites had the great advantage of being able to "stare" at the Earth, this capability comes with challenges in thermal control and the collection of sunlight for power.

FIGURE 7.5 GOES 8–12 satellite (1998–2013).
Image credit courtesy NASA

TABLE 7.1 GOES-8 imager channels.

Central wavelength μm	Spectral interval μm	SNR or NEΔT @ spc input
0.65	0.55–0.75	250 @ 100% albedo
3.90	3.80–4.00	0.11 K @ 300 K
6.75	6.50–7.00	0.14 K @ 300 K
10.7	10.2–11.2	0.09 K @ 300 K
11.95	11.5–12.4	0.70 K @ 300 K

The spinning satellites were in thermal equilibrium since the faces of the satellite alternated between solar exposure and solar shadow. A three-axis stabilized spacecraft would be continuously pointed at the Earth, meaning that one end would be exposed to solar radiation and the other would not. Likewise, the solar panels on the spinners could be mounted on the surface of the spinning spacecraft, but for a three axis stabilized satellite, new arrangements must be made to collect the solar energy to power the spacecraft. These challenges had already been met by India, whose American contractor Space Systems/Loral had built a three-axis geostationary satellite for them, which was operating over the Indian Ocean. NOAA had then contracted with Loral for its geostationary satellites.

Another advantage of going to three-axis stabilization was that the satellite could now carry independent imagers and sounder instruments that could operate simultaneously. The imager was a five-channel (Table 7.1) stepping radiometer.

The spatial resolution was 1 km for the visible channel and 4.0 km for the infrared channels. It could cover the full disk every 30 min and limited areas in a shorter period. This limited area capability was to be able to focus on a line of thunderstorms to be able to identify tornadoes and provide early warning. This capability would be greatly increased in future geostationary weather satellites.

The sounder used in this series had 18 thermal infrared bands plus a low-resolution visible band. The field of view is 8 km, and the sounder is sampled every 10 km. This sounder was patterned after the high-resolution infrared radiation sounder flown on the NOAA polar orbiting satellites. It has six bands in the 15 μm (longwave) spectral region, a split-window pair, three mid-tropospheric water-sensitive bands along with an ozone band (mid-wave), five 4 μm (shortwave) bands, and a low-resolution visible band (Table 7.2).

The goal of the GOES-8 sounder was to provide brightness temperatures with 1.0 K accuracy and a 0.3K precision. The accuracy and performance of the GOES-8 sounder are verified when collocated. GOES-8 sounder and GOES-7 VAS radiances are compared (Fig. 7.6).

TABLE 7.2 GOES-8 sounder channels.

Wavelength μm	Channel number	NEΔT	Purpose
Longwave			
14.71	1	1.02	Stratosphere temperature
14.37	2	0.87	Troposphere temperature
14.06	3	0.60	Upper-level temperature
13.96	4	0.40	Mid-level temperature
13.37	5	0.45	Low-level temperature
Window			
12.66	6	0.20	Total precipitable water
12.02	7	0.13	Surface temperature, moisture
11.03	8	0.10	Surface temperature
Ozone			
9.71	9	0.14	Total ozone
Water vapor			
7.43	10	0.11	Low-level moisture
7.02	11	0.13	Mid-level moisture
6.51	12	0.21	Upper-level moisture
Shortwave			
4.57	13	0.13	Low-level temperature
4.52	14	0.13	Mid-level temperature
4.45	15	0.16	Upper-level temperature
Nitrogen			
4.13	16	0.10	Boundary-layer temperature
Window			
3.98	17	0.17	Surface temperature
3.74	18	0.10	Surface temperature, moisture
Visible			
0.94			Cloud

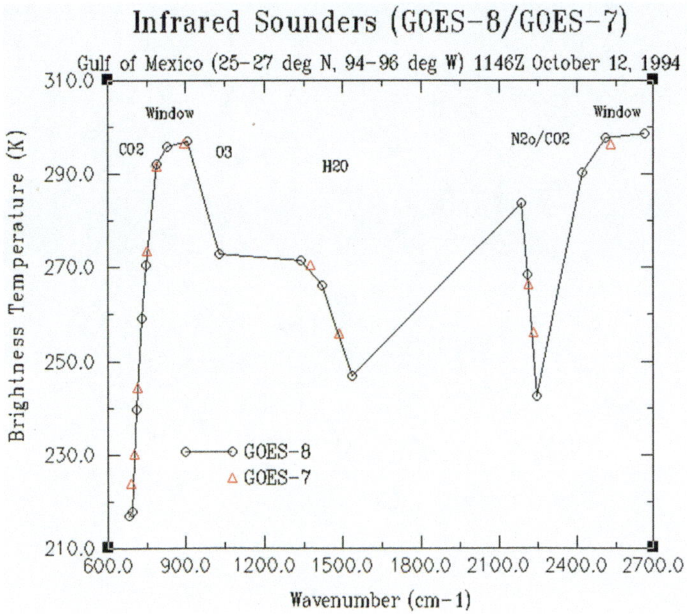

FIGURE 7.6 Comparison of GOES-8 and GOES-7 infrared sounders.
Image credi courtesy NASA

In Fig. 7.6, the two sounders agree within 2.0 K. The additional GOES-8 spectral bands provide more atmospheric detail. The full-time (versus alternate sampling on the GOES-7 spinner) makes operational sounding products possible for the first time, which contributes greatly to mesoscale weather forecasting over the United States. The data can also be used to compute thermal winds over the ocean and supplement the Automated Surface Observing System with upper-level cloud information.

Imagery from the GOES-8 sounder demonstrates the sounder's capabilities for sampling in specific bands. Here we show images from four of the longwave carbon dioxide bands in Fig. 7.7. From the upper left panel (band 2 at 14.4 μm) over to the right (band 3 at 14.2 μm) and then from the lower left (band 4 at 14.0 μm) across to the right (band 5 at 13.4 μm), the imagery shows the process of sensing lower and lower into the atmosphere as the wavelengths of the radiation decrease. One can see how the radiative temperatures increase from band 2 through band 5 in a generally cloud-free region such as the Gulf of Mexico (from gray to green to yellow to red, indicative of the general lapse rate of temperature with height in the troposphere). Note how the clouds along the front, extending from the western Great Lakes to the Texas Panhandle, are seen in the latter three bands but not in band 2. Only the highest cloud tops (along the west coast of Mexico in the central Gulf) are sensed with band 2. Also note that for bands 3–5, the general temperature pattern shows cooler air to the northwest with warmer air in the southeast, while in band 2, the gradient is reversed. Radiances in band 2 are actually sensing the lower stratosphere.

GOES N – P (on orbit) (13–15; Fig. 7.8) further improved the imager and sounder spatial resolutions with an Image Navigation and Registration subsystems which used geographic landmarks and star locations to better correct the geostationary imagery and soundings. The instrument's optics were improved, and more sensitive detectors were employed. Better batteries provided more available power for imaging and sounding.

The imager on these satellites improved the infrared resolution to 2.0 km. A full disk was still available every 30 min with the ability to focus in on smaller target areas with shorter time intervals.

GOES-13 was launched on May 24, 2006 and was initially in a parking orbit waiting for assignment. It was called to duty with the failure of GOES-12 due to a thruster leak. After the leak was fixed, GOES-12 went back to work, and GOES-13 was again deactivated. In April 2010, GOES-13 replaced GOES-12 as GOES-East over the longitude of 75.0°W. On December 14, 2017, GOES-13's direct downlink transmission (GOES VARiable transmission format or GVAR) was disabled. The GOES-13 GVAR and (low rate information transmission) LRIT were relayed to the ground via GOES-14 until January 8, 2018, when it was turned off. GOES-13 ceased transmitting data and began drifting to its storage location at 60.0°W on January 9, 2018, arriving there on January 31, 2018.

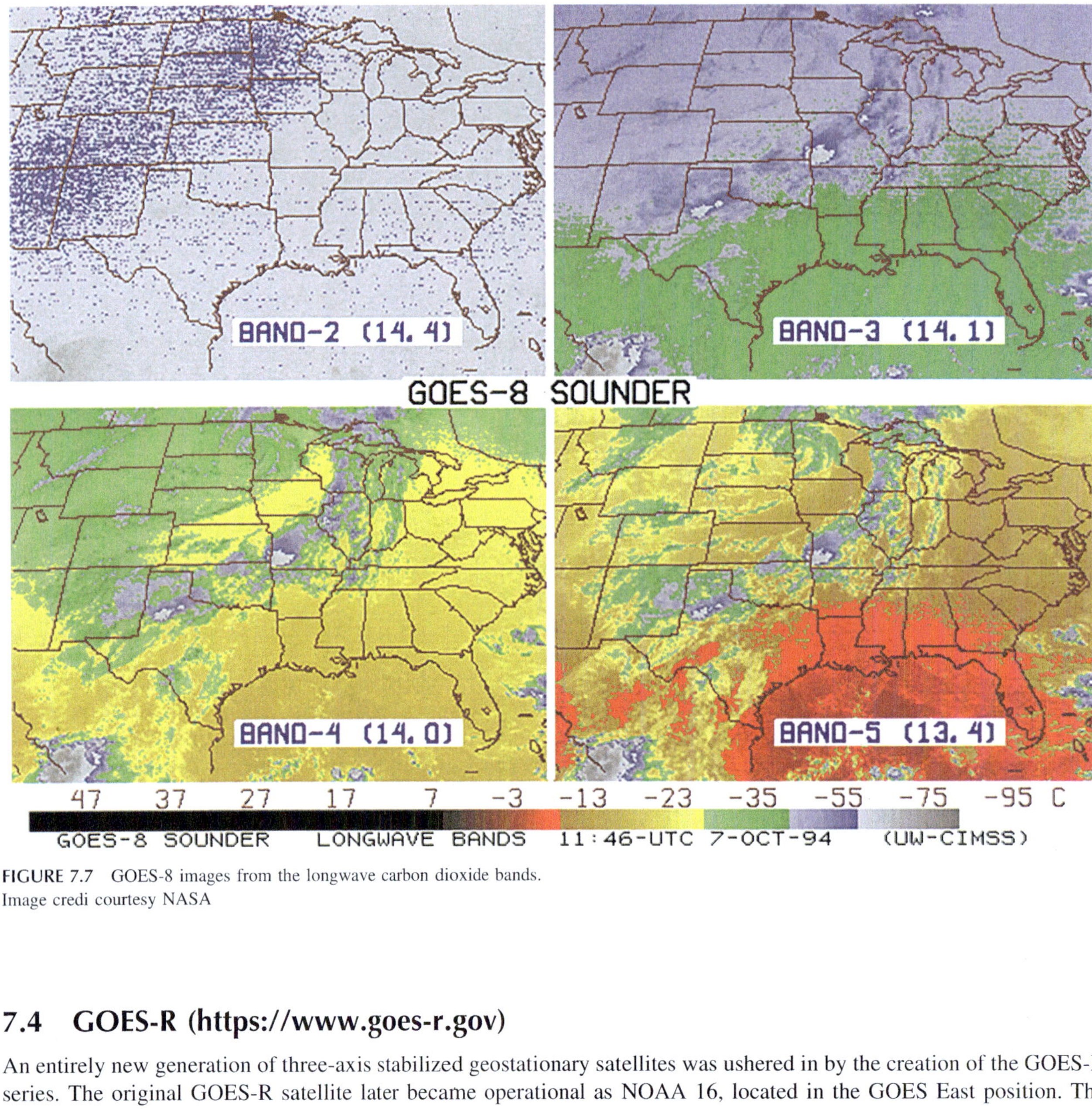

FIGURE 7.7 GOES-8 images from the longwave carbon dioxide bands.
Image credi courtesy NASA

7.4 GOES-R (https://www.goes-r.gov)

An entirely new generation of three-axis stabilized geostationary satellites was ushered in by the creation of the GOES-R series. The original GOES-R satellite later became operational as NOAA 16, located in the GOES East position. The GOES-R series is a four-satellite program designed to extend the GOES satellite system through 2036. The satellite was awarded to Lockheed-Martin, and the author was involved in the proposal that won this bid. The spacecraft is presented in Fig. 7.9 which shows the 6 instruments carried by the satellite. The solar panel is mounted on one side with a very flexible mount, making it possible to reorient the solar panel array to best take advantage of the changing solar illumination.

The satellite was designed for 10 years of on-orbit operation preceded by up to 5 years of on-orbit storage. The spacecraft carries three types of sensors: nadir pointing, solar pointing, and in situ. The satellites are able to operate through periodic station-keeping and momentum adjustment maneuvers, which allow for near-continuous instrument observations. It includes vibration isolation for the Earth-pointed optical bench as well as high-speed spacecraft to instrument interfaces designed to maximize science data collection. This data collection will only be interrupted by all momentum management, station-keeping, and yaw flip maneuvers that will only take 120 min/year. This is nearly a two orders of magnitude improvement over previous three-axis stabilized GOES satellites.

FIGURE 7.8 GOES 13–15 satellite.
Image credi courtesy NASA

FIGURE 7.9 GOES-R satellite. *ABI*, Advanced Baseline Imager; *EXIS*, Extreme Ultraviolet and X-ray Irradiance Sensors; *GLM*, GOES Lightening Mapper; *MAG*, magnetometer; *SEISS*, space environment in situ suite; *SUVI*, Solar Ultraviolet Imager.
Image credit Lockheed-Martin

The satellite driving requirements are

- Spacecraft on-orbit life of 15 years with orbit east-west and north-south position maintained to within +/– 0.1 degree.
- Collect and transmit up to 100 Mbps instrument payload data from each location continuously.
- Continuous rebroadcast function at L-band up to 31 Mbps utilizing dual polarization.
- Provide continuing services search and rescue, data system collection, emergency manager's weather information network (EMWN).

The spacecraft is 6.1 m X 5.6 m X 3.9 m and its dry mass is 2,857 kg. The satellite was launched from Vandenberg on a Falcon Heavy and later all on the Atlas V.

The primary instrument on GOES-R is the Advanced Baseline Imager (ABI), which had its early origin in June 1999 under the direction of Tim Schmitt of the National Environmental Satellite, Data and Information Service (NESDIS). In its earliest inception, it was a 10-channel instrument which was derived from six instruments on other satellites (file:///Users/williamjemery/Desktop/RS%20past%20pres%20futu/background%20off%20the%20web/geosatellites/goes%20r/GOES-16%20-%20Wikipedia.webarchive). In September 1999, NOAA endorsed the continued development of the instrument with the same number of spectral bands. As the development progressed, the number of spectral bands increased from the original ten to twelve. At the same time, NOAA funded research on the advanced baseline sounder (ABS), which would be part of a hyperspectral environmental suite (HES) to go on future geostationary satellites.

In 2001, NOAA planned for the GOES-R generation of GOES satellites to begin with the expected launch of GOES-R in 2012, with the ABI and ABS as the expected instrumentation. Four years later, the number of proposed spectral bands on the ABI instrument increased to 16, covering the visible and thermal infrared portions of the EM spectrum. In September 2006, NOAA dropped plans to include the HES aboard GOES-R due to inadequate testing and cost overruns on the NPOESS system. While GOES-R was expected to cost $6.2 billion, the increased instrument complexity, revised inflation assumptions, and program reserves led the Government Accountability Office (GAO) to estimate a much higher $11.4 billion cost for the program in 2006.

The ABI was built by L3 Harris in Fort Wayne Indiana, independent of the spacecraft contractor and was conceived and built before the spacecraft contractor was even selected. ABI views the Earth's weather in 16 spectral bands (compared with 5 on the previous GOES imager). They include 2 visible channels, 4 near-infrared channels, and 10 thermal infrared channels. The different wavelengths associated with these channels are used by weather models to assess and predict the weather. They are also used to monitor trees, clouds, water surfaces, atmospheric moisture, and smoke. It provides three times more spectral information, four times the spatial resolution, and more than five times faster temporal coverage than the previous system.

The instrument has multiple scan modes. In mode 4, or continuous full disc mode, the ABI produces a full disc image every 15 min, a CONUS (North America latitudes) image (coverage 3000 km by 5000 km) every 5 min, and two mesoscale domains (resolution 1000 km by 1000 km at nadir) every 60 sec or in one subdomain every 30 sec. Mode 6, or 10 min flex mode, which became the default operating mode for GOES-16 and 17 in April 2019, provides a full disc image every 10 m, a CONUS (GOES-16)/PACUS (GOES-127) (here GOES-16 is in the eastern location covering the eastern continental US and slightly offshore, while GOES-17 is in the western position so that it can view the weather coming from the west so half of it covers the eastern Pacific while half of it covers the US west coast) image every 5 min and both mesoscale domains every 60 sec (or one domain in 30 sec). All ABI bands have on-orbit calibration. There are a number of advantages to a 10 min full disc scan. It allows NOAA to match the full disc scanning operations of our partners like the Japan Meteorological Agency (JMA), the Korean EXISUVI Meteorological Administration (KMA), and the next generation of EUMETSAT geostationary satellites. Ten-minute disc imagery will also be critical to our National Weather Service Forecast Offices, National Climate Environmental Prediction Center (NCEP), and the Volcanic Ash Advisory Centers in monitoring weather conditions and allowing for forecasting information from unsampled regions like the eastern Pacific Ocean. This sampling scheme is depicted here in Fig. 7.10.

An example of the GOES-16 full disc is presented here in Fig. 7.11 with the overlaying CONUS region

In Fig. 7.12 the same full disc image is shown with the more intensive mesoscale region indicated.

I always find it amazing that this instrument is able to provide all of this information at the same time. The required complexity of the device must be very high. I'm sure those early NOAA administrators could little expect the success of this instrument and its role in weather forecasting. The rapid sampling in a selected mesoscale region is ideal to follow a line of thunderstorms to detect the formation of a tornado and provide an earlier warning for people to seek shelter in the area. CONUS/PACUS sampling provides the weather forecaster with an image of the cloud distribution that will lead to the upcoming weather in North America. CONUS is able to best detect the formation of hurricanes in the western North Atlantic with the potential to impact the US East Coast. Here, the mesoscale sampling can be focused on the developing hurricane to resolve its structure and monitor its changes over time.

The ABI spectral bands are given here in Table 7.3 From this table, we can see that ABI applications include vegetation, cirrus clouds, snow/ice, cloud particle size, upper-level water vapor, mid-level water vapor, lower-level water vapor, cloud-top phase, ozone, and CO_2. With this complement of channels, the ABI is able to do some of the atmospheric sounding that was lost with the subtraction of the Hyperspectral Environmental Suite (HES) which could profile the atmosphere. The longwave infrared band was used with the mesoscale option to follow a storm on April 5, 2017, that resulted in destructive hail and tornadoes in the Southeastern United States (Fig. 7.13).

The 0.5-km spatial resolution of the ABI in the visible cuts the previous imagery resolution in half. The visible/near infrared has a 1.0-km resolution, and this channel was not previously available. The longer wavelength channels >2 μm

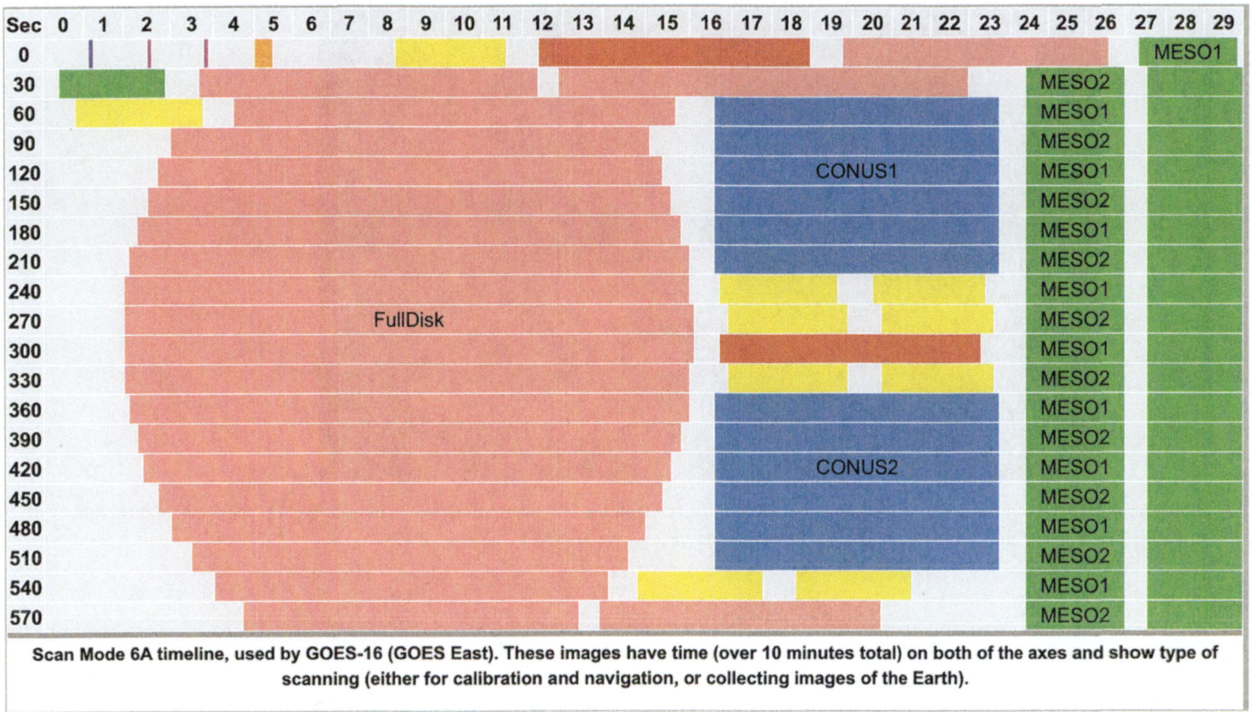

FIGURE 7.10 ABI scan mode 6 for GOES-16.
Image credit courtesy NOAA

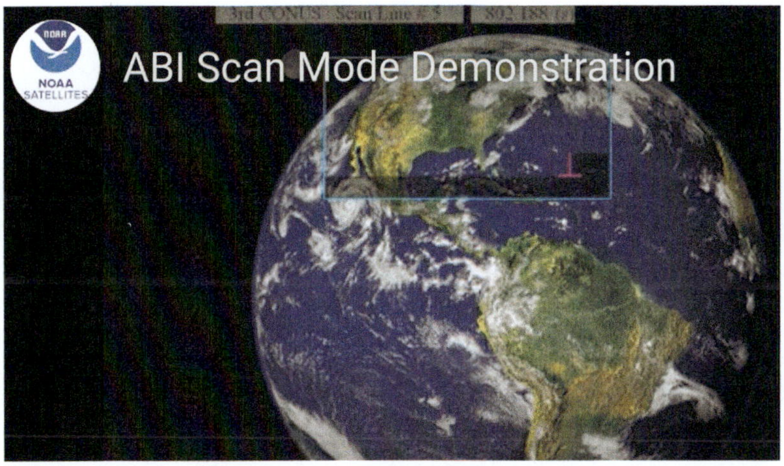

FIGURE 7.11 Upper part of the Advanced Baseline Imager (ABI) full disc image with the CONUS area overlaid.
Image credit courtesy NOAA

have a 2-km spatial resolution, which again is half that of the previous sensors. The CONUS (eastern satellite) and PACUS (western satellite) imagery are available every 5 min providing updated imagery of the west coast weather and the emergence of hurricanes in the east.

GOES-R also carried the first lightning mapper in geostationary orbit, known as the GOES Lightning Mapper (GLM; Fig. 7.14). The GLM is a single-channel, near-infrared optical transient detector that can sense the momentary changes in an optical scene. This change indicates the presence of lightning. GLM measures total lightning (in cloud, cloud-to-cloud, and cloud-to-ground) continuously over the Americas and adjacent ocean regions with a near-uniform spatial resolution of approximately 10 km.

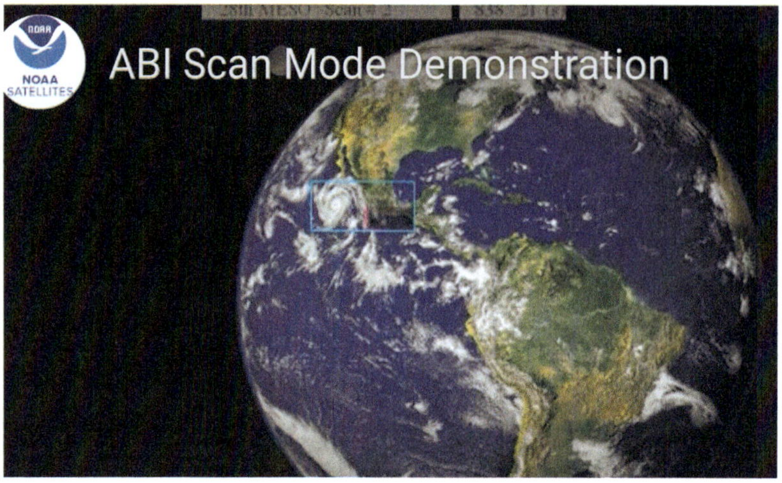

FIGURE 7.12 Upper part of the Advanced Baseline Imager (ABI) full disc image with the 1000 km mesoscale area marked. Note the focus of this rapid repeat sample on the cyclonic feature just off the west coast of the Baja Peninsula. Image credit courtesy NOAA

TABLE 7.3 Summary of the wavelength, resolution, sample use, and heritage instruments of the AGI bands.

ABI Band	Central wavelenght (μm)	Type	Nickname	Best spatial resolution (km)
1	0.47	Visible	Blue	1
2	0.64	Visible	Red	0.5
3	0.86	Near-infrared	Veggie	1
4	1.37	Near-infrared	Cirrus	2
5	1.6	Near-infrared	Snow/Ice	1
6	2.2	Near-infrared	Cloud particle size	2
7	3.9	Infrared	Shortwave window	2
8	6.2	Infrared	Upper-level water vapor	2
9	6.9	Infrared	Midlevel water vapor	2
10	7.3	Infrared	Lower—level water vapor	2
11	8.4	Infrared	Cloud-top phase	2
12	9.6	Infrared	Ozone	2
13	10.3	Infrared	"Clean" longwave window	2
14	11.2	Infrared	Longwave window	2
15	12.3	Infrared	"Dirty" longwave window	2
16	13.3	Infrared	CO_2 longwave	2

ABI, Advanced Baseline Imager.

GLM is a staring CCD imager (1,372 X 1,300 pixels) with a near uniform spatial resolution of 8 km at nadir and 14 km at the edge of the field of view. It covers up to 52°N latitude and has a 70%–90 % flash detection during both day and night (Fig. 7.15).

Severe storms are associated with a significant increase in total lightning, which often occurs many minutes before the radar detects the potential for severe weather. Thus, taken together with radar and other satellite sensors, and ground observations, lightning from the GLM has a great potential to improve the lead time for severe thunderstorm and tornado

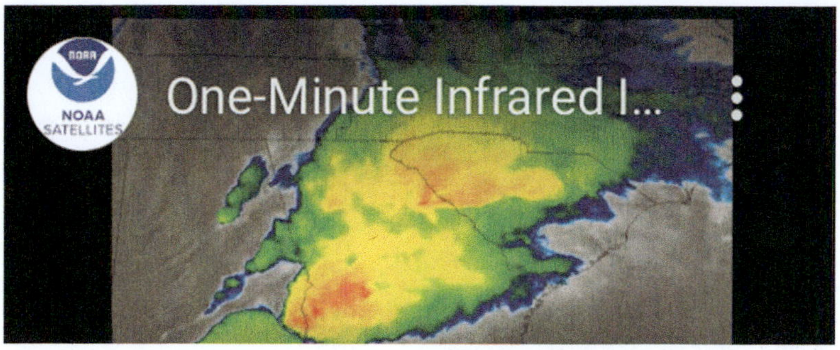

FIGURE 7.13 One minute images of a storm over the Southeast United States on April 5, 2017.
Image credit courtesy NOAA

FIGURE 7.14 GOES Lightening Mapper (GLM). https://www.goes-r.gov/spacesegment/glm.html

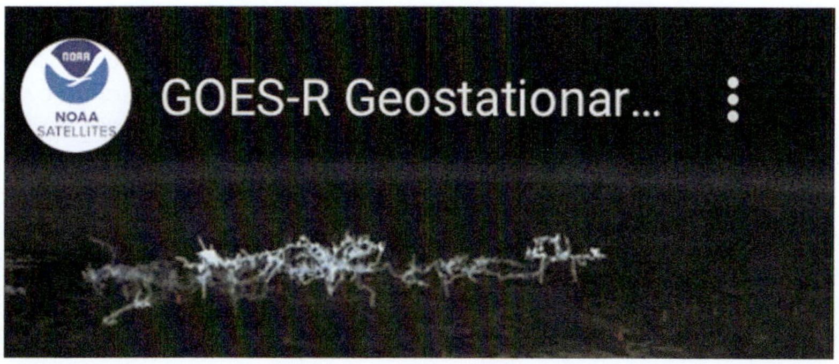

FIGURE 7.15 Snapshot of lightening from GOES Lightning Mapper. https://www.goes-r.gov/spacesegment/glm.html

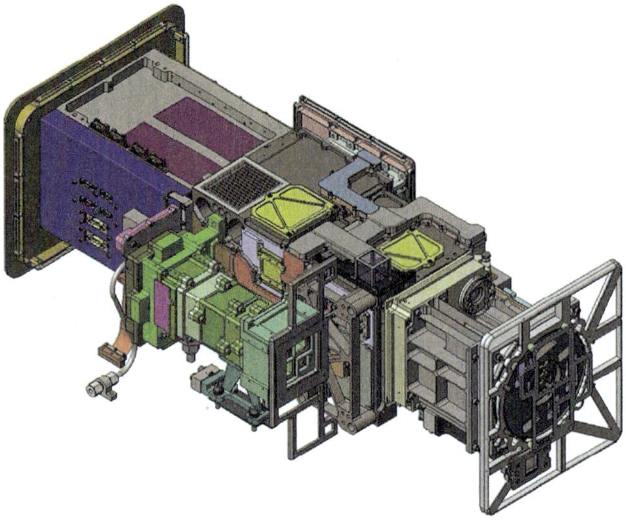

FIGURE 7.16 The Extreme Ultraviolet and X-ray Irradiance Sensors instrument on GOES-R. https://www.goes-r.gov/spacesegment/exis.html

warnings and also to reduce false alarm rates in these other sensors. Data from GLM will be used to create a database to track decadal changes in lightning activity.

The Compact Coronograph (CCOR) carried by GOES-R was provided by the Naval Research Laboratory to augment the Solar Ultraviolet Imager (SUVI) and Extreme Ultraviolet and X-ray Irradiance Sensors (EXIS). The CCOR will image the solar corona (the outer layer of the sun's atmosphere) and help detect and characterize coronal mass ejections (CMEs). CMEs are large expulsions of plasma and accompanying magnetic field distortions from the corona. They can be remotely detected with white light imagery of the upper solar corona. CCOR is designed to capture white light imagery of the upper solar corona. The CCOR is mounted on a sun-pointing platform (SPP) housed on the yoke of the satellite's solar array, along with all the other sun-oriented instruments. Thus SPP tracks the seasonal and daily movements of the sun relative to the spacecraft.

The EXIS, (Fig. 7.16) are also mounted on the SPP. These instruments are designed to monitor the solar irradiance in the upper atmosphere, which is the power and effect of the sun's electromagnetic radiation per unit area. EXIS is able to detect solar flares that could interrupt communications and GPS navigation accuracy. They also interfere with satellite operations (data downlink and commands), high altitude airlines, and power grids on the Earth. EXIS consists of two sensors, the extreme ultraviolet sensor (EUVS) and the X-ray sensor, that together help scientists monitor activity on the sun. NOAA requires the real-time monitoring of the solar irradiance variability that controls the variability of the Earth's upper atmosphere (ionosphere and thermosphere). This is in support of NOAA's space weather operations in Boulder, Colorado.

The EUVS measures changes in the solar extreme ultraviolet, which drives the upper atmospheric variability on all time scales. EUV radiation can severely impact the ionosphere, which, in turn, can result in radio blackouts or dropouts in terrestrial high-frequency communications at low latitudes.

This instrument was built by the Laboratory for Atmospheric and Space Physics (LASP) at the University of Colorado at Boulder. This lab has considerable experience in building and operating instruments for solar measurements. The author was also a member of LASP.

The space environment in situ suite (SEISS) was a collection of 6 different instruments (Fig. 7.17) all designed to make measurements relevant to space weather and its variability. These are the Energetic Heavy Ion Sensor, the Magnetospheric Particle Sensors-High and Low (MPS-HI and MPS-LO), and the solar and galactic proton sensor. The suite also includes the data processing unit. Data from SEISS drive the solar radiation storm portion of NOAA's space weather forecasts, leading to alerts and warnings.

Also on the SPP is the SUVI, which monitors the sun in the extreme ultraviolet wavelength range. Operating both day and night, the SUVI is able to compile full disc solar image around the clock. This instrument replaces the earlier GOES X-ray (SXI) imager and represents both a change in spectral coverage and in spatial resolution over the SXI. The SUVI observes the complex active regions of the sun, solar flares, and the eruption of solar filaments that give rise to CMEs. In this way, the SUVI can provide early warning on possible impacts to the Earth's space environment and terrestrial

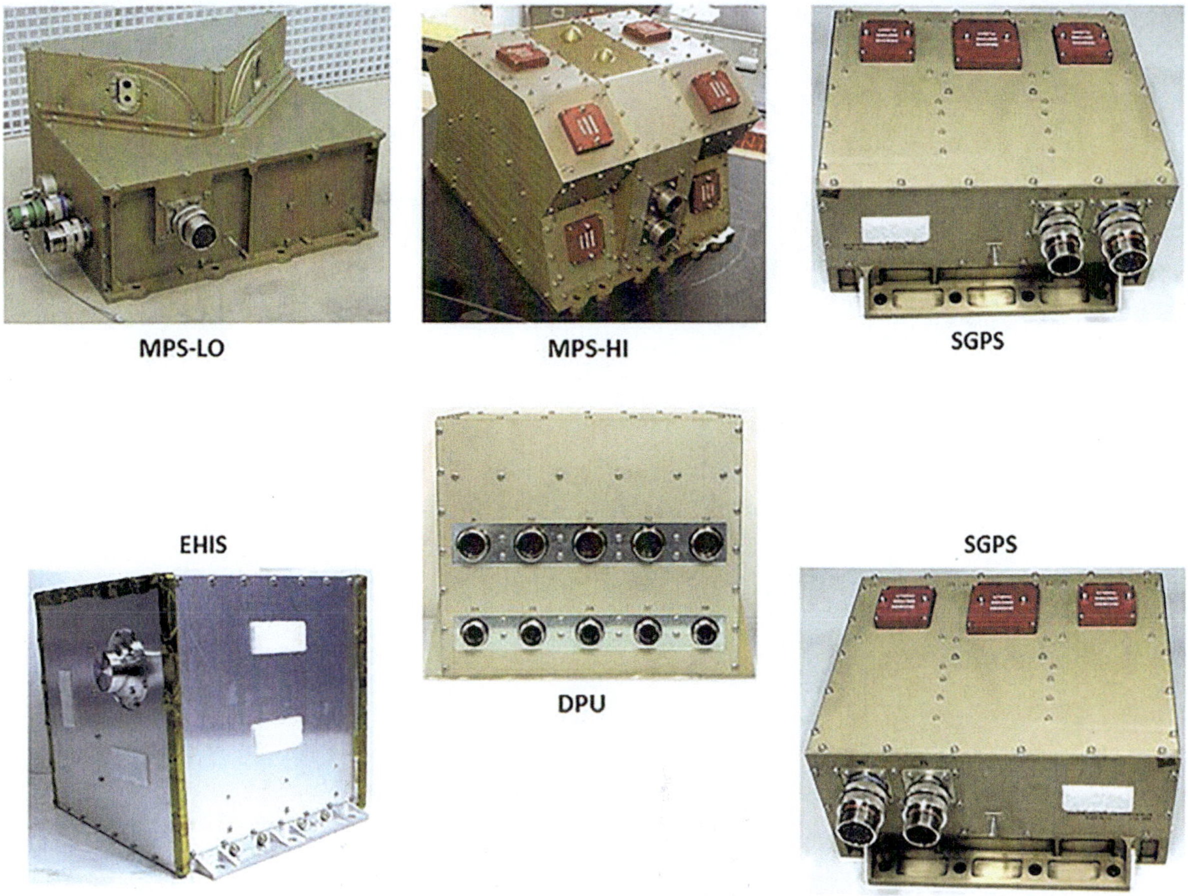

FIGURE 7.17 The instruments making up the space environment in situ suite (SEISS). *DPU*, Data processing unit; *EHIS*, Energetic Heavy Ion Sensor; *SGPS*, solar and galactic proton sensor. https://www.goes-r.gov/spacesegment/seiss.html

activities. SUVI provides data that can be used to estimate coronal plasma temperatures and mass ejections, which are critical to space weather forecasting. Specifically, the SUVI measurements:

1. Locate coronal holes for geometric storm forecasts.
2. Detect and locate flares for forecasts of solar energetic particle events related to flares.
3. Monitor changes in the corona that indicate CMEs.
4. Detect active regions beyond the east limb for activity forecasts.
5. Analyze active region complexity for flare forecasts.

This instrument was built by Lockheed Martin Advanced Technology Center in Palo Alto, CA.

Another space environment monitoring instrument on GOES-R is the magnetometer (MAG, Fig. 7.18). It was also built by the Lockheed Martin group in Palo Alto, CA. This MAG provides measurements of the space environment magnetic field that controls charged particle dynamics in the outer regions of the magnetosphere. These particles present a hazard to spacecraft and human spaceflight. They are also valuable in providing alerts to satellite operators and power utilities.

Note that the boom is an integral part of the MAG. This boom deploys after the spacecraft is in orbit and goes through a series of orbit-raising maneuvers. In other words, the satellite has to be safely in its operational location before the boom is deployed. The GOES-R MAG makes measurements of three components of the geomagnetic field with a resolution of 0.016 nT at a response frequency of 2.5 Hz.

7.5 Land remote sensing

The concept of a land remote sensing satellite program saw its genesis in the mid-1960's from discussions of people from the Office of Naval Research (ONR), NASA, and the US Geological Survey (USGS) under the leadership of its director

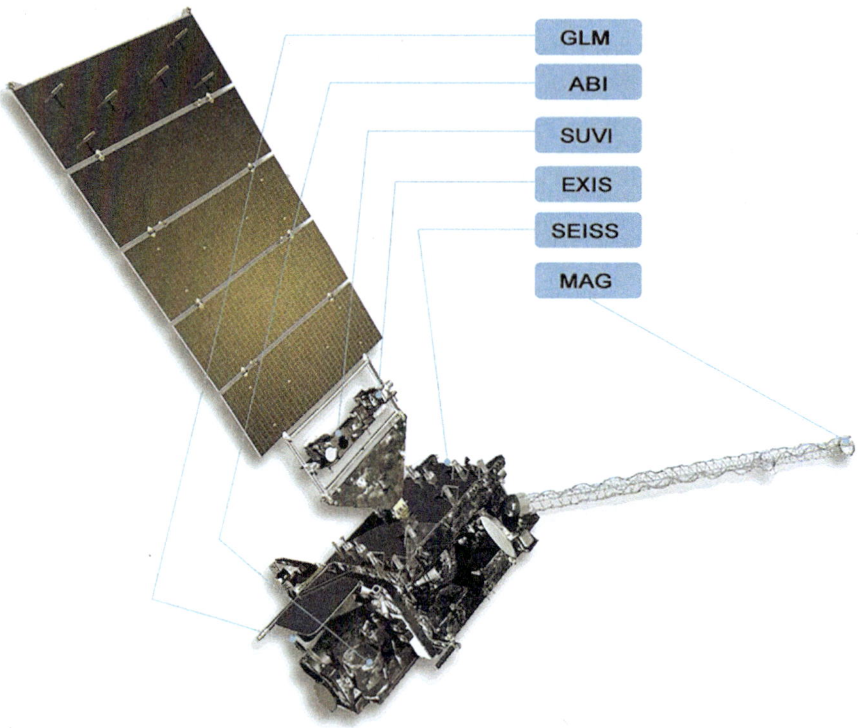

FIGURE 7.18 GOES-R magnetometer.
Image credit courtesy Lockheed Martin

William Pecora (Waldrop, 1982), a scientist within the USGS. Originally, the Department of the Interior (DOI) was to build, launch, and operate an Earth Resources Observation Satellite (EROS). The objectives of the EROS program were (1) construct and fly an Earth observing system by the end of 1969, followed by improved systems as required; (2) provide unclassified remotely sensed data to facilitate assessment of land and water resources of the United States and other nations; (3) design specific systems based on user requirements, distribute data to users and make operational use of the data in resource planning. This was all to be done in the simplest possible way.

Since the development of space technology was NASA's responsibility, the DOI proposal was rejected by the NASA administrator, James Webb, who then met with President Johnson to discuss the DOI proposal. Webb succeeded in creating a NASA program for such a satellite system. In cooperation with DOI, USGS, USDA, and other agencies, NASA designed an Earth-observing satellite, obtained funding for the project, and its eventual successful launch in July 1972 as the first Earth Resources Technology Satellite (ERTS-1), which was later called Landsat-1. This first satellite used the NASA Nimbus bus to carry its sensors, as did Landsat-2. Landsat-1 was launched on July 23, 1972 when it was still known as ERTS-1 (Fig. 7.19) and lasted until January 6, 1978. There were two sensors onboard: a Return Beam Vidicon (RBV) camera and the multispectral scanner (MSS) built by the Hughes Aircraft Company. A TV camera, the RBV was supposed to be the prime instrument, but it was very quickly discovered that the MSS data were superior in content. In addition, the RBV instrument was the source of an electrical transient that caused the satellite to briefly lose altitude control.

The MSS collected data in four spectral bands: the green, red, and two near-infrared bands (Table 7.4). The MSS had a ground resolution of 80 m in all of the spectral bands. The cross-track scanning was achieved by an oscillating mirror; six lines were scanned simultaneously in each of the four spectral bands for each mirror sweep.

Landsats 1–3 were all modified Nimbus satellites. They were all three-axis stabilized using 4 wheels to ±0.7 degree and had 3 hydrazine thrusters. The pixel size was 57 X 79 m, and the scene size was 170 X 185 km. The satellites were in a 15-day repeat orbit, which was good for static ground targets but not for rapidly changing surface or atmospheric phenomena.

Although the DOI did not succeed with their own satellite plans, they still pursued the goals of the EROS program, part of which was to archive and distribute remotely sensed data and support research with satellite data within the DOI. To this end, they set up the EROS Data Center in Sioux Falls, South Dakota, in 1972. This data center became the prime

FIGURE 7.19 Landsat-1.
Image credit courtesy NASA

TABLE 7.4 Multispectral scanner (MSS) channels.

Band 4	**Visible green**	**0.5–0.6 μm**
Band 5	Visible red	0.6–0.7 μm
Band 6	Near-infrared	0.7–0.8 μm
Band 7	Near-infrared	0.8–1.1 μm

repository for Landsat data during all the early phases of the program. NASA was happy to sign over the data handling to the EROS data center as long as they could retain control of the design and deployment of the Landsat spacecraft.

After the launch of Landsat-1, NASA's Goddard Space Flight Center hosted a series of symposia for users to report on their use of Landsat data. The application categories were agriculture, forestry, environmental geology, land use/land cover, water, and marine. Each of the proceedings of these symposia approached 2,000 pages.

In the decades that followed Landsat-1, the program was beset with severe political uncertainty and was casually labeled as "technology in search of an application." Still, the critical mass of support was achieved, and the Landsat program continued with lots of interested users. Landsat 1–3 were operated in near-polar orbits at an altitude of 920 km. They circled the Earth every 103 min, completing 14 orbits per day, producing swaths of imagery 105 km wide. Eighteen days and 251 overlapping orbits were needed to provide nearly complete coverage of the Earth's surface. The amount of overlap varied from 14% at the equator to nearly 85% at 81 degrees north or south. As stated earlier, all three of the first Landsat satellites carried both the RBV and the MSS. For Landsat 3, the MSS was modified to add a thermal infrared channel. Landsat 1 delivered high-quality data for almost 5 years beyond its designed life of 1 year (Table 7.5).

TABLE 7.5 Early Landsat satellite series.

Satellite	Launch date	Termination	Duration	Comments
Landsat 1	23 July 1972	6 Jan 1978	5 years 6 months	Built by RCA, originally ERTS-1
Landsat 2	22 Jan 1975	25 Feb 1982	7 years 1 month	Identical to Landsat 1
Landsat 3	5 Mar 1978	31 Mar 1983	5 years	MSS added thermal IR
Landsat 4	16 July 1982	14 Dec 1993	11 years 4 months	New bus and add thematic Mapper
Landsat 5	1 Mar 1984	5 June 2013	29 years 3 months	Copy of Landsat 4

FIGURE 7.20 Landsat-4 satellite.
Image credit courtesy NASA

As stated earlier, Landsats 1–3 used NASA's Nimbus bus as a platform. With Landsat 4, the program opted for a new spacecraft (Fig. 7.20) built by GE Astro (later to become Lockheed Martin Missile and Space). It used the multimission modular spacecraft (MMS) bus developed originally by Fairchild Industries.

The satellite was designed to be repaired by astronauts in space via the Space Shuttle. This never came to pass. Landsat-4's attitude was controlled with hydrazine thrusters. The spacecraft was three-axis stabilized with reaction wheels. It was powered by three nickel-cadmium batteries (NiCd), which were charged by a single solar array. The satellite was designed to last a minimum of 3 years, much less than the 11 years that it operated.

Landsat-4 carried the same MSS as Landsat-3 with an additional thermal infrared channel. A new instrument, the Thematic Mapper (TM), was added to Landsat-4. The TM had seven bands rather than the four of the MSS with a much higher spatial resolution. Bands 1–5 and 7 each have a spatial resolution of 30 m; band 6 in the thermal infrared channel has a maximum resolution of 120 m.

There was an agreed-upon relationship between NASA and USGS before the launch of Landsat-1 where NASA would continue to develop the satellites and sensors and be responsible for the launch and operation of these satellites. USGS would handle the archival and dissemination of all of the Landsat data through their EROS Data Center. Still, there continued to be tension regarding the Landsat program between NASA, NOAA, DoI, USDA, and USGS. The Carter administration undertook an extensive review of both military and civilian space policies, and by 1979, new policies were formulated in which the civilian program was to be made operational, administered by NOAA, and eventually turned over to the private sector. At about the same time, Congress merged land, ocean, and weather remote sensing systems under the administration of NOAA.

A crisis arose where a burgeoning community of Landsat data users, among them the news media, wanted low-cost imagery. A growing industrial sector believed that privatization of the Landsat data supply would strengthen America's hand in commercial Earth observations. In its effort to reduce the size of the government, the first Reagan Administration acted quickly to move the Landsat program into the private sector. NOAA solicited bids to manage the existing Landsats and oversee the program. Proposals were received from such diverse bidders as aerospace companies, an insurance company in New York, a small geoscience firm in Michigan, and a farmer in North Dakota. In 1985, a contract was signed with EOSAT Company and the transfer of the Landsat program, but not the weather satellites, was complete. It is interesting that the most compelling arguments presented to Congress for Landsat commercialization had to do with data and program continuity, not improved spectral analyzes and fine-resolution, time series data. The result was the Land Remote Sensing Commercialization Act of 1984.

EOSAT's (the winning bidder) first effort was to design and build an altogether new satellite for Landsat-6. On October 5, 1993, the EOSAT-owned Landsat-6 failed at launch after not reaching the velocity to obtain orbit. The satellite carried the Enhanced TM which contained the same seven spectral bands as in the TM instrument on Landsats 4 and 5, with an additional eighth band with a spatial resolution of 15 m instead of the 30 m for the other channels. The eighth band was known as the sharpening band or panchromatic band, sensitive to light from the green through the near infrared. With the failure of Landsat-6, it looked like there might be a gap, except Landsat-5 continued to operate until June of 2013.

The privatization of Landsat data during this period had some very adverse effects. The science community that wanted to use these data in their research could no longer afford them, and many studies failed due to a lack of access to these data. Landsat data had always cost but now the price was so high it was out of the reach of most research budgets. The main data users became the oil companies, insurance companies, and other industrial users who could easily afford the high prices. The discontent of the science community rippled through many agencies and supervisory bodies. The Space Studies Board of the National Academies of Sciences submitted a statement that the commercialization of Landsat data needed to be reversed so that science users could acquire the data.

An extensive study was carried out by NOAA, and as a result, the National Space Council released its National Space Policy Directive #5, establishing new goals and guidelines for the Landsat program. The directive called for a joint DoD/NASA effort to build, launch, and operate Landsat-7. In October 1992, the Land Remote Sensing Policy Act was signed into law, which reversed the 1984 decision to commercialize the Landsat program. The law required NASA to negotiate with EOSAT for a new data policy regarding existing Landsat data that would make data available to the general science user community. NASA was also to assume program responsibility from NOAA.

Hardly a year later, Landsat was again under scrutiny, mainly because of the large budget request for the high-resolution multispectral stereo imager proposed by DoD for Landsat-7. Instead, the National Science and Technology Council (NSTC) recommended that Landsat-7 carry only an improved TM sensor. As a result, DoD withdrew from the program. This resulted in Presidential Decision Directive/NSTC-3, dated May 5, 1994, reconfirming the Administration's support for the program but giving NASA, NOAA, and USGS joint management responsibilities for the overall program. These agencies negotiated with EOSAT for reduced prices for existing Landsat data and proceeded with the planning for Landsat-7.

The government-owned Landsat-7 (Fig. 7.21) was successfully launched on April 15, 1999, from the Western Test Range of Vandenberg Air Force Base, CA, on a Delta-II expendable launch vehicle.

The Earth observing instrument on Landsat-7 was the Enhanced Thematic Mapper Plus (ETM+) that replicated the channels on Landsat-5 with additional capabilities:

- a panchromatic band with 15 m spatial resolution
- onboard, full aperture, 5% absolute radiometric calibration
- a thermal infrared channel with 60 m spatial resolution
- an onboard data recorder

At the time, Landsat-7 was at far the most accurately calibrated Earth-observing satellite, particularly when its measurements are compared with data collected on the ground. Landsat-7's sensor has been called "the most stable, best characterized Earth observation instrument ever placed in orbit." Landsat-7's rigorous calibration standards have made it the validation choice for many lower-resolution sensors. The excellent data quality, consistent global archiving, and reduced pricing ($600) of Landsat-7 led to a large increase in data users. In October 2008, USGS made all Landsat-7 data free to the public, followed in January 2009 when they made all Landsat data free to the public, resulting in a 60-fold increase in downloads. Landsat-7 is now in an extended life mission.

An example of Landsat-7 imagery is presented here compared with a later image from Landsat-9. They are images of the Malaspina Glacier in Alaska (Fig. 7.22).

It should be noted that Landsat-7 had a major failure of one of its scanning subsystems known as the Scan Line Corrector after only 4 years of operation beginning in 1999. This failure reduced the usability for many people requiring precise image locations. Unlike the radiometers of the TIROS series of satellites that simply scanned in one direction, the

FIGURE 7.21 Landsat-7 satellite.
Image credit courtesy NASA

FIGURE 7.22 Landsat-7 and 9 images of the Malaspina Glacier in Alaska. https://science.nasa.gov/earth/earth-observatory/malaspina-glacier-alaska-2-2749/

mirror in the MSS and the TM scanned back and forth, forming what is known as a "bow tie effect" that needed to be corrected to have a continuous image. The lines you see are introduced to correct for this missing capability

The most significant constraint of the Landsat-7 sensor was its limited dynamic range, resulting in high noise levels in dark areas and frequent overexposure in bright areas. The design tried to mitigate that limitation by offering two different amplification settings of the sensor, which were chosen based on the expected brightness of the target Earth surface. As a result, USGS stopped recording areas with particularly high contrast (i.e., polar regions) and concentrated the recording capacity of Landsat-7 on the main lower latitude land masses.

This brings to a close the discussion of Earth remote sensing satellites prior to the year 2000. We will now move into the transition to systems that embody the present-day remote sensing.

Chapter 8

Other major satellite programs

8.1 Satellite altimetry

The earliest satellite altimetry instrument was on GEOS-3 (Geodetic Experimental Ocean Satellite), which was part of a NASA series of satellites to study the Earth's gravity field. The mission objectives were to perform a satellite altimetry experiment on orbit to further support the calibration and position determination of NASA and other agency C-band radar systems. The radar altimeter (ALT) on GEOS-3 provided sub-satellite sea surface height (SSH) measurements with a 0.5 m range precision, which could be improved to 0.2 m in an intensive mode. GEOS-3 orbited in an elliptical orbit 816 by 850 km with an inclination of 115 degrees and a period of 101.52 min. The satellite was gravity gradient stabilized (Fig. 8.1).

The GEOS-3 spacecraft was designed and built by the Applied Physics Lab (APL) of Johns Hopkins University in Laurel, MD. The gravity gradient stabilization required the long boom supporting the small mass away from the main spacecraft. GEOS-3 was launched on April 9, 1975 on a Thor-Delta launch vehicle from Vandenberg, CA. The ALT was a Ku-band instrument designed and built by APL and by the Aerospace and Electronic Systems Department of General Electric Company. The ALT is a multimode radar system with two distinct data-gathering modes (global and intensive). The global mode used a long pulse (200 ns) with six readings per second (since GEOS-3, all altimeters have been using the chirp technique of pulse compression). Both modes operated at 13.9 GHz using a parabolic antenna.

The ALT is the first instrument to fly on a satellite dedicated to mapping the ocean's surface. This instrument was of S-193 heritage on the Skylab mission (1973—the Skylab altimeter recorded a couple of tracks that revealed the existence of trenches in the marine geoid). This satellite was intended to improve the accuracy and provide global coverage. ALT featured two self-rest/calibration modes for use in on-orbit functional test and instrument calibration.

The next big development in satellite altimetry was the launch of the US Navy's Geodetic Satellite (GEOSAT) on March 12, 1985 into an 800 km, 108 degrees inclination orbit (Fig. 8.2).

Obvious from the long neck out to the supplemental mass is the fact that this satellite was gravity gradient stabilized. The first part of the GEOSAT mission data was classified, as the Navy wanted to map the global geoid to better correct their gravity maps for cruise missiles. Called the Geodetic mission, this period lasted for 18 months, which ended on September 30, 1986. At that point the navy had all the data that they needed and due to some heavy lobbying by Jim Mitchell of the Naval Research Lab in Stennis, MS and by George Born of the University of Colorado at Boulder, CO, the navy agreed to move GEOSAT to a near-polar circular orbit with an inclination of 108 degrees, an altitude of 800 km and a period of 101 min. Its repeat period was 17.05 days. This was called the exact repeat mission, which started October 1, 1980, and lasted until January 1990 when it was terminated due to the failure of both onboard tape recorders. The data of the exact repeat mission were not classiified and became well used by the oceanographic community. The Geodetic Mission data were finally completely declassified in 1995.

The gravity gradient attitude control kept the spacecraft to within 1 degree at nadir with a 98.6% accuracy. This attitude control system consisted of a 6-m scissors boom with a 45 kg end mass, redundant momentum wheels, attitude control thrusters, three digital sun-attitude detectors, and a three-axis vector magnetometer. RF communications were achieved through two dual-track high-density tape recorders that would independently record a 10.205 kbits/s telemetry stream. The onboard recording lasted up to 12 h, and playback transmission to the ground occurred at 833 kbits/s on an S-band downlink. The spacecraft was commanded through a VHF uplink from the APL ground station.

The first ocean bottom topography map from altimetry is presented here in Fig. 8.3.

This map was created by David Sandwell and his student Walter Smith, who now works at National Oceanic and Atmospheric Administration (NOAA) on satellite altimetry. They have continued their efforts to map bottom topography with later altimetric satellites.

Satellite Remote Sensing. DOI: https://doi.org/10.1016/B978-0-443-22086-9.00006-X

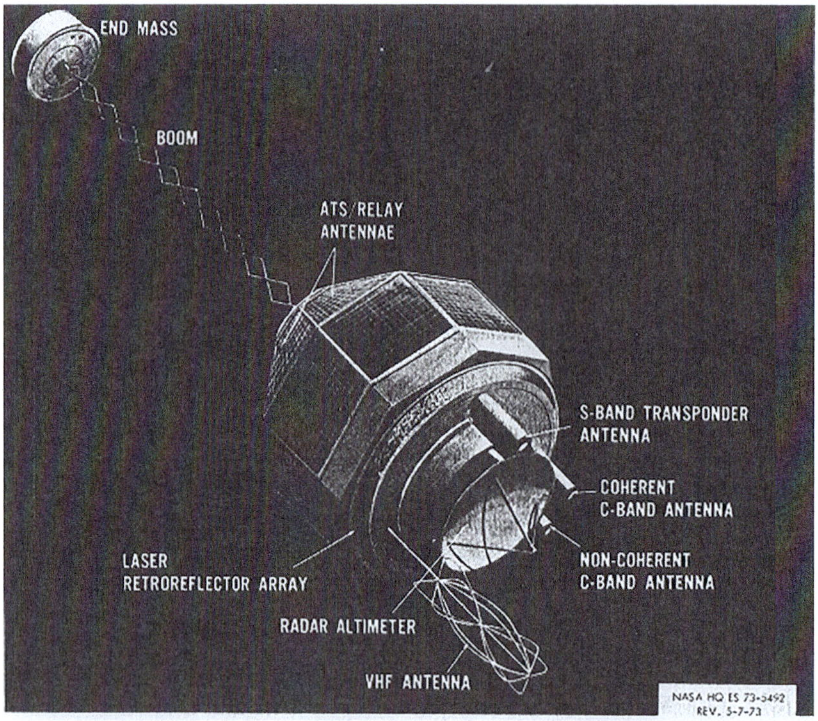

FIGURE 8.1 Geodetic Experimental Ocean Satellite-3 spacecraft. (https://podaac.jpl.nasa.gov/GEOS-3). Image credit: NASA.

FIGURE 8.2 Geodetic Satellite (GEOSAT) spacecraft. (https://en.wikipedia.org/wiki/Geosat). Image credit NOAA JHU/APL.

GEOSAT data were also used by Cheney and Miller (1988) to map sea level variations in the tropical Pacific and Indian Ocean during the 1986-87 El Nino event. Using a combination of crossover and collinear techniques, they were able to generate a time series of sea level anomalies in the tropical Pacific and Indian Oceans for the first years of the GEOSAT exact repeat mission. This time interval contains the 1986-87 El Nino event, which would be the first time that such an event was sensed with satellite ALT data. First, to evaluate the accuracy of the satellite ALT data, they compared the GEOSAT data with coincident tide gauge data from Christmas Island (1.9°N, 157.5°W) as presented here in Fig. 8.4.

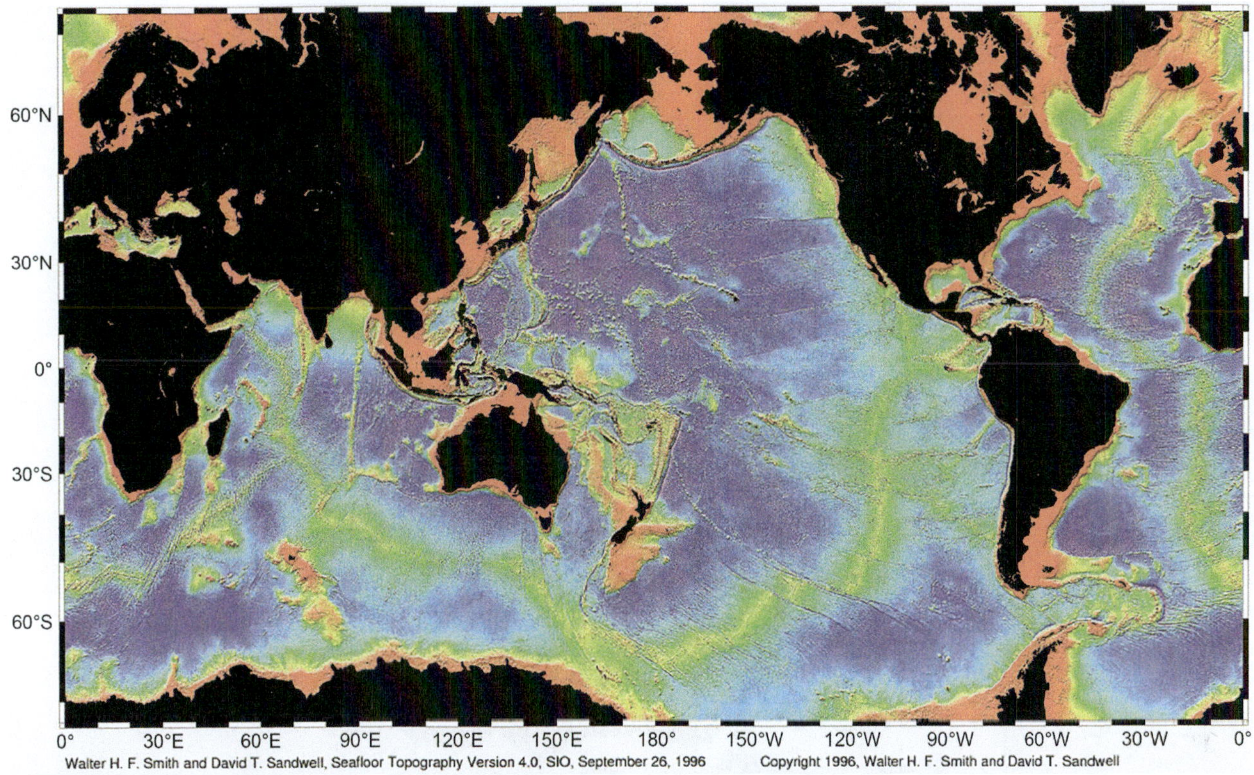

FIGURE 8.3 The ocean seafloor map was created from Geodetic satellite altimeter data. Map lines delineate study areas and do not necessarily depict accepted national boundaries (https://en.wikipedia.org/wiki/Geosat).
Image credit: Scripps Institution of Oceanography, NOAA.

FIGURE 8.4 Comparison of Geodetic Satellite time series (solid line) for a @X1 degree region centered on 1°N, 157°E with tide gauge data collected at Christmas Island (dashed line). The RMS difference is 5.9 cm with a correlation of 0.83. These comparisons are made on monthly means (Cheney and Miller, 1988).

This agreement substantiates the idea that the satellite altimetry really does measure the ocean's surface height and its variations in time and space.

To represent the latitudinal structure of the tropical Pacific, Cheney and Miller (1988) averaged all the GEOSAT data in 1-degree latitude bands (Fig. 8.5).

The question to answer is how, in an El Nino year, water is redistributed within the tropical Pacific. They concentrated on meridional transport as the low frequency El Nino signal is often described as a displacement of water from the equatorial region to higher latitudes. From Fig. 8.5, we can see from the sea level record that the annual and interannual signals are clearly organized into three latitude bands: 20°S–8°S, 7°S–7°N, 8°N–20°N. Of significance to the El Nino is the fact that all three bands display long-term trends. In Fig. 8.6 the GEOSAT data have been replotted as the volume variability for these three zonal bands. The northern band exhibits a strong annual signal, while the Equatorial band also shows a strong annual signal but is almost exactly out of phase with the northern band.

The southern band shows a very weak annual signal.

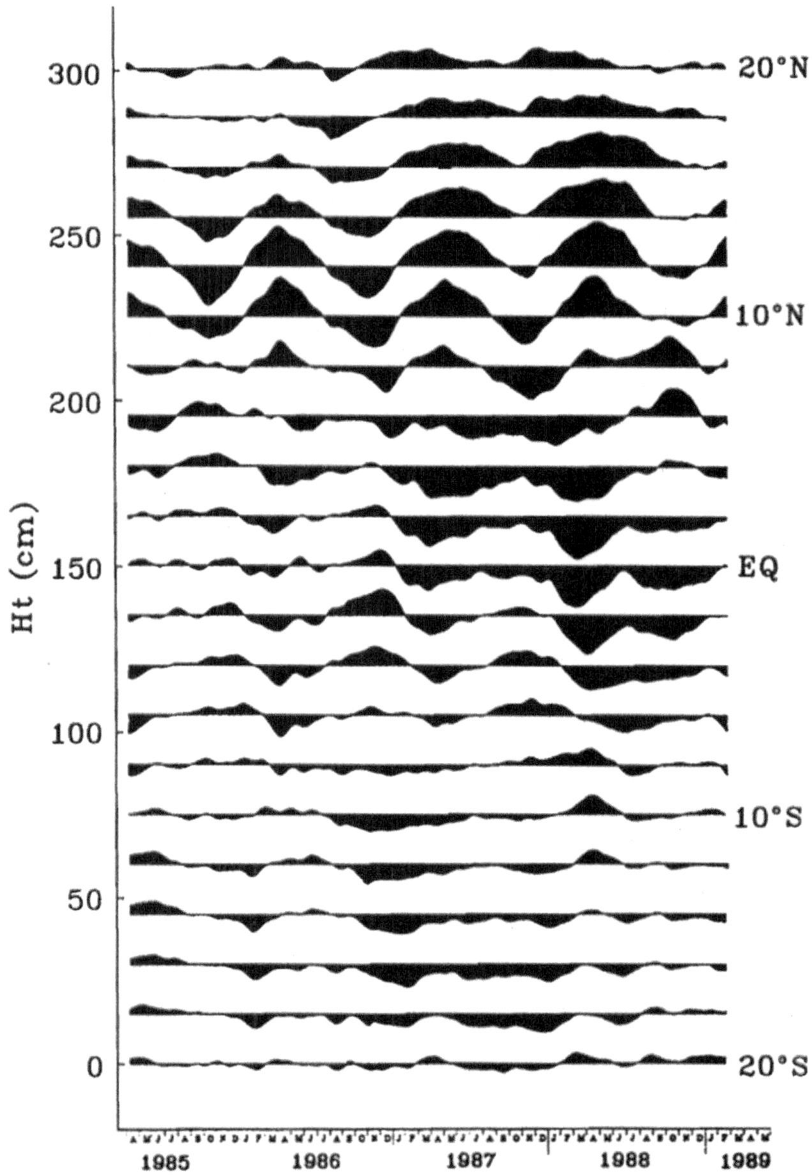

FIGURE 8.5 Zonally averaged sea level determined from Geodetic Satellite time series in alternate 1-degree latitude bands (20°S, 18°S, 16°S....) across the entire Pacific. Vertical offsets between series are 15 cm. The strongest signals are found in the northern hemisphere, principally the result of annual forcing by the wind. A distinct phase change exists between 6°N and 10°N during most years (Cheney and Miller, 1988).

Of greater importance to the El Nino phenomenon is the fact that all three bands display long-term trends. To better display these trends, the GEOSAT data have been plotted as the volume anomalies relative to their respective April 1985–86 cycles in Fig. 8.7.

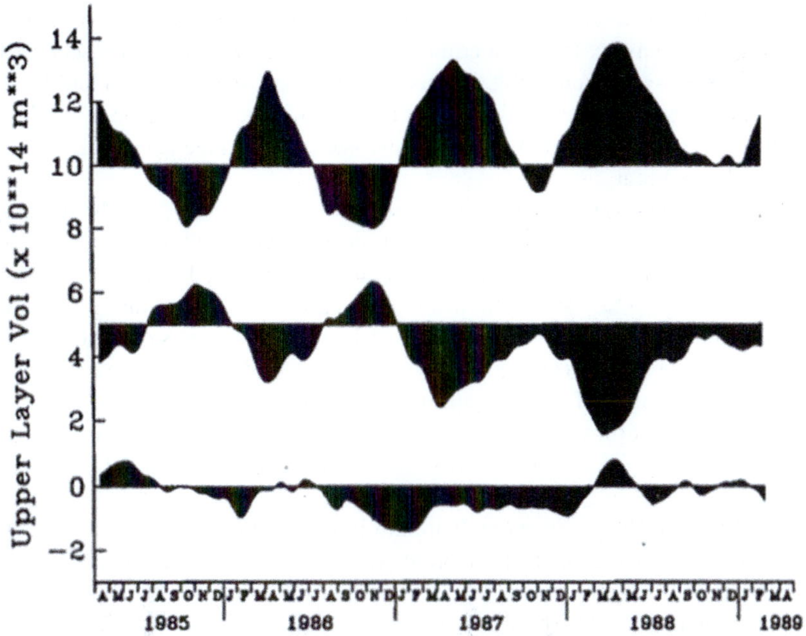

FIGURE 8.6 Upper layer volume determined from all Geodetic Satellite time series in three latitude bands: 20°S–8°S (bottom), 7°S–7°N (middle), and 8°N–20°N (top) (Cheney and Miller, 1988).

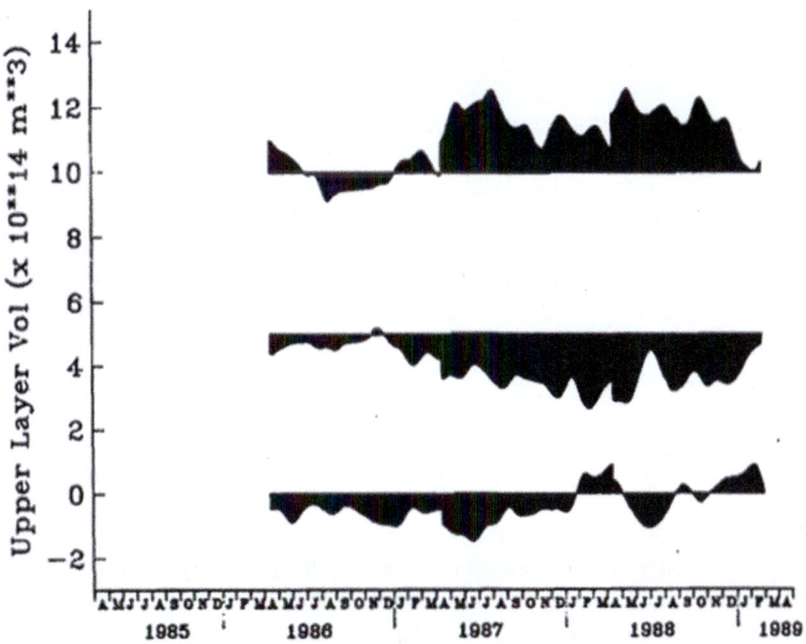

FIGURE 8.7 The volume anomaly with respect to the 1985–86 annual cycle. Long-term trends are now evident (Cheney and Miller, 1988).

The southern band displays negative anomalies starting in mid-1986, while the anomalies in the other two bands do not develop till 1987. Relatively normal conditions return in the south in early 1988, while the equatorial and northern bands remain anomalous until early 1989. All three bands show large anomalies during the third year (April 1987–88) with larger than normal volumes in the northern band, a corresponding deficit in the equatorial region, and continued low volume in the south.

Cheney and Miller (1988) conclude that by integrating GEOSAT-derived sea level time series over the entire width of the Pacific Ocean, they have been able to quantify the net meridional exchange of warm surface water between different latitude bands during the 1986–87 El Nino event. Following the onset of the warm water event in late 1986, the equatorial region in the Pacific lost about 1.5×10^{14} m^3 of water, while the north equatorial region gained a similar amount. These volume anomalies persisted for nearly two years, during which time the warm event gave way to an unusually cold event. Finally, in 1989, all regions returned to normal.

While GEOSAT gave us a glimpse of what was possible with systematic satellite ALT data, the big change came with the launch of TOPEX/Poseidon (TP) on August 10, 1992 a joint project between NASA (United States) and CNES (France). The TP satellite operated for a total of 13 years, 5 months, and 12 days well beyond its design lifetime. The revolution ushered in by TP can be summed up by a statement made by its earliest Program Manager at CNES, Jean-Louis Fellous. "Although I had deep confidence in the fact that TP would reveal unknown features of the ocean circulation, none of us could have imagined that it would revolutionize physical oceanography and climate research to the extent that it did." In a way, TP ushered in the "golden age" of satellite altimetry. "What started as a single satellite launched as a scientific experiment involving just a handful of scientists became the story of many successive satellites and a worldwide community working toward a common good," said Alain Ratier, former TP Program Scientist at CNES and later Director-General of EUMETAT.

The TP satellite (Fig. 8.8) was built in the United States by Fairchild Industries under contract to NASA/Jet Propulsion Laboratory (JPL). The agreement with France was that the satellite would be built by one country and launched by the partner. Thus TP was launched on Ariane 4 P from Kourou, French Guiana. Initially, TP was conceived as two separate satellite ALT missions. NASA was planning the TOPEX mission while CNES was planning the Poseidon mission. Stan Wilson, then Chief of NASA's Oceanic Processes Branch, said, "Our altimetry mission was so expensive that we couldn't afford it on our own." Wilson made an impromptu trip to CNES in Toulouse, France, following a conference in Venice. At the Venice meeting, he had been invited by Michel Lefebvre (who later would become known as one of the leaders of satellite geodesy) to visit CNES. Then Wilson realized that the French were also working on altimetry with a solid science background and purpose. When he realized that he could not sell TOPEX on his own, he wondered if the United States might partner with the French. Meanwhile, CNES had run into an obstacle of its own. While funding was available for their Poseidon program, the French government was not convinced of the value of an altimetry mission. CNES needed to find a partner.

In 1983 a US–French panel was convened in Washington, DC, which recommended exploring a potential collaboration. In 1987 both countries approved a joint program. The joint mission had three main goals: measuring ocean circulation, processing, and distributing the measurements to science investigators, and establishing the basis for these observations to continue long term. Each of these goals produced formidable barriers for the teams on both sides of the Atlantic to overcome. First off, since the joint satellite would carry ALTs from both countries but had only one antenna, the ALTs could not operate simultaneously. An agreement was finally reached that the NASA TOPEX ALT would operate 90% of the time and the CNES Poseidon ALT would only operate 10% of the time. This was likely due to the fact that NASA used a traveling wave tube in its design, used on previous ALTs while the Poseidon ALT was a new design solid-state instrument. The irony of this decision was that the Poseidon ALT operated so well that it became the default instrument for all future NASA-CNES ALT missions.

The next problem was which agency would provide the satellite bus and which would launch the satellite. At a meeting between the NASA administrator James Beggs and CNES president Hubert Curien in Washington DC, Curien offered to provide a SPOT bus for use in the joint mission, leaving NASA to provide the launch. NASA's Wilson recognized that this arrangement would compromise the scientific goals of the program and he was unable to accept the arrangement. The SPOT bus was only capable of flying in a sun-synchronous orbit, which results in the similar illumination of the Earth by the sun each day. Although this orbit is ideal for optical imaging, it is not optimal for determining changes in global sea level. This is due to the fact that changes in sea level are caused in part by tides, including tides caused by the sun. Flying an ALT in a sun-synchronous orbit would result in observing the sea level when the solar component was always in the same phase. Thus,it would not be able to observe the solar component of the tides.

Wilson instead suggested an alternative arrangement where NASA would provide a bus capable of flying in the higher nonsun-synchronous orbit and have CNES provide the launch. Beggs proposed this as a counteroffer, and CNES accepted.

FIGURE 8.8 TOPEX/Poseidon satellite being readied for launch.
Image credit courtesy NASA.

This came with a price since it was viewed as the US endorsing the Ariane launch vehicle. This prejudice soon abated, and the program moved ahead. The earlier Seasat ALT had suggested the utility of a dedicated altimetric satellite, and GEOSAT had further established that value. Still, the ±5 cm accuracy of GEOSAT's sea levels suggested that the next altimetry mission must be much more accurate to be able to resolve the sea level changes associated with variations in the ocean circulation. TP was launched into a geocentric, nonsun-synchronous orbit with a 1,340 km altitude, 66 degrees inclination, and a period of 112 min. This was an orbit that repeated every 10 days. It operated until October 18, 2005 and was decommissioned January 18, 2006.

Fellous said "convincing the other (than CNES) French agencies to participate in a collective effort to develop the relatively small ocean community into a sizeable group of people needed to utilize the altimetry data when they arrived, required some level of diplomacy. In the end the community in both France and the US was formed from some remote sensing people who turned their interest to the study of the ocean and some classic physical oceanographers who saw the benefits they could get from these new satellite data."

Not long after the launch and operation of TP, the data collected as part of the World Ocean Circulation Experiment began to transform our preexisting understanding of the ocean. "One 10-day cycle of TP data showed as good an image of the global ocean circulation as the previous 150 years of ocean measurements from ships had shown," said Fellous. It should be noted here that while the TP data were a dramatic step forward in understanding the geostrophic circulation of the ocean, they by no means covered all of the currents that needed to be observed in physical oceanography studies.

The fundamental principles of radar altimetry are presented here in Fig. 8.9.

The ALT measures the range between it and the sea surface relative to a reference ellipsoid from which the satellite orbit is well known. The quantity of primary interest to the physical oceanographer is the sea level relative to the marine geoid, which then can be related to the geostrophic ocean currents. This SSH can be written as follows:

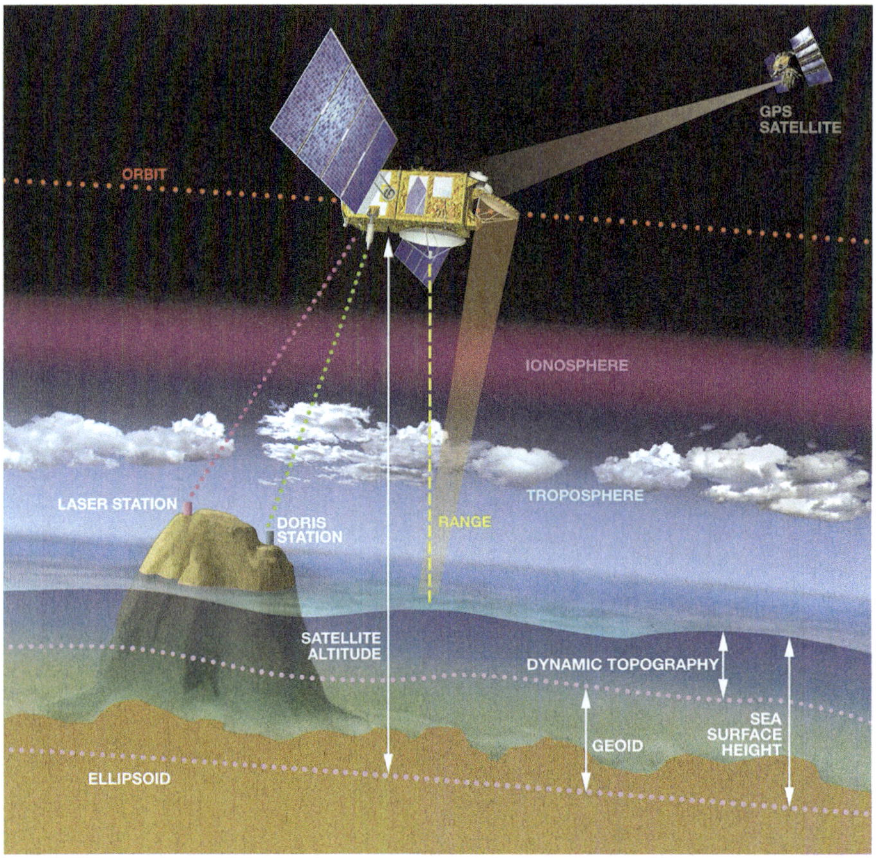

FIGURE 8.9 Satellite altimeter measurement principle.
Credits courtesy CNES and AVISO.

$$\mathrm{SSH}(x,\ t) = G(x) + \eta(x,\ t) + \varepsilon(x,\ t) \tag{8.1}$$

where G is the geoid, η is the dynamic topography of the ocean, and ε is the measurement error (Le Traon, 2007). While many efforts are underway to better define the marine geoid, it is still not well enough known to allow the calculation of absolute dynamic height except at very long wavelengths. However, the interesting variable part of the equation, SSH(x, t), is easily calculated using the "repeat-track method." For a given satellite track (recall that the ALT is a nadir sampling instrument), SSH'(x, t) is obtained by removing the mean profile over several ALT cycles, which contain both the geoid G and the dynamic topography. Thus we can write as follows:

$$\mathrm{SSH'}(x,\ t) = \mathrm{SSH}(x,\ t) - (\mathrm{SSH}(x))_t = (\eta(x,\ t)) - (\eta(x,\ t))_t + \varepsilon(x,\ t). \tag{8.2}$$

To get the absolute SSH, one needs to use a mean dynamic climatology from historical in situ data or to use existing geoid estimates together with an ALT mean surface.

One of the most amazing successes of the TP program was the realization of a ±2 cm error in the satellite's orbit. This was an achievement of people such as George Born at the University of Colorado and the astrodynamics group at JPL. Much of this is due to the many sources of data fed into the orbital models. Born was instrumental in getting a GPS antenna added to TP and the French DORIS Doppler radio tracking added to the standard laser reflector for ranging measurements. This achievement in the reduction of the orbit error is something that really enhanced the TP data for the physical oceanographic community. This orbit error is small compared with the ±10 cm accuracy of the ALTs on ERS 1, 2, and the ±5 cm accuracy of GEOSAT.

Space/time variations in atmospheric pressure result in changes in sea level on the order of a cm but are easily corrected knowing the atmospheric pressure. Ionospheric changes can introduce range errors from 1 to 20 cm, but the dual-frequency ALTs can correct this down to 0.5 cm. One of the largest corrections is for the atmospheric path delay due to atmospheric water vapor. For this reason, starting with TP, all ALT satellites carry a bore-sighted microwave radiometer optimized to measure the water vapor-induced path delay. Called the wet troposphere delay, the correction varies both spatially and temporally, with magnitudes from 5 to 30 cm (highest in the tropical convergence zones where atmospheric convection is strong). The precision of the onboard water vapor correction is about 1 cm.

One of the most difficult errors to correct is known as the "sea state bias" and is due to the fact that the concave form of the ALT wave fronts responds differently to the wave-covered ocean; sea surface troughs reflect better as compared to reflection from the wave crests. This shifts the SSH away from the mean sea surface to a bias that favors the troughs. This error can be as large as 2 cm. The temporal evolution of the ALT error is summarized in Fig. 8.10 taken from LeTraon et al. (2010).

This figure documents the change in error from the early GEOS-3 mission, through Seasat, GEOSAT, and ERS, up to TP, which ushered in the world of modern satellite altimetry. We will introduce some of these later missions in this chapter. One can clearly see the improvement in ALT accuracy over time. The TP mission did not initially have 2 cm orbits, but as ground truth calibration systems came online (like the Harvest platform ground truth site in California), the orbit error reduced. One can see how this improvement in the ALT error was needed to really detect and map the ocean circulation signal.

The ALT waveform (Fig. 8.11) also yields information on the wind speed at that location. The slope of the leading edge is inversely proportional to the significant wave height, while the y-axis from zero indicates the instrument noise level, and the waveform maximum is a measure of the sigma-0 backscatter, which is related to the wind speed.

From the mid 1990's to early 2012, there were multiple ALT satellites operating, giving us the real heyday of ALT data availability. During this period, it became useful to merge these different data together to improve the space-time coverage of the combined ALT data. To merge different ALT data, one must first create homogeneous and intercalibrated data sets. One way to do this is to use the most precise mission as a reference for the other ALTs (Le Traon and Ogor, 1998). The best references were TP or Jason 1, 2, the follow-on satellites to TP. Furthermore, to compute the sea level anomaly (SLA), it is best to use a common reference surface.

One of the main purposes of satellite altimetry was to measure the geostrophic circulation of the ocean. This is derived from the ocean surface topography that functions for the physical oceanographer much as the pressure field does for the meteorologist. The global sea surface from TP is shown here in Fig. 8.12.

The geostrophic currents should flow along the lines of the height contours.

The history of ALT satellites up to 2005 is nicely summarized by Le Traon (2007) and reproduced here as Fig. 8.13. This figure clearly shows the development of satellite altimetry before 2000. This single mode of remote sensing did have a profound effect on research in physical oceanography and transformed the way we approach ocean circulation today.

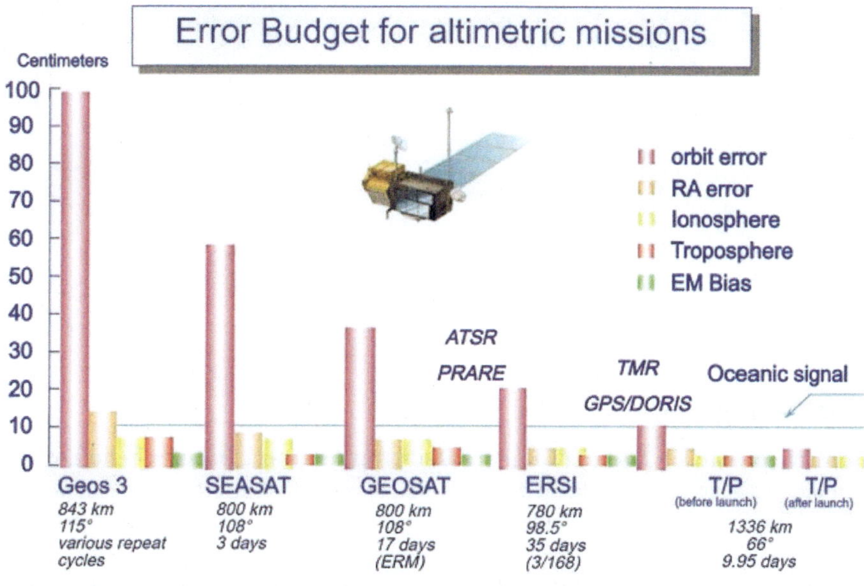

FIGURE 8.10 Evolution of altimeter overall error (Le Traon, 2007).

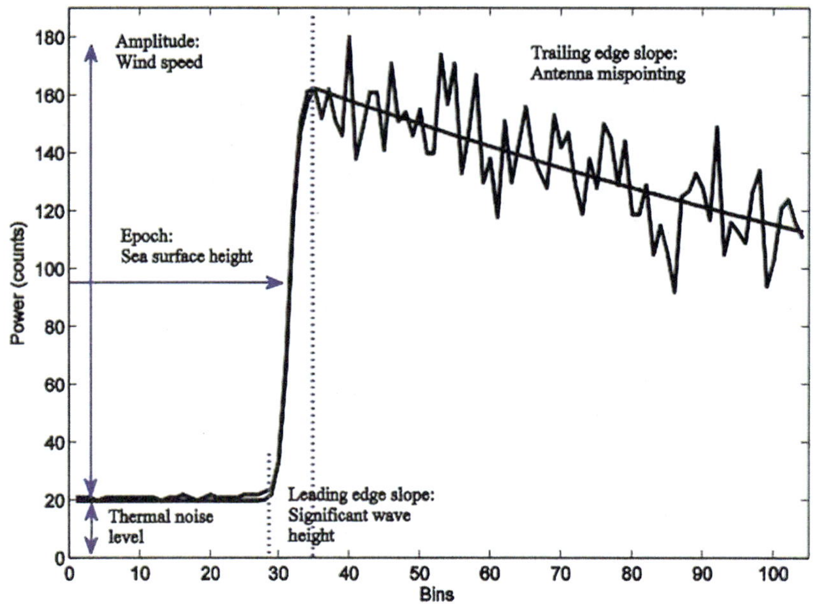

FIGURE 8.11 Altimeter waveform labeled for geophysical parameter retrievals (https://www.aviso.altimetry.fr/en/techniques/altimetry/principle/pulses-and-waveforms.html).

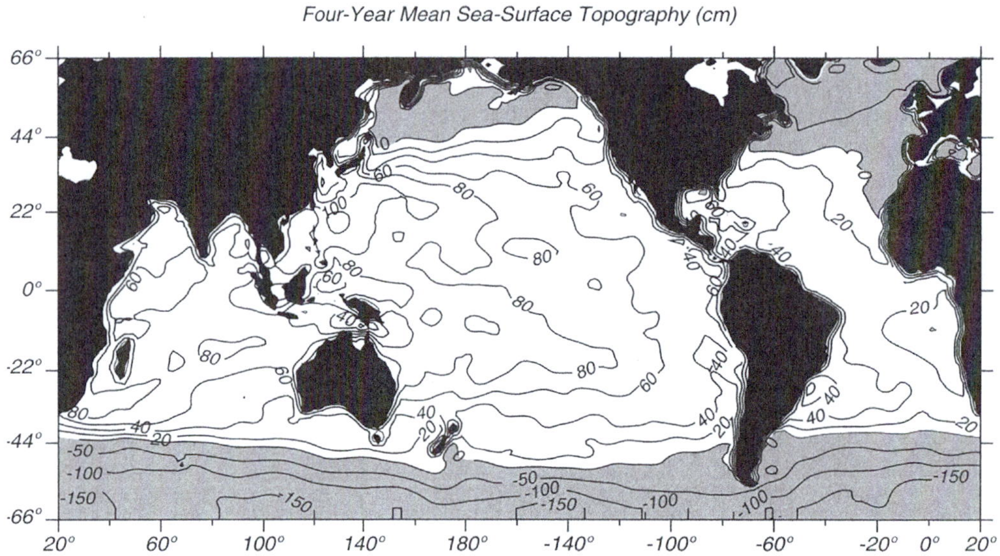

FIGURE 8.12 Global distribution of time-averaged sea surface topography from TOPEX/Poseidon altimeter data between 10/3/92 and 10/6/99 relative to the JGM-3 geoid. (Le Traon and Ogor, 1998). Map lines delineate study areas and do not necessarily depict accepted national boundaries.

8.2 NASA's Earth observing system (EOS)

Around the turn of the century, there were a number of large satellite programs, many of which did not survive in their initially proposed state. They did, however, bring into being some important satellite programs that we have inherited today, and even many of the satellites sponsored under the much-reduced programs have provided valuable data to the science community.

Ideas for comprehensively studying the Earth from space began in the 1980's, and many different ideas were proposed. In the 1990's, Shelby Tilford, then NASA Administrator for Earth Studies, proposed a Mission to Planet Earth of which the most important part was the EOS. EOS began as a series of polar-orbiting spacecraft that were going to be as large as a

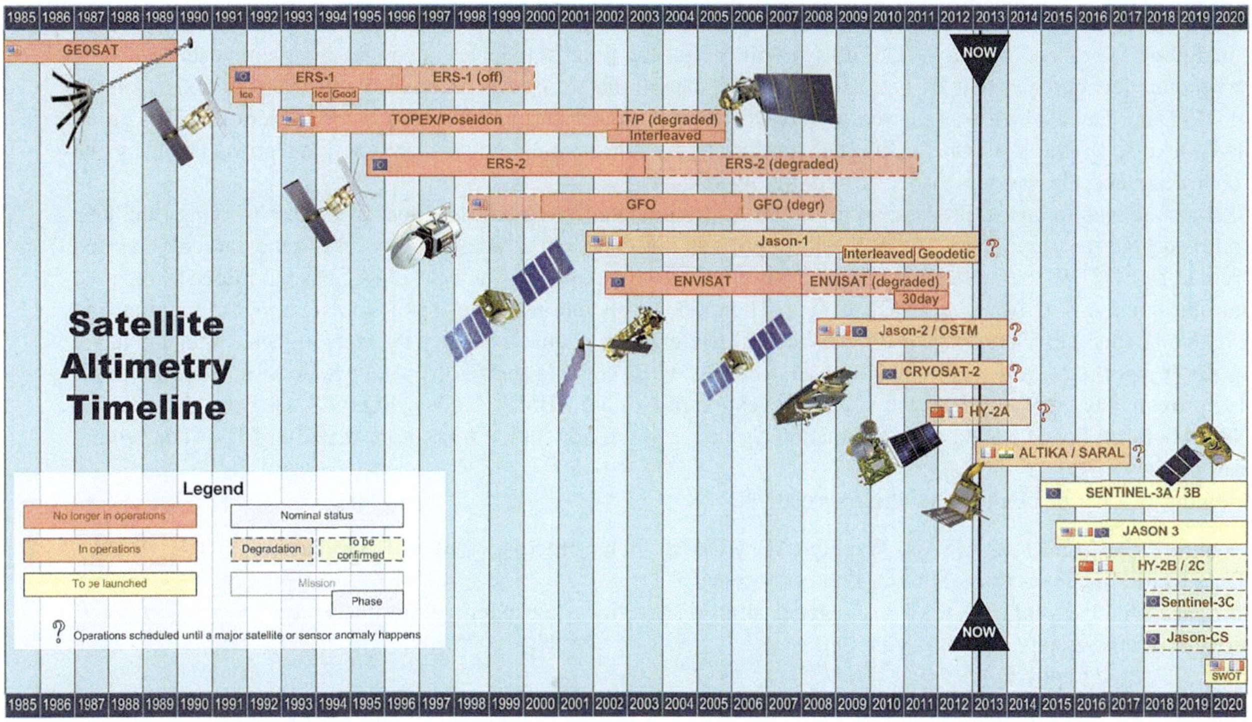

FIGURE 8.13 History of spaceborne radar altimeters (Le Traon, 2007).

Greyhound Bus to carry all of the instruments needed to comprehensively study the Earth. This large program was to cost about $33 billion over the life of the program. There began to be a concern at the Government Accounting Office (GAO) that there were not sufficient remote sensing scientists to utilize the data generated by this very expensive system. In a presentation to the National Research Council (NRC) Committee on Space Studies (CSS, where the author was present), Shelby Tilford explained the need for the large satellite not on scientific grounds but rather on budgetary reasons. He said, "if he used a fleet of smaller spacecraft, he would likely lose a number of them to budget cuts and therefore needed a large satellite."

NASA claimed that the EOS program was needed to detect, understand, model, and predict global climate change. The first EOS satellite was scheduled for launch in 1998; by 2002, a full constellation would be on orbit collecting data from 25 instruments on at least 10 spacecraft. Congressional concerns about EOS's affordability led to a review panel that the author participated in, and the estimated cost was reduced to $11 billion. This was further reduced later to $8 billion in fiscal year 1993 and then to $7.25 billion in fiscal year 1995.

Nevertheless, EOS continued on and was the source of launch and operation of two very successful polar-orbiting satellites. Recognizing the need for more scientists to be involved with EOS data, NASA, in 1995 and 1996, began adding new investigators to the EOS science teams. NASA also embraced an open data policy where any and all EOS data were freely available to anyone interested in them. They also realized that one problem was the lack of interdisciplinary projects that would mix EOS data sets to achieve their objects. So NASA Earth Science solicited interdisciplinary proposals to fill this need. NASA originally selected 29 interdisciplinary science teams in 1989 and 1990, with additional teams selected before the launch of the first EOS satellite then scheduled for late 1995.

There were two NASA missions launched during this period that were not initially part of the EOS system. These were the Upper Atmosphere Research Satellite (UARS) and the TP radar ALT satellite. The GAO found that the science teams for UARS and TP were larger than the EOS teams and that the EOS teams were not going to be adequate to handle the large amount of new data from the EOS satellites. Taken together, the UARS and TP science teams were a little larger than all of the EOS teams, while the EOS data is rather close to 1,000 times greater than the combined data rate of UARS and TP.

8.2.1 EOS data and information system

The data from EOS were supposed to be processed, archived, and distributed by the EOS data and information system (EOSDIS). NASA estimated that the development and deployment of EOSDIS would cost around $3 billion through fiscal year 2000. The US House Committee on Science, Space, and Technology asked GAO to examine several key aspects of EOSDIS, including the extent to which EOSDIS (1) will use prototyping to reduce system development costs and risks, and (2) may require technologies that are beyond the present state of the art. NASA assembled a group of scientists known as the EOSDIS Data Panel, which was chaired by Jeff Dozier, and the author of this book was a member of it. This group met regularly to discuss the state of the EOSDIS project and made visits to review actions planned or taken by the EOSDIS contractor and the various NASA centers involved.

EOSDIS was intended to ingest, process, archive, and distribute data from a large number of Earth observing satellites, from aircraft and field measurements. The goal was to provide the user with a "seamless" access to the data they needed for their research. EOSDIS created a number of Distributed Active Archive Centers (DAACs), distributed across the United States, to archive and distribute EOSDIS data. The mode of operation was to make the data accessible online and for free (Fig. 8.14). Innovative graphical user interfaces were created to enable the query by users to locate data and judge whether or not it met their needs (e.g., amount of cloud cover when viewing the Earth). As of September 2012 EOSDIS contained approximately 10 PB of data with a daily ingestion rate of 8.5 TB/day. Today, EOSDIS has been shortened to ESDIS (NASA's Earth Science Data and Information System), which now holds a lot more than 100 PB of data with a daily ingestion rate of 300 TB/day.

The following is a list of DAACs and their specializations:

- Alaska Synthetic Aperture Radar (SAR) Facility (ASF) DAAC Polar processes and SAR products
- EROS Data Center land processes DAAC. Land processes
- Goddard Space Flight Center (GSFC) DAAC Upper atmosphere, global biosphere, atmospheric dynamics, geophysics, and biological oceanography (ocean color).
- JPL Physical Oceanography DAAC (PO-DAAC)
- Langley Research Center (LaRC) DAAC Radiation budget, tropospheric chemistry, clouds, and aerosols
- National Snow and Ice Data Center (NSIDC) DAAC snow and ice, cryosphere (non-SAR), and climate, University of Colorado
- Oak Ridge National Laboratory (ORNL) DAAC Biogeochemical dynamics
- Socioeconomic Data and Applications Center (SEDAC) at the Consortium for International Earth Science Information Network (CIESIN) human interactions in the environment
- Global Hydrology Resource Center (GHRC) at Marshall Space Flight Center
- National Climatic Data Center (NCDC) of NOAA Climate and weather
- National Geophysical Data Center (NGDC) of NOAA
- Solid Earth geophysics, marine geology and geophysics, solar terrestrial physics, and paleoclimatology
- National Oceanographic Data Center (NODC) of NOAA
  Oceanography
- Satellite Active Archive (SAA) of NOAA

NASA initiated a prototyping project within the early precontract phase of the EOSDIS project known as Version 0, which the author was involved in. It won the Golden Hammer Award from NASA. Still, the GAO did not feel that Version 0 addressed critical areas of technical feasibility that needed to be covered if EOSDIS was to be a success. Of the $83 million that NASA planned to spend on precontract activities in fiscal year 1992, only $19 million was reserved for Version 0 of EODIS. Version 0 was intended to be in place by late 1994 and was to provide some early integration of existing Earth science data systems to demonstrate the feasibility of interconnecting and interoperating distributed, heterogeneous systems and to prototype DAAC data processing and distribution not available with existing data systems. The GAO report states that "According to specialists in data base research and data management, without several significant technological advances EOSDIS cannot reach its goal of providing ready access to the vast amount of EOS and other Earth science data that global change researchers will need."

As a result, NASA developed an evolutionary approach to EOSDIS to satisfy criticisms from the scientific community. NASA adopted the "build a little, test a little" approach starting with Version 0. One of the goals of Version 0 was to use this prototype system to connect existing data centers to share data. The NRC recommended that the early prototypes be developed to cover the following areas:

- data display and user interface,
- browsing capability,
- data formats and media,
- accessibility of data and information,

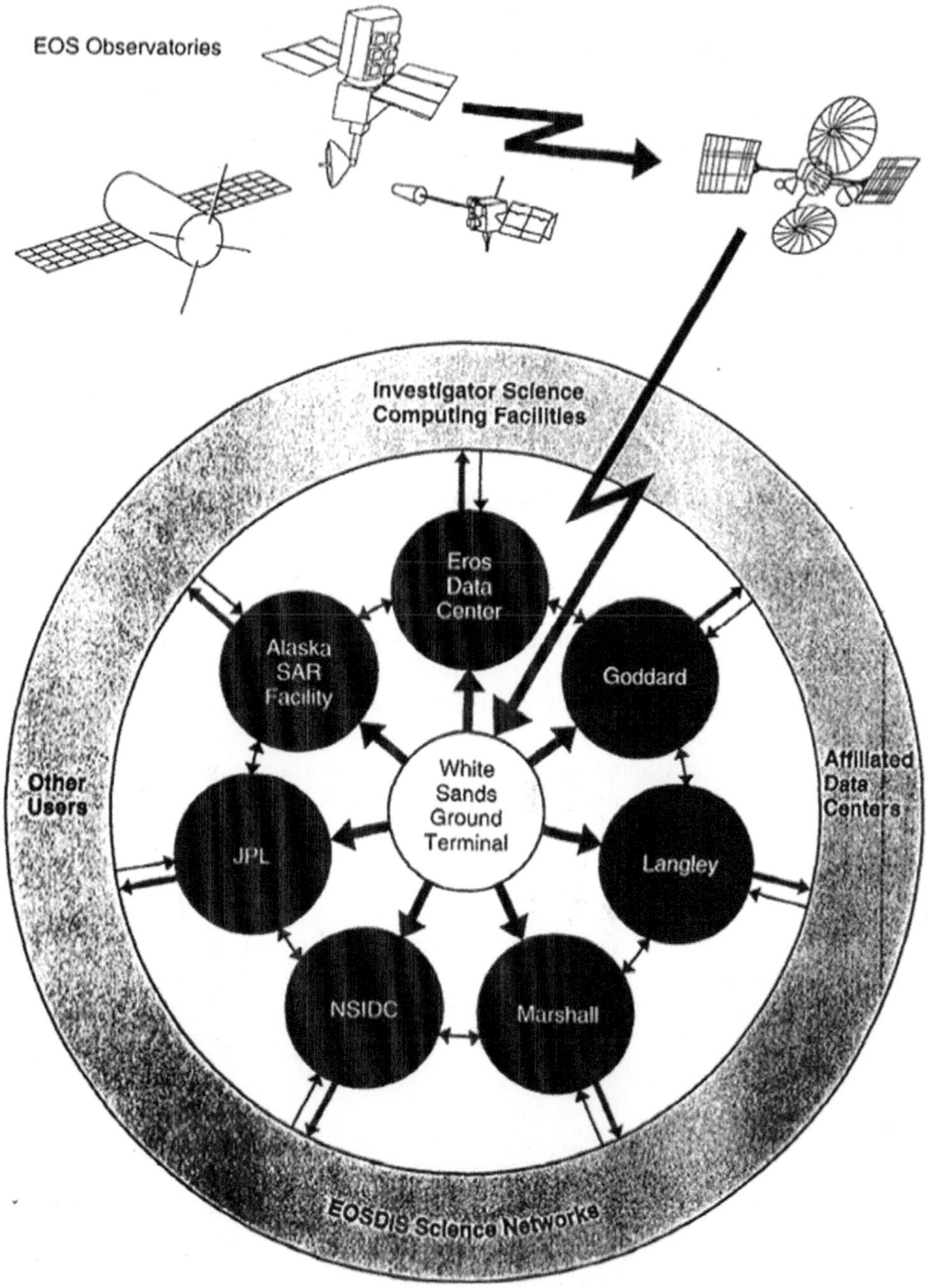

FIGURE 8.14 EOS data and information system (EOSDIS) Data flow. *EOS*, Earth observing system; *JPL*, Jet Propulsion Laboratory; *NSIDC*, National Snow and Ice Data Center.
Image credit courtesy NASA.

- cataloging,
- search and query capabilities,
- model and data interactions,
- data structures,
- data reduction algorithms, and
- networking

The Council stressed that prototyping should address the challenges posed by the immense amount of the planned EOS data sets, which will be larger than any past experience. EOSDIS prototyping should also explore how scientists will interact with the system and its data holdings.

One example of the problems with prototyping was the formulation of data format standards to work with EOS data. This posed a large problem for EOSDIS that was to be interactive between smaller communities that were accustomed to working with their own data in formats that were optimal for their applications. In addition, the large volume of EOS data needed to be translated into standard formats is very large and complex. The Version 0 approach to this problem was very modest, with the equivalent of one quarter of a person at each DAAC being assigned to one or more potential standard formats each.

Rather than address the full range of critical EOSDIS technical challenges, NASA's Version 0 primarily supports and enhances existing Earth science data systems at the seven DAACs. Thus this prototype does not serve to lead into a larger system capable of handling the needs of analyzing the tremendous amount of data envisioned for the original EOS program. According to NASA's development strategy, the EOSDIS Core System contractor will be responsible for initiating and conducting prototyping efforts after the contract is awarded and before the full-scale EOSDIS system is in place. The author recalls a large number of meetings at the GSFC reviewing plans for the EOSDIS system and the frequent dinners at a nearby Chinese restaurant, where we were not sure if the fortune cookie slips didn't foretell the future of the EOSDIS. The chairman, Jeff Dozier, kept an archive of these fortune cookie slips.

8.2.2 NASA's Terra satellite

As a consequence of the dramatic budget cuts, EOS was boiled down to essentially two major polar orbiting spacecraft that in and of themselves broke new ground in Earth observations. The first was known as Terra (Fig. 8.15), named after the "Mother Earth" due to its morning equatorial crossing time allowing for optimal viewing of land surfaces.

It was also referred to as EOS-AM-1 and flew a number of new US instruments, and some contributed by other countries. NASA held a naming contest among US high school students, and the winning essay was submitted by Sasha Jones of Brentwood, Missouri. The satellite was launched from Vandenberg Air Force Base on December 18, 1999 aboard an Atlas IIAS launch vehicle into a near-polar, sun-synchronous orbit with an altitude of 705 km and a 10:30 a.m. descending node.

The satellite carried five instruments:

- ASTER (Advanced Spaceborne Thermal Emission and Reflection Radiometer), which provided high-resolution images of clouds, ice, water, and the land surface using three different subsystems. They are the Shortwave Infrared (SWIR),

FIGURE 8.15 Terra satellite.
Image credit courtesy NASA.

thermal infrared (TIR) and Visible and Near Infrared. Together, they cover 14 multispectral bands at a relatively high resolution. ASTER was provided by Japan's Ministry of Economy, Trade, and Industry (METI).

- CERES (Clouds and the Earth's Radiant Energy System)
- MISR (Multiangle Imaging SectroRadiometer)
- MODIS (Moderate-resolution Imaging Spectrometer)
- MOPITT (measurements of pollution in the troposphere; provided by Canada)

8.2.2.1 Advanced Spaceborne Thermal Emission and Reflection Radiometer

ASTER collects high-resolution images of the Earth's surface in 14 spectral bands (Table 8.1).

ASTER data have been used to create detailed maps of land surface temperature, emissivity, reflectance, and elevation. An ASTER image in Fig. 8.16 shows the 2010 eruption of Mount Merapi in Indonesia, which reveals pyroclastic flow along the Gendol River south of Mount Merapi. A joint project between NASA and the Japanese METI used ASTER data to produce a Global Digital Elevation Model (GDEM)

covering 99% of the Earth's surface. The digital terrain model from NASA's Shuttle Radar Topography Mission (SRTM) covered approximately 80% of the Earth's surface with a global resolution of 90 m and a resolution of 30 m over the United States. The GDEM covers the Earth from 83°N to 83°S, exceeding the SRTM's coverage of 56°S to 60 degrees and becoming the first Earth mapping system to provide coverage of the polar regions. GDEM was created using 1.3 million ASTER visible bands using single-pass stereoscopic correlation techniques, yielding terrain elevation measurements globally at 30 m intervals. GDEM has been criticized for some artifacts in the results and saying that the real GDEM resolution is far less than 30 m. In October 2011 Version 2 of the GDEM was released with improved horizontal and vertical accuracies (Rexer and Hirt, 2014). GDEM version 3 was released on August 5, 2019 and added some stereo pairs while reducing the occurrence of artifacts. Version 3 used 2.3 million ASTER scenes, claiming now to have a 1 m spatial resolution. An image of Mount St. Helens in western Washington State shows the level of detail available with ASTER (Fig. 8.17). ASTER imagery is ideal for studying urban environments such as Las Vegas, Nevada, in Fig. 8.18.

TABLE 8.1 Advanced Spaceborne Thermal Emission and Reflection Radiometer (ASTER) bands.

Band	Wavelength (μm)	Resolution (m)	Nadir or backward	Description
B1	0.520–0.60	15	Nadir	Visible green/yellow
B2	0.630–0.690	15	Nadir	Visible red
B3N	0.760–0.860	15	Nadir	Near infrared
B3B	0.760–0.860	15	Backward	Near infrared
B4	1.60–1.70	30	Nadir	Short-wave infrared
B5	2.145–2.185	30	Nadir	Short-wave infrared
B6	2.185–2.225	30	Nadir	Short-wave infrared
B7	2.235–2.285	30	Nadir	Short-wave infrared
B8	2.295–2.365	30	Nadir	Short-wave infrared
B9	2.360–2.430	30	Nadir	Short-wave infrared
B10	8.125–8.475	90	Nadir	Thermal infrared
B11	8.475–8.825	90	Nadir	Thermal infrared
B12	8.925–9.275	90	Nadir	Thermal infrared
B13	10.250–10.950	90	Nadir	Thermal infrared
B14	10.950–11.650	90	Nadir	Thermal infrared

FIGURE 8.16 ASTER false-color image of the 2010 eruption of Mount Merapi, Indonesia (https://commons.wikimedia.org/wiki/File:2010_Eruption_at_Mount_Merapi,_Indonesia_(ASTER).jpg).

FIGURE 8.17 Advanced Spaceborne Thermal Emission and Reflection Radiometer image of Mount St Helens (https://www.satimagingcorp.com/gallery/more-imagery/aster/aster-mount-st-helens/).

There is a separate ASTER Volcano Archive at JPL, which includes additional data such as the chemical properties of the volcanoes along with information about their state of activity (Pieri and Abrams, 2004; Pieri and Buongiorno, 2010). Moreover, at JPL, the ASTER team has carried out the Global Land Ice Measurements from Space (GLIMS) to monitor the world's largest glaciers primarily using ASTER data. The GLIMS project guided ASTER to acquire imagery of Earth's glaciers that is optimized (best season and best instrument settings) for glacier monitoring. Here, it needs to be added that ASTER was not a constant duty instrument and needed to be activated for targets of interest. For the GDEM project, global "cloud free" imagery was needed, which meant that a lot of ASTER data collected were not suited for the GDEM project. This limited duty cycle restricted the amount of total coverage. Nevertheless, ASTER was able to collect enough imagery to generate the GDEM digital elevation maps.

8.2.2.2 Moderate Resolution Imaging Spectrometer (MODIS)

MODIS was the workhorse instrument of the Terra satellite and was a constant duty instrument. It had heritage in the AVHRR and in the CZCS instruments and was intended to be a major step forward from both of these predecessor instruments. It has a high radiometric sensitivity (12 bit) in 36 spectral bands ranging from 0.4 to 14.4 μm. The various bands are custom tailored to the needs of the community of users and have very low out-of-band response (strong cutoff). Two bands have a spatial resolution of 250 m at nadir, while five other bands have a 500 m resolution, with the remaining 29 bands having a 1 km spatial resolution. A ±55 degrees scan pattern at the Terra 705 km orbit results in a swath width of 2,330 km. The instrument provides global coverage every 2 days.

The scan mirror assembly (Fig. 8.19) is a unique continuously rotating double-sided mirror which scans ±55 degrees driven by a motor encoder built to operate at 100% duty cycle throughout the 6-year instrument design life (which it has greatly exceeded). Commonly referred to as a "paddle wheel," this mirror scan mechanism is totally unique to MODIS and was not repeated on future instruments. The optical system consists of a two-mirror off-axis afocal telescope, which directs the energy to four refractive objective assemblies, one each for VIS, NIR, SWIR/MWIR, and LWIR spectral regions.

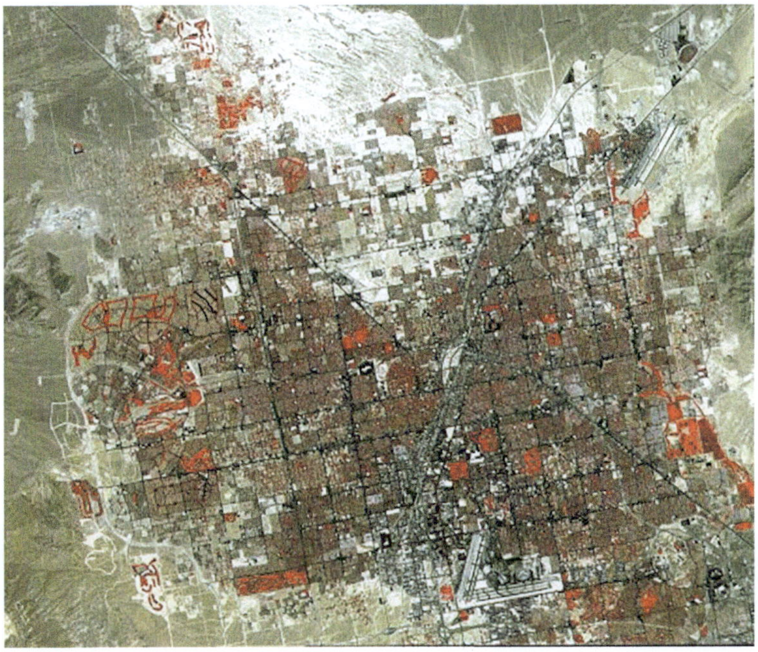

FIGURE 8.18 Advanced Spaceborne Thermal Emission and Reflection Radiometer (ASTER) image of Las Vegas, Nevada (courtesy NASA/METI/AIST/Japan Space Systems and U.S./Japan ASTER Science Team).

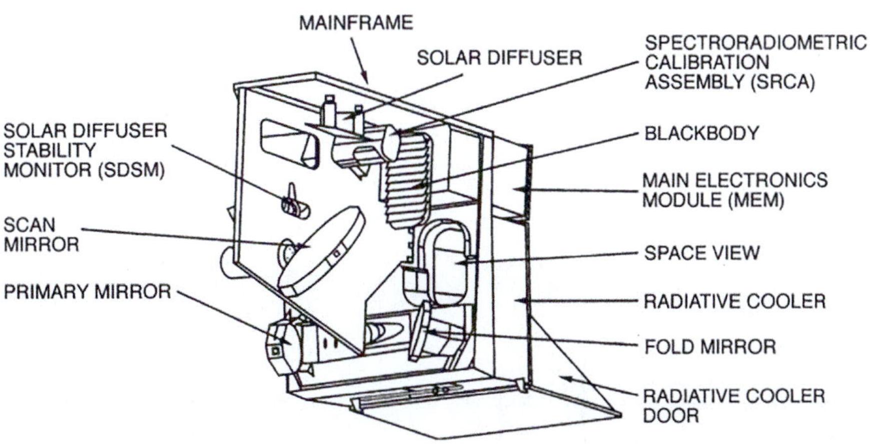

FIGURE 8.19 Cutaway of the Moderate Resolution Imaging Spectrometer instrument (https://ladsweb.modaps.eosdis.nasa.gov/missions-and-measurements/modis/).

A high-performance passive radiative cooler provides cooling to 83 K for the 20 infrared channels on the two HgCdTe Focal Plane Assemblies (FPAs). The visible and NIR use a novel photodiode-silicon readout technology, which has unsurpassed quantum efficiency and low-noise readout with exceptional dynamic range. Analog programmable gain and offset, and FPA clock and bias electronics are in two electronics modules, the Space-viewing Analog Module and the Forward-viewing Analog Module. A third module, the Main Electronics Module, supplies power, control systems, command and telemetry, and calibration electronics. The system also includes four onboard calibrators as well as a space view: a solar diffuser, a v-groove blackbody, a Spectro radiometric calibrations assembly, and a solar diffuser stability monitor.

The MODIS bands are fully described in Tables 8.2 and 8.3.

In Table 8.2, you can clearly see the ocean color bands that are based on the heritage of CZCS and SeaWIFS. There are more water vapor channels for correction of ocean color and sea surface temperature (SST) than were available in any of the heritage instruments. The land, cloud, and aerosol bands expand the sensing capability in these spectral areas. The SST channels are again a split-window, while there are a lot more bands to sense cloud top temperatures and hence altitude. There are some channels dedicated to sensing cirrus clouds and atmospheric water vapor. There are atmospheric temperature bands giving MODIS a modest atmospheric profiling capability.

Typical MODIS images are presented here in Figs. 8.20—8.22. These images provide dramatic new high-resolution images of events on the Earth's surface. While it was designed to last for 6 years, MODIS is still operating some 27 years

TABLE 8.2 Moderate Resolution Imaging Spectrometer (MODIS) bands 1–19.

Primary use	Band	Bandwidth	Spectral radiance	Required SNR
Land/Cloud/Aerosols boundaries	1	620–670	21.8	128
	2	841–876	24.7	201
Land/Cloud/Aerosols properties	3	459–479	35.3	243
	4	545–565	29.0	228
	5	1230–1250	5.4	74
	6	1628–1652	7.3	275
	7	2105–2155	1.0	110
Ocean Color/Phytoplankton/ Biogeochemistry	8	405–420	44.9	880
	9	438–448	41.9	838
	10	483–493	32.1	802
	11	526–536	27.9	754
	12	546–556	21.0	750
	13	662–672	9.5	910
	14	673–683	8.7	1087
	15	743–753	10.2	586
	16	862–877	6.2	516
Atmospheric water vapor	17	890–920	10.0	167
	18	931–941	3.6	57
	19	915–965	15.0	250

Bands 1–19 are in nm, spectral radiance is in W/m^2 μm-sr, SNR=signal-to-noise ratio, and NEΔT=Noise equivalent temperature difference.

TABLE 8.3 Moderate Resolution Imaging Spectrometer (MODIS) bands 20–36.

Primary use	Band	Bandwidth	Spectral radiance	Required NE[Δ]*T* (K)
Surface/Cloud temperature	20	3.660–3.840	0.45(300K)	0.05
	21	3.929–3.989	2.38(335K)	0.20
	22	3.929–3.989	0.67(300K)	0.07
	23	4.020–4.080	0.79(300K)	0.07
Atmospheric temperature	24	4.433–4.498	0.17(250K)	0.25
	25	4.482–4.549	0.59(275K)	0.25
Cirrus clouds water vapor	26	1.360–1.390	6.00	150(SNR)
	27	6.535–6.895	1.16(240K)	0.25
	28	7.175–7.475	2.18(250K)	0.25
Cloud properties	29	8.400–8.700	9.58(300K)	0.05
Ozone	30	9.580–9.880	3.69(250K)	0.25
Surface/Cloud temperature	31	10.780–11.280	9.55(300K)	0.05
	32	11.770–12.270	8.94(300K)	0.05
Cloud top altitude	33	13.185–13.485	4.52(260K)	0.25
	34	13.485–13.785	3.76(250K)	0.25
	35	13.785–14.085	3.11(240K)	0.25
	36	14.085–14.385	2.08(220K)	0.35

after its launch. Many of the channels are degraded, but the fundamental character of the mission continues. As of February 2023 the direct broadcast capability of MODIS data is no longer available. The operators at GSFC continue to nurse the EOS satellites along, and MODIS is the main data generator.

After 2 years of operation of MODIS on the Terra satellite, Solomonson et al. (2002) wrote a summary of the first two years of MODIS operation. They recorded how, from first operation of MODIS on October 31, 2000, a stable configuration using the "B-side" electronics was put into operation. The instrument continued to operate in that mode until June 15, 2001, when the instrument went out of operation due to a high-energy proton event and stayed out of operation until July 2, 2001, when the instrument operations were resumed using the "A-side" electronics. It has continued in this configuration ever since.

From November 2000 to 2002 the instrument showed performance at or above specifications in terms of signal-to-noise in all reflected solar bands (1–9 and 26). The noise-equivalent-delta-temperature (NEΔ*T*) performance in the TIR bands (20–15, 27–35) also showed at or better than specified performance except for anomalies within a band. Some striping, electronic noise, crosstalk, etc., can still be observed, particularly in the shortwave infrared bands and the 1.36 μm band (band 26). In general, however, the Level 1 data for all bands have provided useful information for scientific investigations.

The MODIS team consisted of 28 individuals in addition to several "validation" scientists. This team is divided into groups concerned with studies in land, ocean, and atmosphere problems. The atmospheres group focuses on observing and quantifying cloud microphysical properties such as droplet size, phase (ice or water), optical depth, aerosol temperature, and total column amounts of ozone and water vapor, along with atmospheric profiles of temperature and water vapor. The oceans group produces products for ocean color plus SST. The ocean color products have been shown to merge well with heritage ocean color observations from SeaWIFS and CZCS. There are 13 different land products produced from MODIS data that can be grouped into families of: radiation budget variable including surface reflectance (corrected for the atmosphere), land surface temperature (more difficult than SST due to varying emissivity), snow-cover, surface albedo, and ecosystem variables such as vegetation indices (Fig. 8.23), leaf-area index, fraction photosynthetically absorbed

FIGURE 8.20 Hurricane Igor in the Atlantic Ocean, September 17, 2010 (https://en.wikipedia.org/wiki/Hurricane_Igor).

FIGURE 8.21 Low clouds in central California, January 24, 2015 (https://science.nasa.gov/earth/earth-observatory/an-unrelenting-tule-fog/#:~:text=In%20late%20November%2C%20a%20very,cold%20air%20downslope%2C%20he%20added).

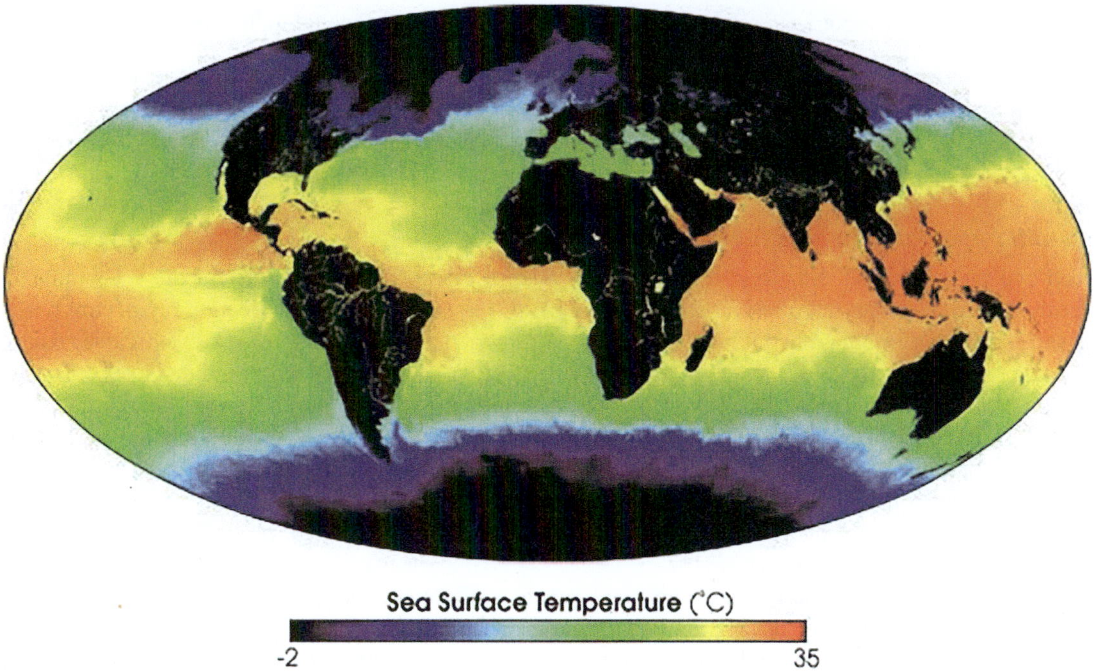

FIGURE 8.22 1-month composite of Moderate Resolution Imaging Spectrometer (MODIS) sea surface temperature imagery for May 2001 from MODIS. Map lines delineate study areas and do not necessarily depict accepted national boundaries.
Image courtesy MODIS Ocean Group, NASA GSFC, and the University of Miami.

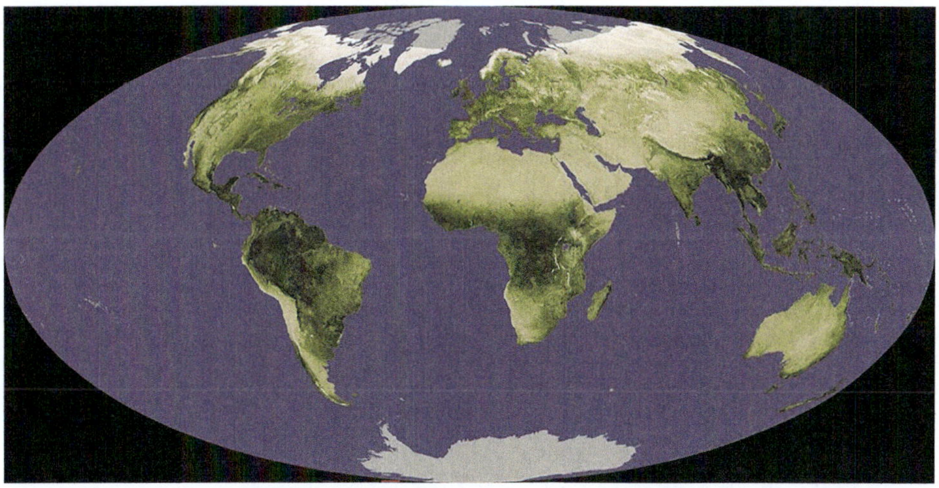

FIGURE 8.23 Moderate Resolution Imaging Spectrometer Normalized Difference Vegetation Index (NDVI) (https://www.google.com/search?q=MODIS+Normalized+Difference+Vegetation+Index+%28NDVI%29&sca_esv=72ed8926a77c5d66&source=hp&ei=YuJlafrdFLan0PEPyNn0yQw&iflsig=AFdpzrgAAAAAaWXwcuoREo78E8gb5lxRKZcaYlkFuoal&ved=0ahUKEwj6luSM7oeSAxW2EzQIHcgsPckQ4dUDCCE&uact=5&oq=MODIS+Normalized+Difference+Vegetation+Index+%28NDVI%29&gs_lp=Egdnd3Mtd2l6IjNNT0RJUyBOb3JtYWxpemVkIERpZmZlcmVuY2UgVmVnZXRhdGlvbiBJbmRleCAoTkRWSSkyBRAhGKABMgUQIRigATIFECEYoAFI0whQAFgAcAB4AJABAJgBhgGgAYYBqgEDMC4xuAEDyAEA-AEC-AEBmAIBoAKMAZgDAJIHAzAuMaAH9gOyBwMwLjG4B4wBwgcDMi0xyAcDgAgA&sclient=gws-wiz).

radiation, net primary productivity, evapotranspiration, and land-cover characteristics, including the location and intensity of wildfires.

We will say more about MODIS when we discuss the Aqua satellite, which also carried a MODIS instrument. One difference is that the Terra MODIS was launched with a known crosstalk in the TIR channels and this crosstalk was corrected in the Aqua MODIS.

8.2.2.3 MISR (multiangle imaging spectrometer)

MISR was designed and built at NASA JPL (Fig. 8.24) with the goal of measuring the intensity of the solar radiation reflected by the Earth system (planetary surface and atmosphere) in various directions and a selection of spectral bands. MISR consists of an innovative combination and configuration of nine separate digital cameras that each gather data in four spectral bands of the solar spectrum. One camera points at nadir, while the others provide forward and backward view angles of 26.1, 45.6, 60.0, and 70.5 degrees. Each of the nine cameras samples in the same four wavelengths (blue, green, reds, and near infrared). MISR is specifically designed to monitor the monthly, seasonal and long-term trends of atmospheric aerosol particle concentrationss including those formed by natural causes and those caused by human activities. MISR can also be used to map upper winds, cloud cover type, and height, as well as characterization of land surface properties, including vegetation canopies, the distribution of land cover types, or the properties of snow and ice fields, along with many other biophysical properties.

Diner et al. (1998) provide a detailed description of the development, calibration, and expected results from the MISR instrument. They report that at the heart of the instrument is the optical bench, which holds the cameras at their light-admitting end with the detector end cantilevered into the instrument cavity. The effect of Earth rotation is taken into account in processing, the 7 min interval between the fore and aft cameras' views of the Earth. The view angles vary slightly with orbital location and position within each camera's field-of-view.

In Table 8.4 are given both, the specified bandwidths and those realized when the instrument was constructed are given. This table shows that the actual channel wavelengths are very close to when they were initially specified, indicating the MISR should be able to fulfill the expectations for which it was built. Electronically, each camera is relatively autonomous, with its own power supply and serial data interface. This power supply provides power to the systems controlling the cameras, as well as the drive and timing signals to the CCD focal plane.

MISR has its own onboard calibrator, which is a pair of deployable diffuser panels that are covered with Spectralon (a pure polytetrafluoroethylene (Teflon) polymer resin), which was compressed and sintered. When not in use, these panels are stowed and protected. They will be deployed at approximately monthly intervals for calibration. Over the North pole, the forward-mounted panel will swing aftward to reflect diffused sunlight into the field-of-view of the aftward-looking and nadir cameras. Over the South Pole, the aftward-mounted panel will swing forward for the calibration of the forward and

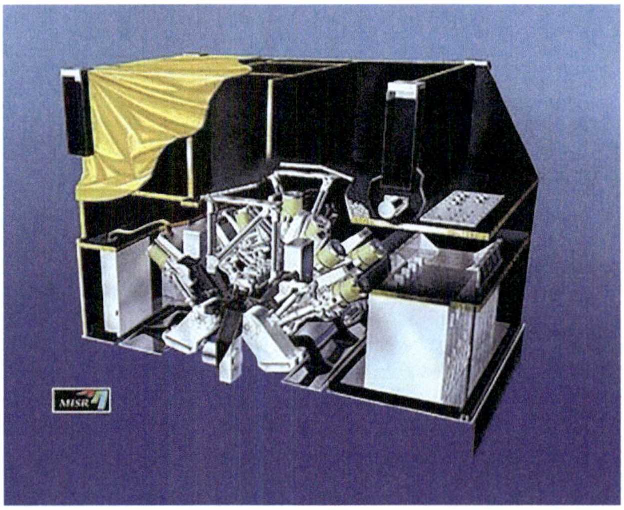

FIGURE 8.24 Artist's rendition of MISR in a cutaway view. The back ends of the nine MISR cameras appear as *yellow* cylinders. In this orientation, MISR is pointed Earthward (https://misr.jpl.nasa.gov/mission/misr-instrument/#:~:text=MISR%20Instrument-,MISR%20Instrument,the%20daylit%20side%20of%20Earth).

TABLE 8.4 MISR radiometric and spectral requirements and as-built specifications.

Parameter	Specification	As-built
Spectral bands (solar weighted, measured in-band response)	Blue: 443±2 nm Green: 555±2 nm Red: 670±2 nm NearIR: 865±2 nm	Blue: 446.4 nm Green: 557.5 nm Red: 671.7 nm NearIR: 866.4 nm
Spectral bandwidths (solar weighted, measured in-band response)	Blue: ≤30 nm Grecn: ≤20 nm Red: ≤20 nm NearIR: ≤60 nm	Blue: 41.9 nm Green: 28.6 nm Red: 21.9 nm NearIR: 39.7 nm
Polarization sensitivity	≤1%	As specified
Absolute radiometric uncertainty	≤3% of absolute radiance at full signal	2%
Signal-to-noise ratio	≥700 at full signal at highest spatial resolution	As specified

nadir cameras. The diffuse calibration targets are monitored in-flight by Radiation-resistant p, intrinsic, and n-doped (PIN) photodiodes and high quantum efficiency diodes. These diodes establish the radiometric scale upon which the instrument calibration is based.

Spatial coregistration of the 36 channels of MISR data is an essential requirement of all MISR geophysical retrievals. This is carried out during ground processing. Instrument-related geometric distortions were accounted for by extensive geometric camera calibration. The coregistration is achieved by projecting the images to a surface defined by a digital elevation model, which then accounts for the angle dependence of the MISR instrument.

MISR was uniquely able to view. clouds due to their large areal extent, high albedo, and variability on many length and time scales. Using its ability to view any scene from multiple directions, MISR is able to improve our ability to model cloud-radiation interaction by taking into account the variable structure of broken cloud fields and will be better able to define cloud-filled and cloud-free scenes. MISR is also capable of classifying clouds and computing regional cloud cover. MISR can use its angular sampling to create 3D structures of clouds that allow modeling of the 3D fields of radiative fluxes that play a critical role in climate feedback. Traditionally, cloud height is inferred from TIR cloud-top temperature, but MISR offers a wholly geometric way to determine cloud-top height without assuming a relationship between infrared temperature and height. Diffusion of radiation through the cloud sides and side illumination causes the angular reflectance of cumuliform cloud fields to differ markedly from stratiform fields, dictating that careful scene classification and angular integration can be implemented by MISR to obtain accurate estimates of top of atmosphere (TOA) albedo.

Cloud screening discriminates between clear and cloudy pixels in an image. Frequently, radiance threshold methods are used on a pixel-by-pixel basis, which may be single channel or multichannel thresholds. These methods are limited in that they represent a single view of the clouds. MISR instead uses thresholds that are a function of view angle, sun angle, relative view-sun azimuth angle, time, and surface type. The result of this process applied to each camera is known as the radiometric camera-by-camera cloud mask.

MISR was intended to be a new way of observing atmospheric aerosols. Aerosols are solid or liquid airborne particulates of various compositions and sizes, found in stratified atmospheric layers. In general, they have sizes between 0.1 and 10 μm, although the size of condensation nuclei is typically about 0.01 μm. Under normal conditions, most of the aerosols reside in the troposphere, where sources such as dust storms, desert and soil erosion, biogenic emissions, forest and grassland fires, and sea spray contribute about 90% of their presence in the atmosphere. The background tropospheric aerosol is temporally and spatially variable. The overall objectives of the MISR aerosol retrievals are as follows:

1. Characterization of aerosol climatic effects
   Aerosols play a direct role in the radiation budget of the Earth on regional and hemispherical scales; however, their net radiative effect, that is, whether they heat or cool the surface, depends on their optical properties and the albedo of the underlying surface.

2. Mapping of aerosol distribution
   The lifetimes of tropospheric aerosols range from about a week to a little over a month. Since these lifetimes are so short, spatial-temporal patterns are often indicators of sources and sinks.
3. Atmospheric corrections

The inherent goal of any atmospheric correction method is the accurate reflectance from the TOA radiance. It is well established that scattering and absorption by aerosols are responsible for dramatic modifications of remotely sensed images of the Earth's surface, resulting in major classification errors.

Substantial improvements in the retrieval of aerosols over the ocean and other dark water bodies are possible with MISR. The retrieval of aerosol optical depth over land is less well developed than the dark water case due to the high brightness and heterogeneity of the land surface. In this case, the simplest approach is to make an assumption about the surface reflectivity or albedo. It should be noted that dense dark vegetation behaves very much like the ocean as a background, greatly enabling the aerosol retrieval.

MISR was also designed to improve surface retrievals using the multiangle sensing as a unique capability. The goals of the MISR surface retrievals are as follows:

1. Surface radiative fluxes
   The bulk of the solar energy provided to the troposphere is absorbed at the lower boundary (oceans and land masses) and then reemitted to the atmosphere through fluxes of sensible and latent heat as well as in the form of thermal radiation. Accurate descriptions of these interactions between the surface and the atmosphere require reliable quantitative information on the fluxes of energy (all types), mass (including water and CO_2), and momentum. MISR is uniquely suited to measuring the surface reflectance and, hence, estimating these fluxes.
2. Land surface classification
   The evolution of terrestrial ecosystems is difficult to monitor at the surface, and satellite platforms offer a unique way to provide an overview of the surface, also sampled in time. MISR's ability to formulate 3D models of the surface greatly enables the construction of physically based models to infer canopy structure.
3. Biological productivity
   While the primary instrument on Terra for assessing ocean productivity is MODIS, MISR acquires images at several angles, and it has excellent SNR at low signal levels. It also provides an excellent correction for atmospheric aerosols, making it possible to retrieve accurate ocean color data unrestricted by sun glitter contamination.

Diner et al. (1998) conclude by saying that MISR provides unique information about clouds, aerosols, and surfaces. This is due to the multiangle sampling of the instrument that enables 3D imaging of clouds and ground surface structures. It also enables stereoscopic height and wind retrievals. Designed for aerosol sensing, MISR is able to permit aerosol composition identification and determine aerosol optical depth. It is excellent for cirrus cloud detection and characterization. It also enables scene-dependent albedo determination and surface classification.

8.2.2.4 Clouds and the Earth's Radiant Energy System

The Terra satellite carried two identical CERES instruments (Fig. 8.25); one operates in a normal cross-track scan mode while the other scans in a biaxial mode (Fig. 8.26). The cross-track mode essentially continues the measurements of the Earth Radiation Budget Experiment (ERBE) and the Tropical Rainfall Measuring Mission (TRMM), while the biaxial scan mode provides new angular flux information to improve the accuracy of angular models used to derive the Earth's radiation balance. The first CERES instrument was a proto-flight module and flew on the TRMM satellite launched in November 1997 from Japan. This instrument failed after only 8 months of operation due to an onboard circuit failure.

As summarized in Smith et al. (2011), CERES objectives were to measure the reflected solar radiance (shortwave) and Earth-emitted (longwave) radiances and, from these measurements, to compute the shortwave and longwave radiation fluxes at the TOA and the surface and radiation divergence within the atmosphere. The TOA fluxes need to be retrieved with an accuracy of 2%. The multiangle sampling improves the calculation of bidirectional distribution functions (BRDFs), which in turn are used to compute the fluxes at TOA from the radiances with errors reduced from the previous ERBE instrument by a factor of 2 or more. A validation protocol, including in-flight calibrations and comparisons of measurements, has further reduced the errors to being less than 1%.

In processing CERES data, a cloud screening is first done using MODIS images, whereby the cloud properties for each CERES pixel are determined to better calculate the BRDF. The cloud information is also used to compute radiation at the surface, through the atmosphere, and to facilitate the study of the relationship between clouds and the radiation budget. The biaxial CERES sampling was used to construct the BRDF (Loeb et al., 2003 a, b), which was then used to correct the

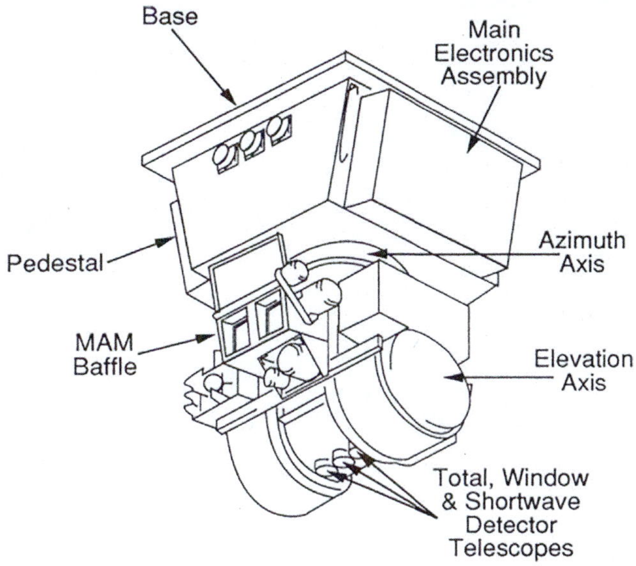

FIGURE 8.25 Clouds and the Earth's Radiant Energy System Instrument (https://ceres.larc.nasa.gov/instruments).

FIGURE 8.26 The two Clouds and the Earth's Radiant Energy System instrument as deployed on the Terra satellite (https://ceres.larc.nasa.gov/instruments/).

cross-track CERES data to further reduce the errors to the 1% level or lower. Fig. 8.27 is a map of outgoing longwave radiation flux for the month of July 2000 based on the two CERES instruments flying on the Terra satellite.

8.2.2.5 Measurement of pollution in the troposphere

The last of the major instruments carried by the Terra satellite was MOPITT (Fig. 8.28), which was intended to measure tropospheric carbon monoxide levels. It was a payload constructed in Canada and contributed to the United States to fly on the Terra satellite of the EOS program. It utilizes gas correlation spectroscopy to measure the presence of different gases in the troposphere.

MOPITT is one of the earliest satellite instruments to use gas correlation spectroscopy. The sensor measures 4.7 μm radiance emitted from the Earth. When it reaches the sensor, it is passed through an onboard sample of carbon monoxide, which is then compared with the radiance that came directly from the Earth. The difference in path length and arrival time

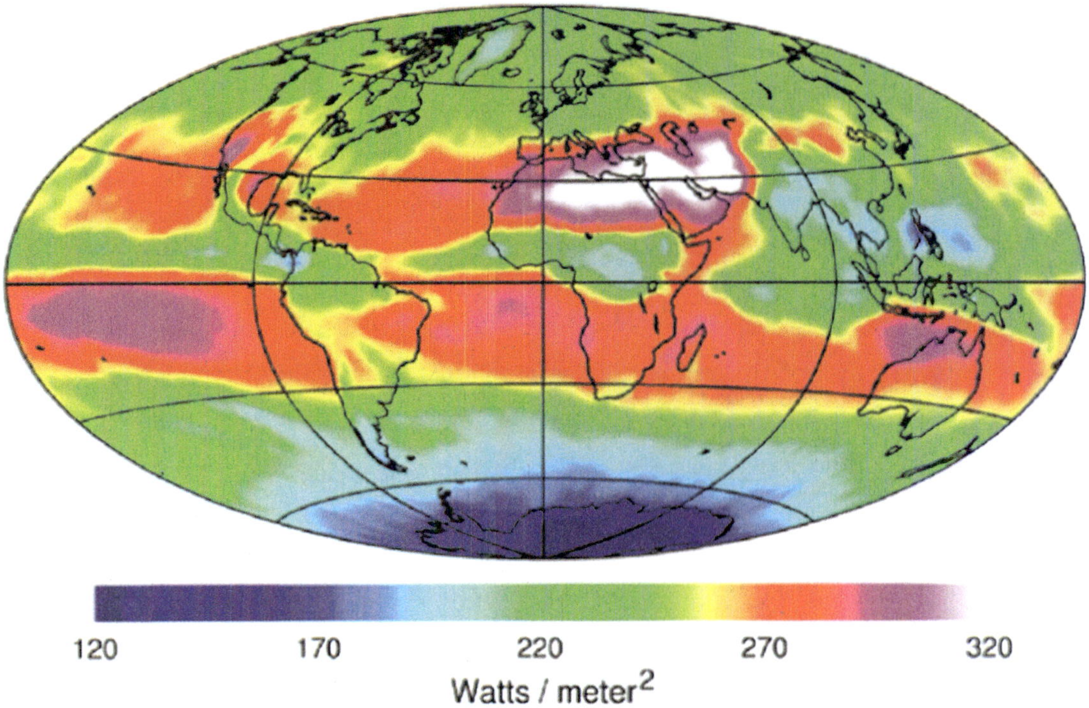

FIGURE 8.27 Longwave flux for July 200 from the two Clouds and the Earth's Radiant Energy System instruments on Terra (https://svs.gsfc.nasa.gov/30604/). Map lines delineate study areas and do not necessarily depict accepted national boundaries.

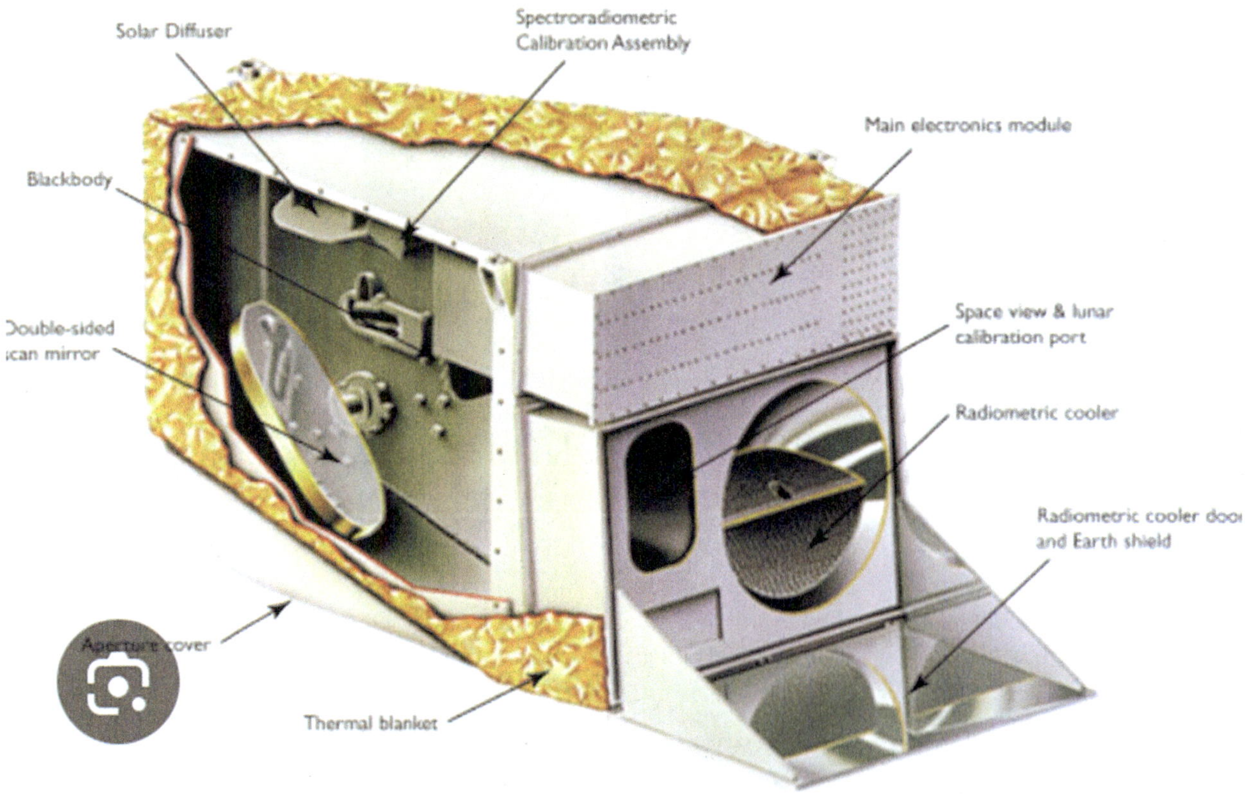

FIGURE 8.28 The measurements of pollution in the troposphere instrument. https://www.google.com/search?q=mopitt+instrument+image&sca_esv=34c521378b1c3854&sxsrf=ANbL-n5ft73M8mZZwG3wpcWz7JyKw2yiNQ%3A1777750701431&ei=rVL2ad-DGqfu0PEP3qHz4Qc&ved=0ahUKEwifjru4rZuUAxUnNzQIHd7QPHwQ4dUDCBE&uact=5&oq=mopitt+instrument+image&gs_lp=Egxnd3Mtd2l6LXNlcnAiF21vcGl0dCBpbnN0cnVtZW50IGltYWdlMgUQIRigATIFECEYoAFI8xxQ6QVYyxJwAXgAkAEAmAGaAaABvAaqAQMxLja4AQPIAQD4AQGYAgegAoUGwgIKEAAYRxjWBBiwA8ICBhAAGBYYHsICBRAAGO8FwgIFECEYqwKYAwCIBgGQBgSSBwMxLjagB7wPsgcDMC42uAf_BcIHBTAuNC4zyAcSgAgB&sclient=gws-wiz-serp#sv=CAMSZxowKg4zZlJfNWxkZ3ZrTktNTTIOM2ZSXzVsZGd2a05LTU06DkRMQ21XTWQxUG54UzRNIAQqLwobX0pGUDJhWS0wRE1DcDBQRVAxOW5NLUFrXzQ2Eg4zZlJfNWxkZ3ZrTktNTRgAMAEYByDUtaKFBUoIEAIYASACKAE.
Image credit NASA.

is an estimate of tropospheric carbon monoxide. A similar process takes place at 2.3 μm, which measures radiation reflected from the Earth but again is compared with an onboard carbon monoxide reference sample. Finally, the 2.2 μm channel is designed to measure methane rather than carbon monoxide. The path length differences cause a frequency shift of the spectral lines, which is then used to measure the atmospheric gas.

MOPITT's spatial resolution is 22 km horizontally and about 5 km vertically with a 640 km wide swath width (Fig. 8.29). MOPITT can view different layers of the troposphere by varying modulated cells. Pressure-modulated cells are used to observe upper layers of the atmosphere, while length-modulated cells are able to view lower layers of the troposphere. In the two optical systems, eight channels measure carbon monoxide or methane. Channels 1, 2, 3, 5, 6, and 7 target carbon monoxide. These six channels vary in cell pressures from 25 to 800 mb with cell lengths from 2 to 10 mm. Channels 4 and 8 measure methane. All channels operate at 300 K.

Before MOPITT data on carbon monoxide were very limited, and there was no global map of its distribution. MOPITT made it possible to calculate monthly mean maps of global carbon monoxide distributions (Fig. 8.30).

Some areas of the application of MOPITT data are as follows:

1. Better understanding of the increase in ozone concentration in the troposphere allows scientists to carry out interim pollution mitigation methods.
2. An assessment of antipollution practices, such as automobile emission reduction standards.
3. Analysis of environmental pollution caused by human activities.
4. The generation of maps of pollution transport, such as the direction of carbon monoxide emissions after large fires.
5. The confirmation of newly developed climate models that include areas of carbon monoxide production or reduction.

8.3 NASA's Aqua satellite

With Terra in the morning orbit, NASA's other main EOS satellite was put in the afternoon orbit, which later became known as the "A-TRAIN" orbit (more on that later). Since the AM satellite had been named Terra, it seemed only fitting that this second EOS satellite be named Aqua (Fig. 8.31), representing the ocean that covers 2/3 of the Earth's surface. In addition, the six instruments carried by Aqua were capable of observing all the elements of the Earth's water cycle.

The six instruments onboard Aqua are (Fig. 8.32) as follows:

1. Atmospheric infrared sounder (AIRS).
2. Advanced Microwave Scanning Radiometer for the EOS (AMSR-E)
3. Advanced microwave sounding unit-A (AMSU-A)
4. CERES; two copies
5. Humidity sounder for Brazil (HSB)
6. MODIS

The satellite was launched on May 4, 2002 into a near polar, sun-synchronous orbit at a 705 km altitude and an equatorial crossing time of 1:30 p.m. (south to north) and 1:30 a.m. (north to south). Launched with a 6 year design life, the satellites and their instruments have continued operation well over 20 years. Aqua is an international project combining efforts from the United States, Japan, and Brazil. Aqua ran out of station-keeping fuel in December 2021 and is now in a free-drift mode, slowly descending below the A-TRAIN and drifting to later equatorial crossing times and lower altitudes. In the fall of 2026 the passive destruction of the satellite will begin.

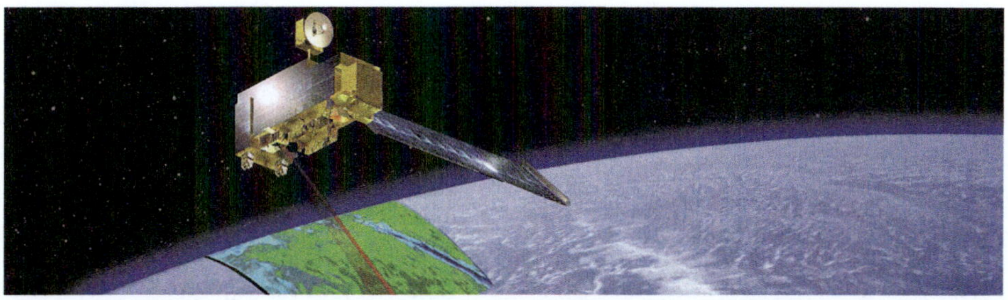

FIGURE 8.29 Measurements of pollution in the troposphere sampling on Terra (https://terra.nasa.gov/areas/mopitt).

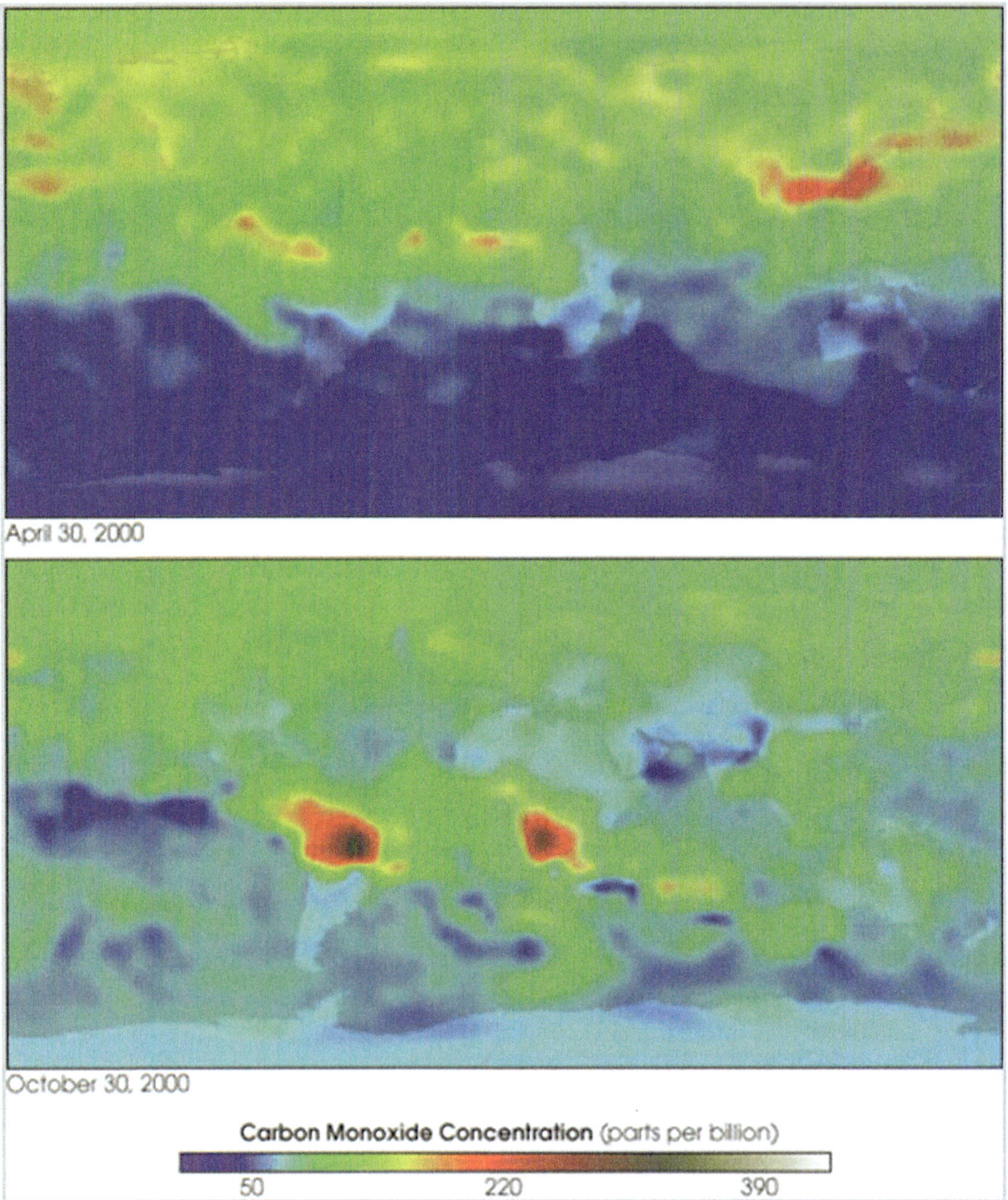

FIGURE 8.30 Global carbon monoxide for 2000 (top April 30; bottom October 30) (https://terra.nasa.gov/areas/mopitt). Map lines delineate study areas and do not necessarily depict accepted national boundaries.

We will not treat CERES and MODIS instruments as they are the same as those that were carried by the Terra satellite (with the exception that the crosstalk in the TIR channel was corrected) but will discuss those instruments unique to the Aqua satellite.

8.3.1 Atmospheric infrared sounder (AIRS)

AIRS was a complex and expensive grating spectrometer, infrared sounder (Fig. 8.33), intended to provide detailed profiles of the atmosphere. The heart of the instrument was a grating spectrometer that could observe various gases in the Earth's atmosphere (Fig. 8.34). This spectrometer operated over the range of 3.7–15.4 μm wavelengths with a spectral resolution of

FIGURE 8.31 Artist's rendition of the Aqua satellite.
Image credit courtesy NASA.

FIGURE 8.32 The Aqua satellite showing the instruments unique to Aqua. *AMSU-A*, Advanced microwave sounding unit-A ; *HSB*, humidity sounder for Brazil.
Image credit courtesy NASA.

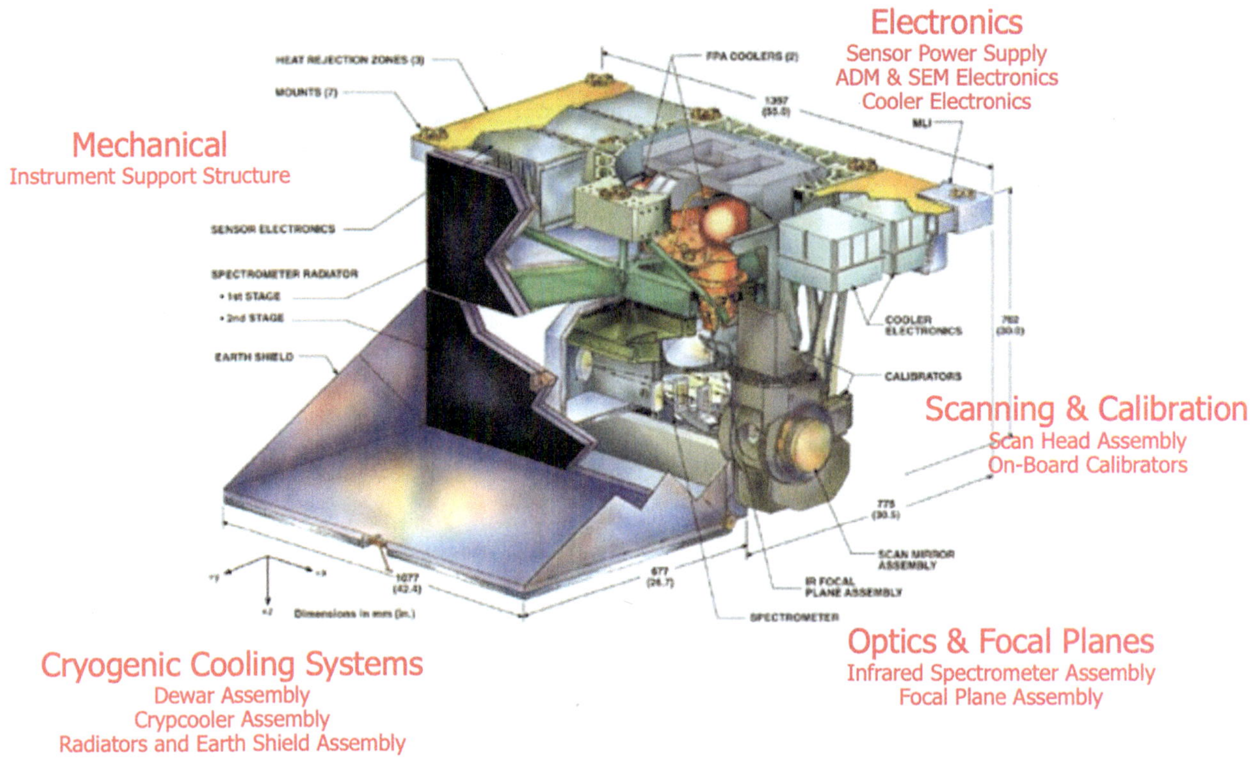

FIGURE 8.33 The atmospheric infrared sounder instrument.
Image credit courtesy JPL.

$\Delta\lambda$=1200 with 2,378 HgCdTe detectors in 17 arrays. The grating spectrometer optics are presented here in Fig. 8.35. Here we can see that the incoming radiance goes through a series of reflections and passes through an optical "slit" before being sent to the focal collimator. There, the radiation interacts with the grating, a fold mirror, and a "Schmidt" mirror before being sent to the focal plane, where the array of detectors generates the instrument signals for the various wavelengths.

The instrument has a 1,650 km swath with a basic 13.5 km resolution at nadir and a 1 km vertical resolution. It uses redundant 60 K Stirling cycle coolers to maintain a 20°C–25°C operating temperature for the infrared detectors. It has a 1.0-K atmospheric temperature retrieval accuracy per 1 km layer in the troposphere. The sensitivities (NEΔT) are 0.14 K at 4.2 μm, 0.20 K from 3.7 to 13.6 μm, and 0.35 K from 13.6 to 15.4 μm. The instrument was designed and built at NASA's JPL in Pasadena, CA.

As an infrared sounder, AIRS operated in concert with the NOAA-supplied microwave sounder AMSU-A.

AIRS provided measurements of atmospheric carbon dioxide as seen here in Fig. 8.36.

AIRS can also measure global maps of methane to provide comprehensive measurements of the two most important "greenhouse" gases. More traditional atmospheric profile measurements are of atmospheric temperature, water vapor, ocean surface temperature, land surface temperature, land surface emissivity, humidity, and clouds. AIRS can also detect carbon monoxide emissions from the burning of plant materials, such as the large fires in Brazil. AIRS can provide a 3D view of Earth's ozone layer and reveal how it is transported over the Earth. AIRS provides the best view of the ozone hole over the Antarctic during the polar winter. AIRS is also able to measure concentrations of Sulfur dioxide and dust in the atmosphere.

AIRS collects data in 6 min chunks, and there are 135 scan lines times 90 footprints for a total of 12,150 pixels per granule (~125 MB). So AIRS generates a lot of data. To get an idea of the daily coverage, we present the global atmospheric temperature (K) at 1,000 cm^{-1} in Fig. 8.37.

AIRS operated in conjunction with the AMSR-A microwave sounding unit provided by NOAA. Fig. 8.38 shows how the pixels of AMSU-A and AIRS overlapped (upper right) and how the AIRS pixels were a function of scan angle. Finally, this figure shows the daily scan pattern of AIRS.

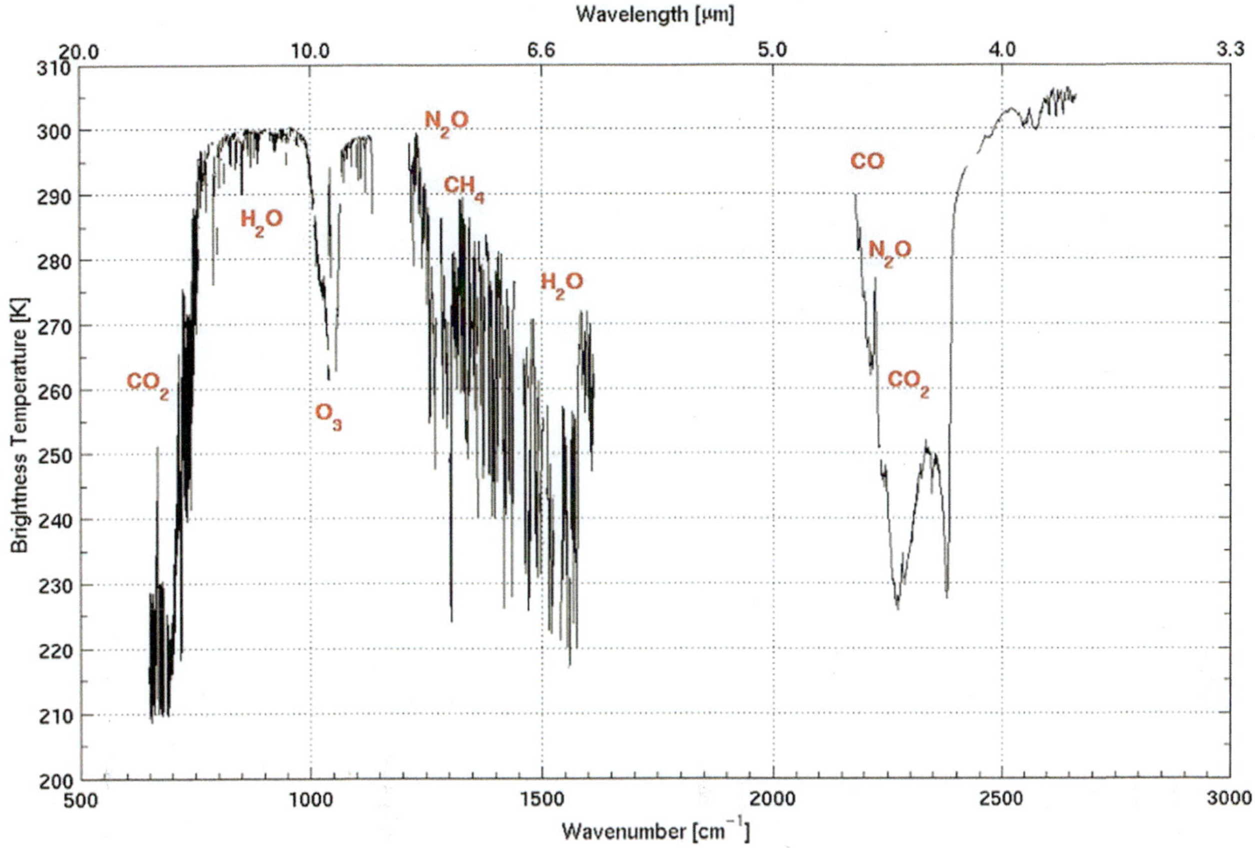

FIGURE 8.34 Characteristic brightness temperatures for atmospheric gases as a function of wavenumber (1/cm). Image credit courtesy JPL.

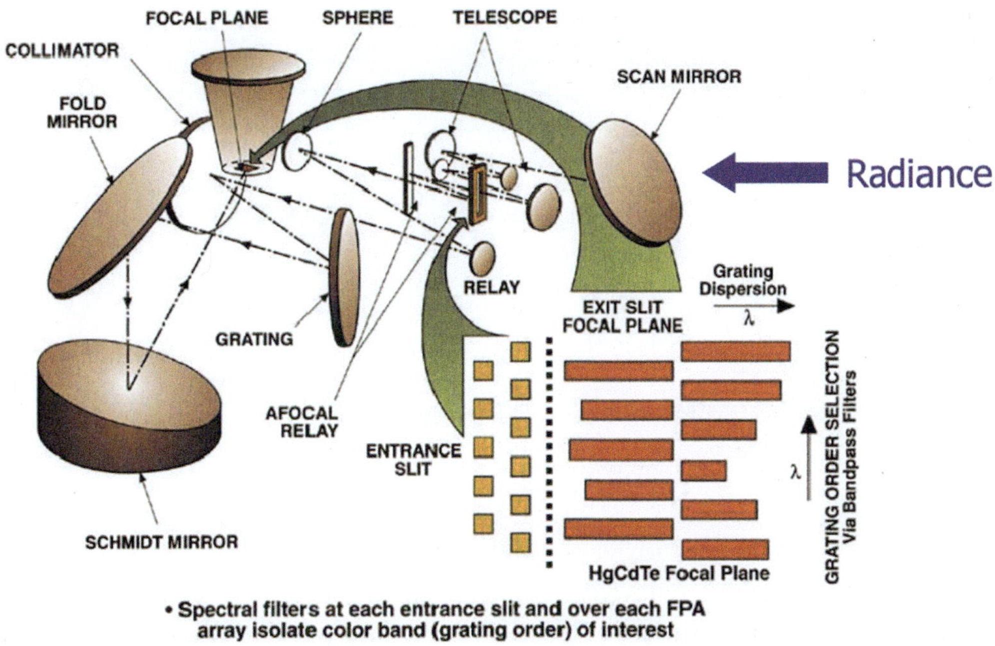

FIGURE 8.35 The atmospheric infrared sounder grating spectrometer optics. Image credit courtesy JPL.

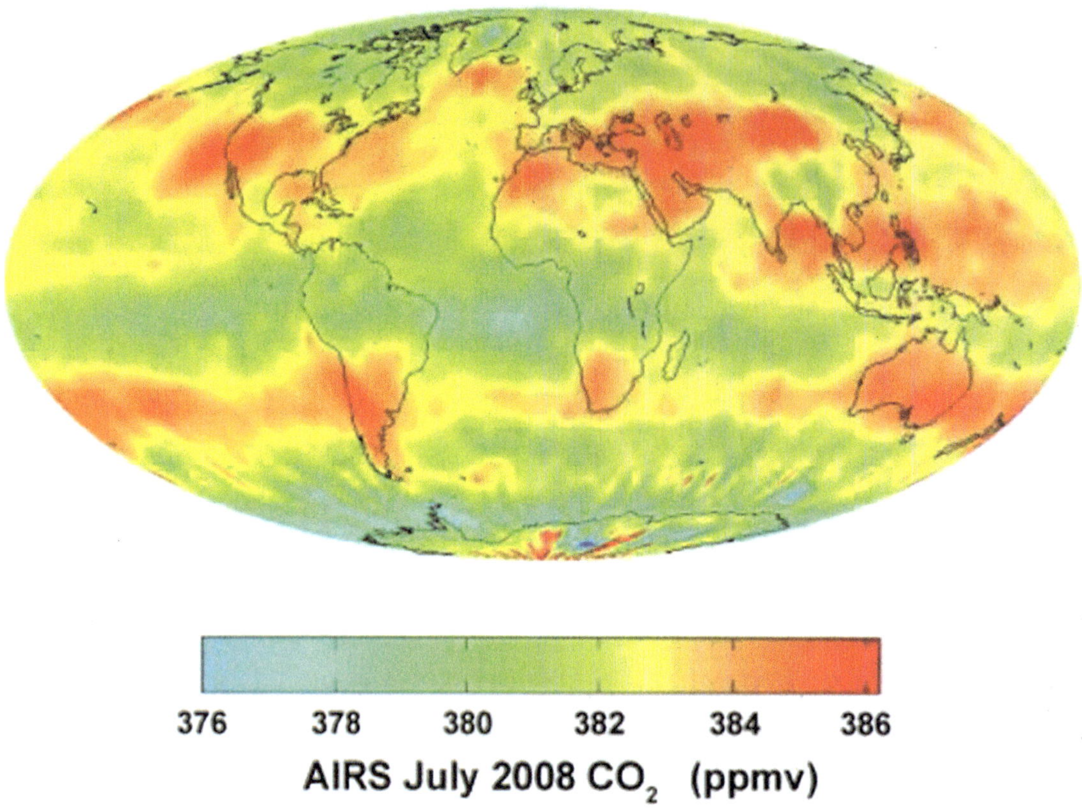

FIGURE 8.36 AIRS. Average CO_2 for July 2008. Map lines delineate study areas and do not necessarily depict accepted national boundaries. Image credit courtesy JPL.

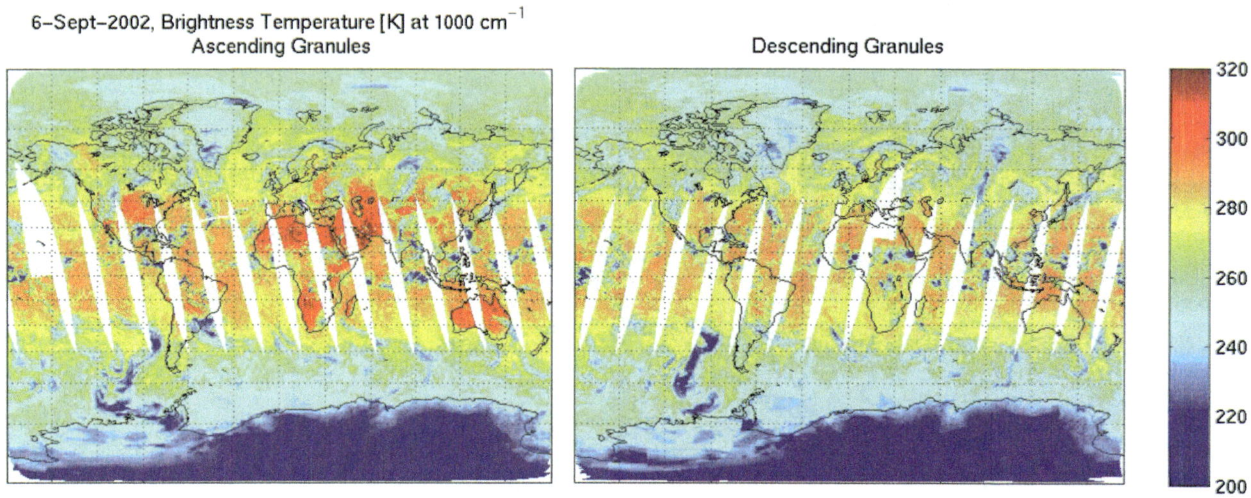

FIGURE 8.37 AIRS atmospheric temperature (K) for ascending and descending passes on 6 September 2002. Map lines delineate study areas and do not necessarily depict accepted national boundaries.
Image credit courtesy JPL.

From this figure, it is clear how the pixel size increases with scan angle and how the AMSU-A pixel overlaps completely with three AIRS pixels and partially with another 6 AIRS pixels. The gaps between scans in Fig. 8.39 will be filled in successive orbits.

Lambrigtsen et al. (2003) discuss the alignment of AIRS with AMSU-A and HSB. They make it clear that the four atmospheric sounding instruments (AIRS, HSB, AMSU-A1, and A2) need to work in concert, requiring a very accurate

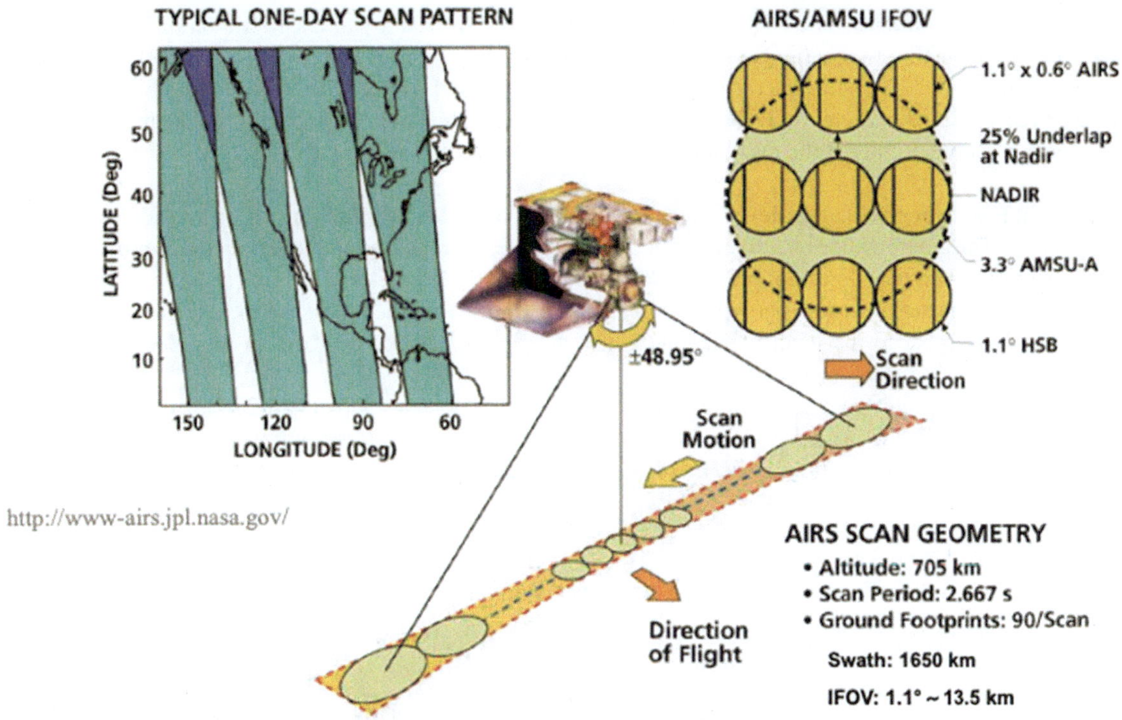

FIGURE 8.38 Atmospheric infrared sounder (AIRS) sampling and coverage, including overlap with Advanced Microwave Sounding Unit-A (AMSU-A).
Image credit courtesy JPL.

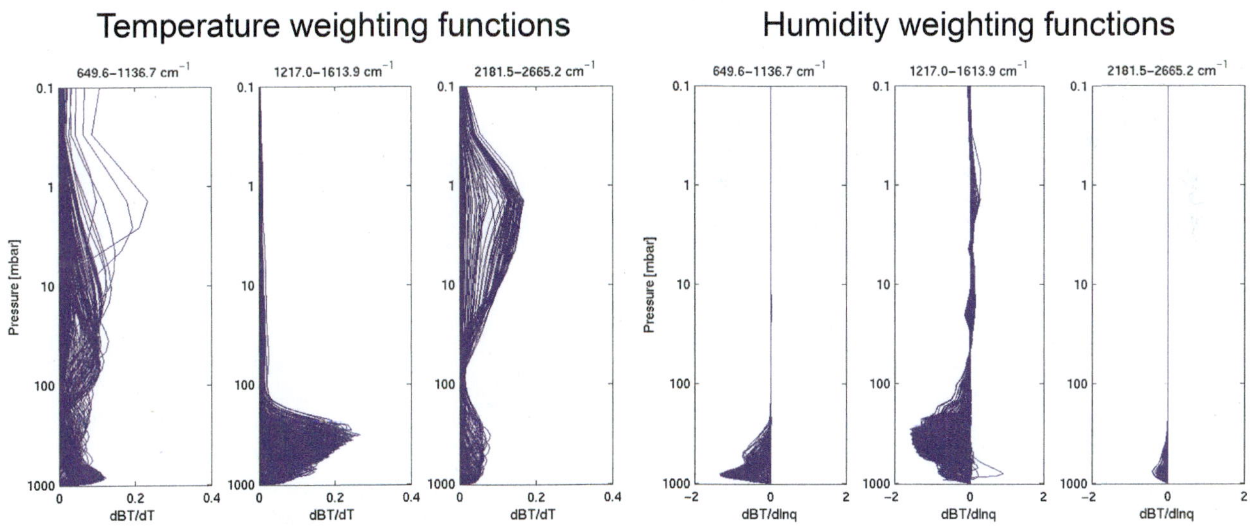

FIGURE 8.39 Temperature (left) and humidity (right) Atmospheric infrared sounder weighting functions.
Image credit courtesy JPL.

and well-maintained coalignment. They describe this process and examine possible errors in coalignment. They also point out that the AIRS instrument has as its goal to produce radiosonde-quality soundings from a satellite instrument. To do this, AIRS has 2,378 infrared channels with an expected temperature accuracy from ARIS of 1 K in 1 km vertical layers in the troposphere. Although the misalignment error between sensors is small, it is an error that can be corrected. The alignment requirements fall into two categories. The first is an instrument-level requirement for the two AMSU instruments, which is that they are within ±0.2 degrees for all scan positions. The second has to do with inter-sensor coalignments. It should be noted that AIRS and HSB have essentially the same pixel resolution and hence the same matchup with two AMSU-A instruments. The AIRS and HSB pixels should overlap by at least 90% on the ground. This alignment is maintained by adjusting two parameters: the instrument yaw angle and the time of start of each scan cycle. Finally, they point out that due to a scan step discrepancy between AMSU-A and AIRS/HSB, it is not possible to obtain a perfect set of AIRS/HSB/AMSU clusters as described, even if the scan line clustering is perfect. Near nadir, the cluster match can be near perfect but out at the swath edges, the center of the AMSU field of view (FOV) will be separated from the AIRS/HSB cluster by about 0.5 degrees.

The AIRS/AMSU combined geophysical retrieval method uses an iterative, least-squares physical inversion of cloud-cleared infrared radiances. This combination yields a total of 324,000 air temperature and specific humidity profiles every day. This combination has a horizontal resolution of 45 km at nadir that increases to 100 km at large scan angles. The vertical weighting functions are presented here in Fig. 8.39, where $K=dB/dx$ is the matrix of partial derivatives of the brightness temperature with respect to the input parameter X. The weighting functions (=rows of K) reflect the relative contribution from each level to the total measured radiance.

AIRS proved useful in mapping hurricanes, and an example is Hurricane Helene on September 27, 2024, shown here in Fig. 8.40. This image clearly shows the intensity of the storm at its center but also similarly high values in its outer bands.

AIRS was also able to map the concentration of carbon dioxide and methane on a global basis. Fig. 8.41 is a global map of monthly mean carbon dioxide from AIRS for July 2008.

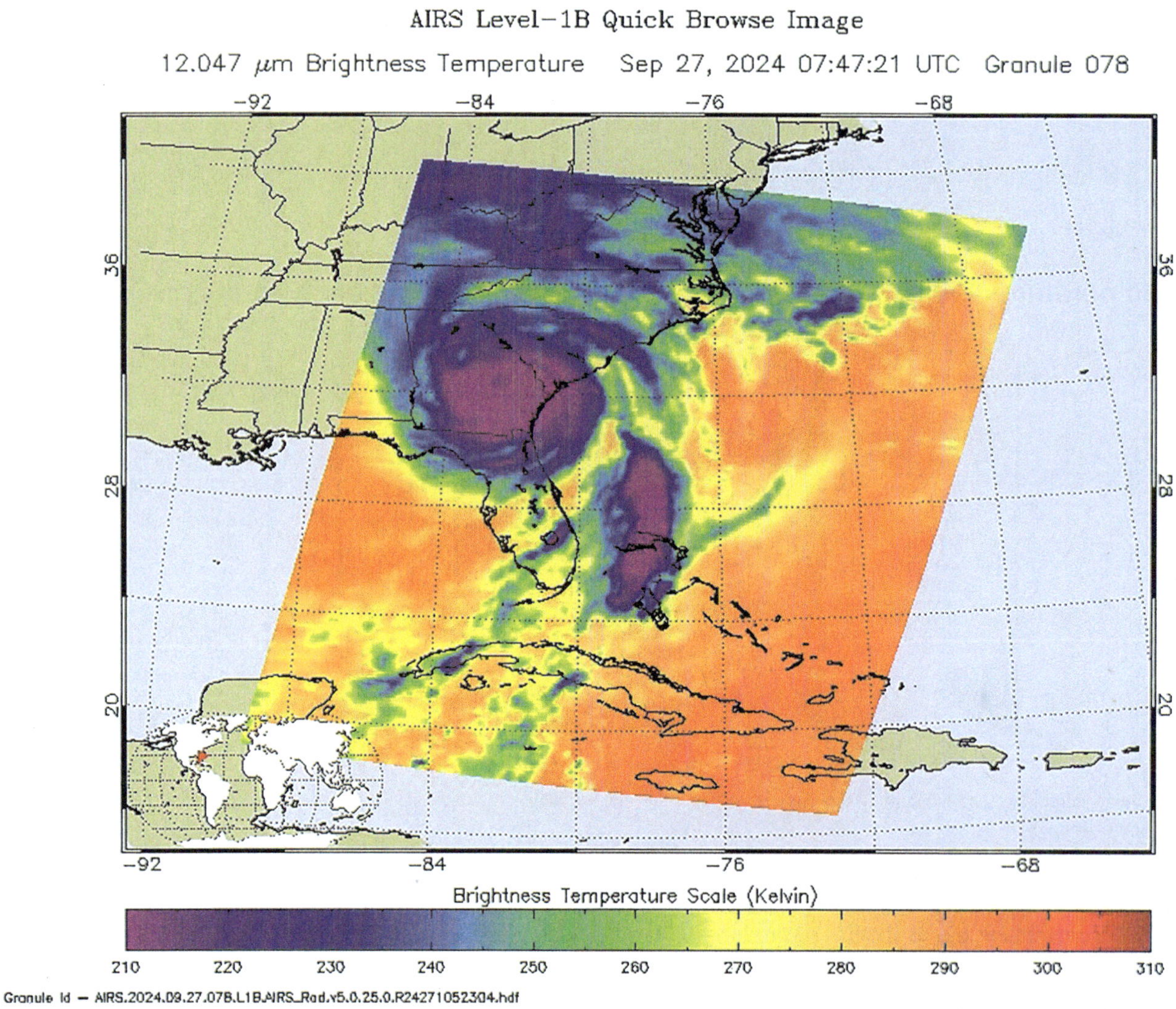

FIGURE 8.40 Infrared image of Hurricane Helene September 27 (NASA/Jet Propulsion Laboratory-Caltech). Map lines delineate study areas and do not necessarily depict accepted national boundaries. *AIRS*, Atmospheric infrared sounder. Image credit courtesy JPL.

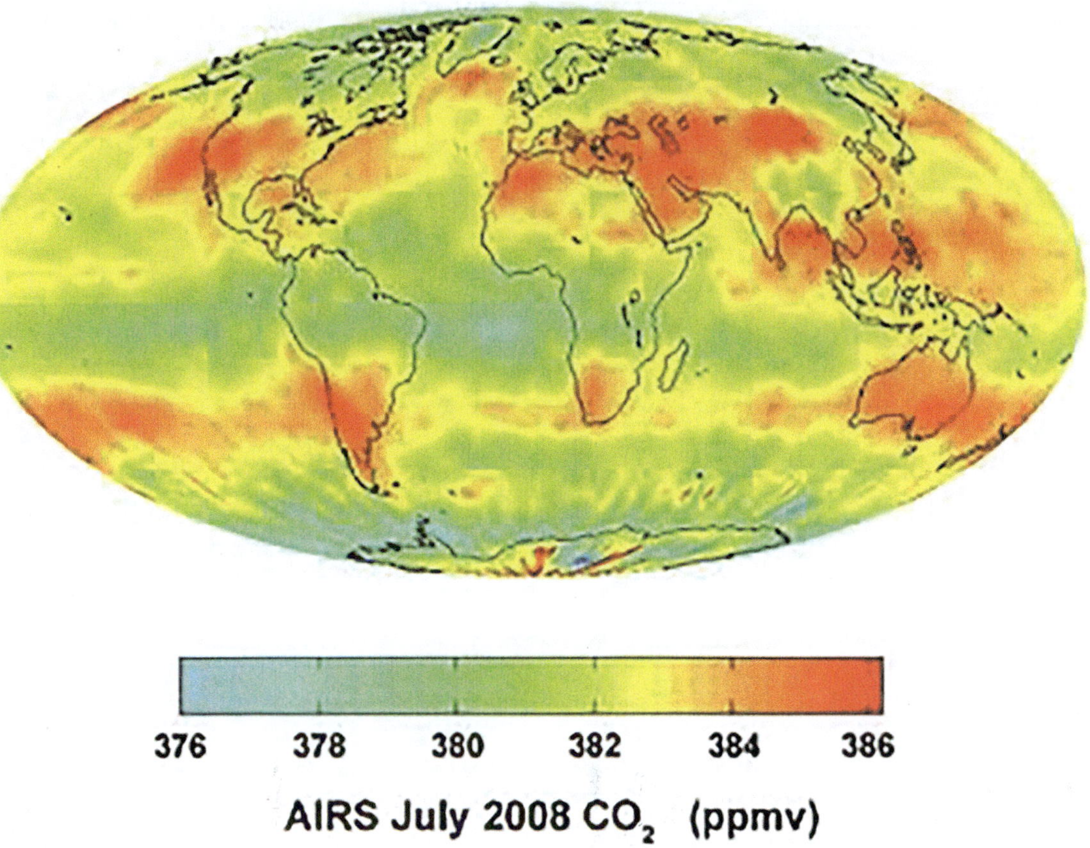

FIGURE 8.41 July 2008 mean carbon dioxide from AIRS. Map lines delineate study areas and do not necessarily depict accepted national boundaries. *AIRS*, Atmospheric infrared sounder.
Image credit courtesy JPL.

8.3.2 Advanced microwave sounding unit (AMSU-A1, A2)

AMSU is an advanced microwave sounding radiometer that normally flies on NOAA polar orbiting weather satellites. In the case of NASA's AQUA satellite, it was decided to fly two copies of the AMSU instrument to work in concert with AIRS to augment the detailed infrared atmospheric profiles from AIRS with a set of passive microwave soundings. The role of the AMSU-A is to provide temperature and humidity soundings under completely overcast conditions and to aid the cloud detection and clearing for the companion IR sounder (AIRS). The temperature sounding exploits mainly the oxygen band at 50 GHz. The 3 dB instantaneous footprint of AMSU-A is 3.3 degrees (45 km) and the calibration precision is 0.3 K. AMSU-A scans across track in a "stop and stare" mode at a scan rate of 8 s. The swath width is about ±1,027 km, and there are 30 mirror positions per scan. Both AMSU instruments were built by Northrop Grumman in Los Angeles, CA.

The AMSU-A frequencies, passbands, bandwidths, NEΔT, calibration accuracy, and polarization angle are presented here in Table 8.5.

The instruments are calibrated each data cycle by viewing both warm and cold calibration targets. Once every 8 s the AMSU-A measures 30 Earth views, the cold space view twice and the warm internal blackbody target twice.

A good example of the AMSU profiling capability is seen in Fig. 8.42 which shows a cross section of Hurricane Floyd on September 11, 1999. On the right of the figure is a three-channel image of Hurricane Floyd at approximately the same time. The anomalies in Fig. 8.43 were calculated by taking the anomaly relative to conditions outside the storm. The warming in the eye is correlated to a reasonable extent with a minimum surface pressure (warming decreases mean column density, which results in a decrease in column weight, which in turn, is closely related to surface pressure). The darkest red and orange correspond to the largest positive temperature anomalies, which appear in the lower troposphere. The cool anomalies appear in the lower troposphere in the cross section. The two symmetric anomalies on either side of Floyd's eye correspond to the stormy eyewall, and the other cold anomaly to the left of the eye likely coincides with a

TABLE 8.5 Advanced Microwave Sounding Unit-A (AMSU-A) channel characteristics.

Channel	Central freq (GHz)	No. pass bands	Bandwidth (MHz)	NEΔT (K)	Calib acc (K)	Pol ang (degree)
AMSU-A2						
1	223.8	1	±135	0.3	2.0	90-q
2	31.4	1	±90	0.3	2.0	90-q
AMSU-A1						
3	50.3	1	±90	0.4	1.5	90-q
4	52.8	1	±200	0.25	1.5	90-q
5	53.596±0.115	2	±85	0.25	1.5	nospec
6	54.40	1	±200	0.25	1.5	nospec
7	54.94	1	±200	0.25	1.5	nospec
8	55.50	1	±165	0.25	1.5	nospec
9	F_{LO}=57.290344	1	±165	0.25	1.5	nospec
10	F_{LO}±0.217	2	±39	0.4	1.5	nospec
11	F_{LO}±0.3222±0.048	4	±18	0.4	1.5	nospec
12	F_{LO}±0.3222±0.022	4	±8	0.6	1.5	nospec
13	F_{LO}±0.3222±0.010	4	±4	0.8	1.5	nospec
14	F_{LO}±0.3222±0.0045	4	±1.5	1.2	1.5	nospec
15	89	1	±3000	0.5	2.0	90-q

NEΔT, Noise-equivalent-delta-temperature.

FIGURE 8.42 (Left) Cross section of temperature anomalies derived from Advanced Microwave Sounding Unit-A for Hurricane Floyd at 2332Z on September 11, 1999. The contour interval is two degrees Celsius, with red color indicating the "warmest anomalies" and blue the "coldest." (Right) A three-channel (multispectral) image of Hurricane Floyd at approximately the same time.
Image credit courtesy NOAA.

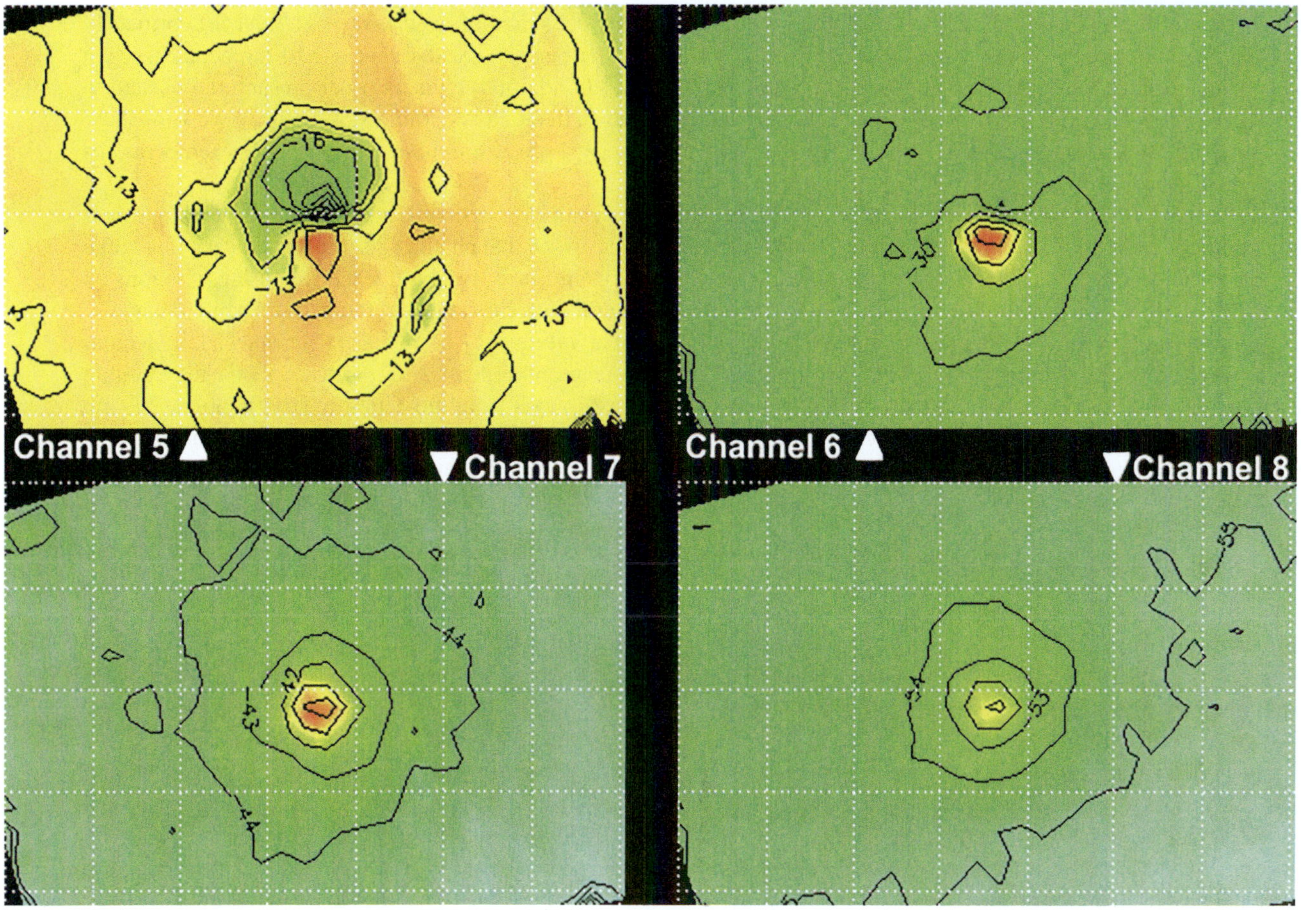

FIGURE 8.43 Channels 5–8 advanced microwave sounding unit (AMSU) images of Hurricane Fabio taken by AMSU-A1 on NOAA-17 on September 5, 2003. The top row of images is channel 5 (left) and channel 6 (right), while the lower row of images is channel 7.
Credit courtesy CIMSS.

thunderstorm in a spiral band coiling inward toward the eye. In reality, however, you should ignore these three cold anomalies, as they are likely false. Heavy rain in the eyewall and spiral-band thunderstorms greatly attenuates microwaves (raindrops scatter and absorb microwaves), resulting in unrealistically weak upwelling that is accidentally interpreted as a large cool anomaly. The attenuation of microwaves by heavy rain is one of the limitations of these kinds of remote sensors.

AMSU can also be used to view tropical cyclones in different layers. Forecasters generally look at four specific AMSU channels to evaluate the temperature in the upper half of the troposphere and in the lower atmosphere. These are as follows:

- Channel 8 (55.5 GHz) approximately 100 mb (about 15 km)
- Channel 7 (54.9 GHz) approximately 200 mb (about 12 km)
- Channel 6 (54.5 GHz) approximately 350 mb (about 10 km)
- Channel 5 (53.6 GHz) approximately 550 mb (about 5 km)

As an example, we present the images from channels 5–8 from Hurricane Fabian from 02Z on September 5, 2003 (Fig. 8.43). The warm core is particularly apparent in the images for channels 6 and 7 (350 and 200 mb) as marked by the yellow, orange, and red colors. Historically, the maximum warming over the eye of a hurricane was thought to occur about 200 mb, which coincides with channel 7 on AMSU. Recent research has suggested that AMSU does not have sufficient vertical resolution to truly pinpoint the exact altitude of the maximum warm anomaly. In fact, the maximum warm anomaly may meander between the middle and the upper troposphere at various times during a storm's life cycle and be located most frequently toward the middle troposphere. Thus it is best to monitor all four of these channels to view the behavior of the eye of the storm.

To demonstrate the profiling capability of AMSU, we will look at the behaviors of channels 3 and 7. AMSU primarily senses microwave radiation emitted by molecular oxygen, which has the benefit that the decrease of molecular oxygen with increasing altitude is roughly the same at any place and any time. In addition, clouds do not present an obstacle to microwave radiation, meaning that AMSU can profile the atmosphere even in cloudy conditions. Thus since we know how oxygen is vertically distributed in the atmosphere, it allows us to assign specific altitudes to brightness temperatures measured at microwave frequencies.

Between 50 and 60 GHz (AMSU band coverage), molecular oxygen absorbs strongly at some frequencies and less strongly at others. If we look at AMSU channels 3 and 7 (Fig. 8.44), we can see that channel 3 samples at a very low value of the oxygen curve while channel 7 is at a peak of the oxygen absorption/emission. Looking at the weighting functions on the right of Fig. 8.44, we can see how channel 3 senses a microwave emission maximum at the surface while channel 7 senses the emission maximum at about 12 km altitude. This means that at a frequency of 54.9 GHz (channel 7), molecular oxygen strongly absorbs microwave radiation and emissions from the ground at 54.9 GHz do not reach the satellite because they are absorbed higher up. In the final analysis, microwave emissions from molecular oxygen at about 200 mb (about 12 km) provide the greatest contribution to the upwelling radiation that reaches the satellite at this frequency. This same process is carried out for the other 11 channels on AMSU-A1 and on the 2 channels of AMSU-A2.

8.3.3 Humidity sounder for Brazil

The HSB rounded out the sounders flying on the Aqua satellite. It was a four-channel passive microwave radiometer/sounder, with one channel at 150 GHz and three channels at 183 GHz (Fig. 8.45).

It was very similar to the earlier AMSU-B instrument except without the 89 GHz surface sounding channel. The exact channel characteristics are presented here in Table 8.6.

HSB was a heritage instrument of AMSU-B, except that the 89 GHz channel was dropped. It was intended to study profiles of atmospheric water vapor and provide improved input data to the cloud-clearing algorithms in the AIRS and AMSU-A atmospheric profiling suite. It was built by Matra Marconi Space Ltd. in the United Kingdom under a contract with the Brazilian National Institute for Space Research (INPE). HSB suddenly and without warning stopped scanning on February 5, 2003, at 21:39 UTC. The most likely cause was a failure in the scan electronics. By design, the HSB had very little redundancy in its hardware.

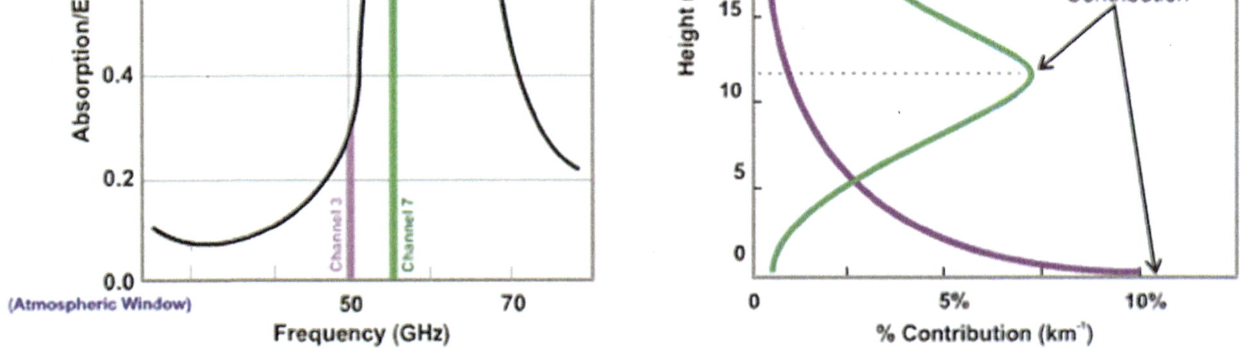

FIGURE 8.44 (left) The microwave spectrum between 50 and 60 GHz, with the frequencies of channel 3 in purple and channel 7 in green. (Right) The vertical weighting functions for channels 3 and 7, showing where each responds to an oxygen maximum.
Credit: courtesy David Babb.

FIGURE 8.45 The humidity soundeer for Brazil (HSB).
Image credit courtesy NASA.

TABLE 8.6 Channel characteristics of the humidity sounder for Brazil (HSB).

Channel number	Frequency (GHz)	Bandwidth (at nadir)	NEΔT (K)
1	89.9±0.9	Deleted	Deleted
2	150±0.9	4000	0.68
3	183.31±1.0	2×500	0.57
4	183.31±3.0	2×1000	0.39
5	183.31±7.0	2×2000	0.30

NEΔT, Noise-equivalent-delta-temperature.

8.3.4 Advanced microwave scanning radiometer–Earth observing system

Another instrument unique to the Aqua satellite was the AMSR-E (Fig. 8.46), which was a Japanese instrument with lots of heritage in earlier AMSR instruments (Kawanishi et al., 2003).

Typical of microwave radiometers Fig. 8.47 shows the large reflector dish that spun around a vertical axis to produce the scan. Major modifications were made on AMSR-E over the earlier AMSR instrument. One was scaling down this reflector from 2.0 m diameter to 1.6 m to fit into the launch vehicle fairing. AMSR-E had a deployable antenna system and the 50 GHz channel of AMSR was deleted. Since Aqua flew in a lower orbit than the Japanese ADEOS-II satellite (that carried the AMSR) the reduction in antenna size had little effect on the ground spatial resolution of AMSR-E relative to the previous AMSR instruments.

The characteristics of the AMSR-E channels are presented in Table 8.7.

AMSR-E is a total power microwave radiometer with dual polarization capability for all frequencies. The conical scan rate is 40 rev/min at a constant incidence angle of 55 degrees. The various frequencies are collected by multiple feed horns. Calibration measurements are made every scan by viewing an onboard high temperature source (300 K) and a cold sky view mirror to give a deep space view close to 0.0 K. An overview of the mechanical aspect of AMSR-E is given in Fig. 8.47.

Covering a wide range of frequencies with a single antenna poses some real challenges. You want a large reflector to improve high spatial resolution on the ground, but you also need to have an accurate mirror surface to realize high main-

FIGURE 8.46 The advanced microwave scanning radiometer–Earth observing system instrument. Image credit courtesy JAXA.

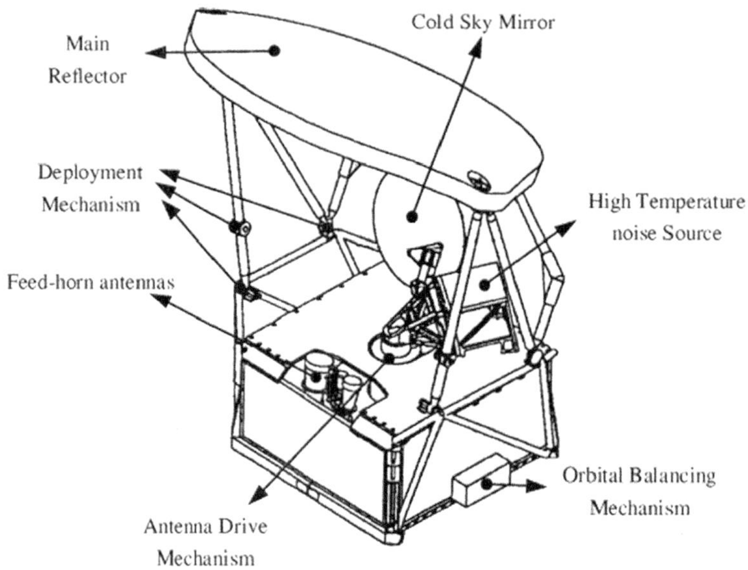

FIGURE 8.47 Advanced microwave scanning radiometer–Earth observing system mechanical structure. Image credit courtesy JAXA.

beam efficiency, especially in the 89 GHz channels. To reduce weight and minimize thermal deformation on orbit, the main reflector is made of an aluminum honeycomb panel sandwiched between carbon-fiber-reinforced plates. The reflector surface is deposited with aluminum coated with silica dioxide. There are six dual-mode feed-horns illuminated by the reflector. The 18.7 and 23.8 GHz channels use the same feed horn (Fig. 8.48) while two feed horns are used for the 89 GHz channel. The largest horns are the 6 and 10 GHz low-frequency channels. Staggered integration timing for each radiometer compensates for differences in beam direction for the different feed horn antennas. Horizontal and vertical polarizations are created by splitting the signal with orthogonal-mode transducers.

The excellent sensitivity of AMSR-E is displayed in its NEΔT values and is a consequence of the highly performing low-noise amplifiers that make a full-power microwave radiometer possible. The scan geometry of the AMSR-E is shown here in Fig. 8.49.

The reflector continuously rotates counterclockwise about an axis parallel to the geocentric direction, with an off nadir view angled at 47.5 degrees (55 degrees incidence angle on the equator). Thus the scan direction is from right to left when

TABLE 8.7 Advanced microwave scanning radiometer–Earth observing system (AMSR-E) channel characteristics for a scene temperature of 150 K.

Parameter	Performance and characteristics						
Center frequency (GHz)	6.925	10.65	18.7	23.8	36.5	89.0(A)	89.0(B)
Bandwidth (MHz)	350	100	200	400	1000	3000	3000
Polarization	Vertical and horizontal polarization						
NEΔT (K)	<0.34	<0.70	<0.70	<0.60	<0.70	<1.20	<1.40
NEΔT (K), as built, V-pol/H-pol	0.32/0.34	0.49/0.57	0.55/0.47	0.56/0.54	0.51/0.41	0.89/1.01	1.18/0.91
Dynamic range (K)	2.7–340						
Main beam efficiency (%), as built	95.1	94.8	95.8	94.8	93.9	94.5	93.7
Cross polarization (%), as built	0.34	0.29	0.32	0.89	0.43	0.77	0.68
Feed-horn spillover (%), analysis	3.6	2.8	2.2	3.2	2.4	2.4	2.6
Incidence angle (deg.)	55.0						54.5
Off-nadir angle (deg.)	47.5						47.0
Beam width (deg.)	2.20	1.50	0.80	0.92	0.42	0.19	0.18
IFOV (km) cross-track x along-track	43x75	29x51	16x27	18x32	8.2x14	3.7x6.5	3.5x5.9
Sampling interval (km), along-track	10.1					4.1/6.0 (B→A/A→B)	
Sampling interval (km), cross-track	9.0					4.5	4.4
Swath width (km)	1450						
Integration time (msec)	2.5					1.2	
Digital quantization (bits)	12	10					

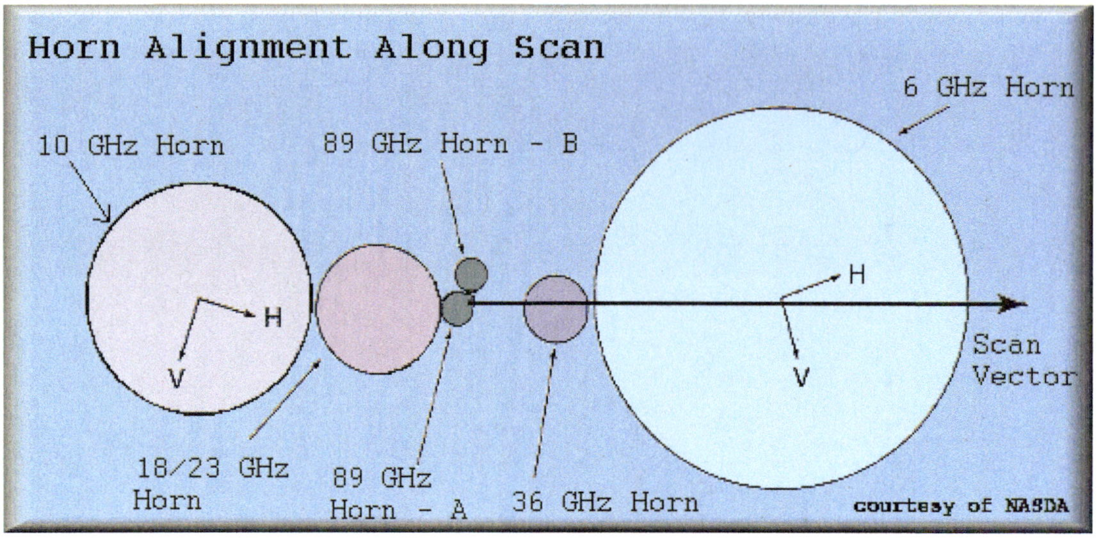

FIGURE 8.48 Advanced microwave scanning radiometer–Earth observing system feed horns. Image credit courtesy JAXA.

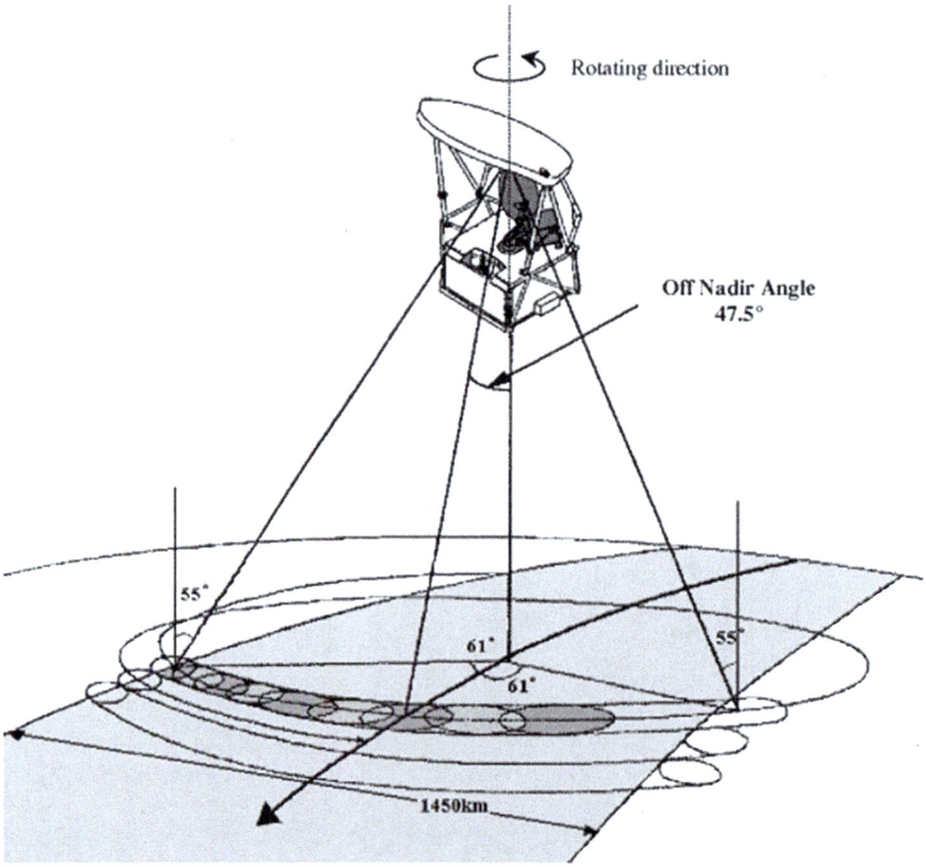

FIGURE 8.49 Advanced microwave scanning radiometer–Earth observing system scan geometry. Image credit courtesy JAXA.

looking in the satellite orbit direction. Only ±61 degrees is used in the forward direction, which is reduced in ground processing, resulting in a swath width of 1,450 km. During each rotation, the subsatellite point advances by 10.1 km, and a sampling period of 2.6 ms ends up with a "cross-track" sampling resolution of 8.9 km except for the 89 GHz channels. For 89 GHz, the sampling period is 1.3 ms, resulting in a spatial resolution of 4.5 km on the ground. On each scan, 196 samples (392 for 89 GHz) are obtained within ±61 degrees scan angle.

Kawanishi et al. (2003) conclude that the major hardware improvements with AMSR-E have provided finer spatial resolution and improved the capability of measuring SST and soil moisture on a global basis. An important application of AMSR-E data has been its assimilation into numerical weather forecast models. Frequent sampling of transient phenomena, such as severe tropical storms, can be resolved by AMSR-E under totally cloudy conditions. The calibration systems of AMSR-E have contributed greatly to its success.

High resolution (0.25 degrees) SST has been retrieved by people at remote sensing systems and distributed by the National Center for Atmospheric Research (https://climatedataguide.ucar.edu/climate-data/sst-amsr-e-sea-surface-temperature-remote-sensing-systems).

Monthly mean AMSR-E derived SST for July 2006 is presented here in Fig. 8.50 along with the number of observations and the standard error. One can see the excellent coverage in the middle panel except for an equatorial band in both the Pacific and the Atlantic. This is likely due to heavy tropical precipitation that will mask out the microwave SST signal. The same is true around the UK and western coast of Europe. The SST map is limited by the lower spatial resolution of the AMSR-E sensor compared to infrared SST maps. The standard error maps exhibit largest values in the western boundary current regions, where strong mesoscale variability contributes to space-time variations of SST. There are again higher values just north of the equator in both major oceans, but this is likely due to SST variability in the equatorial wave guide. There is relatively strong SST variability in known upwelling areas and in the Southern Ocean.

An interesting comparison was made between the AMSR-E SST and the SST from the UK Hadley Center (Fig. 8.51). The EOF1 patterns are quite similar with the Hadley SST, being smoother, which reflects the smoothing used by the Hadley Center to combine both satellite and in situ SST data sources. In contrast, the AMSR-E SST exhibits much more variability, particularly in the warm center of the SST field. Still, the time series for the first EOF between June 2006 and December 2010 appears very similar and have a correlation of 0.97. This is encouraging as it says that the AMSR SST is representative of this much more processed and smoother Hadley SST.

The AMSR-E antenna stopped spinning at the required speed on October 4, 2011. This is most likely due again to the lack of lubricant in the spin mechanism. The AMSR-E continued to function but at a greatly reduced capacity. It did, however, clearly demonstrate the value of such an instrument for Earth remote sensing.

8.4 The A-Train

NASA's Aqua satellite was placed into an orbit that became known as the A-Train for Afternoon Train (Fig. 8.52). Most of the satellites in this orbit were launched in the early to mid 2000's. Launched in May of 2002, Aqua was one of the first in this orbit. It was followed by NASA's Aura satellite, launched July 15, 2004. That same year, the French PARASOL mission was launched by CNES into the A=TRAIN orbit on December 2, 2004. NASA's CloudSat followed with a launch on April 28, 2006. It was launched together with CALIPSO, which was a joint CNES and NASA project. It followed CloudSat by no more than 8.5 sec. In the following decade, GCOM-W1 (SHIZUKU) was launched by JAXA on May 18, 2012, and NASA's OCO-2 satellite was launched into the lead position on July 2, 2014. OCO-2 replaced OCO, which was destroyed in a launch failure on February 24, 2009.

CloudSat was moved to a lower orbit on February 22, 2018 and followed soon after by CALIPSO in September 2018. This was called the C-TRAIN that only contained the two satellites.

A good example of combining data from these dedicated missions with other satellites is the combination of an image of Hurricane Bill from Aqua MODIS with cloud heights from the CALIOP lidar (on CALIPSO) on 19 August 2009 (Fig. 8.53).

Stephens et al. (2002) provided us with a table (Table 8.8) that best describes the A-TRAIN satellites and their capabilities. In this table, Aqua is labeled as the lead constellation spacecraft, which just recognizes that it was first in this orbit. The table lists all of the instruments on Aqua that we have introduced earlier. The table goes on to specify the character of CloudSat (Stephens was the CloudSat project scientist), noting that it lags Aqua by less than 120 s. The primary instrument is a 94 GHz cloud radar with 500 m range gates and a FOV of 1.4 km. CALIPSO, a joint NASA and CNES satellite, was launched with CloudSat into an orbit that lagged CloudSat by about 15±2.5 s. The principal instrument on CALIPSO was the CALIOP lidar, but it also carried a 3-channel infrared radiometer with an FOV of 1 km and a swath of 64 km. PARASOL was a French (CNES) satellite that lagged CALIPSO by about 2 min. It carried a nine-

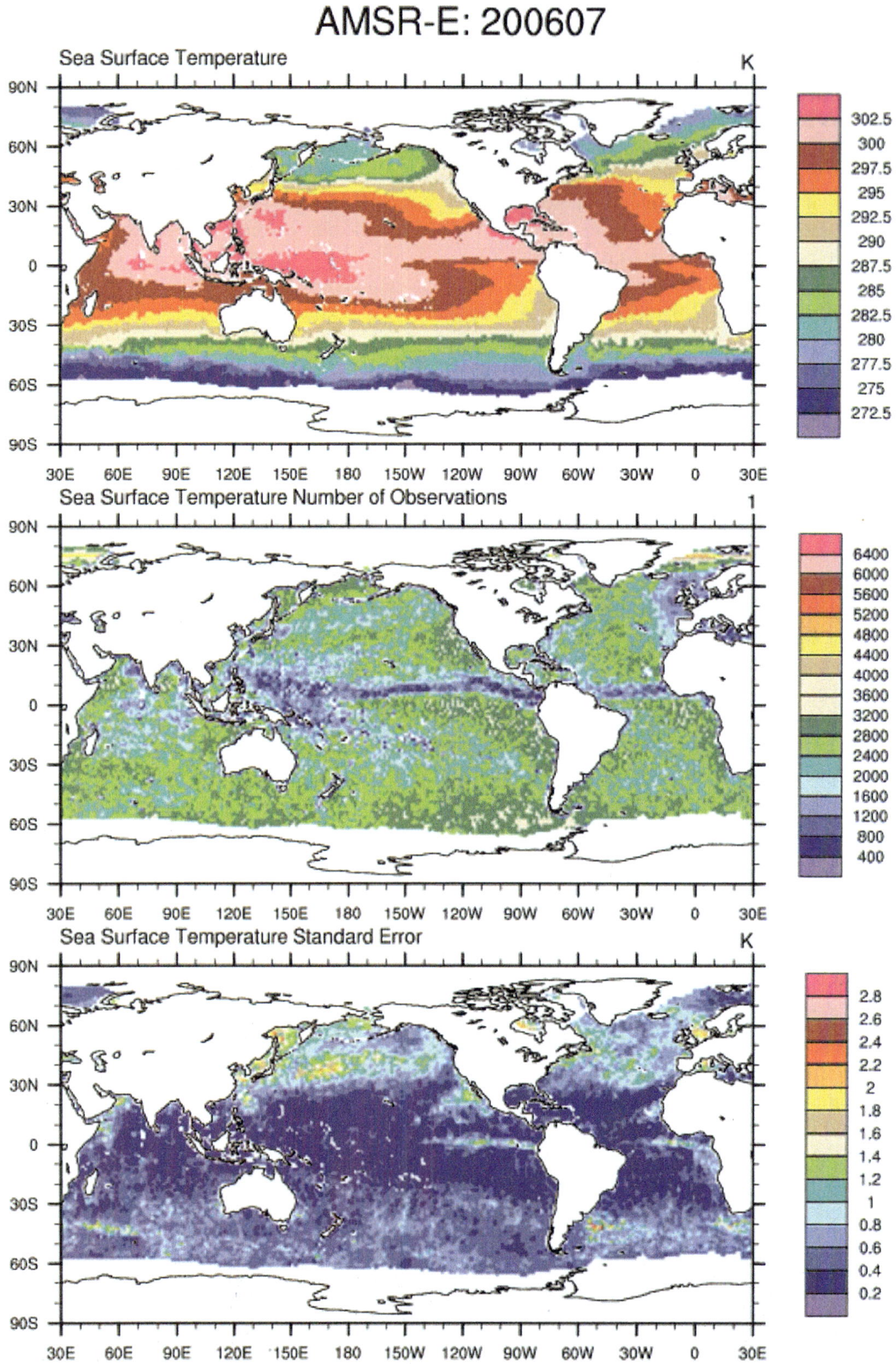

FIGURE 8.50 Advanced microwave scanning radiometer–Earth observing system (AMSR-E) sea surface temperature (SST), (top) mean monthly SST for July 2006, (middle) number of AMSR-E SST observations, (bottom) standard error of the AMSR-E SST in July 2006 (courtesy Hadley Center, UK). Map lines delineate study areas and do not necessarily depict accepted national boundaries.

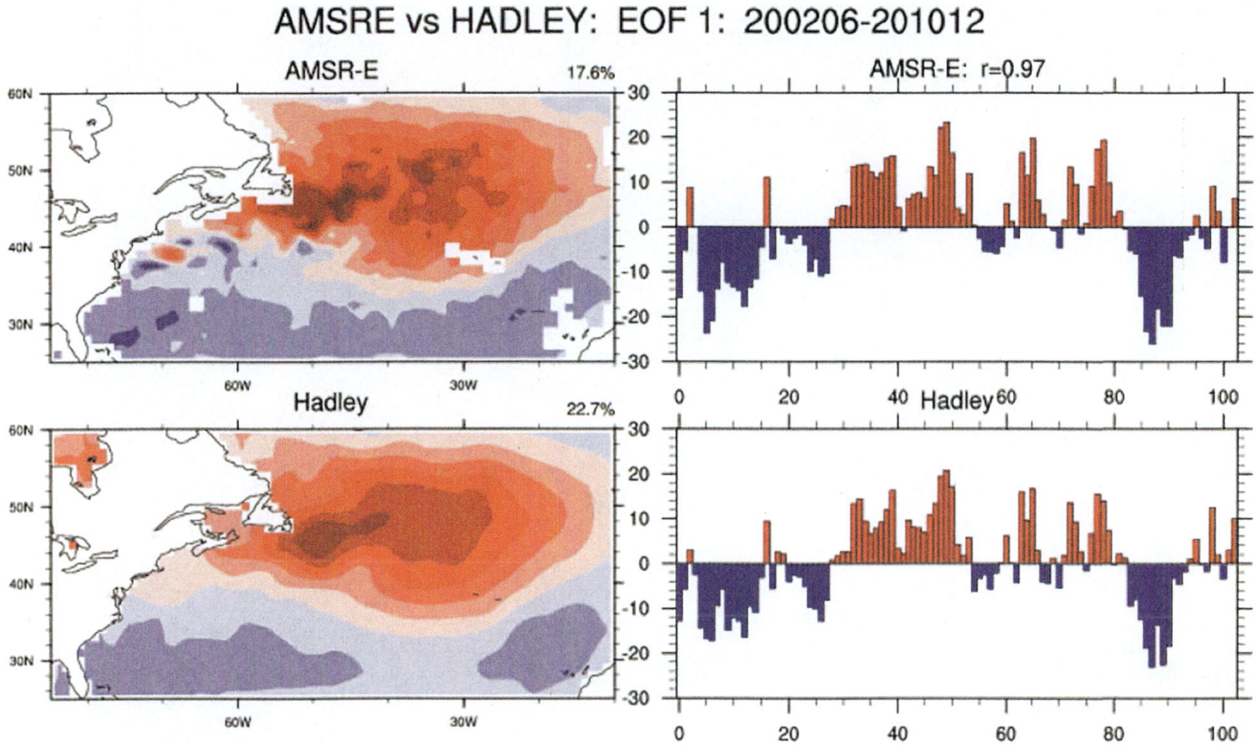

FIGURE 8.51 (left) The first orthogonal function for both Advanced microwave scanning radiometer–Earth observing system (AMSR-E) and the Hadley sea surface temperature (25°N–60°N, 275°E–330°E), (right) time series (6, 2002–12, 2010) of the first EOF. Correlation is 0.97. Image credit courtesy Hadley Center, UK.

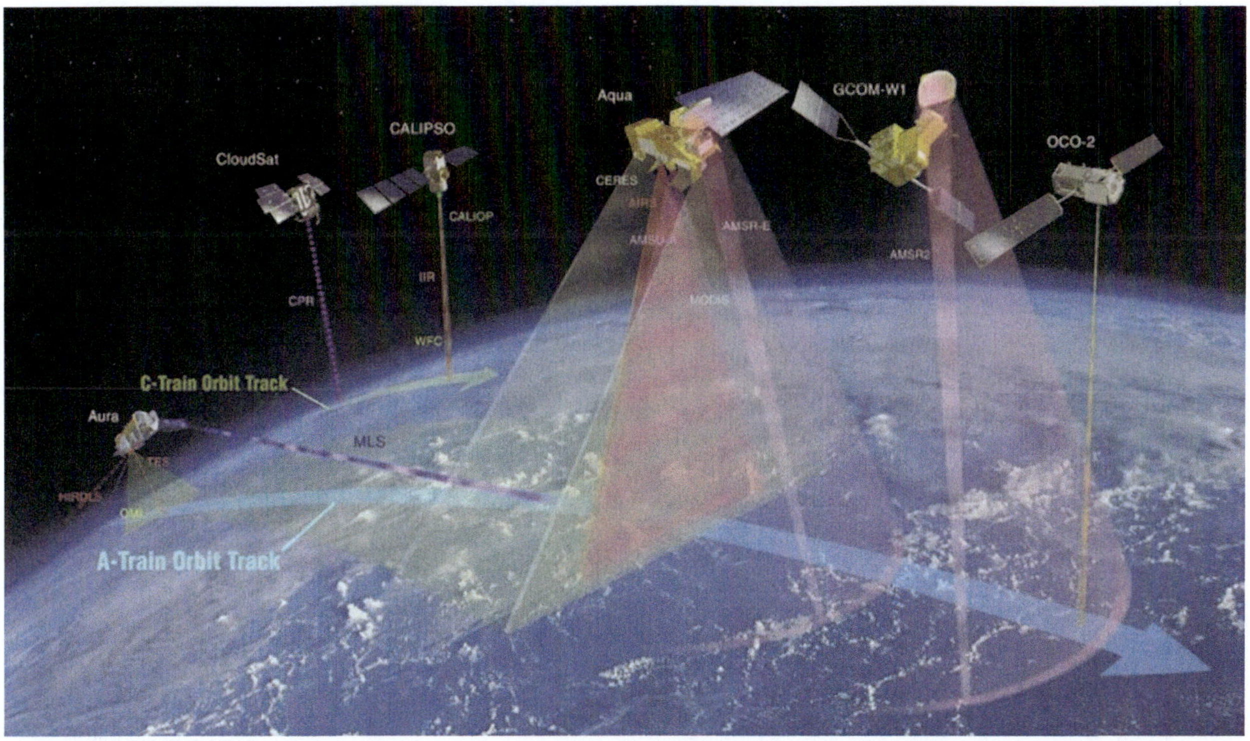

FIGURE 8.52 The international Afternoon Constellation or A-Train. Includes the C-Train satellites (CALIPSO and CloudSAT). Credit NASA.

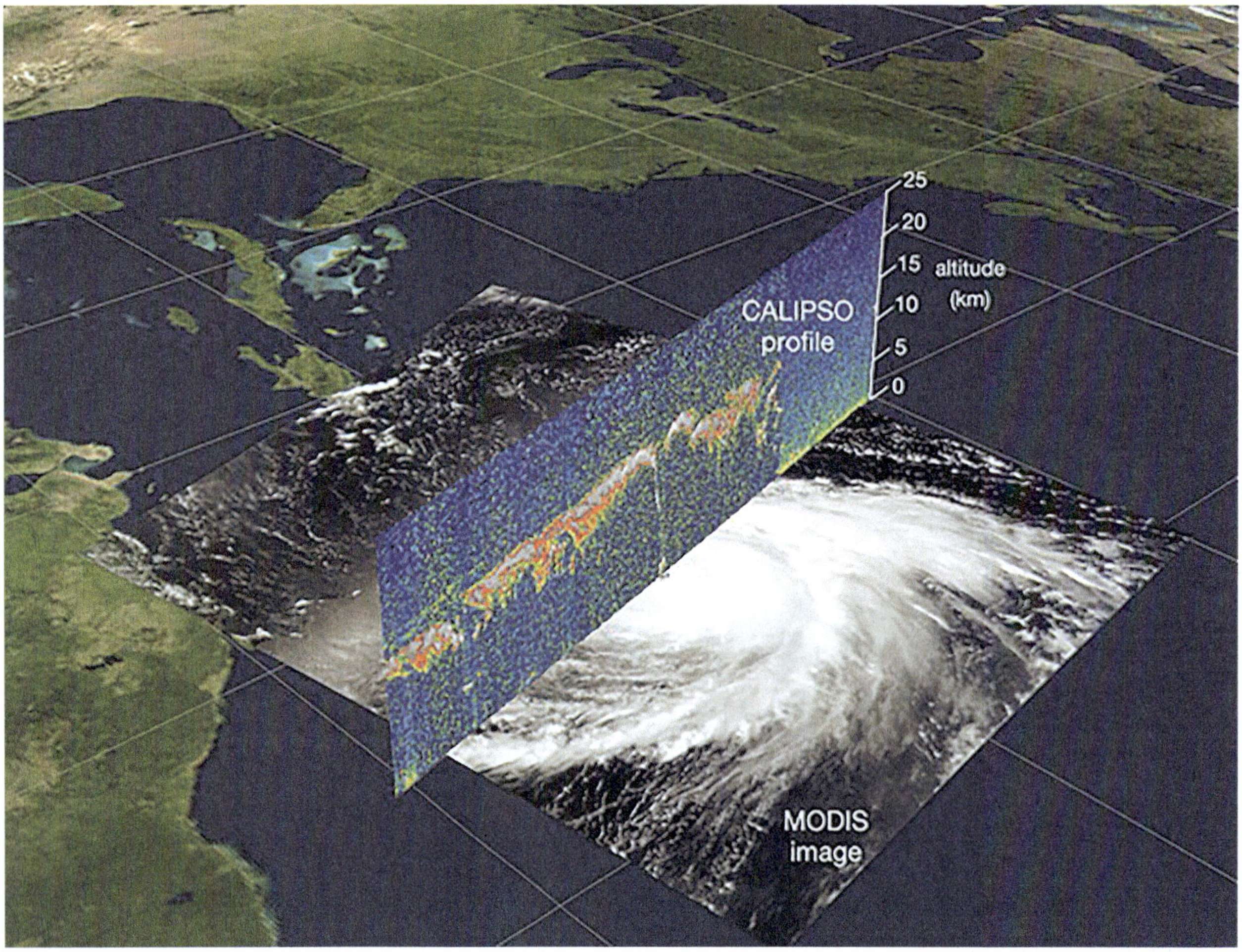

FIGURE 8.53 Image of Hurricane Bill as seen by Moderate Resolution Imaging Spectrometer (MODIS) on the Aqua satellite with cloud heights from the CALIOP lidar on the CALIPSO satellite, taken on August 19, 2009 (https://www.wral.com/weather/image/20240060/).

channel polarimeter with channels in the visible and the near infrared. CloudSat, CALIPSO, and PARASOL were all designed to give some redundant and some independent information on clouds. NASA's Aura satellite lagged Aqua by about 15 min and carried an infrared limb sounder, a microwave limb sounder, and infrared imaging spectrometer (TES) with a 0.5×5 km resolution, a narrow swath, and variable pointing capability. Finally, Aura had a UV grating spectrometer (OMI) with a 12×24 km resolution. All of the Aura instruments were designed to monitor trace gases and aerosols.

Thus all of the A-TRAIN satellite instruments focused on a better understanding of processes in the atmosphere that affect the Earth. To better appreciate the importance, we present (Fig. 8.54) the response of numerical weather prediction models to changes in atmospheric CO_2. Note that the cloud feedback on the weather far outweighs the water vapor effect. The top diagram shows the response terms of precipitation, while the bottom shows the change in surface temperature.

The reason that precipitation and surface temperature are shown as these are the primary consequences of global warming rather than higher atmospheric temperature.

8.5 National Polar-Orbiting Operational Environmental Satellite System (NPOESS)

In its inception NPOESS was to be the US next-generation polar-orbiting satellite system that would monitor the Earth's weather, atmosphere, oceans, land, and near space environments (file:///Users/williamjemery/Desktop/RS%20past%20pres%20futu/background%20off%20the%20web/NPOESS/NPOESS%20-%20Wikipedia.webarchive). NPOESS satellites (Fig. 8.55) were to host proven technologies and operational versions of sensors that were under operational prototyping by NASA at the time. This was a tri-agency program led by an Integrated Program Office (IPO) with staff from NOAA, the US Department of Defense (DoD) and NASA. The lead agency was to be NOAA who would eventually

TABLE 8.8 **Sensor complement and related products of the A-TRAIN (https://science.nasa.gov/mission/a-train/).**

Spacecraft	Payload	Characteristics	Cloud and aerosol products
Aqua lead constellation spacecraft	MODIS	36-channel visible radiometer, 2300-km-wide swath, variable resolution from 0.25 to 1 km.	Land, ocean, and atmospheric products. The latter includes cloud and aerosol optical depths and particle size information, as well as cloud emissivity and cloud-top height.
	AIRS/ AMSU-A/ HSB	Combination of IR and microwave sounders. Swath of ±50 degrees, resolution of IR sounder ~10 km.	Temperature and moisture profiles in clear atmosphere. Some cloud properties.
	AMSR-E	Six-channel microwave radiometer. 1445-km swath, asymmetric FOV with variable resolution from ~6×4 km (89 GHz) to 43×75 km (6 GHz).	LWP, column water vapor, liquid precipitation, principally confined to ocean regions.
	CERES	Broadband and spectral radiances converted to fluxes, resolutions at nadir −20 km.	TOA radiation budget. Primary product is time mean fluxes, but instantaneous fluxes are also produced.
CloudSat Lags Aqua by a variable amount but less than 120 s	94-GHz radar (CPR)	500-m vertical range gates from surface to 30 km. High sensitivity, FOV approximately 1.4 km.	Cloud profile information, liquid and ice water content profiles, and precipitation. The information is obtained by combining the radar measurements with *Aqua* measurements, including MODIS and AMSR-E, as well as with the CALIPSO lidar.
CALIPSO Separation is maintained by CloudSat. Lags CloudSat by 15±2.5 s	Lidar (CALIOP)	532- and 1064-nm channels with depolarization. FOV of approximately 300-and 70-m resolution.	Cloud profile information primarily of upper-tropospheric clouds. Optical depth of thin cirrus. Aerosol profiles with attached optical depth estimates. Aerosol information requires averaging over 10s of kms especially in daylight.
	IIR	Three-channel IR radiometer with a FOV of 1 km, swath 64 km.	Cirrus cloud optical properties.
PARASOL Lags CALIPSO by ~2 min	POLDAR	Nine-channel polarimeter with channels in the visible and near-infrared. Resolution of 5 m, swath of 400 km.	Cloud and fine mode aerosol optical depths and particle sizes.
Aura Lags Aqua by about 15 min	HIRDLS	IR limb sounder.	Trace gases and stratospheric aerosol.
	MLS	Microwave limb sounder.	Trace gases, ice content of thin upper-tropospheric cloud.
	TES	IR imaging spectrometer, 0.5×5 km resolution, narrow swath, and variable pointing.	Trace gases could also provide high spectral resolution data on clouds.
	OMI	UV grating spectrometer, 13×24 km resolution.	Ozone and aerosol index.

AIRS, Atmospheric infrared sounder; *AMSR-E*, Advanced microwave scanning radiometer–Earth observing system; *AMSU-A*, Advanced Microwave Sounding Unit-A; *CERES*, Clouds and the Earth's Radiant Energy; *FOV*, field of view; *HIRDLS*, infrared limb sounder; *HSB*, humidity sounder for Brazil; *MLS*, microwave limb sounder; *MODIS*, Moderate Resolution Imaging Spectrometer; *TOA*, top of atmosphere.

operate all of the spacecraft. Initially, the DoD was put in charge of procurement and NASA was strictly advisory. The plan was for this system to replace both the NOAA polar orbiter program and the DoD DMSP program, which was something that had been tried before but never saw fulfillment. The launch of the first NPOESS satellite was planned for

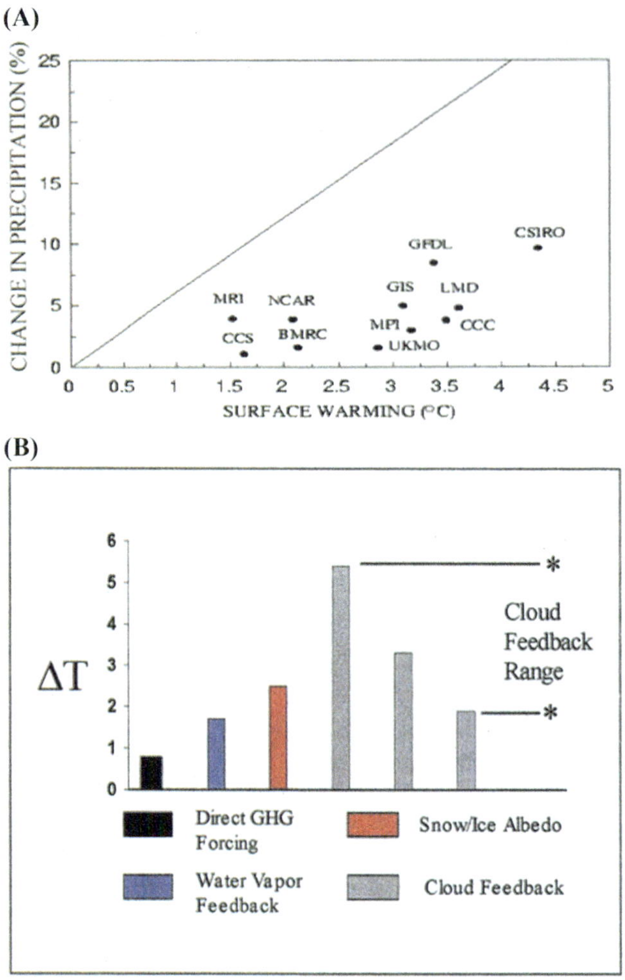

FIGURE 8.54 (A) The responses of various coupled ocean-atmosphere Global Climate Models to an imposed doubling of CO_2. The letters refer to different institutions that run these models. (B) The response of a single climate model to an imposed doubling of CO_2, showing the different feedback as they are added to the model (from left to right). Note the different treatments of cloud feedback (Souffer and Manabe, 1999).

2013. Note, this broke with the previous practice of NASA developing the spacecraft and sensors and then turning them over to NOAA for their operation.

NPOESS did not use instruments that had been prototyped by NASA but rather selected a set of brand new instruments that had never flown before. These are as follows:

1. Visible/infrared radiometer suite (VIIRS) was intended to replace heritage instruments like the AVHRR but also bring into operations capabilities on NASA MODIS instrument. It was intended to be the main imager for atmosphere, ocean, and land applications.
2. Conical microwave imager/sounder (CMIS) was to collect microwave imagery and soundings of the atmosphere.
3. Crosstrack infrared sounder (CrIS) was the primary sounding instrument to provide vertical profiles of temperature, moisture, and pressure in the atmosphere.
4. Global positioning system occultation sensor was intended to provide data for atmospheric occultation measurements of atmospheric properties.
5. Ozone mapping and profiler suite (OMPS) was to collect data on the vertical and horizontal distribution of ozone in the Earth's atmosphere.
6. Space environment sensor suite was actually an instrument with significant heritage. It collected data on the neutral and charged particles, electron and magnetic fields, and optical signatures of aurora.

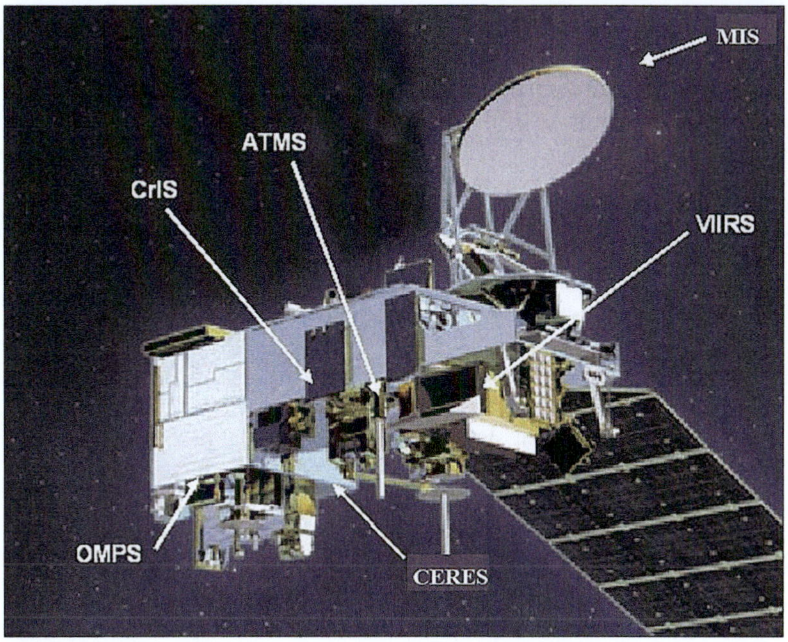

FIGURE 8.55 NPOESS satellites.
Image credit NOAA/NASA.

7. Aerosol polarimeter sensor was designed to retrieve specific aerosol and cloud parameters using multispectral photo polarimetry.
8. Advanced Technology Microwave Sounder (ATMS) (under development by NASA), in conjunction with CrIS, was to provide global observations of temperature and moisture profiles at high temporal resolution (~daily).
9. Data Collection System (DCS) also had considerable heritage and was to replace the earlier Argos DCS then flying on NOAA polar orbiters.
10. Earth Radiation Budget Sensor (ERBS) was intended to measure the Earth's radiation budget parameters with similar to ERBE and CERES heritage instruments.
11. Radar ALT similar to that flying on the Jason series of satellites.
12. Search and Rescue Satellite Aided Tracking works in conjunction with other NOAA polar orbiting and geostationary satellites to locate downed aircraft and ships in distress.
13. Total Solar Irradiance Sensor (TSIS) is a total solar irradiance (TSI) monitor plus a 0.2–2 μm solar spectral irradiance monitor.

A design review in 2006 removed several of the climate instruments from the program. Removed were TSIS, ERBS, ALT, and OMPS. This last instrument was later assigned to fly on what was then known as NASA's NPOESS Preparatory Project (NPP). This was a spacecraft that was due to fly in 2011. During this design review, the number of NPOESS satellites was reduced from six to four. Concerns remained about the development of the NPOESS, many sensors, particularly the primary imager the VIIRS. The author was involved as a consultant on the original design of the VIIRS. When the design was finalized, all of the consultants were let go, and some of the in-house Raytheon engineers who had been working on VIIRS were also let go.

The author also contributed to the main NPOESS contract that was won by then TRW, which was subsequently bought by Northrop-Grumman. As a consequence, I was also a member of the NPOESS Advisory Panel, but in 2003, I was working for a year at NASA Headquarters, and people thought I should not participate in the advisory panel at least for that year. I then engaged with the panel again in 2004 and was witness to all the problems that NPOESS had. The biggest problems were very large cost overruns. This did not surprise me with a program led by NOAA and a procurement assigned to the DoD. As I said earlier NASA had a strictly advisory role and did not make any direct decisions other with NASA as part of the development of their own NPP platform. The one real accomplishment of this advisory panel was that we were able to convince NASA and NPOESS that NPP should fly in the afternoon rather than the morning orbit. It needs to be mentioned here that NPP was subsequently renamed Suomi NPP to honor Vern Suomi, the man who had designed the spin-scan camera that made the earliest geostationary weather satellites possible.

NPP (Fig. 8.56) was initially scheduled to launch in October 2006 with a life cycle cost of $560 million. While NASA met its schedule for the spacecraft and the instruments, they were dependent on the other NPOESS IPO partners, who were unable to deliver their three instruments, leading to a 5-year launch delay and a 54% increase in cost. NPP was finally launched in October 2011, and the life-cycle costs had grown to $864 million.

The satellite was built by Ball Aerospace and Technologies Corporation of Boulder, Colorado (contract awarded in May 2002). The bus is a variation of Ball's standard BCP 2000 platform that was used earlier on ICESat and CloudSat.

Since the project involved the DoD, NPOESS was obliged (it overran original cost estimates by more than 25%) to undergo a review by the US House of Representatives under the Nunn-McCurdy Act. They found that the program "has a troubled history of cost increases and schedule delays and has been the subject of several Science Committee hearing." The Department of Commerce Inspector General raised concerns about NOAA's program management and award fees paid to the prime contractor, Northrop Grumman. Even after the 2006 revamped program, it was estimated to have acquisition costs of $11.5 billion, which is an increase of about 50% over the official baseline of $7.4 billion. Earlier cost estimates assumed a first launch in 2010.

On February 1 of 2010 the US White House terminated the NPOESS program and renamed the follow-on program the Joint Polar Satellite System (JPSS), which would be managed by NASA's GSFC, returning to the earliest satellite days in the United States, where NASA developed the satellites and instruments and turned them over to NOAA for operations. The idea was that these JPSS satellites would be much like the Suomi NPP satellite. The instruments would be a subset of the NPOESS list. JPSS requested $1.06 billion for 2011 and an additional $3.47 billion over the 4 following years.

The first JPSS-1 satellite (Fig. 8.57) was launched on November 18, 2017, 3 years later than originally expected. When JPSS-1 reached its orbit NOAA followed tradition and renamed the satellite NOAA-20 to signify that this is a continuation of the earlier NOAA polar orbiting satellites. VIIRS, OMPS, CERES, and TSI were the instruments on JPSS-1. It also carried a new RBI that would measure reflected sunlight and emitted thermal radiation from the Earth. This would be a continuation of the ERBE record and would supplement CERES. Finally, another instrument was installed on JPSS-1, which was Libera, which was a follow-on to CERES, and was named Libera, who was the daughter of CERES, the Roman goddess of agriculture. Libera will observe the Earth's radiation budget, analyzing solar radiation entering the Earth's atmosphere and the amount of it that is either absorbed, reflected or emitted. Libera will maintain the 40-year record of the solar radiation balance.

The TSI (also called the ECP for energetic charged particle) sensors are government-supplied instruments for JPSS. They are designed and prototyped by the US Air Force, which contracted with Applied Technology Associates of Paso Robles, CA to build the flight units. The Air Force also contracted with Teledyne Brown Engineering of Huntsville, AL, to assemble ECP-Lite, which was developed by the Aerospace Corp. These two sensors fly on different spacecraft in the JPSS series. The intent is to dramatically improve the ability to monitor and forecast space-weather.

The basic payload for the JPSS satellite was the Visible Infrared Imaging Radiometer Suite (VIIRS), which collects global visible and infrared observations of land, ocean, and atmosphere as mentioned earlier. They also carried the CrIS designed to retrieve high-resolution, three-dimensional temperature, pressure, and moisture profiles of the atmosphere. These profiles will be used to enhance numerical weather forecast models to facilitate better short and long term weather forecasts. Next is the ATMS, which is a 22-channel cross-track sounder to retrieve atmospheric temperature and moisture

FIGURE 8.56 Suomi NPOESS Preparatory Project satellite.
Courtesy Ball Aerospace.

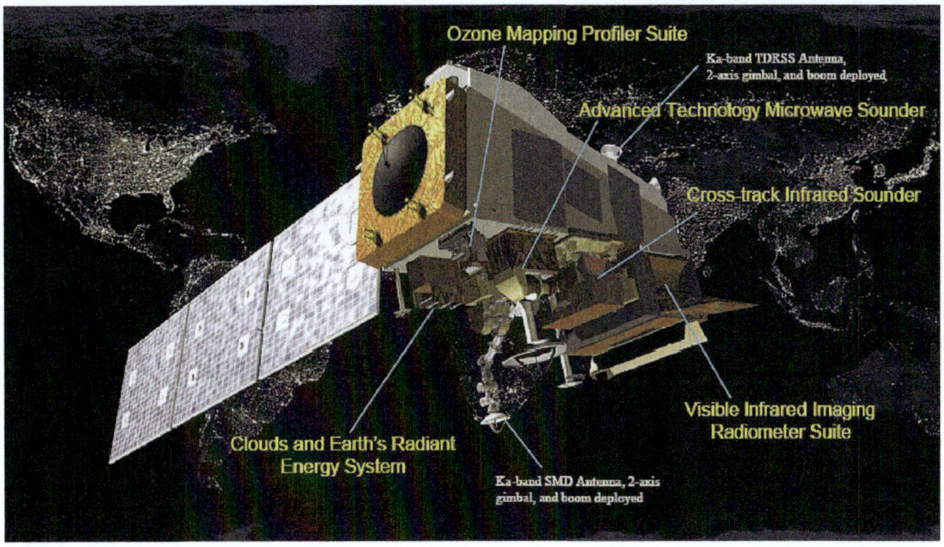

FIGURE 8.57 Joint Polar Satellite System-1 or NOAA 20.
Credit courtesy NASA GSFC.

profiles. ATMS data is used in conjunction with CrIS profiles to improve weather and climate forecasting. JPSS also carried the OMPS, which is a combination of three hyperspectral instruments that extends the 25+ year record of total ozone and ozone profile records. The improved resolution of OMPS allows for a better testing and monitoring of the complex chemistry involved in ozone destruction near the troposphere. OMPS products, when combined with cloud predictions, also help produce better ultraviolet index forecasts.

The next instrument, the CERES, was discussed earlier in terms of the NASA Terra and Aqua satellites. The Radiation Budget Instrument (RBI) carried by the JPSS satellites measured reflected sunlight and thermal emissions radiated from the Earth. RBI was not on JPSS-1 and started with JPSS-2. The instrument was canceled by NASA on January 26, 2018, citing cost and schedule issues. So, this particular instrument was not flown on any of the JPSS satellites. The TSI calibration transfer experiment (TCTE) measures the sun's energy output. The TCTE was originally flown on a US Air Force Space Test Program Satellite-3 that was launched on November 19, 2013. JPSS was able to continue these measurements into the future. Libera is a follow-on to the CERES instruments. It will observe the Earth's radiation budget, analyzing solar radiation entering the Earth's atmosphere and the amount that is either absorbed, reflected, or emitted. Thus, it is important to determine whether or not the Earth is warming, cooling, or balanced.

JPSS-2 was launched on November 10, 2022, and carried the same instruments as JPSS-1/NOAA-20 to continue the series of measurements of atmospheric, terrestrial, and oceanic condition for weather forecasting and long-term climate change. JPSS-3 is scheduled for launch in 2026. It will carry the same complement of instruments as on the previous JPSS satellites. JPSS-4 will carry updated versions of the instruments carried by earlier JPSS satellites and is scheduled for launch in 2031.

The JPSS Common Ground System converges the NOAA–NASA civil polar environmental satellite program with the NASA Suomi spacecraft and the Air Force's Defense Weather Satellite System (DWSS) ground systems into a single common system that satisfies both US and international partner environmental monitoring satellite needs from polar orbit.

The original plan was for the JPSS satellites to fly in the afternoon orbit while the DoD satellites covered the morning orbit in its DWSS. However, DWSS was canceled in April of 2012, leaving the military to rely on the remaining DMSP satellites discussed earlier in this text. More recently, the DoD has had Ball Aerospace (now BAE Systems) to build the Weather system follow-on microwave (WSF-M), the first in the DoD's Weather System Follow-ons following the demise of NPOESS (Fig. 8.58).

This satellite was launched on April 11, 2024 into a sun-synchronous orbit with an altitude of 854 km. It makes a substantial contribution to ocean surface winds, tropical cyclone intensity, and low-Earth-orbit energetic charged particles. WSF-M was designed to fill three high-priority gaps in the DoD Space-based environmental monitoring system. These are ocean surface vector winds (as discussed for WindSat), tropical cyclone intensity, and low Earth orbit energetic charged particles. This satellite leveraged knowledge from the Ball-built Global Precipitation Measurement (GPM) Microwave

Imagery (GMII), which became the on-orbit reference standard for calibrating precipitation measurements in the GPM constellation. WSF uses the Ball Configurable Platform, a proven, agile spacecraft with 50 years of on-orbit performance.

WSF-M channels' characteristics are spelled out here in Table 8.9.

The instrument had a conically scanning 1.8 m reflector that fed feeds for the six different frequencies. Three channels have full polarimetric capability for the wind vector retrievals. The third and fourth stokes vectors are obtained by digitally sampling and cross-correlating v-pol and h-pol amplitude signals to yield the ocean surface wind vectors. This passive microwave approach to measuring the ocean surface wind vector was pioneered by the Windsat satellite.

The wind speed and wind direction errors are given in Fig. 8.59.

Ball's partner Harris (EDO) built the 37 GHz polarimetric noise source and down-converter assemblies. These are combined with the Ball-built Digital Receiver Assembly (DRA) and the existing 37 GHz receivers developed for GMI to provide and end-to-end receivers at 37 GHz. Ball's Main Reflector partner, AASC, built and tested the 1.8 m reflector, building on experience gained with the 1.2 m reflector used in GMI. Ball was responsible for the instrument electronics controller.

A precursor to WSF was the US Navy "WindSat." This was an instrument (Fig. 8.60) developed by the Naval Research Lab in Washington, DC, and launched on a free flyer satellite known as the Coriolis satellite, built by Spectrum Astro of Gilbert, Arizona. It became known as WindSat. The Navy wanted to measure wind direction from a passive microwave instrument rather than an active microwave instrument like the NASA scatterometer. To do this, the WindSat instrument was fully polarimetric, and the primary goal was to measure the ocean wind vector from a passive microwave instrument. This wind vector retrieval used the polarization out to the fourth Stokes parameter to get direction. Coriolis was launched on June 6, 2003 into an 840 km circular sun-synchronous orbit and operated until December 31, 2020. Like other passive microwave radiometers the WindSat instrument could be used to measure a variety of ocean parameters such as SST, total precipitable water, cloud liquid water, sea ice, snow, soil moisture, vegetation water content, and rain rate.

Gaiser et al. (2004) provide a full description of the WindSat instrument and its development. They note that the ocean surface vector wind is a key parameter for short-term weather forecasting, which is a critical factor in ocean operations. The WindSat instrument was designed to demonstrate the capability of a polarimetric microwave radiometer to measure the ocean surface wind vector from space. They quote that "recent research in polarimetric radiometry has demonstrated that usable wind direction information can be obtained by combining the vertical and horizontal polarization with the cross correlation of those same two polarizations. The cross correlation terms represent the third and fourth parameters of the

FIGURE 8.58 Weather system follow-on microwave satellite.
Courtesy of Ball Aerospace.

TABLE 8.9 Weather system follow-on microwave (WSF-M) Channel characteristics.

Channel	Center frequency (GHz)	Bandwidth (MHz)	Polarizations	Beamwidth (deg)	3dB footprint(km)
1	10.85	500	V, H, 3rd, 4th	1.06	38×23
2	18.85	500	V, H, 3rd, 4th	0.66	23×15
3	23.8	370	V, H	0.58	21×13
4	36.75	500	V, H, 3rd, 4th	0.43	15×10
5	37.3	<2500	V, H	0.43	15×10
6	89	<6000	V, H	0.43	15×10

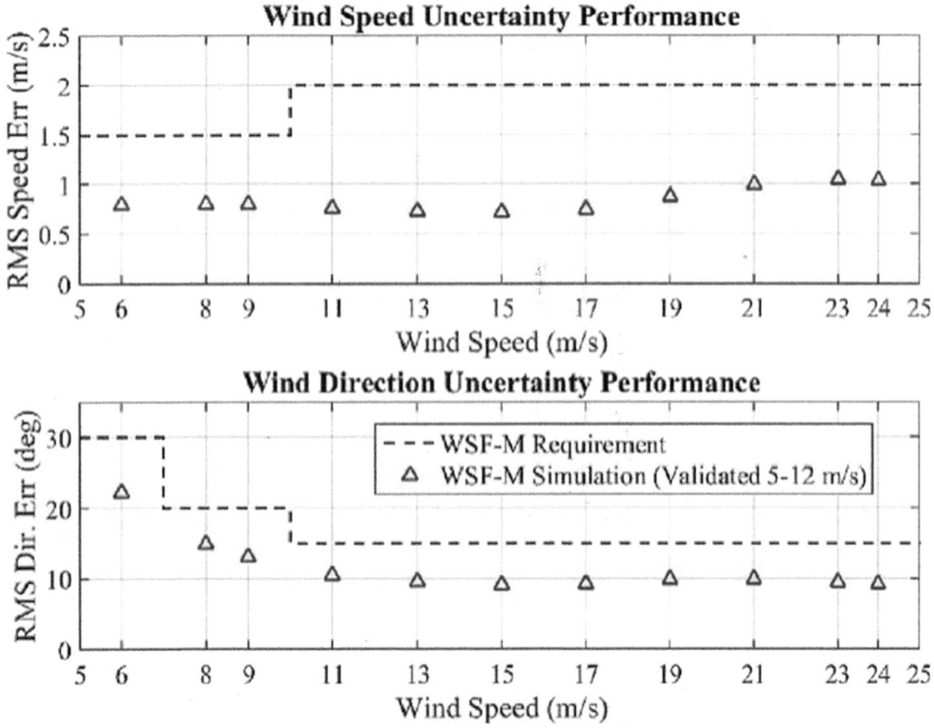

FIGURE 8.59 Speed and direction uncertainty.
Image credit courtesy Ball Aerospace.

modified Stokes vector. The Stokes vector provides a full characterization of the electromagnetic signature of the ocean surface and the independent information needed to uniquely determine the ocean surface wind direction."

The requirements for wind vector measurements can be broken into requirements on wind speed and sea surface wind direction. These requirements are given here in Table 8.10. Here for comparisons are the requirements for the CMIS on JPSS and the interagency operational requirements document (IORD) refers to earlier Table 8.10 requirements for the now nonexistent NPOESS. Experience has now shown that the passive microwave wind vector retrievals are only accurate at wind speeds greater than 7 m/s, while the scatterometer wind vectors are valid for winds greater than 3 m/s. Still, if you are only interested in strong wind events, WindSat data would do the trick. Unfortunately, much of the ocean is not occupied by winds above 7 m/s.

The WindSat frequencies and polarizations are given here in Table 8.11.

FIGURE 8.60 The WindSat instrument in the thermal/vacuum chamber.
Image credit courtesy NASA.

In the polarimetric bands (10.7, 18.7, and 37.0 GHz), three dual-polarization feed horns are used: V/H, +45 degrees/-45 degrees linear, and left/right circular. The third and fourth Stokes parameters are formed by taking the brightness temperature difference between the +45 degrees and −45 degrees polarizations, and the left-hand and right-hand circular polarizations as described in the following equation:

$$I_s = \begin{bmatrix} I \\ Q \\ U \\ V \end{bmatrix} = \begin{bmatrix} T_v \\ T_h \\ T_{45} - T_{-45} \\ T_{lc} - T_{rc} \end{bmatrix} = \begin{bmatrix} \langle E_v E_v^* \rangle \\ \langle E_h E_h^* \rangle \\ 2\ \mathrm{Re}\langle E_v E_h^* \rangle \\ 2\ \mathrm{Im}\langle E_v E_h^* \rangle \end{bmatrix} \tag{2.45a}$$

where T_v, T_h, T_{45}, T_{-45}, $T_{lc,}$ and T_{rc} are the brightness temperatures (radiances) at vertical, horizontal, +45 degrees, −45 degrees, left-hand circular and right-hand circular polarizations respectively. This is the modified Stokes vector equation. These different polarizations are collected by different feed horns in the WindSat instrument (Fig. 8.61).

In addition, the antenna/calibration subsystem consists of the main reflector (Fig. 8.60), the 11 feed horns (Fig. 8.61), a warm calibration load, and a cold sky reflector. The main reflector is a 1.8 m diameter offset parabolic reflector fabricated of graphite epoxy with vapor-deposited aluminum coating. The reflector and most of system rotate at a rate of 31.6 rpm. The circularly polarized feeds use polarizers based on the design of Simmons (1955), while the other WindSat polarizers achieved the necessary phase shifts with tuning screws rather than dielectric vanes. This arrangement of feed horns resulted in the Earth projected beams as presented here in Fig. 8.62.

For this unique instrument, the antenna performance requirement was very tight, and the efficiency was set at 95% to reduce the contamination from sidelobes. To maintain, the polarization purity requirement was set at 99.9% for all the polarimetric bands to enable the accurate computation of the fourth Stokes parameter and hence the ocean wind vector. For the vertical and horizontal bands, this requirement was reduced to 99.5%. This results in residual cross-polarization coupling of −30 and −23 dB for the polarimetric and linear bands, respectively.

TABLE 8.10 Windsat performance requirements.

Parameter	Accuracy			Range			Spatial resolution		
System	WindSat	CMIS	IORD	WindSat	CMIS	IORD	WindSat	CMIS	IORD
Wind SPEED	±2 m/s or 20%			3-25 m/s			~30 km	20 km	
Wind direction	±20 degrees (3-25 m/s)			0-360 degrees			~30 km	20 km	

CMIS, Conical microwave imager/sounder.

TABLE 8.11 Performance requirements for WindSat. The absolute accuracy requirements are 0.75 K for the vertical and horizontal polarizations and 0.25 K for the third and fourth Stokes parameters.

Freq (GHz)	Polarization	BW (MHz)	Absolute accuracy (K)	Integration time (msec)	NEDT (K IFOV)	Dynamic range (K)
6.8	V, H	125	0.75	5.00	0.63	3–330
10.7	V, H,U,4	300	0.75/0.25	3.93	0.44	3–330
18.7	V,H,U,4	750	0.75/0.25	2.19	0.44	3–330
23.8	V,H	500	0.75	1.63	0.60	3–330
37.0	V,H,U,4	2000	0.75/0.25	1.08	0.42	3–330

The NEDT requirements assume scene brightness temperature of 250 K. *FOV*, Field of view.

The most challenging part of this polarization measurement is to minimize and characterize the cross polarization in the antenna and in the system itself. In addition, there is coupling introduced when the spacecraft's polarization basis is rotated relative to the Earth. The characterization of the antenna led to a rigorous alignment campaign at an antenna range. The calibration using the hot load and sky values NEDT is compared with the requirement, the prelaunch estimate, and the on-orbit performance are shown in Fig. 8.63.

The big question with WindSat was how well it could measure the ocean wind vector? Since WindSat overlapped with the QuikSCAT scatterometer, it is possible to compare the two. In Fig. 8.64 we present a wind speed comparison between WindSat and QuikSCAT. You can see that they agree very well up to about 32 m/s where they diverge, with WindSat speeds being greater than QuikSCAT. The dashed lines in this figure refer to an earlier version of the WindSat wind speed algorithm. These algorithms continually mature with ongoing space instruments. The divergence at low wind speeds shows that neither instrument performs well at very low wind speeds, as mentioned earlier.

The big question was how did WindSat do with wind direction, and a similar comparison between QuikSCAT and WindSat wind directions is given here in Fig. 8.65.

We see from Fig. 8.66 that both sensors appear to agree very well at about 6 m/s wind speed. QuikSCAT is thought to be accurate down to wind speeds greater than 3 m/s. Again, there is increased variability and divergence at the highest wind speeds.

WindSat has the following limitations:

- Cannot detect winds within 30–50 km of a coast (land contamination in a pixel).
- Temporal resolution is only twice per day, and data latency issues exist between 5 and 15 UTC because of NPP downlink conflict.
- Both scatterometer and passive microwave radiometers are contaminated by precipitation and tend to have a high bias in these regions, although the wind direction appears to be reasonable.
- Passive microwave sensors like WindSat are affected by cloud liquid water (CLW, >0.2 kg/m^2) while scatterometers are not, hence WindSat will be missing data in areas of thick clouds.
- Sea ice is also a contaminant.

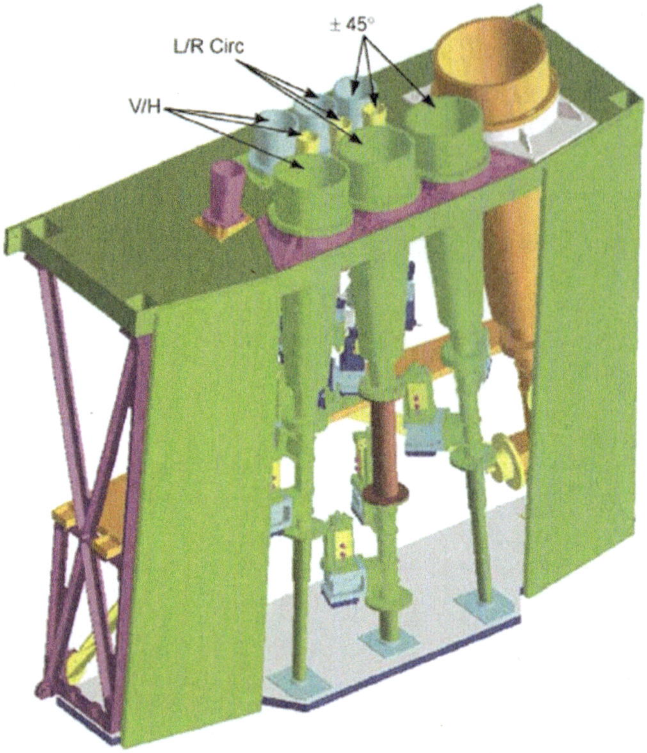

FIGURE 8.61 The WindSat feed bench. Green horns are the 10.7 GHz, blue horns are the 18.7 GHz, yellow horns are the 37 GHz, the orange horn is 6.8 GHz, and the magenta horn is 23.8 GHz (NASA/GSFC).
Image credit courtesy Naval Research Lab.

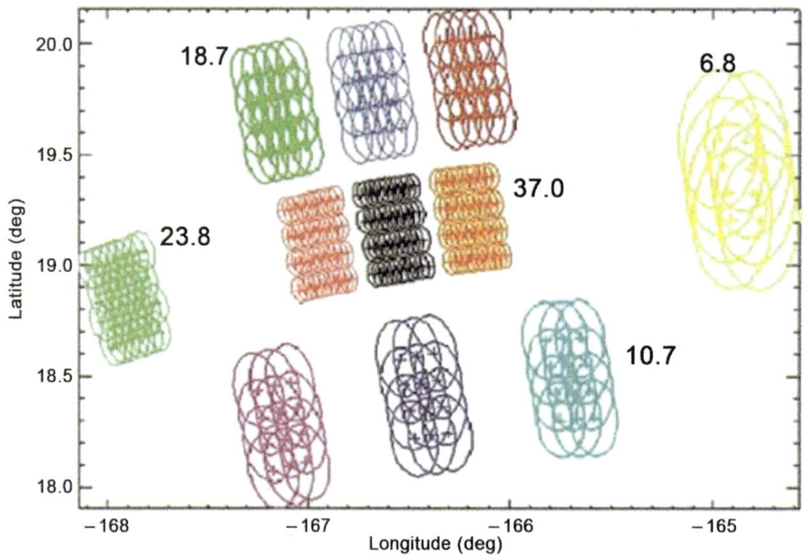

FIGURE 8.62 WindSat antenna footprints projected on the surface of the Earth. Note, the colors correspond to the feed horns, with center dark color being the intersection of two feed patterns. All frequencies are Nyquist sampled in the along-scan direction. One can also see the relative sampling rate for each of the frequencies.
Image credit courtesy NASA.

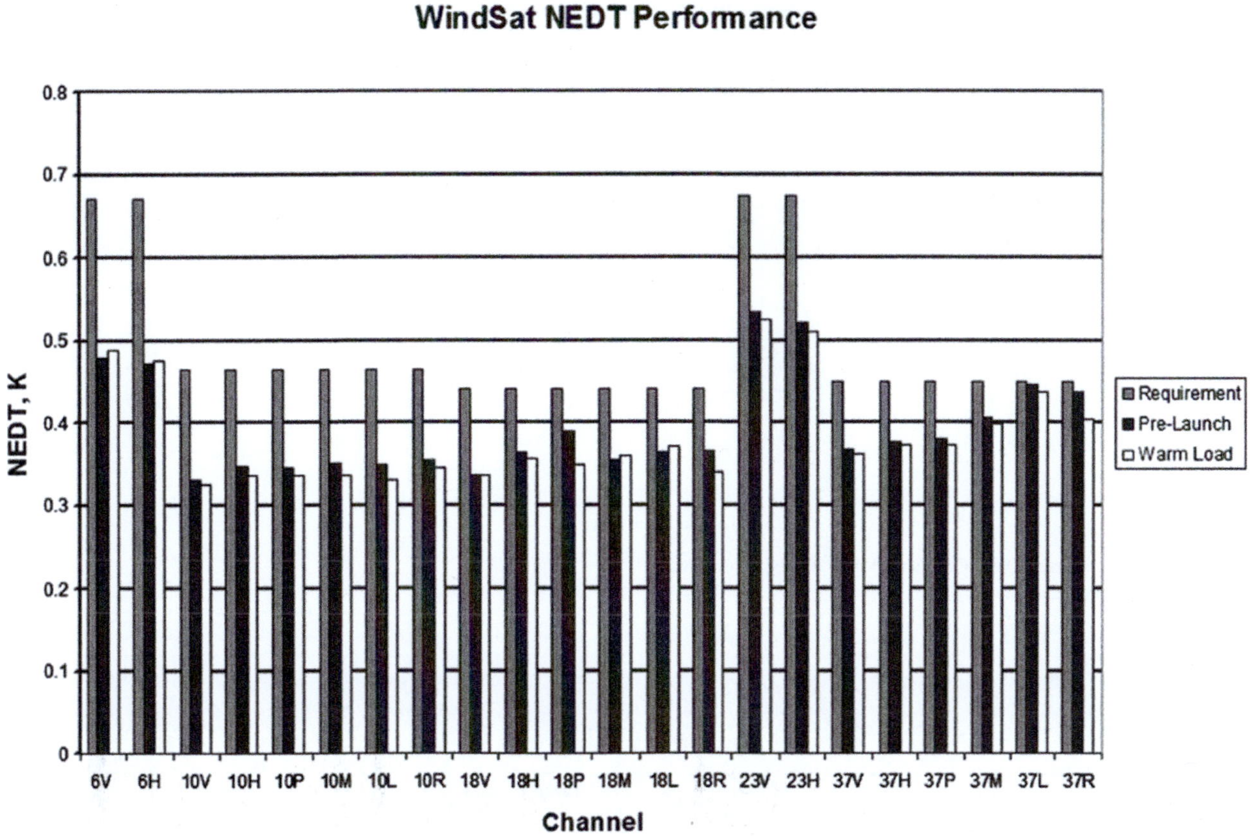

FIGURE 8.63 WindSat NEDT performance for requirement, prelaunch, and on-orbit values. Image credit courtesy NASA.

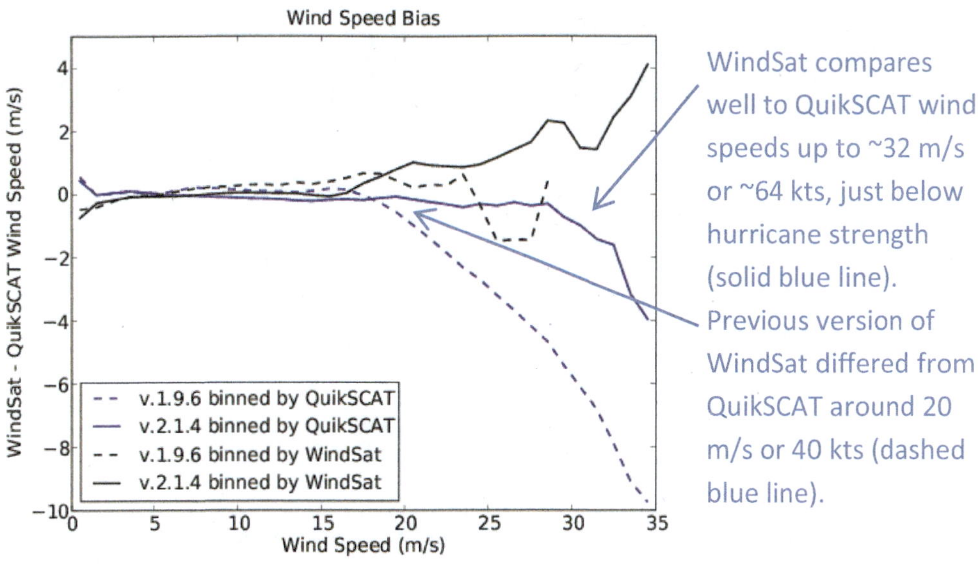

FIGURE 8.64 Comparison of wind speed between that from QuikSCAT and WindSat. Graph provided by the Naval Research Laboratory.

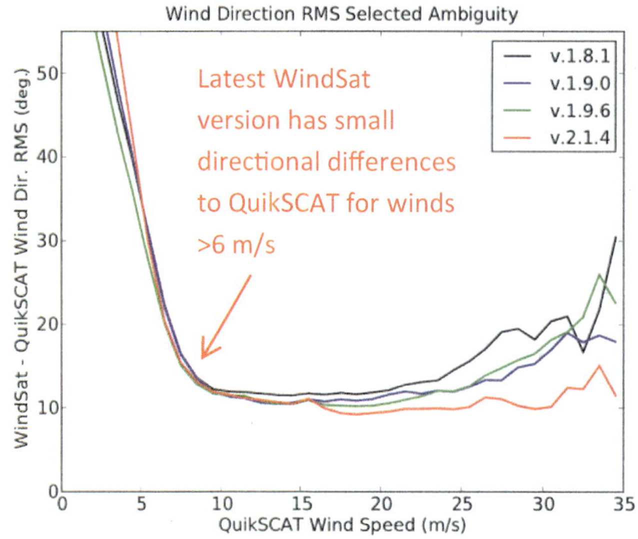

FIGURE 8.65 Comparison between QuikSCAT and WindSat wind directions.
Graph from the Naval Research Laboratory.

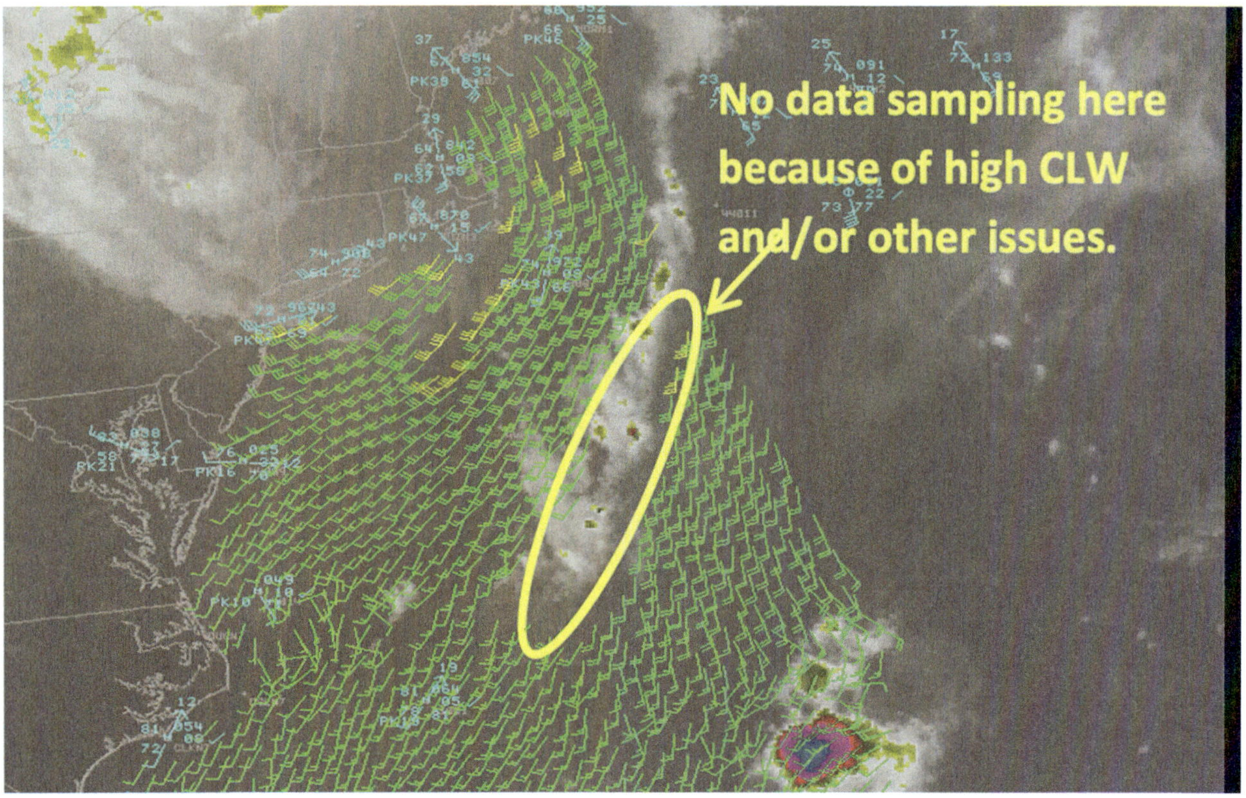

FIGURE 8.66 WindSat wind barbs taken after the passage of Hurricane Irene on the US East Coast.
Acknowledgment goes to Dr. Tom Lee and Mr. Michael Bettenhausen of the Naval Research Laboratory in Monterey, CA.

The limit of CLW can be clearly seen in Fig. 8.66, which are the WindSat wind barbs taken just after the passage of Hurricane Irene. Note, the loss of wind barbs in the high CLW area. The wind barbs are overlain on an infrared image, which reflects the high CLW content of the area missing wind barbs.

Chapter 9

Non-US satellite programs

9.1 European research satellites

9.1.1 Satellite Pour l'Observation de la Terre satellites

Actually, the first commercial European remote sensing system was the SPOT (Satellite Pour l'Observation de la Terre) series of satellites created under the French Space Agency CNES (Centre National d'Etudes Spatiales) in 1986 and run by the Spot Image company in Toulouse, France. The SPOT series operated five satellites between 1986 and 2015, creating a 29-year time series of high-resolution satellite imagery. The SPOT-1 satellite is shown in Fig. 9.1. Airbus Defense and Space continued the SPOT series after CNES ended the program in 2015. SPOT 6 and 7 are commercial satellites owned by Airbus, providing continuity through 2024. Fig. 9.2 shows the growth of this program from the first launch in 1997 up to 2024.

SPOT-2 quickly followed SPOT-1, launched on January 22, 1990, and was then followed by SPOT-3, launched on September 26, 1993, SPOT-4 on March 24, 1998, SPOT-5 on May 4, 2002, SPOT-6 on September 8, 2012, and SPOT-7 on June 30, 2014. SPOT-4 stopped operating in July of 2013. SPOT-5 brought upgrades but stopped functioning on March 31, 2015. Up to SPOT-5, the imager on the satellites provided 10 m panchromatic images and 20 m multispectral resolution. SPOT-1 was decommissioned on December 31, 1990, and SPOT-2 was deorbited in July 2009. SPOT-5 carried two high resolution geometrical instruments with a resolution of 2.5–5 m in panchromatic mode and 10 m in the multispectral mode (20 m in the shortwave infrared (1.58–1.78 μm). SPOT-5 also carried a high-resolution spectrometer that could point forwards and backwards, making it possible to collect stereo image pairs that could be used to compute vertical relief. This timeline of SPOT satellites is presented here in Fig. 9.2.

While SPOT-1–5 were all launched on Ariane rockets, SPOT 6 and 7 were launched by India's Polar Satellite Launch Vehicle (PSLV), with SPOT-6 being launched on September 9, 2012 and SPOT-7 on June 30, 2014. Their architecture had a centrally mounted optical instrument, a three-axis star tracker, a fiber-optic gyro, and four control momentum gyros. They were both in the same orbit as the Pleiades 1A and 1B satellites at an altitude of 694 km, forming a constellation of 2 by 2 satellites 90 degrees apart from one another. Their optical instrument provided 1.5 m resolution panchromatic (450–745 nm) images and 1.5 m color (blue 450–525 nm; green 530–590 nm; red 625–695 nm) and near-infrared (760–890 nm) images with a 6-m resolution multispectral image. Each scene had a footprint of 60 km×60 km.

SPOT was developed due to increasing Landsat data prices in 1984, which made SPOT images an economically viable alternative. SPOT 1–3 carried two identical high-resolution visible (HRV) push-broom radiometers operating in either panchromatic or multispectral modes. The pan band covered 0.51–0.73 μm, while the multispectral had a green band (0.50–0.59 μm), a red band (0.61–0.68 μm), and a near-infrared band (0.79–0.89 μm). The HRV has a swath width of 60 km and a spatial resolution of 20 m. SPOT used the CNES tracking network supplemented by the ground station in Kiruna, Sweden, and the Norwegian ground station in Svalbard. An image collected by SPOT-6 is of Uluru (Ayers Rock) in the outback of Australia and is presented here in Fig. 9.3.

9.1.2 The Earth remote sensing series of satellites

It must be remembered that other countries could receive and process weather imagery from US satellites. As a result, most of the early efforts of the European Space Agency (ESA) were satellites aimed at studying processes beyond strictly meteorological applications. On July 17, 1991, ESA launched their first Earth observation satellite, known as ERS-1 (Fig. 9.4). Built by the Airbus Space Division, this satellite carried a suite of instruments led by a synthetic aperture radar (SAR) and a passive microwave radiometer (MWR). Orbiting at 785 km in a sun-synchronous polar orbit, the 5.3-GHz or C-band was VV polarized, and the SAR scanned a 100-km wide strip of the Earth's surface with a 30-m spatial resolution.

Satellite Remote Sensing. DOI: https://doi.org/10.1016/B978-0-443-22086-9.00001-0

FIGURE 9.1 SPOT-1 satellite.
Image credit courtesy: ESA.

FIGURE 9.2 Growth of the SPOT satellites up to 2024.
Image credit courtesy: ESA.

The "side-looking" instrument was oriented 23 degrees off nadir. Global coverage of imagery (sea height) could be generated as a 9-month average, while the wave global coverage could be accomplished in 1.5 months.

Another active instrument was the Active Microwave Instrument Scatterometer (AMI-SCAT). This instrument was a multimode radar which could operate in three modes: (1) Imaging mode where C-band measurements were used to collect imagery of the Earth, (2) Wave mode, where the reflectivity of the sea surface caused by surface waves could be used to identify the length and direction of wave systems, and (3) Wind scatterometer modes, which use three different antennas to measure wind speed and direction over the ocean. It was also a C-band (5.3 GHz) instrument with side-looking antennas. The scatterometer was to measure the sea surface wind vector and also large-scale soil moisture variations. On a 500-km swath, there were 3 looks per pixel (45, 90, and 135 degrees in azimuth). Global coverage was possible in 3 days. ERS-1 also carried the Along Track Scanning Radiometer (ATSR), which was designed, as mentioned earlier in this book, to measure very accurate sea surface temperatures (SSTs). The ERS-1 (Fig. 9.4) mission terminated on March 10, 2000 after completing 45,000 orbits and transmitting 1.5 million SAR images to Earth. A similar satellite, ERS-2, was launched in April 1995 to continue this observation series. The instantaneous field of view (IFOV) was 20 m along track. The MWR, also flown on ERS-1, was used for the atmospheric water vapor correction of the altimeter measurement.

FIGURE 9.3 SPOT-6 image of Uluru, Australia.
Image credit courtesy: SPOT/Imag.

FIGURE 9.4 ERS-1.
Image credit courtesy ESA.

ERS-1 also carried a single-frequency nadir-pointing radar altimeter operating in the Ku band. The optical sensor on ERS-1 was the ATSR built by Rutherford Labs in the United Kingdom. It was a four-channel infrared radiometer and microwave sounder for measuring temperatures at the sea surface and the tops of clouds. Finally, it also carried an MWR that was used to measure atmospheric water vapor, as well as provide a water vapor correction for the radar altimeter. To accurately determine its orbit, the satellite included the PRARE (Precision Range and Range-rate Equipment) and a laser reflector. The PRARE failed soon after launch, and the laser reflector was used to calibrate the altimeter to within 10 cm.

ERS-1 had two different mission phases using 3- and 35-day repeat cycles. During the geodetic mission, a 168-day repeat cycle was used to map the Earth's bathymetry and geoid over the seas using the radar altimeter. On 10 March, the ERS-1 attitude control system failed, terminating the mission. ERS-2 was launched on April 21, 1995. In addition to the instruments carried by ERS-1, it also had the GOME (Global Ozone Monitoring Experiment), and the ATSR was upgraded to ATSR-2, which added three visible bands specialized for ocean chlorophyll and terrestrial vegetation. ERS-2 was launched into the same orbit as ERS-1, allowing an intercalibration of the shared instruments. ERS-2 operated without gyroscopes from February 2001, resulting in some degradation of the data. The tape drive failed on June 22, 2003 limiting the mission to data collected only in line of sight of the ground stations. ERS-2 (Fig. 9.5) was basically a copy of ERS-1, which was launched on April 21, 1995 into the same orbit as ERS-1.

The satellite also experienced improvements in data handling and onboard storage.

In the EOS program, NASA had plans to build a polar orbiting satellite that was as big as a Greyhound bus. That never came to pass at NASA, but ESA decided to build one that large for their research satellite operations. As a result, their ENVISAT (ENVIronmental SATellite) became the largest Earth observation satellite ever built at a weight of 8 t. It was 26 m long, 10 m wide, and 5 m deep. It carried 10 different instruments with a goal of continuing the time series established by ERS1 and 2. Here, Fig. 9.6 clearly shows the size of ENVISAT as depicted in this artist rendition. ENVISAT cost 2.3 billion Euro (including 300 million Euro for 5 years of operation of the satellite. The Sentinel program has taken over the measurement responsibilities of ENVISAT. The satellite operated fully for 10 years and ended on April 8, 2012.

FIGURE 9.5 The ERS-2 satellite.
Image credit courtesy ESA.

FIGURE 9.6 The ENVISAT satellite is in orbit.
Imager credit courtesy ESA.

It was launched on March 1, 2002, and weighed 8,211 kg (including 319 kg of fuel) at the time of launch. It carried an Advanced ATSR (AATSR), an Advanced SAR (ASAR), the Doppler Orbitography and Radiopositioning Integrated by Satellite (DORIS), the Global Ozone Monitoring by the Occultation of Stars (GOMOS), the laser retro reflector (LRR), the medium resolution imaging spectrometer (MERIS), the Michelson Interferometer for passive atmospheric sounding (MIPAS), the MWR, a Radar Atimeter-2 (RA-2), a Scanning Imaging Absorption Spectrometer for Atmospheric Cartography (limb-scanning component, SCIAMACHY-limb) and the Scanning Imaging Absorption Spectrometer for Atmospheric Cartography nadir-scanning component, SCIAMACHY-nadir. Most of the instruments on ENVISAT were improved versions of those flown on ERS-1,2. The only instruments unique to ENVISAT are the pair of SCIAMACHY instruments designed to measure and monitor a series of important gases as well as aerosols. The main mission of ENVISAT was to be studies of atmospheric chemistry, ozone depletion, biological oceanography, ocean temperature and color, wind waves, hydrology, arboriculture, natural hazards, monitoring of maritime traffic, atmospheric pollution, cartography and snow and ice.

The AATSR was an improved version of the earlier ATSRs 1 and 2, with a primary goal of infrared measurements of SST. Like its predecessors, it was provided by the United Kingdom with assistance from Australia. The ASAR was built upon the experience gained with the Active Microwave Instrument (AMI) and was also supplied by the United Kingdom. The Global Ozone Monitoring by Occultation of Stars (GOMOS) had a main scientific goal to monitor ozone in the stratosphere and mesosphere with high accuracy and high vertical resolution. Together with SCIAMACHY and MIPAS, it contributed to the atmospheric sounding of ozone. GOMOS measured atmospheric parameters related to stratospheric ozone chemistry like NO_2, NO_3, H_2O, and aerosols, as well as ozone dynamics like temperature, air density, and turbulence. These were determined by measuring stellar spectra in the ultraviolet to the near infrared range while stars are being eclipsed by the Earth. It was regarded as a demonstrator for the stellar occultation technique and a precursor for other occultation instruments. Fig. 9.7 presents a GOMOS assimilated ozone field centered on the Antarctic.

The MERIS was a programmable, medium-spectral resolution, imaging spectrometer covering the solar reflective spectrum. Initially dedicated to ocean color measurements, the MERIS scope was expanded to include atmospheric and

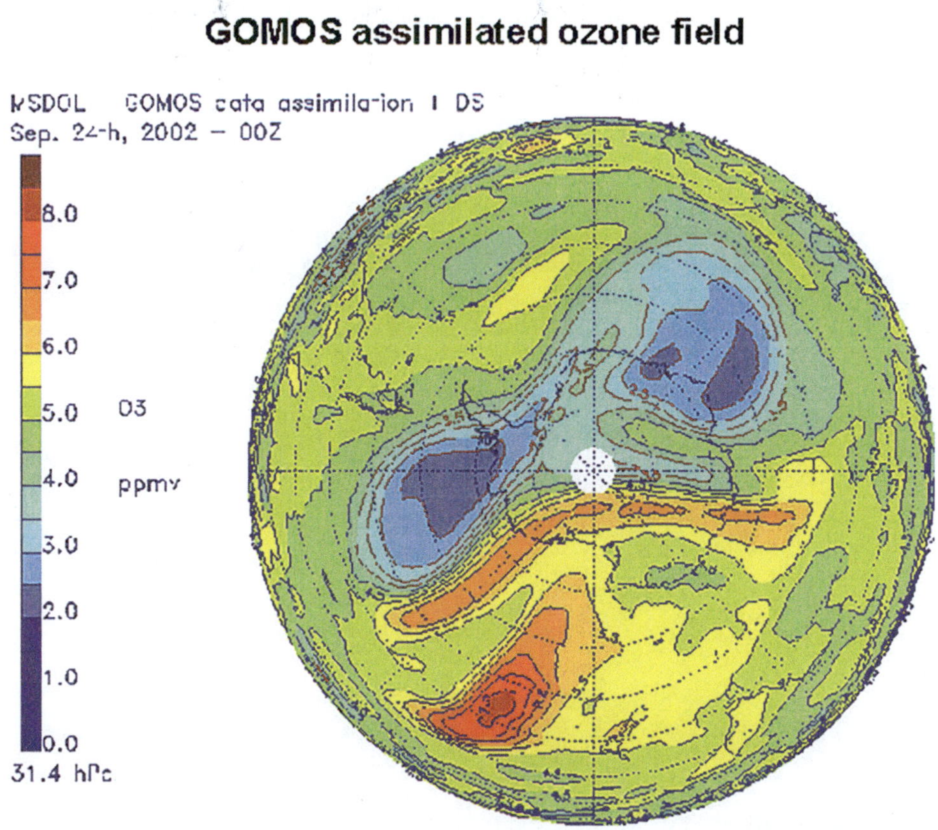

FIGURE 9.7 Assimilated GOMOS ozone field centered on the Antarctic. Map lines delineate study areas and do not necessarily depict accepted national boundaries.
Image credit courtesy ESA.

land surface measurements. MERIS had the very high spectral and radiometric resolution necessary for the study of ocean color. It had a dual spatial resolution to differentiate between global (lower resolution) and coastal oceans (higher resolution). An outstanding aspect of MERIS was the ability to program its spectral bands in terms of their width and position.

MIPAS was the Michelson Interferometer for Passive Atmospheric Sounding, which was a Fourier transform spectrometer to measure at high resolution gaseous emissions at the Earth's limb. It provided data to examine these emissions in the middle and upper atmosphere. It operated over a wavelength range from 4.15 to 14.6 μm. The measured gaseous emission features are known to play a major role in atmospheric chemistry. The high resolution and the low noise performance of the instrument made these precise measurements possible.

The MWR measured the integrated atmospheric water vapor column and cloud liquid water content, surface emissivity, and soil moisture, as well as the surface energy budget to support atmospheric studies, and ice concentrations. The water vapor and cloud liquid water measurements were used as corrections for the radar altimeter measurements. RA-2 was a dual-frequency, nadir-pointing, pulse-limited radar altimeter operating at 13.575 (Ku-band) and 3.2 GHz (S-band) using a single dish antenna. It was based on the heritage of the RA on ERS-1,2. It operated over all surfaces: ocean, land, and ice. RA-2 was the first high-latitude altimeter aimed to improve the precision of earlier altimeters over open oceans, which would also enhance measurements over ice sheets.

The Scanning Imaging Absorption spectrometer for Atmospheric Cartography (SCIAMACHY) was an entirely new instrument with the goal of measuring trace gases in the troposphere and in the stratosphere. It had a very large wavelength range between 240 and 2400 nm and still retained its high resolution. An example of the SCIAMACHY forecast of the Antarctic ozone hole is presented here in Fig. 9.8.

Some of the specific objectives of this instrument were

- Stratospheric ozone with focus on the Antarctic ozone hole and the midlatitudes.
- Tropospheric pollution arising from industrial activity and biomass burning.
- Troposphere–stratosphere exchange.
- Special events such as volcanic eruptions, solar proton events, and related regional and global phenomena.

The instruments had three different viewing geometries: nadir, limb, and sun/moon occultations. The latter yielded total column values as well as distribution profiles in the stratosphere and in some cases the troposphere for trace gases and aerosols.

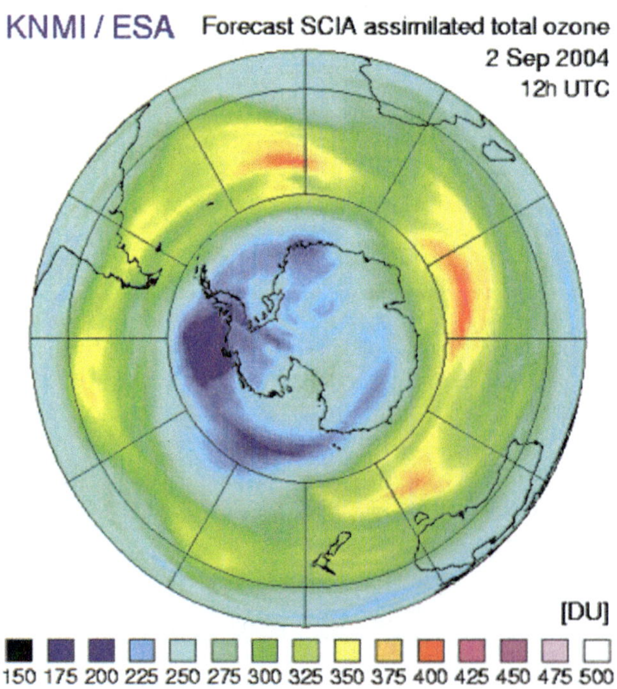

FIGURE 9.8 Ozone hole forecast by SCIAMACHY. Map lines delineate study areas and do not necessarily depict accepted national boundaries. Image credit courtesy ESA.

The satellite carried a couple of non-scientific instruments that improved the performance of the spacecraft. One was the LRR, which reflected high-energy infrared laser emission to improve the orbital calculations of the satellite required by the satellite altimeter. The other instrument used to improve the orbit was the DORIS. It measured the Doppler shift of radio frequency (RF) signals transmitted from ground beacons. This system was developed by the French Space Agency (CNES) to improve the orbits of all altimeter satellites.

9.1.3 Copernicus and the European Union Space Program

The Copernicus program was established in 2014 to follow on to previous space programs of the European Union (EU). Earlier efforts had made great achievements, but the results were limited to research and development (R&D) activities. Copernicus was designed to move from strictly R&D activities using a phased approach. Activities feeding into the formal Copernicus Program started in 1998 with the Global Monitoring for Environmental Security (GMES) program. In October of 2002, the nature and the scope of the "security" component of GMES were defined as addressing prevention of and response to crises related to natural and technological risk, humanitarian and international cooperations, monitoring of compliance with international treaties for conflict prevention, humanitarian and rescue tasks, peacekeeping tasks, and surveillance of EU borders.

In November 2011, the European Commission presented its plans for the future GMES for the time 2014 and onwards. In October 2014, ESA and the European Commission established a budget of for the new Copernicus Program covering the years 2014–2020. The Commission proposed to set up a GMES fund, similar to the European Development Fund, with financial contributions from all member states, based on their gross national income. In October 2014, a budget of 4.3 billion euros was set up to cover the construction of the satellites and their operation.

The series of satellites in the Copernicus Program was known as the Sentinel satellites (Fig. 9.8). Sentinel-1 satellites provide all-weather, day and night radar imaging of land and ocean surfaces. Both satellites were launched by Soyuz rockets. Sentinel-1A was successfully launched on April 3, 2014 and Sentinel-1B was launched on April 25, 2016. These satellites shared the same orbit. All Sentinel-1 spacecraft are designed to carry a C-band SAR (C-SAR) along with a Data Storage and Handling Assembly. The C-band SAR provides 1 dB radiometric accuracy with a central frequency of 5.405 GHz. This instrument has a ground spatial resolution of 5 m. The SAR has both horizontally and vertically polarized data steams, which are broadcast to Earth on two separate data streams.

The C-band SAR has four different operational modes:

- Strip Map Mode, with a 5-m resolution and an 80-km swath. The primary use of this mode is to monitor small islands and act as an emergency management for disaster events. Data products are offered in single (HH or VV) or double (HH +HV or VV+VH) polarizations.
- Interferometric Wide Swath Mode, with a 5×20 m^2 spatial resolution and a 250-km swath. This is the main operational mode over land and accomplishes interferometry through burst synchronization. Data products are available in single (HH or VV) or double (HH+HV or VV+VH) polarizations.
- Extra Wide Swath Mode with a 20×40 m^2 resolution and a 400-km swath width. This mode is used mainly to monitor wide coastal areas for ship traffic and environmental hazards such as oil spills or changes in sea ice.
- Wave Mode, which again has a 5×5 km^2 spatial resolution and a 20-km swath width. This is the main operational mode over the ocean, and data are only offered in a single (HH or VV) polarization.

Sentinel-1C was launched on December 5, 2020 and is due to be followed by Sentinel-1D in December of 2025 (Fig. 9.9).

There are a great many applications of Sentinel data. Here, in Fig. 9.10 is a composite of the land cover type for Ireland derived from Sentinel-1A data. One can clearly see the differences of the coastal areas from the interior. The mountains that define the island are marked by the blue areas. Farm and pastureland are colored green.

Sea ice motion is shown here in Fig. 9.11.

The image in Fig. 9.11 is a composite of three Sentinel-1 radar images collected in December, January, and February of 2024. The subject area is the Nares Strait off the northeastern tip of Ellesmere Island (gray shaded area in the lower left of the image). Each monthly image has been given a different color to show ice movement in Nares Strait as it empties into the Lincoln Sea in the Canadian Arctic. The difference between the red and yellow depicts how the ice cover has changed over the three months. The maximum extent of Arctic sea ice hit another record low this past winter (2024–25). The decrease in sea ice cover can be attributed to a very warm autumn and winter, made more severe by a series of extreme winter "heat waves" in the Arctic Ocean. Alert is the northernmost settlement in the world and is inhabited mostly by

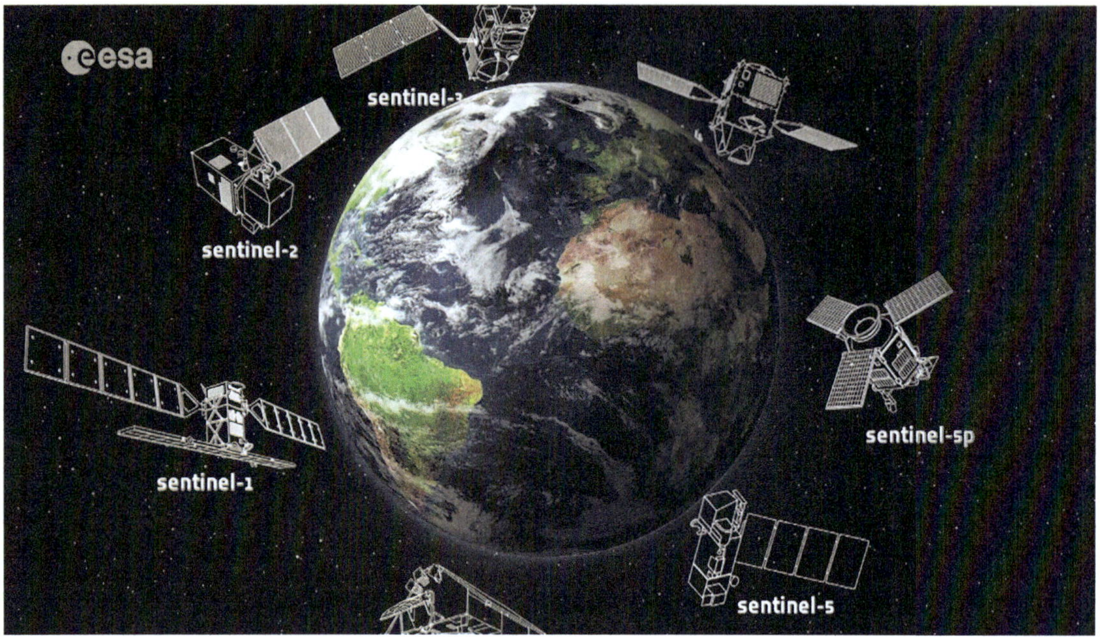

FIGURE 9.9 Copernicus satellite constellation.
Image credit courtesy ESA.

military and scientific personnel on rotation. Alert is 800 km from the North Pole. Instrument measurements made by teams from Alert will be used to validate and verify the satellite images.

One rather unique aspect of the Sentinel series SARs was their ability to collect interferometric images. The image in Fig. 9.12 shows the ground deformation after the eruption of Calbuco volcano in Chile. This volcano, on April 22–23, 2015, experienced significant explosive eruptions.

This was the first eruptive activity in Calbuco since 1972. This interferometric SAR image was created by Paul Lundgren of NASA's JPL in Pasadena, California, from Sentinel-1A C-SAR.

This image shows the relative deformation of the Earth's surface using SAR data acquired on April 14 and April 26, 2015 that span the eruption. The concentric pattern of fringes (color cycles) lies to the west of the Calbuco volcano and represents subsidence of the surface due to deflation of a buried magma reservoir. The black areas are water, with the north being Lago Lianquihue and the south being Puerto Montt Bay (Pacific Ocean).

9.1.4 EUMETSAT and European weather satellites

9.1.4.1 Meteosat First Generation

In November 1977, Europe, via the EUMETSAT (European Organization for the Exploration of Meteorological Satellites) office, launched their first geostationary weather satellite, Meteosat 1 (Fig. 9.13). This was the first of the Meteosat First Generation series of seven satellites that concluded in 2017 with the retirement of Meteosat-7.

Following the early GOES satellites, Meteosat 1 was a spinner carrying a Visible and Infrared Imager (MVIRI), which was a high-resolution radiometer and the primary payload. It also carried the Laser Synchronization from Stationary Orbit (LASSO), which was a series of laser reflectors. It was also equipped with a Space Environment Monitor-2 (SEM-2) and an electron spectrometer flown only on Meteosat 3 and onwards. The MVIRI operated in three spectral bands: the thermal infrared (10.5–12.5 μm), water vapor absorption band (5.7–7.1 μm), and the visible (0.5–0.9 μm). The instrument had a field of view (FOV) of 18 degrees and an IFOV of 0.065 mrad in the visible band, and 0.14 degrees in the water vapor and thermal infrared bands. The LASSO was capable of time standard synchronization to an accuracy of one nanosecond, which made it possible to measure the orbital distance of the satellite accurately to within 0.1 m. The SEM-2 covered the energy range from 42.9 to 300 keV with data provided in two temporal resolutions, high at 500–600 seconds averages, and low at 30 minutes averages.

The Meteosat buses were built by Alcatel Space, France, as the prime contractor along with a consortium of European contractors. The bus design life was five years, and the bus contained the sensor payload, the RF communication

FIGURE 9.10 Composite of Ireland's land cover derived from Sentinel-1A data. Map lines delineate study areas and do not necessarily depict accepted national boundaries.
Image credit courtesy ESA.

systems, and the Meteosat data collection system (DCS). RF communications were in the S band for transmission of data and for uplink of command and control. It also used L band for high-resolution digital image dissemination to meteorological users.

Meteosat started in March 1983 when an intergovernmental conference at ESA decided that they would construct and build a series of 3 geostationary satellites for operational weather forecasting. On May 24, 1983, 16 countries signed the EUMETSAT (European Organization for the Exploitation of Meteorological Satellites) convention, which came into force in June 1986. On January 1, 1987, the responsibility for the operation of the Meteosat satellites was transferred from ESA to EUMETSAT. EUMETSAT is the owner of the Meteosat series beginning with Meteosat 4. EUMETSAT was responsible for the MSG (Meteosat Second Generation [SG]) program, which saw its first launch in 2002 and operational services beginning in 2004. Meteosats were all operated over the 0.0 degree longitude (Fig. 9.14).

Like most other geostationary weather satellites, Meteosat served as a relay system for meteorological observations. This system was known as MOSAIC (Meteosat Operational System for Data Acquisition and InterChange). Since 1991, users have been able to easily merge satellite imagery with other conventional meteorological observations. A list of the early Meteosat satellites is given here in Table 9.1.

FIGURE 9.11 Ice movement in Alert, Canada. False color RGB image of scans for 3 different months from Sentinel-1A. Image credit courtesy ESA.

FIGURE 9.12 Sentinel-1A InSAR image showing ground deformation of Calbuco volcano, Chile. Image credit courtesy ESA.

Meteosat-5 lasted 16 years, while Meteosat-6 operated for 18 years, and Meteosat- 7 lasted for 20 years. The satellites were manufactured by a consortium called COSMOS with Aerospatiale's Cannes Mandelieu Space Center as prime, working with Matra, MBB, Selenia Spazio, and Marconi Company. The satellite spins at 100 rpm. Meteosat 1, 2, and 3 were developed and operated by the ESA. Meteosat 4, 5, and 6 were initially managed by ESA, which then turned them over to the newly formed EUMETSAT. Meteosat-7 was developed and operated by EUMETSAT.

The first image from Meteosat-1 is shown here in Fig. 9.15.

9.1.4.2 Meteosat Second Generation (MSG)

EUMETSAT wanted to improve on the data from the Meteosat satellites and thus began the development of MSG. While other countries like the United States were moving from spinning geostationary satellites to three-axis stabilized

FIGURE 9.13 Meteosat 1.
Image credit courtesy EUMETSAT.

FIGURE 9.14 Visible image from Meteosat 4.
Image credit courtesy EUMETSAT.

TABLE 9.1 List of early Meteosat satellites (EUMETSAT).

Satellite	Launch date	Mission end
Meteosat-1	November 23, 1977	Imager failed in November 1979; data collection ended in 1984
Meteosat-2	June 10, 1981	Moved to graveyard orbit in December 1991
Meteosat-3 (*Metosat-P2*)	June 15, 1988	Retired in 1995
Meteosat-4 (*MOP*-1)	April 19, 1989	Deactivated in November 1996
Meteosat-5 (*MOP*-2)	March 02, 1991	Decommissioned and placed into graveyard orbit in February 2007
Meteosat-6 (*MOP*-3)	November 20, 1993	Continued data transmission service until late 2010 or early 2011
Meteosat-7 (*MTP*)	September 03, 1997	Placed into graveyard orbit in April 2017

FIGURE 9.15 First image from Meteosat-1.
Image credit courtesy EUMETSAT.

spacecraft, EUMETSAT decided to improve on the spinner. Crucial to this effort was the Spinning Enhanced Visible and Infrared Imager (SEVIRI), which was a multispectral radiometer that is used for the monitoring of both weather and climate. It introduced full color images in geostationary orbit and had channels to measure atmospheric humidity and temperature, to compute wind fields from image displacements, could observe cloud type and amount, measure precipitation rate, ozone, the radiation budget, and vegetation. MSG 1 and 2 also carried the Geostationary Earth Radiation Budget (GERB) absolute radiometer. The main goal of GERB is to measure Earth's radiation budget in terms of climate change. It measures the top layer of the atmosphere (continuous temporal sampling) or, more specifically, measures the shortwave and emitted longwave portions of the EM spectrum. The second-generation satellites also carry a Geostationary Search and Rescue (GEOS&R) communications payload with a relay function for search and rescue operations.

The SEVIRI (Fig. 9.16) is the primary instrument of the MSG.

SEVIRI samples the full disk every 15 minutes. The instrument has 12 spectral channels (Table 9.2).

FIGURE 9.16 SEVIRI instrument.
Image credit courtesy EUMETSAT.

TABLE 9.2 SEVIRI channels (EUMETSAT).

Central wavelength	Spectral interval (99% encircled energy)	SNR or NEΔ*T* @ specified input
N/A (broad bandwidth channel)	0.6–0.9 μm	4.3 @ 1% albedo
0.635 Mm	0.56–0.71 μm	10.1 @ 1% albedo
0.81 μm	0.74–0.88 μm	7.28 @ 1% albedo
1.64 μm	1.50–1.78 μm	3 @ 1% albedo
3.92 μm	3.48–4.36 μm	0.35K @ 300K
6.25 μm	5.35–7.15 pm	0.75K @ 250K
7.35 μm	6.85–7.85 μm	0.75K @ 250K
8.70 μm	8.30–9.10 μm	0.28K @ 300K
9.66 μm	9.38–9.94 μm	1.50K@ 255K
10.8 μm	9.80–11.8 μm	0.25K @ 300K
12.0 μm	11.0–13.0 μm	0.37K @ 300K
13.4 μm	12.4–14.4 μm	1.80K@ 270K

SEVIRI, Spinning Enhanced Visible and Infrared Imager.

Spatial resolutions were 1 km for the HRV channel and 3 km for the thermal infrared and the three other visible channels. The focal plane contained 42 detectors and was cooled to 85/95K.

SEVIRI had a very unique sampling scheme (Fig. 9.17) which is made up of two different scans. The rapid scan (line scan) goes from east to west as the spacecraft spins at 100 rpm. The spin axis is perpendicular to the orbital plane and is nominally oriented in the north-south direction. The slow scan goes from south to north by means of a separate scanning mechanism that rotates the scan mirror in steps of 125.8 microradians. A total scan range of ±5.5 degrees (1527 scan lines) covers the 22 degrees Earth imaging range in the south-north direction, and 1,249 scan lines to cover the whole Earth in the baseline repeat cycle. The full Earth's disc image is collected in about 12 minutes, and the scan mirror goes back to its initial position where a flip-flop mechanism is used to insert the reference black body into the optical path for calibration. The black body is then moved out of the optical path after about 2 seconds, and the Earth observation is resumed, resulting in an overall repeat cycle of 15 minutes.

One unique aspect of SEVIRI is the multiple thermal infrared channels and the precise onboard calibration procedure. This made it interesting to compute thermal infrared observations of SST. The multiple infrared channels could also be used for improved cloud clearing. This resulted in improved estimates of SST from SEVIRI in spite of its poorer spatial resolution.

The sequence of MSG satellites is given here in Table 9.3.

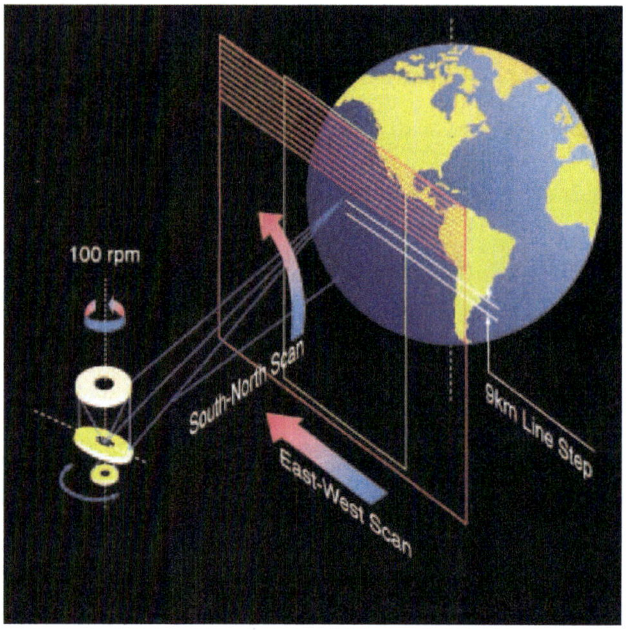

FIGURE 9.17 SEVIRI's scanning principle.
Image credit courtesy EUMETSAT.

TABLE 9.3 Meteosat Second Generation (MSG) Satellites.

Satellite	Launch date	Mission end
Meteosat-8 (MSG-1)	28-08-2002 22:45 UTC	Retired July 1, 2022
Meteosat-9 (MSG-2)	22-12-2005 22:33 UTC	Availability lifetime is until 2025
Meteosat-10 (MSG-3)	05-07-2012 21:36 UTC	Availability lifetime is until 2030
Meteosat-11 (MSG-4)	15-07-2015 21:05 UTC	Availability lifetime is until 2033

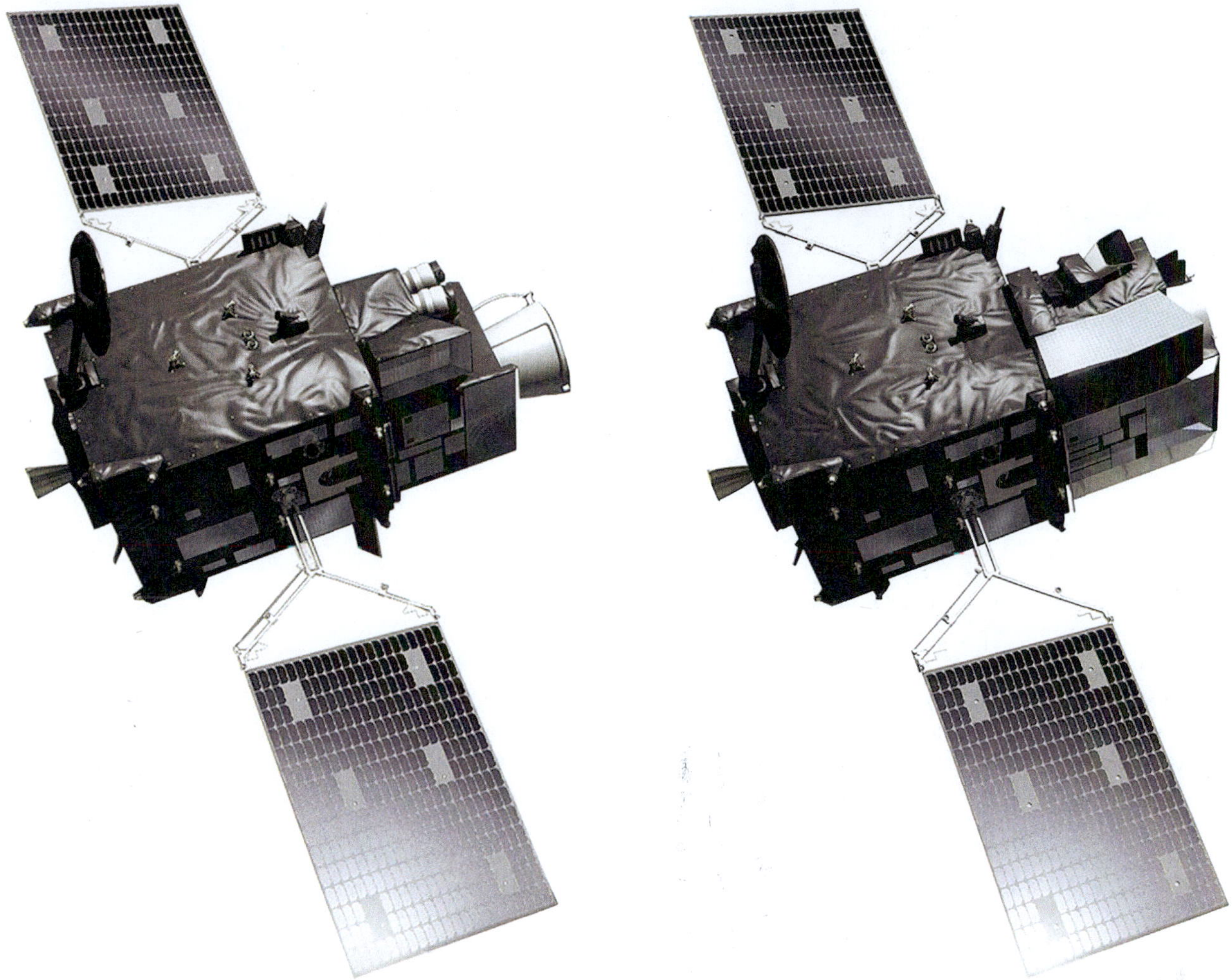

FIGURE 9.18 Meteosat third generation satellites: sounder (left) and imager (right).
Image credit courtesy EUMETSAT

Notice that EUMETSAT followed the NOAA practice of renaming their satellites once they were on orbit. Thus MSG-1 became Meteosat-8 when it entered operation. Most of the MSG satellites are expected to last well past 2030. MSG-1 lasted almost 10 years, and the others are expected to last 20 years or more.

9.1.4.3 Meteosat Third Generation

Like other meteorological satellite organizations around the world, Europe decided the next round of geostationary weather satellites would be three-axis stabilized, and thus Meteosat Third Generation (MTG) satellites would be three-axis stabilized. In addition, the decision was taken to follow the example of MetOp and divide the sensing capabilities into two separate satellites (Fig. 9.18). The MTG-I (for imager) will carry the Flexible Combined Imager (FCI) and Lightning Imager (LI), while the MTG-S satellites will be equipped with an infrared sounder (IRS) and an Ultraviolet Visible Near-infrared (UVN) sounder. There will be a total of six satellites in this series. MTG is again a collaboration between ESA and EUMETSAT, with ESA being responsible for the development and implementation of the space segment technologies and the first pair of MTG satellites. ESA will also procure all recurring satellites, which will be operated by EUMETSAT.

The FCI has 16 channels in the visible and infrared portions of the spectrum, of which eight are in the spectral domain between 0.4 and 2.2 μm. Durand et al. (2015) give a complete description of the FCI, its operation, and applications. There will be six MTG satellites, four imaging satellites (MTG-I) and two sounding satellites (MTG-S). The FCI is designed to provide full disc images of the Earth every 10–2.5 minutes in its 16 spectral channels between 0.44 and 13.3 μm, with a

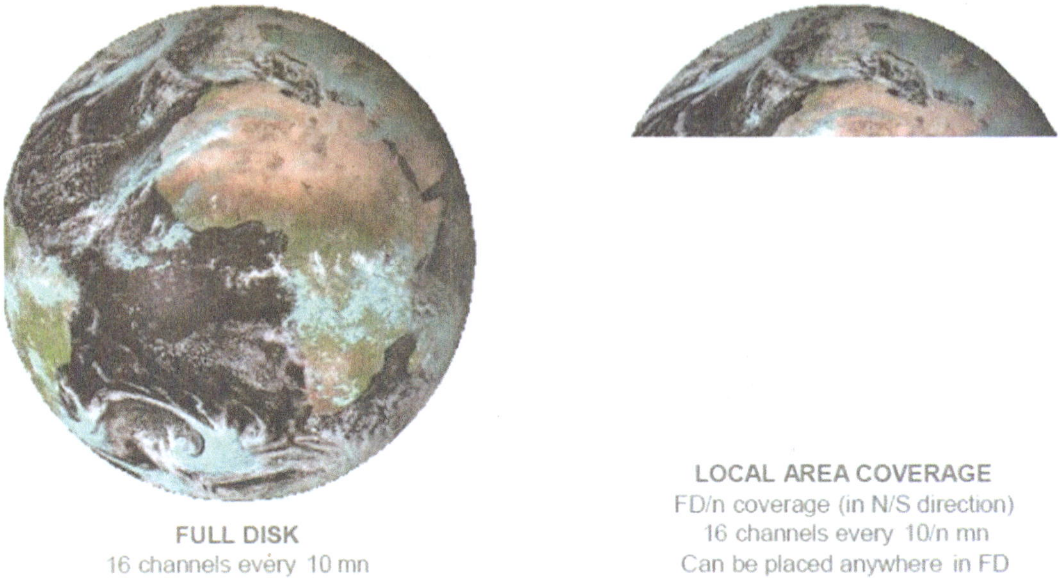

FIGURE 9.19 Meteosat Flexible Combined Imager full disk and Local Area coverage. Image credit courtesy ESA.

ground resolution ranging from 0.5 to 2 km. Onboard sensor calibration is carried out with a Metallic Neutral Density (MND) filter for the VIS/NIR channels and a thermally controlled black body for the thermal infrared channels.

The FCI has the ability to image a full disc every 10 minutes, but it can also collect a more limited sample known as the Local Area Coverage (LAC) scenario. LAC data can be repeated every 2.5 minutes, making it possible to focus on target cloud formations that may be associated with severe weather events. This is demonstrated in Fig. 9.19.

This capability is similar to the Advanced Baseline Imager (ABI) on the GOES-R series of NOAA satellites that can sample a full disc every 5–15 minutes but also a 1000 × 1000 km^2 box with a temporal resolution of 30 seconds and a spatial resolution of 0.5–2 km. This rapid sampling is needed to detect cloud overshoots that indicate the presence of a tornado. The FCI channels are all listed in Table 9.4.

The FCI optical telescope includes a double-gimbaled scan flat mirror that images the full Earth disc. At nadir, the average incident angle of the optical beam on the mirror is 30 degrees, which resulted from a need to minimize the polarization ratio induced by the reflective coatings at high incidence angle. In addition, this allowed tilting of the Optical Bench Assembly on which the telescope is mounted. This yielded a volume that accommodated the cryostat integration below the optical bench. After being collected by the scan mirror, the light coming from the Earth reaches a three-mirror anastigmat telescope. Due to the high temperature of the entrance cavity reached under solar illumination, the telescope aperture stop is physically located at exit pupil level, where the environment is thermally regulated and stable.

The thermal design of the FCI was critically important considering all the various temperature requirements and environmental conditions. Thus the FCI was split into five thermal enclosures:

- The warm optics enclosure, including all the elements, is fine-tuned to 20°C in the operation mode.
- The entrance cavity (including the E/W scan component, M1 and M2 mirrors), which was subjected to solar intrusions around midnight.
- The electronics enclosure is dedicated to the dissipative video electronics.
- The scan enclosure is dedicated to the N/S actuators and encoder.
- The cold assembly, including IR and NIR focal planes at cryogenic temperature.

Each area had its own thermal control and is thermally isolated from the other components. With the exception of the active coolers of the cold detection assembly, the thermal control is passive and uses conventional hardware. Most of the radiators are located on the dark side of the spacecraft, and a sunshade is implemented around these radiators to protect from the sun when the satellite is flipped twice a year at equinox.

There are 16 spectral channels that are split into five groups with spectral separation carried out by dichroic beam-splitters. The selected polarization-oriented design leads to the three-dimensional arrangement offering an optimized

TABLE 9.4 Flexible Combined Imager (FCI) spectral channels with their ground sampling distance.

Spectral channel	Central wavelength, λ_0 (µm)	Spectral width, $\Delta\lambda_0$ (µm)	On-ground spatial sampling distance (km)
VIS 0.4	0.444	0.060	1.0
VIS 0.5	0.510	0.040	1.0
VIS 0.6	0.640	0.050	1.0/0.5
VIS 0.8	0.865	0.050	1.0
VIS 0.9	0.914	0.020	1.0
NIR 1.3	1.380	0.030	1.0
NIR 1.6	1.610	0.050	1.0
NIR 2.2	2.250	0.050	1.0/0.5
IR1 3.8	3.800	0.400	2.0/1.0
IR1 6.3	6.300	1.000	2.0
IR1 7.3	7.350	0.500	2.0
1R2 8.7	8.700	0.400	2.0
IR2 9.7	9.660	0.300	2.0
IR3 10.5	10.500	0.700	2.0/1.0
IR3 12.3	12.300	0.500	2.0
IR3 13.3	13.300	0.600	2.0

polarization balancing between in-plane and out-of-plane polarization states. The VIS focal plane corresponds to the five visible channels (0.4–0.9 µm) with a dedicated detector assembly located at the entrance telescope focal plane. This direct implementation enables an increased transmission using only silver-coating mirrors and a suitable image quality between bands due to an achromatic optical path. The common NIR/IR focal plane combines 11 spectral channels from 1.3 to 13.3 µm in four groups with dedicated band-pass filtered detector arrays. This favors the interchannel coregistration performance between specified groups of channels. The cold optics operate at a temperature of 80K.

The FCI is considered an absolute radiometer and must therefore be equipped with calibration references. The calibrated accuracy was specified to be below 5% for the VNIR channels and below 0.7 K for the IR channels. In addition, the FCI must maintain its radiometric stability over a day to be below 0.1% for the VNIR channels and less than 0.1K for the IR channels. The radiometric stability over the entire instrument lifetime is supposed to be 2% for the VNIR channels and 0.3K for the IR channels. Unlike SEVIRI, which bases its radiometric accuracy on a vicarious calibration process, the FCI VNIR channels calibration is based on an end-to-end full entrance aperture solar calibration. It is performed by inserting a MND filter at the telescope exit pupil. For the IR channels, an approach similar to SEVIRI was followed, and a well-characterized radiance source (black body) is introduced to calibrate the gain fluctuations that are due to instrument aging and thermal fluctuations. Unlike SEVIRI, the black body (Fig. 9.20) is cross-calibrated with a large black body before launch, and only FCI drifts contribute to the calibration budget.

The black coating is a highly specular residual reflective coating from Acktar, allowing for volume and mass reduction in the black body. The FCI uses two black bodies placed in front of its entrance. The one black body will be allowed to fluctuate with the ambient temperature, while the other will be set at a fixed cold temperature to simulate deep space. An in-flight calibration phase, which consists in acquiring every 10 minutes two images: one on the internal black body with a floating (but well-known) temperature (between 15°C and 22°C) and the other when the instrument entrance is illuminated by deep space (dark conditions). These will be taken together to update the calibration coefficients. As for VNIR, the deep space acquisition is done at the end of each swath, which is then used to update the calibration coefficients.

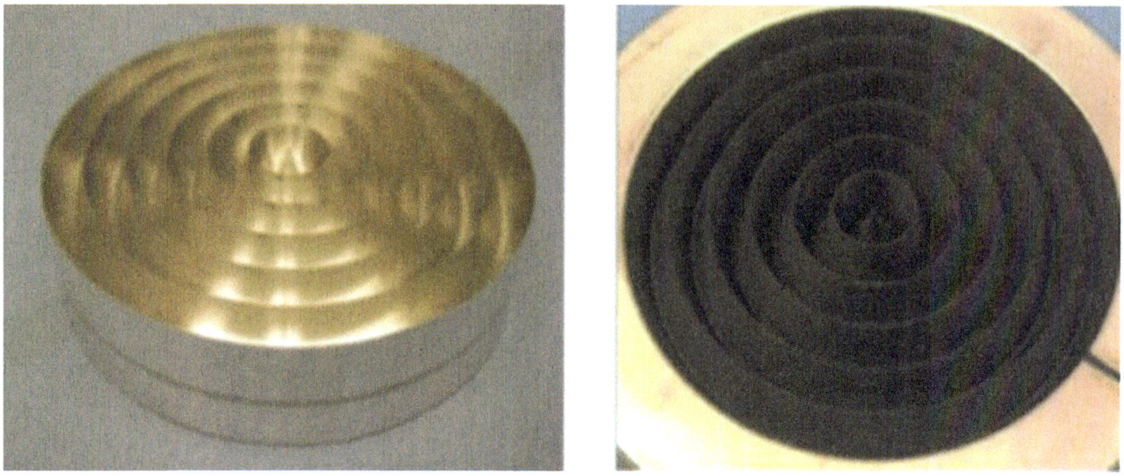

FIGURE 9.20 Emissive plate of the black body before (left) and after (right) black coating.
Image credit courtesy ESA.

Another new sensor on MTG was the Lighting Imager (LI), which measures total lightning (cloud-to-cloud lightning [IC] and cloud-to-ground lightning [CG]). The main objectives of the LI are to detect, monitor, track, and extrapolate in time:

- Development of active convective areas and storm lifecycle
- Lightning climatology
- Chemistry (NO_x production)

The object is to deliver information on location and strength of lightning flashes with a timeliness of 30 seconds. Lightning is not recognized by its bright radiance alone, but by its transient short pulse character. Variable adapting thresholds are used for each pixel to take into account changes in the background radiance.

The LI is designed as a four-camera unit (Fig. 9.21) with each camera having a 1,170 X 1,000 pixel resolution. The detectors are CMOS back-thinned, backside illuminated with integrated analog-to-digital converters.

Following the terminology developed earlier for the lightning mapping instrument on TRMM (the LIS), the data products are broken into the following:

Events: What the instrument measures directly, the pixel is triggered in the detector grid.

Group: A collection of neighboring triggered events in the same integration period (1 ms), representing a lightning stroke in nature.

Flashes: A collection of groups in a temporal and spatial vicinity representing a "geophysical" flash.

There were two kinds of products from the LI, first, the LI initial processing, which were point data in the LI grid space. Also in the initial processing were groups (strokes) and flashes with geographical coordinates. Second came the accumulated products, which were product densities shown in the fixed MTG-FCI imager grid (same grid as for the FCI IR channels with their 2 km resolution. This enabled the merging of LI data with FCI data in such a way as to be most useful to the users. The system to merge individual events into groups and flashes is presented here in Fig. 9.22.

EUMETSAT users are mostly interested in:

1. Flashes
2. Understanding of "what kind of flash" they are getting (strength, duration, extent).

The real-time users would be best served with the flashes (groups) and a supporting accumulated product coming from events. The periodicity of the product should be short enough that it fits well for any further applications, allowing the stacking of data (30 seconds).

The accumulation of events is shown for the time $t = 30$ seconds in Fig. 9.23 which shows the diversity of events in different "flashes."

An example of how this data is applied is presented here in Fig. 9.24 with comparisons shown from more traditional measurements of lightning. This comparison was produced by the Belgium Royal Meteorological Institute (RMI). It compares the LI retrievals with ground measurements made by electrodes in the ground to observe lightning strikes. In addition, RMI provides the operational weather radar composite, which reveals the cloud structure associated with the lightning strikes. Note the higher spatial resolution of the LI measurements compared to the somewhat blocky pattern of

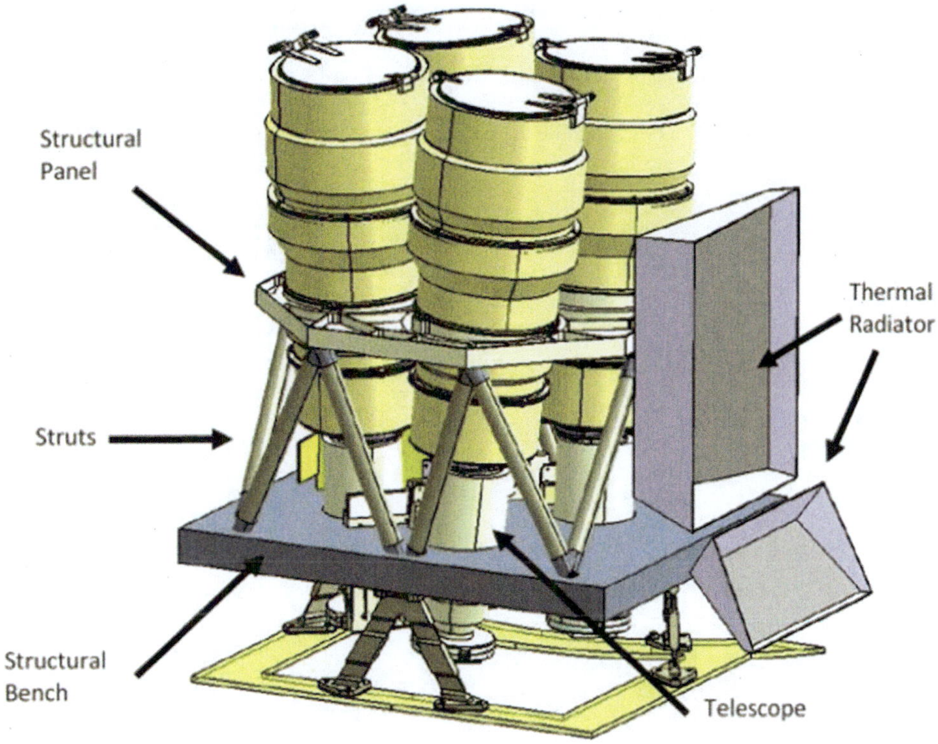

FIGURE 9.21 The Lightning Imager on Meteosat Third Generation.
Image credit courtesy EUMETSAT.

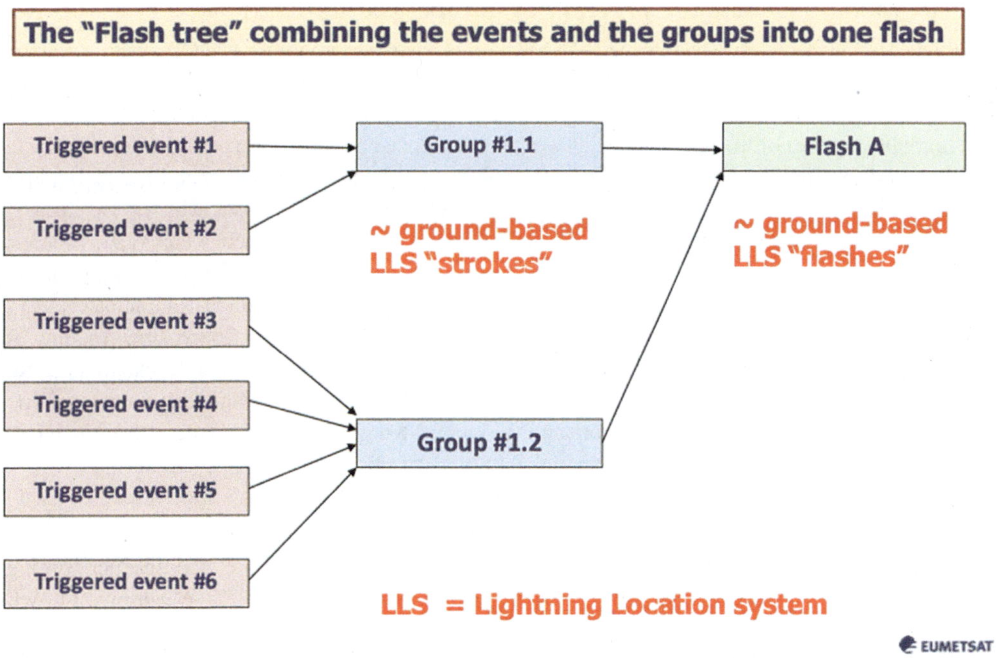

FIGURE 9.22 Schematic showing how individual events are combined into strokes and flashes.
Image credit courtesy EUMETSAT.

the ground-based network. A large number of these comparisons were carried out to verify the success of the LI in clearly retrieving lightning strikes.

A comparison of MSG and MTG capabilities is given here in Table 9.5.

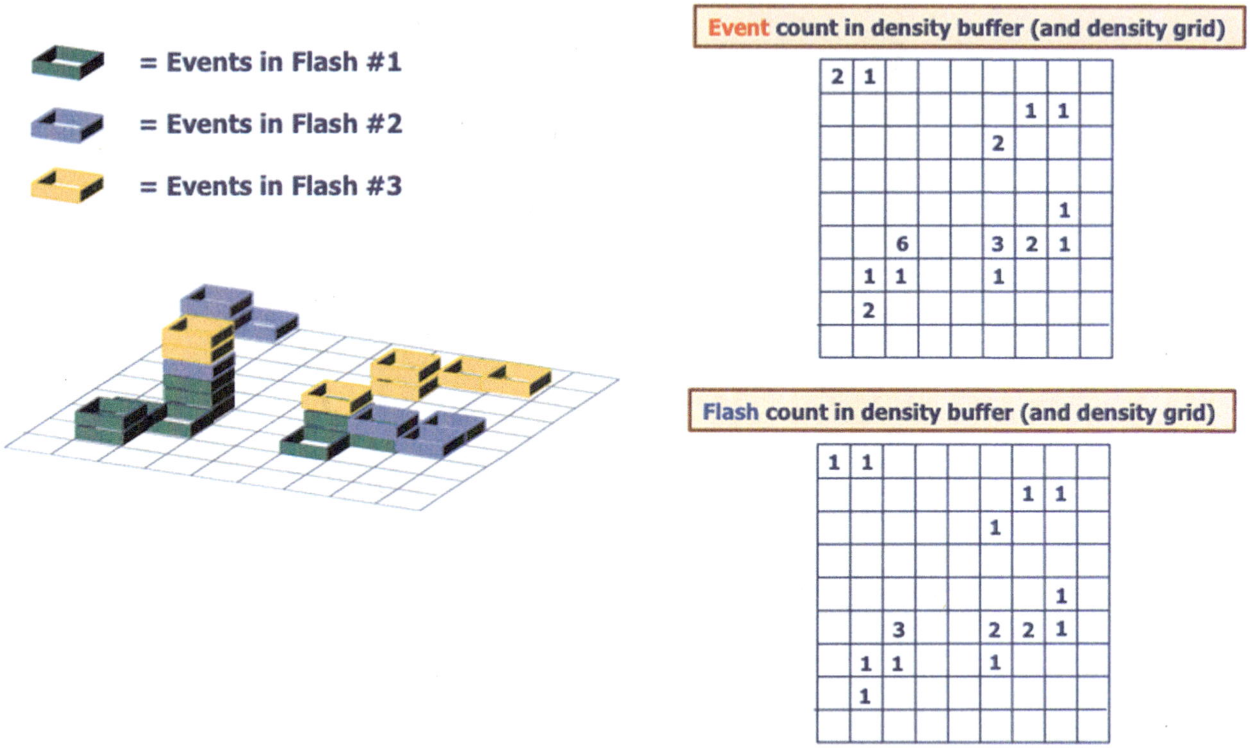

FIGURE 9.23 Accumulation status at t=30 s.
Image credit courtesy EUMETSAT.

9.1.4.4 Meteorological Operational Satellite Program of Europe (MetOp)

The European polar orbiting satellite program Meteorological Operational satellite program (MetOp) was a collaboration between the ESA and the European Organization for the Exploitation of Meteorological Satellites (EUMETSAT) with the main aim to provide polar orbiting satellite services for weather prediction and climate monitoring. Among other observables, MetOp provides temperature and humidity profiles of the atmosphere, as well as wind speed and direction over the ocean to be used by numerical weather prediction models. The first satellite MetOp-A (Fig. 9.25) was launched on October 19, 2006 and operated until November 2021. It carried the Advanced Microwave Sounding Units (AMSU-A1/AMSU-A2), which were carried by the NOAA polar orbiter spacecraft. It provides temperature profiles from 3 mbar (45 km) to the surface, water vapor profiles, snow and ice coverage, cloud liquid water, and rain rate. MetOp-A also carried the High-resolution Infrared Radiation Sounder, which was not carried on MetOp-C. The primary imager was the Advanced Very High Resolution Radiometer (AVHRR/3) provided by NOAA. An advanced DCS (Argos A-DCS) was also carried by MetOp-A. The Space Environment Monitor (SEM-2) was also provided by NOAA. The Search and Rescue Processor (SARP-3) and the Search and Rescue Repeater (SARR) were also part of the MetOp-A payload but not carried by MetOp-C. The Microwave Humidity Sounder (MHS) was very similar to AMSU-B but was built by Astrium EU in Portsmouth, United Kingdom, under contract to EUMETSAT.

There were four instruments unique to MetOp-A. An Infrared Atmospheric Sounding Interferometer (IASI) provided much the same kind of data as did AIRS for the NASA Aqua satellite. It is a Fourier transform spectrometer based on the Michelson interferometer. In total, there were two IASI instruments deployed, one on MetOp-A and one on MetOp-C. IASI is a nadir-viewing instrument recording infrared emissions from 645 to 2,760 cm^{-1} with a spectral resolution of 0.25 cm^{-1}. While the focus of IASI is on atmospheric temperature and water vapor profiles, it is also capable of retrieving information on various trace gases. The IASI bands are listed here in Table 9.6.

As an interferometer, IASI in reality had 8,461 channels with one embedded imaging channel. IASI was contributed to MetOp by the French Space Agency (CNES). Its mission objectives were to measure atmospheric temperature profiles, the height of the planetary boundary layer, the height of the tropopause, the integrated water vapor, specific humidity, and the temperature of the tropopause.

The Advanced Scatterometer (ASCAT) was also introduced on MetOp-A. It measured the ocean surface wind vector using fan beam antennas. It was an evolution of the AMI-SCAT flown on ERS-1 and ERS-2. It was a C-band (5.255 GHz) scatterometer with a 1-MHz bandwidth and looked both left and right with three antennas on each side. This led to two

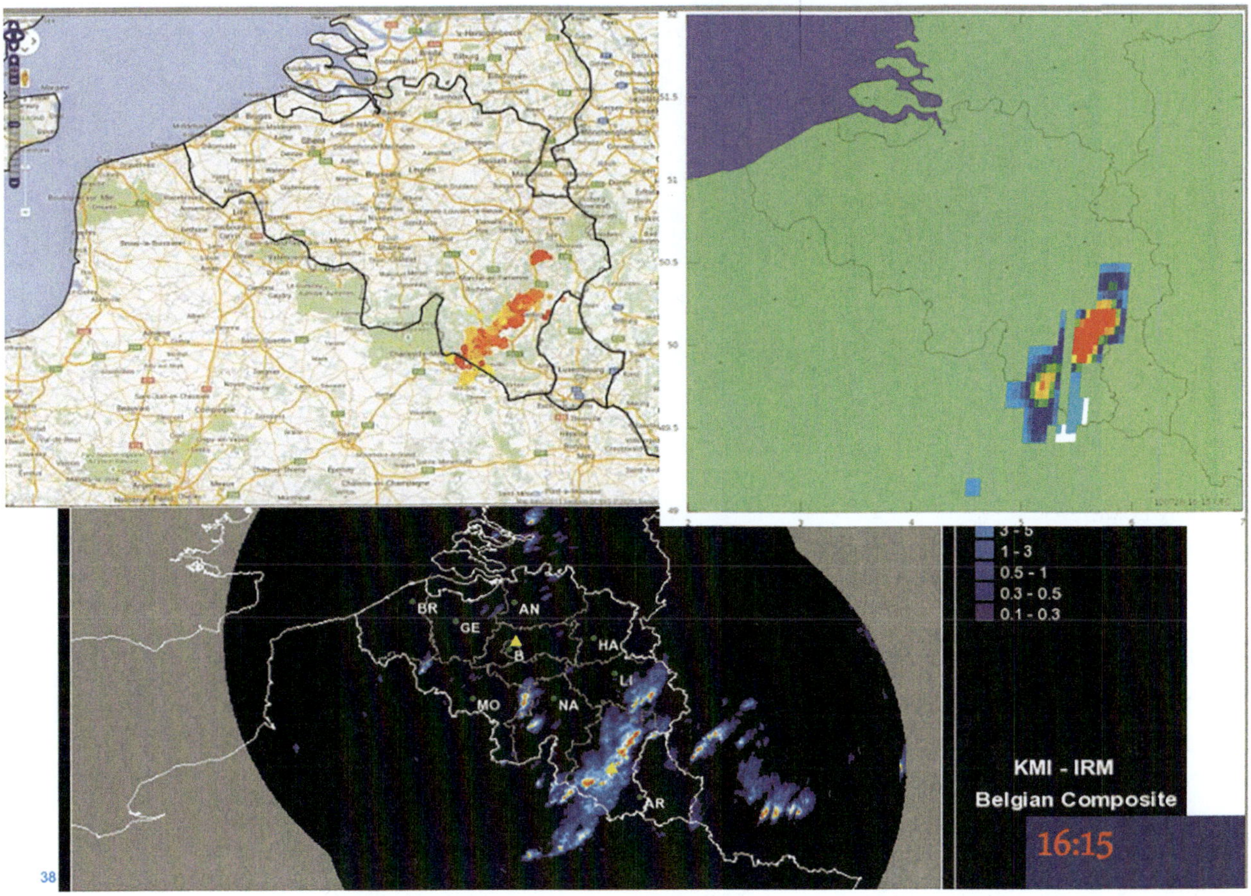

FIGURE 9.24 Comparisons of Lightning retrievals. Top left is the LI accumulated product, top right is the Royal Meteorological Institute ground-based data sensors in Belgium and France, and the bottom panel is a weather radar composite. Map lines delineate study areas and do not necessarily depict accepted national boundaries.
Image credit courtesy EUMETSAT.

TABLE 9.5 Comparison between Meteosat Second Generation (MSG) and Meteosat Third Generation (MTG) (EUMETSAT).

MSG	MTG
Operated by EUMETSAT HQ in Darmstadt	Operated by EUMETSAT HQ in Darmstadt
Geostationary orbit at 36,000 km	Geostationary orbit at 36,000 km
Two-satellite constellation, both imagers	Three-satellite constellation with two imagers and an operational infrared sounder
Spinning satellite views Earth 5% of rotation	3-Axis stabilized views Earth 100% of time
Images of Europe every 5 min, Images of Europe and Africa every 15 min	Images of Europe every 2.5 min. Images of Europe and Africa every 10 min
Spatial resolution 1 to 3 km rel to inst	Spatial resolution 0.5–2 km rel to inst
Constellation delivers 2 Mbps every day	Constellation delivers 110 Mbps every day
No lighting mapper	Lightning imager for detecting storms
No infrared sounder	Infrared sounder 3D profiles every 30 min
No UV, visible or near infrared sounder	UV, visible+near infrared sounder 60 min

FIGURE 9.25 Artist conception of MetOp-A in orbit.
Image credit courtesy EUMETSAT.

TABLE 9.6 Infrared Atmospheric Sounding Interferometer (IASI) bands and applications (EUMETSAT).

Band	Name of region	Spectral region (cm^{-1})	Absorption band	Usage
B1	R1	650–770	CO_2	Temperature profile
B1	R2	790–980	Atmospheric window	Surface and Cloud properties
B1	R3	1000–1070	O_3	O_3 sounding
B1	R4	1080–1150	Atmospheric window	Surface and Cloud properties
B2	R5	1210–1650	H_2O	Humidity profiles; CH_4 and N_2O properties
B3	R6	2100–2150	CO	CO column amount
B3	R7	2150–2250	N_2O and CO_2	Temperature profile; N_2O column amount
B3	R8	2350–2420	CO_2	Temperature profile
B3	R9	2420–2700	Atmospheric window	Surface and cloud properties
B3	R10	2700–2760	CH_4	CH_4 column amount

550 km swaths separated by a 700-km gap. Each pixel had three looks (45, 90, and 135 degrees azimuth). The other two unique instruments were the SEM/Total Energy Detector (SEM/TED) and the SEM/MEPED or Medium Energy Proton Detector. The SEM/TED measured proton and electron fluxes at the platform level with eight electrostatic analyzers and pulse height discriminators operating in two ranges: 0.05–1 and 1–20 keV. The instrument was provided by NOAA. The SEM/MEPED also operated at the platform level to provide directional (nadir and 90 degrees from nadir) and omnidirectional measurements. Directional protons are measured in six bands in the energy range 0.003–6.9 MeV, electrons in three bands between 30 and 300 keV. Omnidirectional proton energy is measured in the ranges >16 MeV, >35MeVm, >70 MeV, and >140 MeV.

The EUMETSAT Polar System (EPS) was planned as a series of three satellites to be launched sequentially over a period of 14 years starting in 2006 with MetOp-A (later 2010, 2014). MetOp is viewed to be complementary to the NOAA POES system. In the cooperation between NOAA and EUMETSAT (agreement signed in November of 1998) EUMETSAT is committed to taking over the morning polar orbit from NOAA, while NOAA continues to

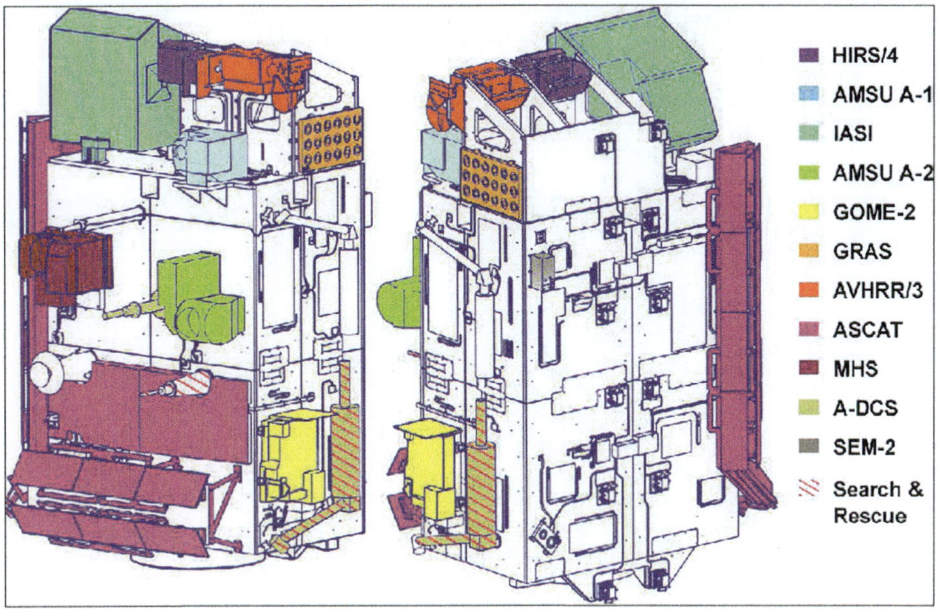

FIGURE 9.26 Schematic overview of the MetOp payload.
(Image credit courtesy ESA).

provide the afternoon satellites. Both services were coordinated and integrated with the full exchange of data, instruments, and operational services. Many of the instruments flown on MetOp were provided by NOAA, giving a continuing record, in particular, of the AVHRR imager. The primary EPS objective was to continue the continuity and availability for operational purposes, polar satellite observation in the morning orbit. In addition, EPS planned to provide enhanced monitoring capabilities by deploying new sensors to provide increased measurement capabilities. The EPS system is an end-to-end system comprising both a space segment and a full ground segment to distribute the data to interested users.

Like NOAA, MetOp had a naming convention that was the reverse of NOAA's, and when the satellite was built, it was designated by a number and later transitioned to a letter. Thus the first satellite built was MetOp-1, which then became MetOp-A. MetOp-2 transitioned to MetOp-B, and MetOp-3 to MetOp-C. As with NOAA polar orbiter data, all MetOp data were downlinked on HRPT/LRPT stations. The configuration of the payload on MetOp is presented here in Fig. 9.26.

The collaboration between ESA and EUMETSAT is planning for MetOpSG and the first satellite has already been built (Fig. 9.27) is now launched and operating. MetOp-SG is planned as a series of six meteorological satellites flying in pairs, MetOp-SG-A and MetOp-SG-B, each carrying complementary sensor payloads. They will continue to fulfill the mission of measurements in the afternoon polar orbit, but will greatly expand the measurement capabilities of the program. They will all have an orbital altitude of 824 km at an inclination of 98.7 degrees and a repeat cycle of 29 days.

The primary instruments on MetOp-SG-A1 are

3MI. Multiviewing, Multichannel, multipolarization Imager

IASI-NG. Infrared atmospheric sounder interferometer—next generation (CNES)

METimage—Multispectral imager (DLR)

Microwave Imaging Radiometer (MWI, ESA)

MWS Microwave Sounder (ESA)

RO Radio Occultation Atmospheric Sounder (RO, ESA)

UVNS (Sentinel-5) Ultraviolet Visible Near-infrared Shortwave-infrared Spectrometer

Space Environment Monitor (NOAA)

Low Light Imager (NOAA)

Radiation Energy Radiometer (NOAA)

Ice Cloud Imager (ICI, ESA)

Scatterometer (SCA, ESA)

Argos-4 (Data Collection Service (NOAA/CNES)

Search and Rescue (COSPAS-SARSAT)

FIGURE 9.27 MetOp-SG-A1.
Image credit courtesy EUMETSAT.

The MetOp-SG program is once again a collaboration of ESA and EUMETSAT. ESA develops the prototype MetOp-SG instrument and procures these instruments on behalf of EUMETSAT, which is responsible for the overall program, operates the satellites, and funds the recurrent satellites. This is called EUMETSAT SG or EPS-SG. On October 16, 2014, contracts were signed to build three pairs of MetOp-SG satellites with Airbus Defense and Space of France taking the lead position as contractor. On 15 December, Europe and the United States signed off on a plan of cooperation of EUMETSAT and NOAA in a Joint Polar System (JPS). This continued the interchange of instruments and data but shifted the afternoon orbit to EUMETSAT and the morning orbit to NOAA. This agreement extends these observations into the mid-2040s timeframe.

The MetOp-SG plan is to have two satellites, A and B, on the same orbit (Fig. 9.28).

The A and B satellites carry somewhat different payloads, which is apparent from a diagram of the two satellites (Fig. 9.29).

The A satellites carry METimage, IASI-NG, MWS, Sentinel-5, 3MI and the RO. The B satellites were instrumented with the SCA, the MWI, ICI, RO, and Argos-4. The only overlapping instrument is the RO, which is designed to use GPS signals to profile the atmosphere. The B satellites had a greater need for stability in orientation and carried 6 reaction wheels, whereas the A satellites had only 5. Taken together, data from the A and B satellites provide a great deal of information on the Earth and its atmosphere. The satellites are expected to exceed their nominal 8.5-year lifetime, and the overall program has a 21-year lifetime.

9.2 Japanese Weather Satellites

9.2.1 The Japanese Geostationary Weather Satellites

In 1977, Japan launched their first geostationary weather satellite called GMS-1 (Geostationary Meteorological Satellite) or Himawari-1. It was built by Hughes Space and Communications Company (HSC) for Japan. Nippon Electric Corporation (NEC) was the prime contractor for Japan's National Space Development Agency (NASDA), and HSC was the major subcontractor. The satellites were built in El Segundo, CA. The principal instrument was the VISSR instrument developed by Hughes Santa Barbara Research Center for the US GOES series. In the GMS, the satellite spun at 100 rpm, while the antennas did not spin and remained pointed Earthward. GMS also carried a space environment monitor

FIGURE 9.28 Artist's rendition of the two MetOp satellites in orbit.
(Image credit courtesy ESA).

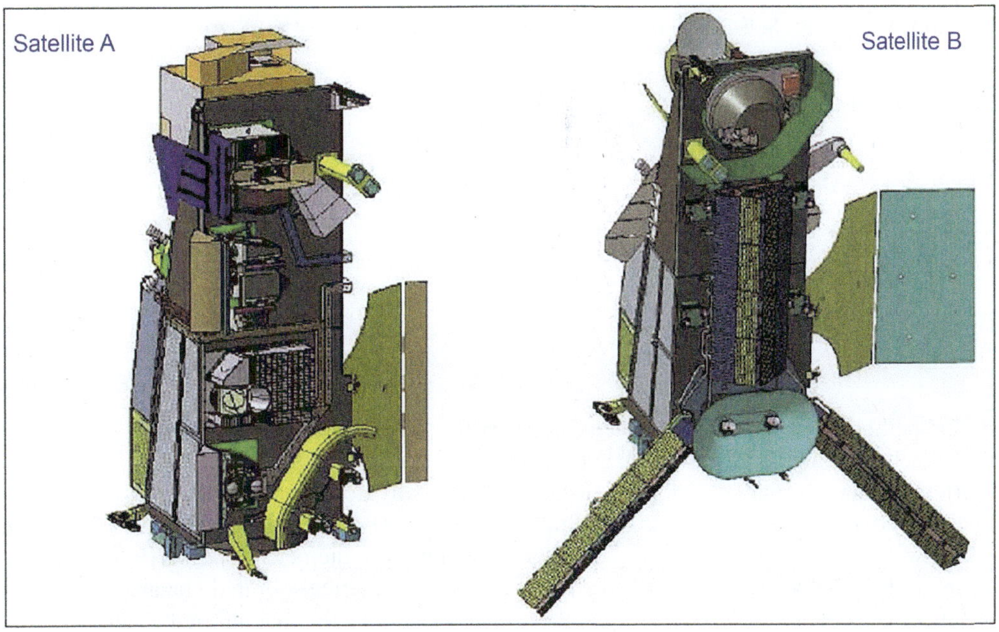

FIGURE 9.29 Schematic of the MetOp A and B satellites.
(Image credit courtesy ESA).

developed by NEC, which was used to investigate energetic particle activity emanated by the sun and to assess the effect of extreme solar activity on terrestrial communications. These satellites also provided a data collection and dissemination system. This system was very important for collecting data from remote locations. For example, information collected from remote buoys can provide early warning of impending storm surges or Tsunamis.

In February 1987, Japan launched its first Earth observation satellite MOS-1 (Marine Observation Satellite-1) designed to study the ocean (Fig. 9.30). It was operated under the auspices of NASDA. It was in a 900 km polar orbit. It carried a Multi-Spectral Electronic Self-Scanning Radiometer (MESSR), designed to provide information on coastal zone dynamics, water resources, vegetation dynamics, and geological applications; a Microwave Scanning Radiometer (MSR), which provided information on sea ice, snowfall, and water vapor content at the ocean and land surfaces; and a Visible and Thermal Infrared Radiometer (VTIR) to measure SST.

FIGURE 9.30 MOS-1 Satellite.
(Image credit courtesy JAXA).

TABLE 9.7 Multi-Spectral Electronic Self-Scanning Radiometer (MESSR) bands Japan Aerospace Exploration Agency (JAXA).

Band	Wavelength (μm)	Resolution (m)	Swatch Width (km)
Band 1 (VIS)	0.51–0.59	50	100
Band 2 (VIS)	0.61–0.69	50	100
Band 3 (NIR)	0.72–0.8	50	100
Band 4 (NIR)	0.8–1.1	50	100

MESSR was a four-band pushbroom radiometer, with the characteristics in Table 9.7. These channels are similar to the Landsat MSS spectral channels, making them useful for both marine and land applications.

The MSR had two different channels, one at 23.8 GHz that was horizontally polarized and one at 31.4 GHz that was vertically polarized. The 23.8 GHz channel had a 32-km resolution, while the 31.4 GHz channel had a 23-km resolution. They both had a 317 km swath width. The VTIR was a mechanical-scanning radiometer. The VTIR was a 4-channel visible/infrared radiometer with a visible channel at 0.5–0.7 μm, a short-wave thermal infrared channel at 6.0—7.0 μm, one long-wave thermal infrared channel at 10.5–11.5 μm, and the second long-wave thermal infrared channel at 11.5–12.5 μm. These satellites were operated by the Japan Aerospace Exploration Agency (JAXA) the successor of NASDA.

Japan also had its own Earth Resources Satellite known as EORC or JERS (Japan Earth Resources Satellite). It carried two instruments: the OPS (Optical Sensor) which was an 8-channel VIS/NIR/SWIR radiometer designed to study the Earth's land surface. The OPS channels are described in Table 9.8.

It was a push-broom scanner with a 75-km swath width and an 18-m spatial resolution for all channels. It could provide global coverage in 1 month in daylight.

JERS also carried a SAR for high-resolution, all-weather sensing of ocean, land, and ice. It was an L-band SAR with a frequency of 1.275 GHz and HH polarization. It was side-looking at 32 to 38 degrees with a swath of 75 km, 400 km off-nadir. It had a spatial resolution of 18 m and could cover the globe in 6 months, assuming a 10% duty cycle.

Starting in 1996, Japan inaugurated a new series of Earth-observing satellites originally known as Midori (green) but later changed to ADEOS (Advanced Earth Observing Satellite). ADEOS carried a large suite of instruments: The Advanced Visible and Near-infrared Radiometer, which was a 5-channel VIS/NIR radiometer with one band being a panchromatic band. It had a 16-m resolution and an 80-km swath width; the Improved Limb Atmospheric Spectrometer-1

TABLE 9.8 Visible and Thermal Infrared Radiometer (VTIR) channels Japan Aerospace Exploration Agency (JAXA).

Central wavelength (μm)	Spectral interval (μm)
0.56	0.52–0.60
0.66	0.63–0.69
0.81	0.76–0.86
0.81 (fore pointing)	0.76–0.86
1.65	1.60–1.71
2.07	2.01–2.12
2.19	2.13–2.25
2.33	2.27–2.40

(ILAS-1) consisting of two grating spectrometers one in the 6.21–11.77 μm range with a spectral resolution of 0.129 μm, and the other in the NIR range of 0.753–0.784; the Interferometric Monitor for Greenhouse gases, which was an IR interferometer over the spectral range of 3.3–14.0 μm split in 3 bands with an 8-km spatial resolution; NASA scatterometer for sea surface wind measurements, a Ku-band (13.995 GHz) side looking instrument with 3 antennas right and left and a 50-km spacing; Ocean Color and Temperature Sensor was a 12-channel radiometer with 8 narrow bands in the range 0.40–0.89 (for ocean color and vegetation), 4 bands in the range 3.5–12.7 for SST with a 700-m resolution; Polarization and Directionality of the Earth's Reflectance to measure aerosols, ocean color, vegetation and bidirectional distribution function with a 6.5-km resolution; Retroflector in Space which reflected ground laser beams to study atmospheric gases; Total Ozone Mapping Spectrometer, a 6-channel grating spectrometer 312.5, 317.5, 331.3, 339, 360.0, and 380 nm with a 50-km resolution. Unfortunately, ADEOS died after less than a year of operation.

The development of geostationary weather satellites in Japan followed the pattern set by the UnitedStates and the Europeans. The first was the Geostationary Meteorological Satellite-1 (GMS-1) or Himawari (meaning sunflower) 1. This satellite was a spinner very similar to the US GOES 4–7 satellites (Fig. 9.31).

This series of geostationary satellites lasted from the launch of GMS-1 July 14, 1977, to the retirement of GMS-5 July 21, 2005 (Table 9.9). It was supposed to be replaced by MTSAT-1, but a launch failure on November 15, 1999 eliminated this satellite, and Japan used GOES-9 over the Pacific for its operations until the new Multifunctional Transport SATellites MTSAT-1R was launched on February 26, 2005, which lasted till December 4, 2015. It was followed by MTSAT-2 with a launch on February 18, 2006 and lasting till March 10, 2017. All of these spacecraft had a Himawari designation as well, with MTSAT-2 corresponding to Himawari 7. Next came a new series and the decision to name the satellites as Himawari, so the next satellite was Himawari-8, which entered in the era of three-axis stabilized satellites into the Japanese system.

MTSATs 1 and 2 were created as three-axis stabilized and are considered second generation (SG) meteorological satellites, which are different from the European SG meteorological satellites, which continued to be spinners. The MTSAT imager had five spectral bands, one for the visible (0.675 μm) and four for the infrared (3.75, 6.75, 10.8, and 12.0 μm), which was an improvement over the GMS-5 imager in detecting low-level cloud or fog at night. The visible channel has a spatial resolution of 1 km, and the infrared channels have resolutions of 4 km. The MTSAT satellites were owned and operated by the Japanese Ministry of Land, Infrastructure, Transport, and Tourism along with the Japanese Meteorological Agency (JMA). They provided coverage for the hemisphere centered on 140°E, which includes Japan and Australia who are the primary users of the MTSAT imagery. As listed in Table 9.10 the MTSAT series was retired in 2017.

Himawari 8, 9 (Fig. 9.32) ushered in a new era of three-axis stabilized geostationary weather satellites. Suzuki et al. (2016) provides a comprehensive review of Himawari-8, which carries the Advanced Himawari Imager (AHI), which has capabilities similar to those of the ABI onboard the GOES-R satellites. The AHI was built by Harris/Excelis for the Himawari satellites. The spacecraft itself was constructed by Mitsubishi Electric with some assistance from Boeing. Bessho et al. (2016) provides a comparison of the Himawari-8/AHI imager with the previous MTSAT-2 imager (Table 9.10).

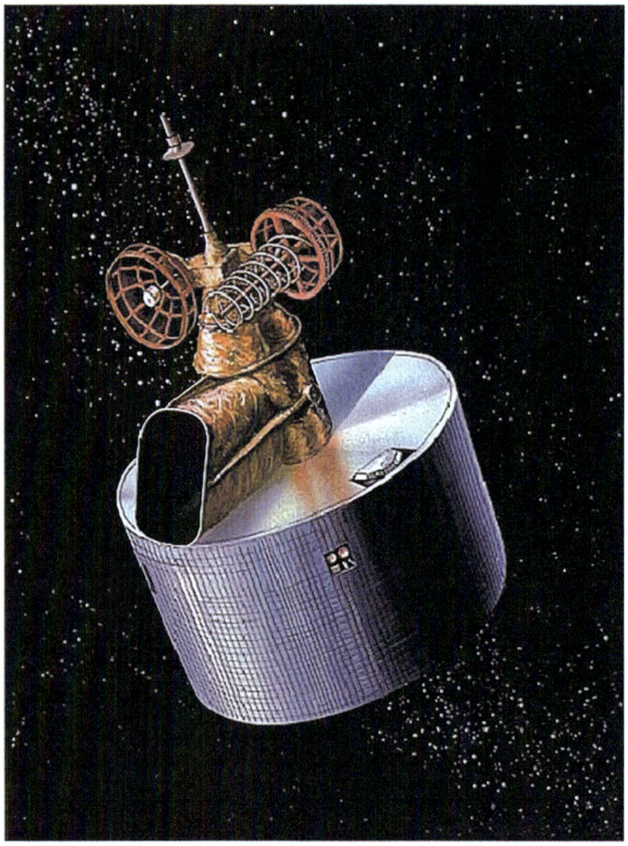

FIGURE 9.31 The first generation of Himawari.
Image credit courtesy JAXA.

TABLE 9.9 The list of Himawari meteorological satellites Japan Aerospace Exploration Agency (JAXA).

Name	Launch date (UTC)	Retirement	Rocket	Launch site
GMS 1 [ja;pt] (Himawari 1)	Jul 14, 1977	Jun 30, 1989	Delta 2914	Cape Canaveral
GMS-2 [ja; pt] (Himawari 2)	Aug 11, 1981	Nov 20, 1987	N-II (N8F)	Tanegashima
GMS-3 [ja;pt] (Himawari 3)	Aug 3, 1984	Jun 23, 1995	Nil (N13F)	Tanegashima
GMS-4 [ja;pt] (Himawari 4)	Sep 6, 1989	Feb 24, 2000	H-I (H20F)	Tanegashima
GMS-5 [ja; pt] (Himawari 5)	Mar 18, 1995	Jul 21, 2005	H-II (F3)	Tanegashima
MTSAT-1 (Mirai 1)	Nov 15, 1999	Launch failure	H-II (F8)	Tanegashima
GOES-9 (Pacific GOES)	May 23, 1995	Jun 14, 2007	Atlas I (AC-77)	Cape Canaveral
MTSAT-1R (Himawari 6)	Feb 26, 2005	Dec 4, 2015	H-IIA (F7)	Tanegashima
MTSAT-2 (Himawari 7)	Feb 18, 2006	Mar 10, 2017	H-IIA (F9)	Tanegashima
Himawari 8	Oct 7, 2014	Stand-by	H-IIA (F25)	Tanegashima
Himawari 9	Nov 2, 2016	Operational	H-IIA (F31)	Tanegashima
Himawari 10	2028	2045	H3	TBD

TABLE 9.10 Himawari-8/AHI versus MTSAT-2 imager Japanese Aerospace Exploration Agency (JAXA).

	Himawari-8/AHI			MTSAT 2/Imager		
	Band no.	**Wavelength (μm)**	**Spatial resolution (km)**	**Band**	**Wavelength (μm)**	**Spatial resolution (km)**
Visible	1	0.47	1			
	2	0.51				
	3	0.64	0.5	VIS	0.68	1
Near-infrared	4	0.86	1			
	5	1.6	2			
	6	2.3				
Infrared	7	3.9		IR4	3.7	4
	8	6.2		IR3	6.8	4
	9	6.9				
	10	7.3				
	11	8.6				
	12	9.6				
	13	10.4		IR1	10.8	4
	14	11.2				
	15	12.4		IR2	12.0	4
	16	13.3				

AHI, Advanced Himawari Imager.

FIGURE 9.32 Hunawari-8/9 satellites.
Image credit courtesy JAXA.

The spatial resolution of 4 km in the infrared in the far right column applies to both MTSAT and AHI. The additional channels in the AHI enable the analysis of new phenomena not available with the MTSAT-2 imager. The utility of these channels compared with the MTSAT channels is shown here in Fig. 9.33 which shows the spectral response functions of the various channels against the EM spectrum. One can clearly see how the visible channels cover the red, blue, and green portions of the spectrum. The near-infrared channels are beyond the visible light. The thermal infrared channels provide the

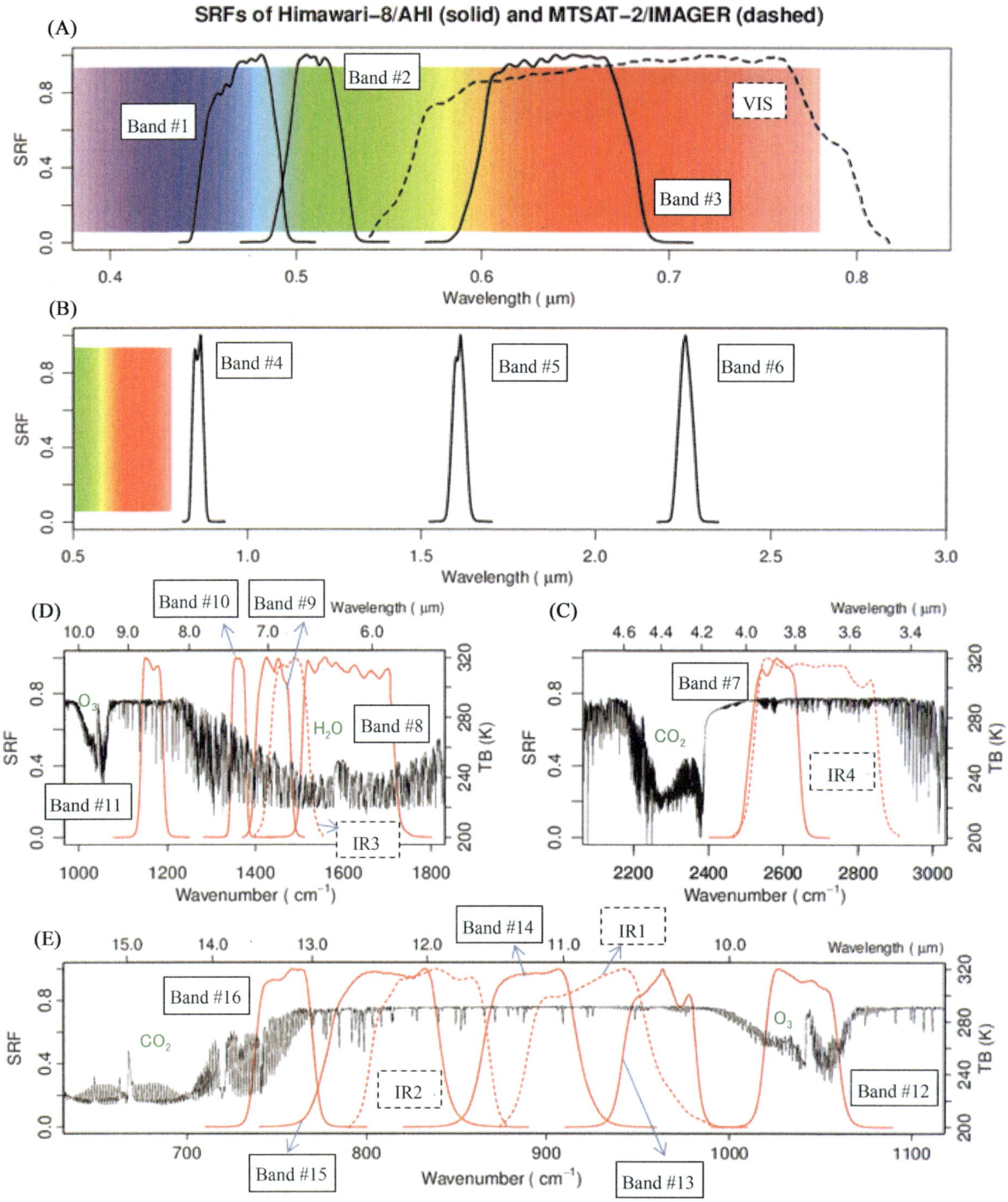

FIGURE 9.33 Spectral response functions for (A) visible bands, (B) near-infrared bands, (C–E) thermal infrared bands of the AHI (solid lines). Also shown are the SPFs for the MTSAT imager channels (broken lines). *AHI*, Advanced Himawari Imager.
From Bessho et al. (2016).

traditional longwave split window for SST calculations, but also a large number of shorter wave infrared channels that are useful for forest fire and flood detection and mapping. Again, the channels are very similar to those on the ABI of GOES-R.

Also similar to the ABI, the AHI is able to produce a number of different scans. The five scans are the Full Disc, Japan Area, Target Area, and two Landmark Areas. The scan ranges for the Full Disc and the Japan Area are preliminarily fixed (corresponding to full disc and conus on the ABI), while the Target Area and Landmark area are flexible to enable the prompt response to developing meteorological conditions (similar to the rapid scan of the ABI). The Landmark Area data are exclusively used for navigation and moon observations for calibration. Full disc images are taken every 10 minutes; Japan Area images are collected every 2.5 minutes, and Landmark Area images every 30 seconds.

To support accurate image navigation and registration, the Himawari-8/9 satellites have star trackers, inertial reference units (IRU), and angular rate sensors (ARSs) installed on the optical bench together with the AHIs for attitude determination. These sensors are important to correct for the thermal strain on the optical benches. To mitigate the thermal strain, a carbon fiber-reinforced plastic panel with low thermal expansion is used for the optical bench. In addition, the IRUs and the ARSs are installed close to the AHI to ensure its directional accuracy. The typical residual error after the coregistration and AHI optical system temperature corrections varies diurnally. The variation is about 2.5 pixels for band no. 5 and about 0.5 pixels for the VIS and NIR bands. For the IR bands, the variation is 0.1 pixels at the maximum.

For the thermal channel calibration, the AHI has a blackbody as the Internal Calibration Target (ICT), and for the visible channel, there is a solar diffuser as a Solar Calibration Target (SCT). The ICT is temperature-controlled using heaters and generates a constant radiance. The SCT is the first solar diffuser installed on a JMA satellite and is made of Spectrolon (manufactured by Labsphere). MODIS on NASA's Aqua/Terra has the same system for its onboard calibrator to reflect and diffuse sunlight for measurement by the detectors as part of the calibration process. The required accuracy of the solar calibration is within 5% of the radiance. The ICT is used together with a deep space view for the cold reference in the IR bands. Himawari 8/9 data are available in a variety of formats and from different data systems. JMA also provides Level 1b data to external users via multiple channels, including the Himawari Cloud service and the Himawari Cast service. There is significant potential for the application of Himawari 8/9 data to various areas outside of meteorology. The JMA is dedicated to working with analysts in other national and international Earth science fields to enable their use of Himawari 8/9 data.

In October 2022, the prime ministers of Australia and Japan announced an intention to strengthen cooperation between their governments to enhance Australia's extreme weather and multihazard monitoring capabilities, with a focus on the successor to the Japanese geostationary meteorological satellite Himawari-9. In addition to extending a partnership that began between the two agencies in 1977, this agreement includes the opportunity to be part of the new Japanese Himawari-10 program. Himawari-10 (Fig. 9.34) will carry a visible infrared imager and a hyperspectral infrared sounder, both built by the L3 Harris Technologies of the United States.

FIGURE 9.34 Himawari-10.
Image credit courtesy Misubishi Electric Corp.

FIGURE 9.35 The Global Precipitation Measurement Core Satellite.
Image credit courtesy JAXA.

It will also be equipped with a space environment sensor from the National Institute of Information and Communications Technology of Japan (NICT). The imager will be an improvement in terms of wavelength bands and resolution over Himawari 8/9. Himawari-10 will be the first Japanese satellite to use a hyperspectral infrared sounder to gather 3D information about water vapor and atmospheric temperature profiles. The space environment sensor will measure the intensity of proton and electron beams at geostationary orbit. Himawari's design life is 15 years.

9.2.2 The Global Precipitation Measurement Mission

Another program that Japan was heavily involved in was the Global Precipitation Measurement (GPM) program. It is a collaboration between the US (NASA) and JAXA (Japanese Aerospace Exploration Agency). The goal of this program is to observe Earth's precipitation every 2–3 hours. The constellation that makes up the GPM is contributed to both by NASA and JAXA. The project is managed by NASA's Goddard Space Flight Center (GSFC) and consists of a Core Observatory satellite (Fig. 9.35) assisted by a constellation of spacecraft contributed by other agencies.

The supporting constellation consists of eight smaller satellites. The GPM core satellite was assembled and tested at GSFC and then launched from the Tanegashima Space Center, Japan, on a Mitsubishi Heavy Industries H-IIA rocket. The satellite flies in a low Earth orbit at an altitude of approximately 400 km and an inclination of 65 degrees.

One instrument carried by the core satellite is the Dual-frequency Precipitation Radar (DPR), which provides three-dimensional maps of storm structure across its swath, including the intensity of rainfall and snowfall at the Earth's surface. The DPR has two frequencies, allowing researchers to estimate the sizes of precipitation particles and detect a wider range of precipitation rates. The Ku-band radar, similar to the PR on TRMM, covers a 245-km swath. The Ka-band radar initially covers a 120-km swath, which was later increased to 245 km. The spatial resolution of both frequencies is 5 km at nadir. The DPR was developed by JAXA in conjunction with Japan's National Institute of Information and Communications Technology. Data is relayed to the ground via the TDRSS relay satellites.

The core satellite is also equipped with the GPM Microwave Imager (GMI) built by Ball Aerospace in Boulder, Colorado (Fig. 9.36).

The GMI collects passive emissions in 13 different frequency/polarization channels. These data can be used to create quantitative maps of precipitation across its 885 km swath. The GMI uses a large reflector that rotates on an axis perpendicular to the spacecraft. Rainfall is sensed by vertical and horizontal channels at 10.6 GHz. The 18.7-GHz channel is also dual-polarized and used to retrieve rainfall. Finally, a vertically polarized 23 GHz channel is added to the rain retrieval. Heavy rain and hail are sensed with a dual-polarized 37 GHz channel together with a dual-polarized 89 GHz channel. Finally, snow uses the 89 GHz horizontally polarized channel together with a dual-polarized 166 GHz channel. The snow sensing adds a vertically polarized 183±3 GHz channel and a vertically polarized 183±7 GHz channel. The data from GMI is used as a reference standard for an international network of partner precipitation-measuring satellites known as the GPM constellation. The data from the GPM Core Observatory and the partner satellites are merged into a single global precipitation dataset called IMERG, which is updated every 3 hours.

The scan geometry of the GMI follows the scan pattern set by earlier passive microwave instruments (Fig. 9.37) where the rotating reflector scans out a circular path along the satellite ground track. The off nadir angle is 48.5 degrees, which

FIGURE 9.36 The GMI is being built at Ball Aerospace.
Image credit courtesy Ball Aerospace.

represents an Earth-incidence angle of 52.8 degrees. This angle was chosen to be the same as the TMI (TRMM microwave imagery) to continue the earlier time series. The reflector spins at 32 RPM, and the GMI collects microwave radiometric brightness temperature measurements over a 140-degree sector centered on the spacecraft ground track. The rest of the scan is used for calibration measurements of both cold space and an internal hot load calibration target. This 140 degree swath translates to an 885-km swath on the Earth's surface. For comparison, the DPR (radar) on the GPM core satellite has a cross-track swath width of 245 km, so that only the central part of the GMI swath overlaps the radar swath. This combination is used to improve the algorithm to compute geophysical parameters from GMI brightness temperatures.

The DPR is one of the primary instruments on the GPM core satellite. As stated earlier, this radar system has both Ku-band and Ka-band radars. These radars provide rain-sensing capabilities over land and ocean, both day and night. Both radars have a beam width of 0.71 degrees and pulse repetition rates between 4,100 and 4,400 Hertz. This radar made it possible to see inside Hurricane Maria two days before it devastated Puerto Rico (Fig. 9.38). This is a 360-degree visualization of the hurricane where green and yellow correspond to low rates of precipitation and red and purple colors indicate high rainfall rates. At the top of the hurricane, where temperatures are colder, the blue and purple indicate both light and heavy frozen precipitation.

The GPM constellation of satellites involves other United States and other countries' satellites. Again, the GPM project is led by the United States and Japan, but other nations have contributed to this constellation. This constellation is clearly depicted in Fig. 9.39 which also shows the flags of the participating countries. The United States contributes seven satellites between NOAA and NASA, 2 of which are joint between Japan (JAXA) and NASA, in addition to its own GCOM-W1 satellite. France contributes to a satellite joint with India, and EUMETSAT represents Europe with MetOp B/C satellites. Thus this combination of satellites means that in addition to the DPR and GMI on the core satellite, GPM includes observation from:

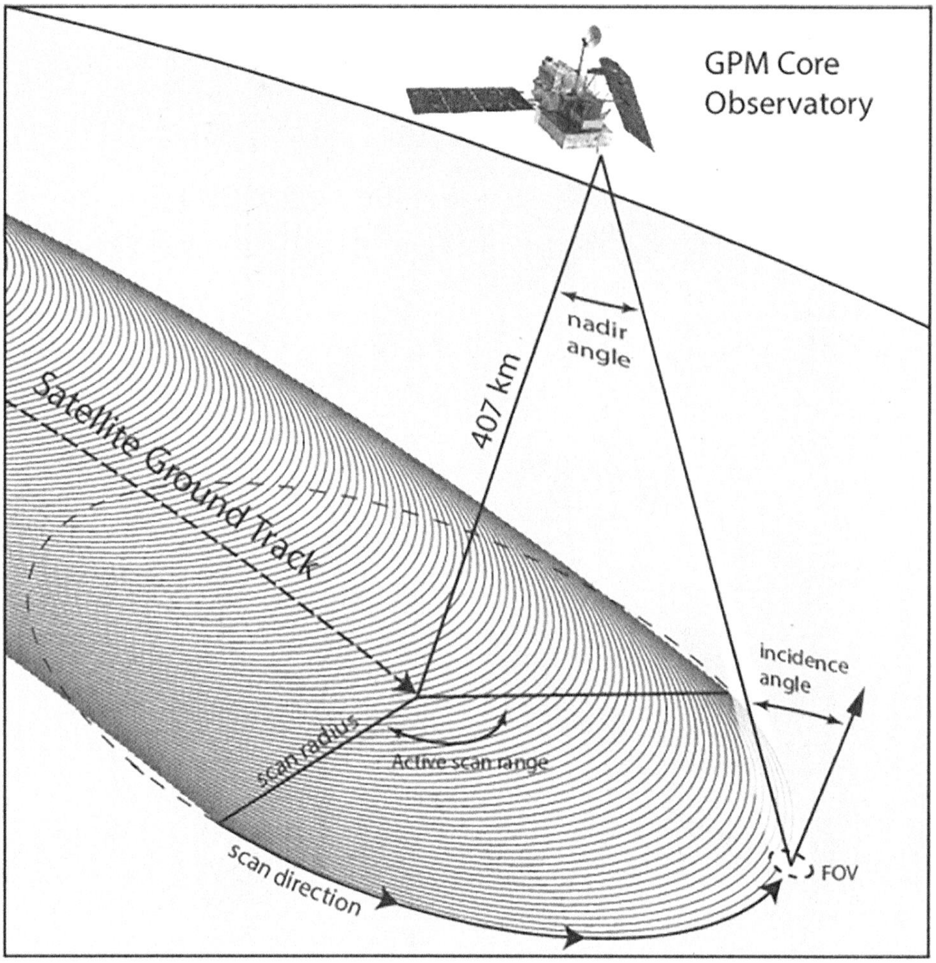

FIGURE 9.37 The GMI scan geometry.
Image credit courtesy Ball Aerospace.

- Special sensor microwave imager/sounder (SSMIS)
- Advanced MSR-2 (AMSR-2) on JAXA's GCOM-W1
- Multifrequency Microwave Scanning Radiometer (MADRAS) and the Multichannel Microwave Humidity Sounder (SAPHIR) on the Mega-Tropiques satellite provided by France (CNES) and India (ISRO).
- MHS on NOAA-19 satellite
- MHS on the MetOp series of satellites
- Advanced Technology Microwave Sounder (ATMS) on NPP
- ATMS on the JPS satellites
- Microwave Sounder on WSF, the Defense Weather Satellite System (DWSS)

Agreements have been signed between the cooperating countries to continue this cooperation to retrieve and map global precipitation. The United Nations has identified the GPM program as an outstanding example of the peaceful utilization of space.

9.3 Germany and TerraSAR-X Program

We can't leave the subject of SAR satellites without mentioning Germany's TerraSAR-X program. TerraSAR-X is a public–private partnership between the German Aerospace Center (DLR) and EADS Astrium. The exclusive commercial rights are held by Astrium. The first TerraSAR-X satellite (Fig. 9.40) was launched on June 15, 2007 and has been in operation since January 2008. The satellite uses a phased array SAR antenna operating with an X-band (wavelength of 31

FIGURE 9.38 Hurricane Maria two days before hitting Puerto Rico as imaged by the DPR on the. GPM core satellite (https://svs.gsfc.nasa.gov/13079/). Image credit courtesy NASA.

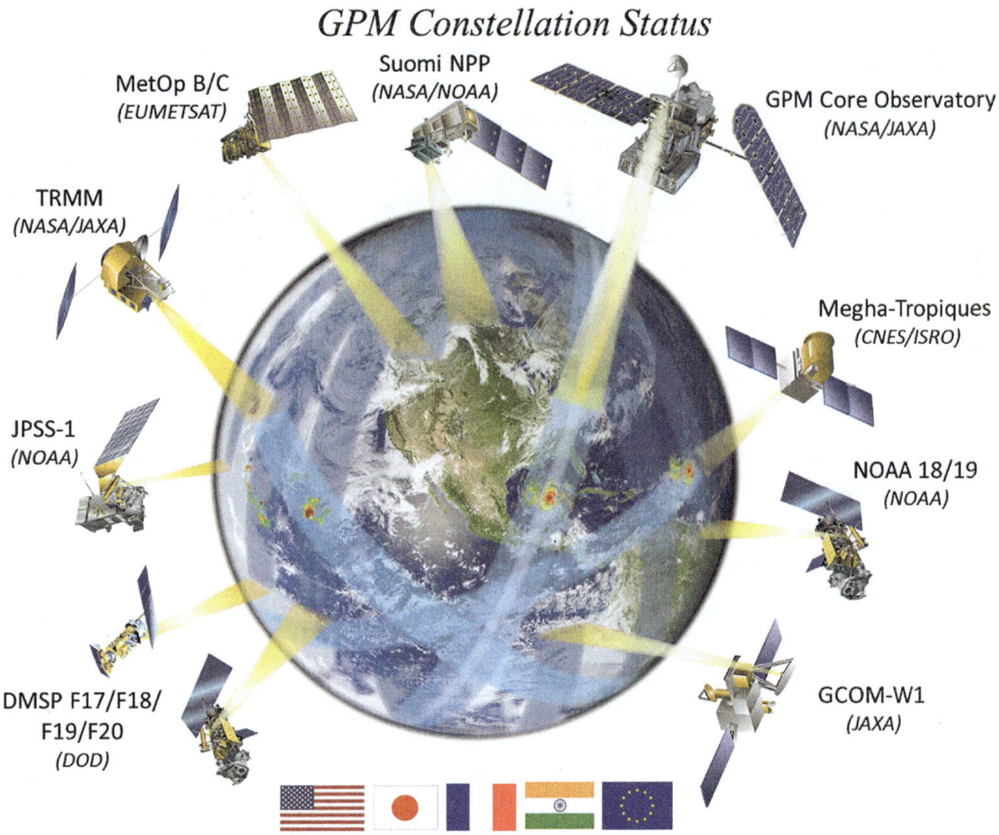

FIGURE 9.39 The GPM satellite constellation with the flags of the participating countries. Image credit courtesy NASA.

mm and a frequency of 9.65 GHz). With processing, the SAR images have a spatial resolution of 1 m. The satellite flies in a sun-synchronous polar orbit with an altitude of 514 km, which was selected to provide the best solar power to the satellite.

TerraSAR-X acquires SAR data in three modes:

- SpotLight mode: up to 1 m resolution, scene size 10 km (width)×5 km (length);

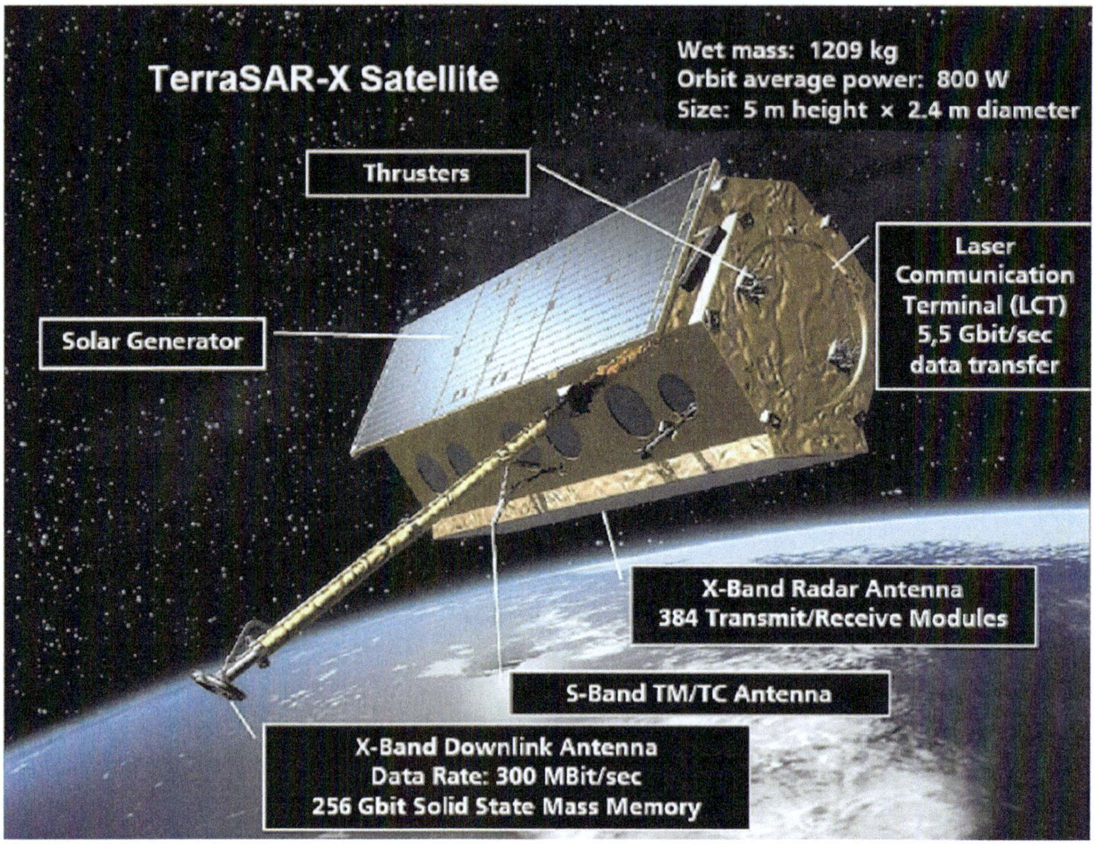

FIGURE 9.40 TerraSAR-X satellite.
Image credit courtesy DLR, EADS Astrium GmbH.

- StripMap mode: up to 3 m resolution, scene size 30 km (width)×50 km (length);
- ScanSAR mode: up to 16 m resolution, scene size 100 km (width)×150 km (length).

The TerraSAR-X antenna allows a variety of polarimetric combinations: single or dual polarization, or full polarimetric. Depending on the desired application, one of four different processing levels can be selected:

- Single Look Slant Range Complex (SSC)
- Multi Look Ground Range Detected (MGD
- Geocoded Ellipsoid Corrected (GEC)
- Enhanced Ellipsoid Corrected (EEC).

The SAR instrument on TerraSAR-X was called the SAR-X and was a side-looking, swath from 10 to 100 km, depending on operation mode as mentioned earlier. Data from all of the operation modes were available in HH, W, HV, or VH polarizations. The primary goals of this instrument were to determine:

- The fraction of vegetated land;
- Glacier motion;
- Glacier topography;
- Land cover type;
- Land surface topography;
- Soil moisture at the surface;
- Soil type;
- Vegetation type.

TerraSAR-X introduced a number of innovations, like the resolution and scan area being changeable in a 1:10 relationship. Larger areas with lower spatial resolutions or smaller areas with greater spatial resolution. In addition, the

FIGURE 9.41 Tandem-L satellites.
Image credit courtesy DLR.

patch antenna can be electronically steered so that the point of view is adjustable. Using the adjustable angle of the radar sensor (along with the precession by the Earth), the SAR can observe any spot on the Earth within 1–3 days. In general, TerraSAR-X has an 11-day repeat cycle at the equator. As with any polar orbit, this revisit time decreases as you move towards the poles.

Since June 21, 2010, TerraSAR-X has been joined ty the TanDEM-X satellite, which is almost identical to TerraSAR-X in design. TanDEM-X, together with TerraSAR-X, forms the first reconfigurable SAR interferometer in space. Its purpose was to be able to make precise land surface height measurements to create a new and accurate digital elevation model (DEM). Thus TerraSAR-X is now part of two missions: its original TerraSAR-X mission and, together with the TanDEM-X satellite.

Scientists and engineers at DLR are now working on the next generation of SAR satellites for climate research and environmental monitoring. It will take the form of Tandem-L, which will be an interferometric radar mission. It will consist of two identical L-band radar satellites used to investigate dynamic processes in the biosphere, cryosphere, geosphere, and hydrosphere (Fig. 9.41).

Tandem-L satellites will carry a single instrument, an L-band SAR (L-SAR). To meet expected user requirements, 26 geophysical products have been identified. To satisfy these challenging requirements, the mission consists of a 3D structure mode fully polarized SAR interferometry to derive tomographic images with fine vertical and horizontal resolutions. This capability is needed for 3D forest and ice mapping and the generation of digital terrain and surface models. Using repeat pass interferometry, the deformation mode will be able to measure and monitor displacements of the Earth's surface with an accuracy capable of detecting earthquakes and evaluating risk conditions. L-SAR modes will also be used to measure the variations of forest biomass, glacier motion and melting, near-surface soil moisture, and ocean surface dynamics. L-SAR will have a 7-m spatial resolution, which can be increased to 1 m for spot images. It will have a 350-km swath width in single/dual polarization mode and a 175-k swath width in quad polarization mode. Tandem-L satellites will fly in sun synchronous, dawn-dusk orbits with a repeat cycle of 16 days. Up to four global data acquisitions will be performed during each repeat cycle. The Tandem-L satellites are designed to last for 10 years with consumables for 12.

The data products, their resolution, and accuracy are given here in Table 9.11.

Fig. 9.42 shows these applications in a graphical form, which includes the time scales of the measurements.

The challenges posed by the need to resolve and monitor this large variety of parameters dictate the need for new approaches to observing the Earth from space. In response, Tandem-L will have two modes:

1. The 3D structure mode that employs fully-polarimetric single-pass SAR interferometry to measure structural parameters of semitransparent volume scatterers. Combining multiple interferometric acquisitions with varying cross-track baselines (Fig. 9.43), it is possible to derive tomographic images with fine vertical and horizontal resolutions to depict accurate measurements of 3D forest and ice structures and to be able to compute accurate DEM and surface models.
2. The deformation mode uses repeat-pass interferometry to measure small displacements of the Earth's surface, achieving accuracies of cm down to mm size. In an effort to minimize errors introduced by atmospheric variations, the

TABLE 9.11 The Products Plan for the Tandem-L Interferometric SAR Mission.

Biosphere	Products	Coverage	Resolution	Accuracy
	Forest height		50 m global 30 m regional	~10%
	Vertical forest structure		50 m global 30 m regional	~20% for 10 m layers
	Forest structure change	All forest areas	50 m global 30 m regional	~15%% for each layer
	Above-ground biomass		100 m global 50 m regional	~20% or 20 t/ha
	Biomass change		100 m global 50 m regional	~10% or 10 t/ha
Geo-Lithosphere	LOS deformation (tectonics)	High strain areas	50 m	1 mm/yr (E/V), 10 mm/yr after 10 years
	3D deformation (tectonics)	High strain areas	50 m	1 mm/yr (E/V), 10 mm/yr after 10 years
	Subsidence and landslide	Urban risk areas	7 m	1 mm/yr after 10 years
	LOS displacement (volcanoes)		50 m	10 mm
Cryosphere	Glacier velocity maps	Most glaciers	50–500 m	1–50 m/yr
	Sea ice type and thickness	Arctic and Antarctic	5–50 km	5%-20%/0.5–1.0 m
	3D ice structure	Greenland and selected areas	100 m	10 m Vertical resolution
	Ice sheet elevation change	Ice sheets worldwide	50 m	0.5–1.0 m
	Permafrost	Selected Arctic regions	10 m (quad)	1 cm LOS displacement/season
Hydrosphere	Soil moisture	Selected areas	50 m	0.05–0.1 mm^3/m^3
	Agriculture mapping	Selected areas	20 m (16 looks, quad pol)	1 dB rad res., NESZ ≤-28 dB
	Wind speed and wave height	Coastal regions	4 km	Speed 2 m/s, height 0.1 m
	Ocean currents	Selected areas	4 km	0.05 m/s
DEM	DEM and surface model	Global	~12 m (bare), 25 m (forest)	2 m (bare), 4 m (vegetated)
	Global basemap and landcover	All land surfaces	10–20 m	Single (2/yr), quad (2/yr)
	Emergency	Local	1 m	Single/dual/quad

DEM, Digital elevation model.

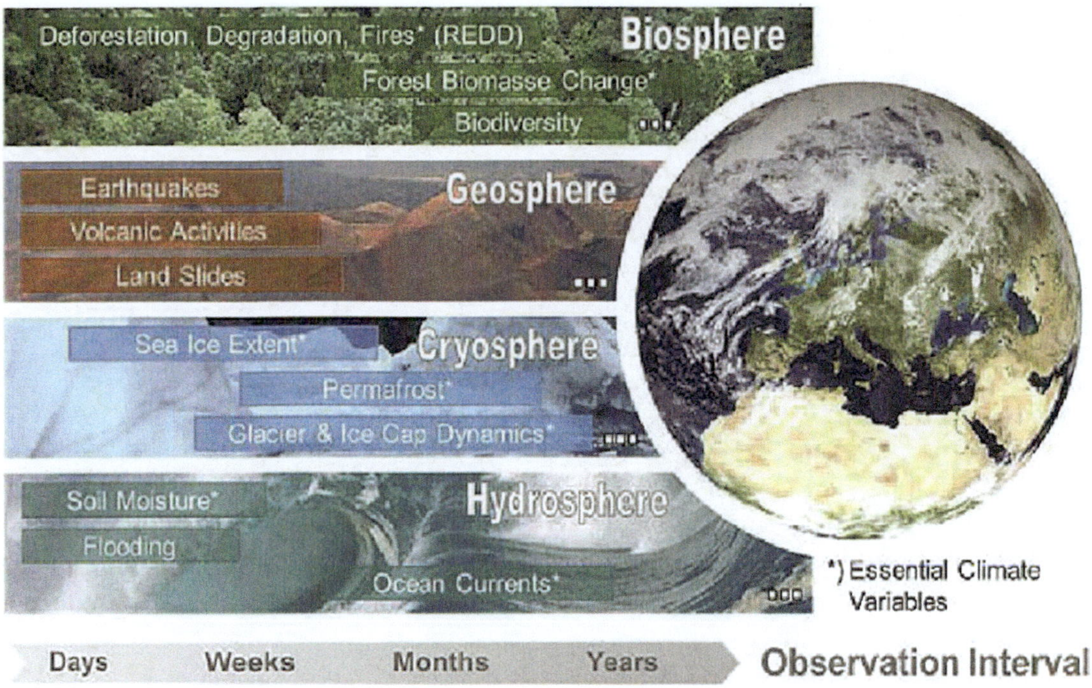

FIGURE 9.42 Examples of the dynamic processes in the bio-, geo-, cryo-, and hydrosphere, along with the observation intervals required to resolve and monitor these conditions. Processes with an * denote critical climate variables.
Image credit courtesy DLR.

processing maximizes the number of image acquisitions. A special SAR imaging mode is instituted with a 350-km wide swath and a 7-m spatial resolution.

During each 16-day repeat cycle, up to 4 global data collections can be performed from different viewing directions in single and dual polarization modes. Surface deformation measurements are further supported by flying the master satellite in a closely controlled orbital tube with a radius of 250 m. To obtain the required cross-track baselines for single-pass interferometry and tomography, the inclination of the slave satellite will be periodically adjusted. This results in a natural drift of the ascending mode and allows for large periodic baseline variations with a minimum amount of satellite fuel expenditure.

Tandem-L presents a large challenge in the amount of data that has to be transferred to the ground. To accommodate this large volume of data, a high-performance Ka-band downlink with a net data rate of up to 2.6 Gbits/s will be used. Thus, the appropriate ground station network will need to download eight terabytes of data each day. To demonstrate the feasibility of such a mission, a data acquisition plan has been developed, and Fig. 9.43 shows how the available data volume is distributed among the different applications.

We can't leave the discussion of TerraSAR-X and all of its successors without presenting an image from it. A TerraSAR-X image of the city of Las Vegas, Nevada, is presented here in Fig. 9.44.

The fourth flight unit of the TerraSAR-X program was the High Resolution Wide Swath SAR, or HRWS-SAR. Its primary goal is all-weather land observations, including surveillance and emergency response applications. Yu et al. (2023) discuss how the HRWS-SAR requires the separation of range-ambiguous echoes with the space-time coding (STC) array. To separate the echoes corresponding to different range-ambiguity regions, the equivalent transmit beamforming is performed in the two-dimensional space-frequency domain. In addition, to compensate for the loss of range resolution during the beam forming process, a frequency splicing method is proposed. Finally, the distributed target simulation is used to demonstrate the effectiveness of obtaining HRWS-SAR images in the STC radar.

9.4 Italy and COSMO-SkyMed

Another SAR program is the COnstellation of small Satellites for the Mediterranean basin Observation (COSMO-SkyMed) that was funded by the Italian Ministry of Research and the Italian Ministry of Defense and carried out by the

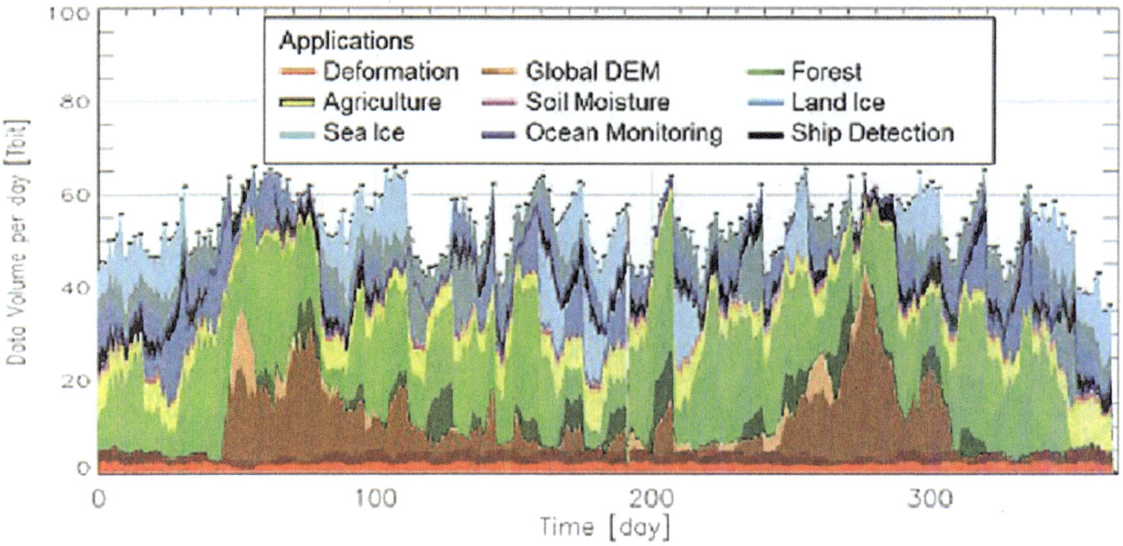

FIGURE 9.43 Daily data volume acquired for the different application areas over one year. Note that most data acquisitions serve more than one application.
Image credit courtesy DLR.

FIGURE 9.44 TerraSAR-X image of Las Vegas.
Image credit courtesty: DLR.

Italian Space Agency (ASI). This constellation was intended for both civilian and military use. The primary contractor was Thales Alenia Space. This constellation contains four satellites (Fig. 9.45) with a SAR as their main payload.

Each satellite carries the SAR-2000 SAR, which, like TerraSAR-X, is an X-band SAR. Also following the example of TerraSAR-X the instrument has a phased array antenna known as the SAR Active Antenna (SAA). The satellite is also equipped with an SAR Internal Electronics (SIE) unit to provide control of the programmable phased array antenna. Also like TerraSAR-X the programmable antenna enables the instrument to provide Stripmap, ScanSAR and Spotlight modes of sampling. The programmable antenna also allows for Wide Area Surveillance, Continuous Imaging, and High Resolution Imaging modes. This level of flexibility allows the system to fulfill both military and civilian requirements. The Spotlight mode has a spatial resolution of less than 1 m for a spot size of 10×10 m^2, while the Stripmap mode has a spatial resolution of 3–15 m and a swath width of 40 km. The two ScanSAR modes (wide region and huge region) allow for a large view region of swaths of 100 and 200 km, respectively, with a corresponding decrease in spatial resolution to 30 and 100 m, respectively. In a 24-hour period, the maximum acquisition capacity is 1,800 images, either 75 Spotlight and 365 Stripmap or 150 ScanSAR images for each satellite. The data acquisition quantity depends on the user requests.

FIGURE 9.45 COSMO-SkyMed Satellite.
Image credit courtesy ASI.

The satellites are all in a sun-synchronous dawn-dusk orbit with a period of 97 minutes and at an altitude of 619 km. The satellites in this constellation can be operated in two different modes. The "nominal orbital configuration" where the satellites are equally spaced in the same orbit, or the interferometric mode. The latter mode can feature two satellites in the same plane in close proximity or two satellites in phase on slightly different orbital planes. This second mode allows for real-time interferometry that can produce 3D images of the Earth's surface. The first COSMO was launched on June 8, 2007, followed soon after by COSMO-2 on December 9, 2007. Both were launched from Vandenberg AFB on a Delta II rocket. The third launch waited till October 25, 2008 and the fourth was launched on November 6, 2010. So all four of the planned satellites are operating and providing the world with satellite imagery and also satellite interferometric imagery (Fig. 9.46).

In Fig. 9.46 COSMOSkyMed images have been used to assess the damage caused by a magnitude 6.6 Earthquake in central Italy on October 30, 2016. The color variation from yellow to red indicates increasingly more significant ground surface change. The damage proxy map (DPM) in Fig. 9.47 was produced by the Advanced Rapid Imaging and Analysis (ARIA) team at JPL and Caltech. This DPM image compared favorably with the optical images of damage provided by Maxar and others.

Another significant application of COSMOSkyMed data has been the interferograms produced of the Kilauea volcano on the island of Hawaii (Fig. 9.48).

The image shows ground deformation during an intrusive activity at Kilauea volcano. This particular interferogram is restricted to a single 1-day timeframe from 6 pm Honolulu Standard Time (HST) on January 31 through February 1, 2024. It therefore reveals the volcano signals with reduced interference of data noise. Colored fringes denote areas of ground deformation, with more fringes indicating more deformation. Each color cycle represents 1.5 cm of ground motion toward or away from the satellite. The patterns indicate overall deflation of the summit area as magma moved underground to the southwest, where the patterns show an uplift (up to about 50 cm) and spreading (along with subsidence) due to the intrusion of a dike (a vertical sheet of magma). Comparing this to an earlier interferogram shows that the dike intrusion moved farther to the SW consistent with the patterns of seismicity. (Images from the Italian Space Agency provided through the Hawai'i supersite.)

In the years to come COSMOSkyMed will continue with an SG of satellites. In 2000, ASI signed a cooperative agreement with CONAE Argentina's Space Agency referred to as SIASGE which is a system combining the four Italian satellites with two Argentinian satellites, SAOCOM (L-band SAR) in the same orbital plane. This will integrate the four X-band SARs with the two L-band SARs. The ASI has plans to go beyond X-band as well and is looking at the benefits of the various radar bands.

9.5 Taiwan and the COSMIC Satellites

The FormoSat-3/COSMIC-1 (Constellation Observing System for Meteorology, Ionosphere and Climate) is an international collaborative program between the National Space Program Office (NSPO) of Taiwan and the University Corporation for Atmospheric Research (UCAR) of the United States started in December 1997; the project has launched a

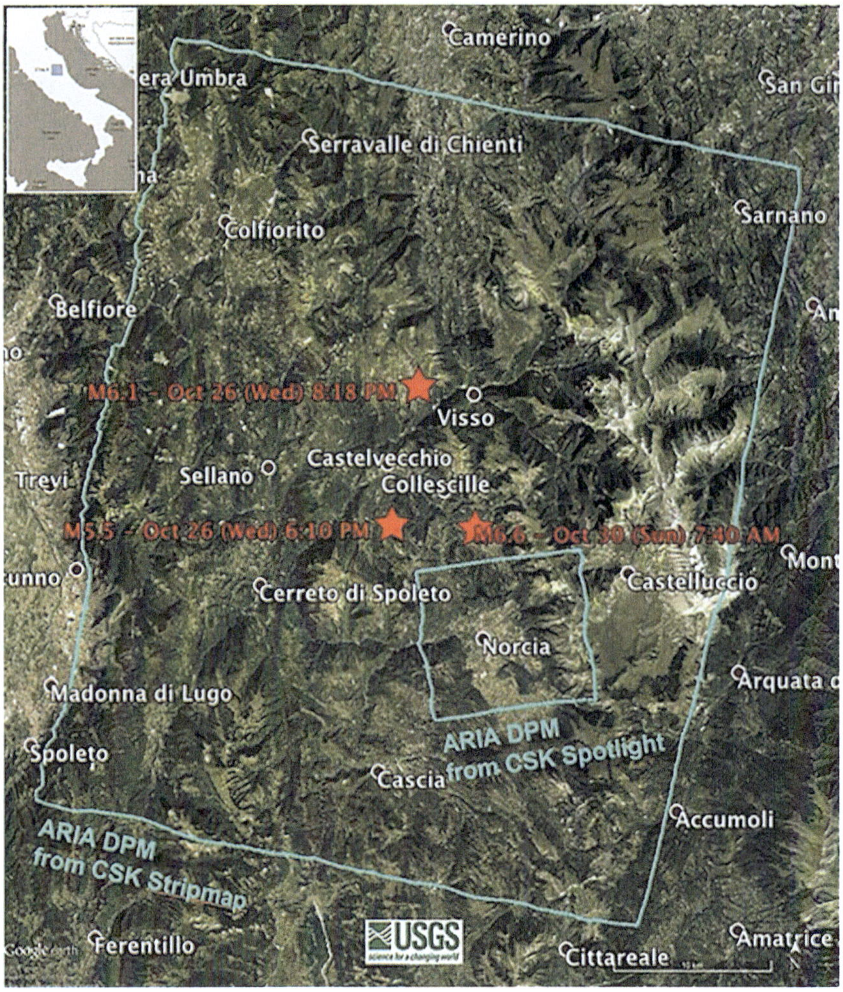

FIGURE 9.46 Damage proxy map (DPM) derived from COSMO-SkyMed Oct. 30-31, 2016. Map lines delineate study areas and do not necessarily depict accepted national boundaries.
Image credit courtesy ASI.

Low Earth Orbit (LEO) constellation of six microsatellites (Fig. 9.49) to collect atmospheric profiles using GPS signal occultations. NSPO is the prime sponsor and owner of the satellites, while UCAR (which administers the National Center for Atmospheric Research [NCAR] in Boulder, Colorado) will be responsible for using the atmospheric profile data in numerical weather/climate prediction models. An overall goal is to demonstrate the utility of atmospheric limb soundings in operational weather prediction. Each satellite carries an advanced version of the JPL-developed GPS receiver to use for occultation measurements. In addition, each satellite carries two tiny, simple secondary instruments (tri-band beacon and photometer).

Orbital Sciences Corporation (OSC) of Dulles, VA, United States is the prime contractor for the microsatellites as selected by NSPO. A joint team of NSPO and OSC engineers designed and built the satellites, with the early work done at OSC and integration and test performed at NSPO. Several companies of Taiwan were selected to participate in the project by providing satellite components as listed here:

- Satellite computer (Acer Technologies Inc.)
- Mission interface unit (Acer Technologies Inc.)
- Solar sensor (Shihlin Electric & Engineering/eBright Corp.)
- Rechargeable storage battery (Shihlin Electric & Engineering/eBright Corp.)
- Current converter (Shihlin Electric & Engineering/eBright Corp.)
- Satellite antennas (Victory Industrial Corp.)

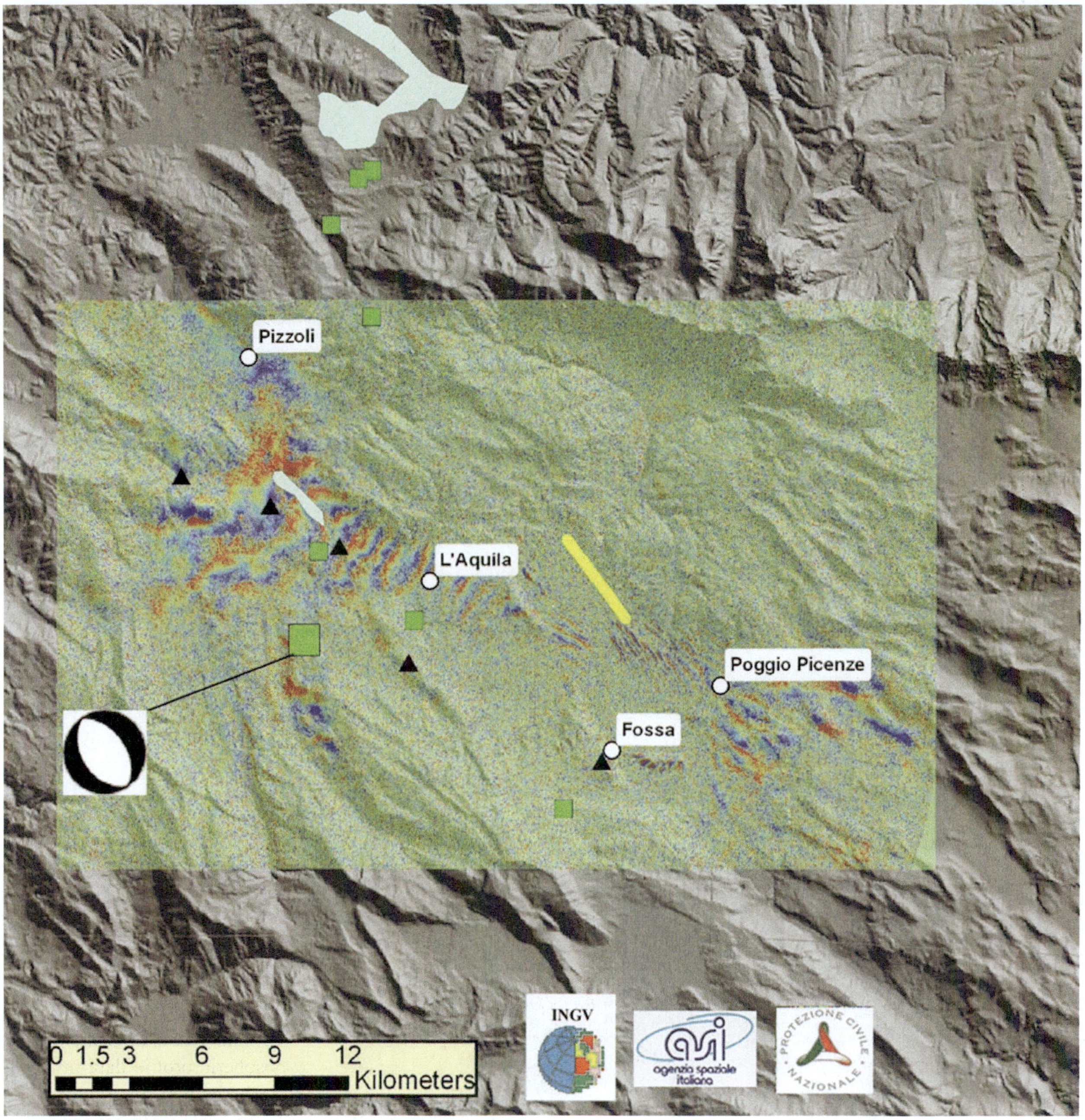

FIGURE 9.47 COSMO SkyMed interferogram over the L'Aquila area.
Image credit courtesy JPL/Caltech.

- Receiving coupler (Victory Industrial Corp.)
- Transmitting filter (Victory Industrial Corp.)
- Satellite heating elements (Yung Tien Industrial Co.)

The GPS RO technique is based on the following: As a GPS signal travels through the atmosphere it is retarded and bent by the structure of the atmosphere it penetrates (Fig. 9.50).

The deviation of the signal leads to a phase and Doppler shift, which can be measured very accurately by the GPS receiver onboard the LEO COSMIC-1 satellites. Since the GPS transmitter and COSMIC receiver positions and velocities are very accurately known the signal bending angle can be calculated from the Doppler shift at the COSMIC-1 receiver. Using this bending angle versus atmospheric impact parameter data, vertical profiles of refractivity as a function of tangent

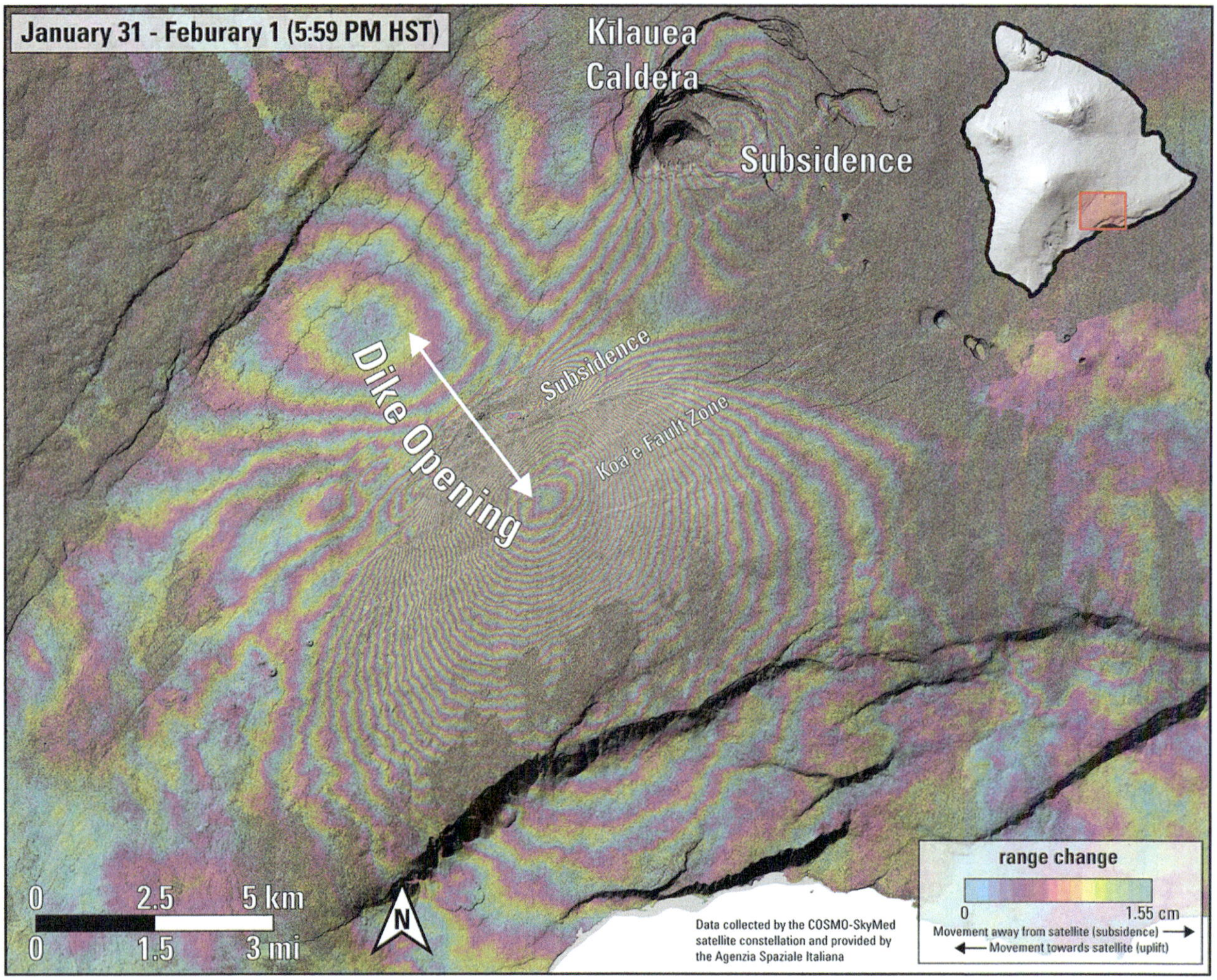

FIGURE 9.48 Interferogram of the Kilauea Volcano area from COSMOSkyMed.
Image credit courtesy ASI.

FIGURE 9.49 Artist's rendition of FormoSAR-3/COSMIC-1 satellite.
Image credit courtesy NSPO.

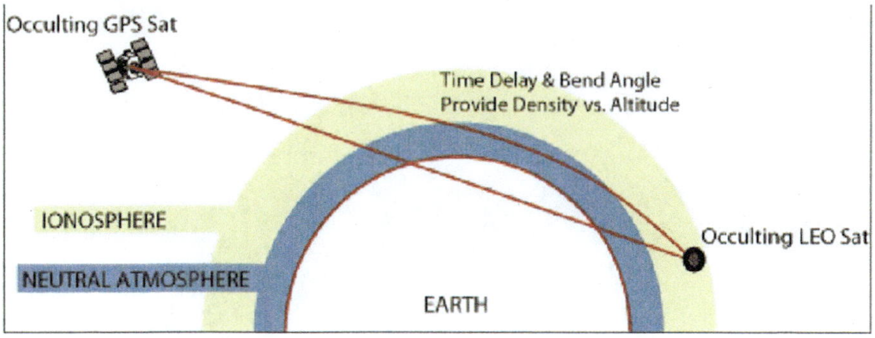

FIGURE 9.50 Occultation scheme of GPS signals to Low Earth Orbit (LEO) satellites.
Image credit courtesy: Broad Reach Engineering.

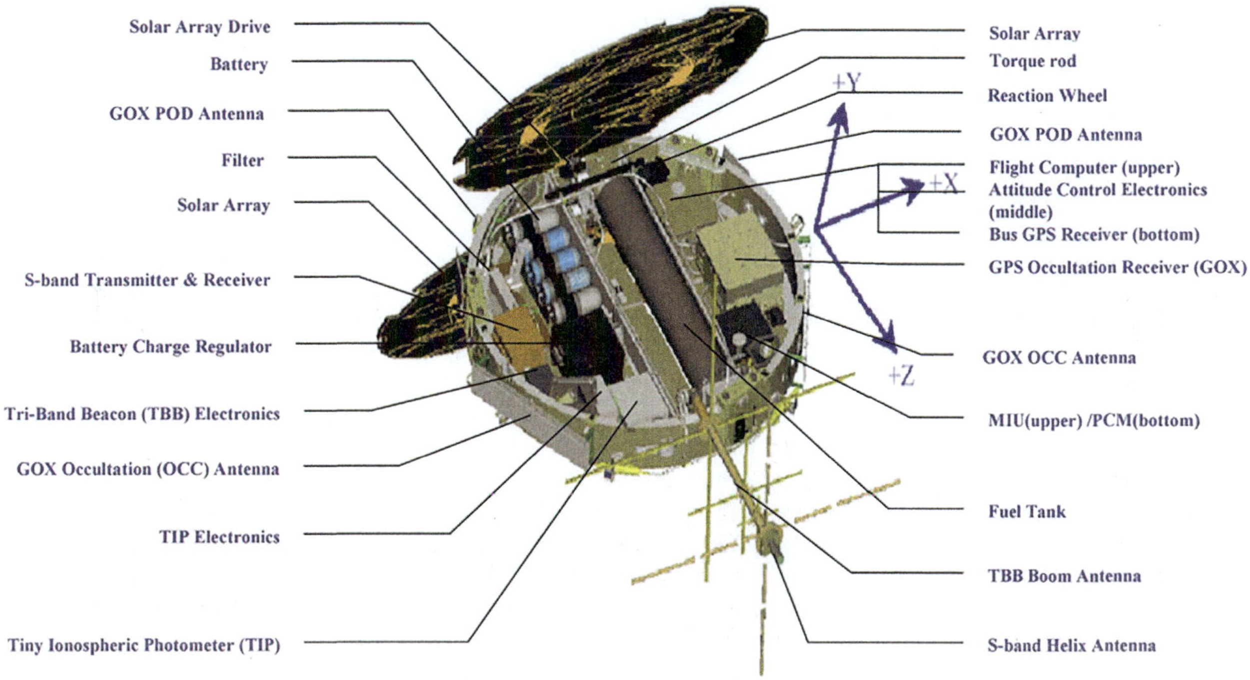

FIGURE 9.51 Cutaway view of the COSMIC-1 satellite.
Image credit courtesy NRL, UCAR.

point radius can be derived. This can be processed to derive both atmospheric profiles and electron density in the ionosphere.

COSMIC-1 is built as a "plate-like" unit (Fig. 9.51) so that multiple satellites can be stacked and launched together. The first launch was on April 15, 2006 from Vandenberg AFB, CA on a Minotaur-1 launch vehicle. All six spacecraft were launched into a single orbital plane. This was a "parking" orbit at 500 km altitude and an inclination of 72 degrees. This initial launch was followed by a 13-month constellation deployment sequence where each satellite is separated from the launch vehicle individually. After an on-orbit checkout, each satellite was boosted by onboard thrusters to different altitudes ranging up to 800 km. The different rates of orbital nodal precession will then gradually drift the orbit planes apart, until an approximately even distribution of six orbit planes is achieved. The satellites will begin collecting occultation atmospheric soundings during the orbit-adjustment period. The total deployment time is approximately 387 days. The sixth satellite moves first and during the transition phase, there are 121 days for the fifth and fourth satellites to operate in a tandem mode. There are also 198 days for the third and second satellites to be in tandem flight. These

FIGURE 9.52 COSMIC-2 in the clean room.
Image credit courtesy UCAR.

temporary configurations are being used for a gravity study to determine the high-order harmonics of the geopotential from the GPS data.

In May of 2013, the COSMIC-1 constellation provided 1,600 daily RO atmospheric profiles in the neutral atmosphere, along with 1,300 daily electron density profiles, even after 7 years of operation. The data are processed at the COSMIC Data Analysis and Archival Center in the United States and at the Taiwan Analysis Center for COSMIC (TACC) in Taiwan. 90% of the RO profiles are delivered to operational weather forecasting centers within 3 hours of observation. These RO profiles have proven to be significant in the Earth and Space Weather Communities.

On August 29, 2016, the United States agencies and Taiwan agreed to move ahead with a follow-on RO mission known as COSMIC-2. On June 25, 2019 COSMIC-2 satellites were launched from the Kennedy Space Center in Cape Canaveral, Florida, and initially deployed into a low Earth orbit at 720 km altitude with an inclination of 24 degrees. The spacecraft were subsequently lowered into orbits of about 550 km altitude, with the last of the six COSMIC-2 satellites achieving its orbit in February of 2021. The COSMIC-2 satellites (Fig. 9.52) were roughly of the size between a standard kitchen oven and a refrigerator. They all carried the following instruments:

- Tri-Global Navigation Satellite System Radio Occultation Receiver (TGRS). This receiver provides high-quality cross sections of the atmosphere. It remotely senses temperature, pressure, moisture, and lightning formation in the atmosphere. It also measures electron density and flashes of light produced by the passage of particles through the upper atmosphere in a phenomenon known as scintillation. This instrument was developed at NASA's Jet Propulsion Lab (JPL).
- Ion Velocity Meter (IVM) This instrument measures ion velocity in the upper atmosphere, which is a key metric for space weather predictions as well as ion concentrations, composition, and temperature. These observations are used to assess the effects of geomagnetic and solar storms on our atmosphere. This instrument was provided by the US Air Force.
- RF Beacon: This instrument measures the total electron content in the Earth's upper atmosphere, which is used in space weather prediction.

COSMIC-2 data are received by a system of ground stations located throughout the tropics (Fig. 9.53).

Their array of global downlink stations (Fig. 9.42) makes it possible to quickly download all of the data from each of the six satellites. After reception the data are transferred to data processing centers in the United States and Taiwan. Once processed the atmospheric profile data are then provided to the World Meteorological Organization Global Transmission System (GTS), which then distributes the data products to meteorological agencies and weather prediction centers around the world. They are used to improve forecasting, especially of short-term severe weather phenomena that are difficult to retrieve from traditional atmospheric sampling methods. The water vapor profile information improves forecasting for

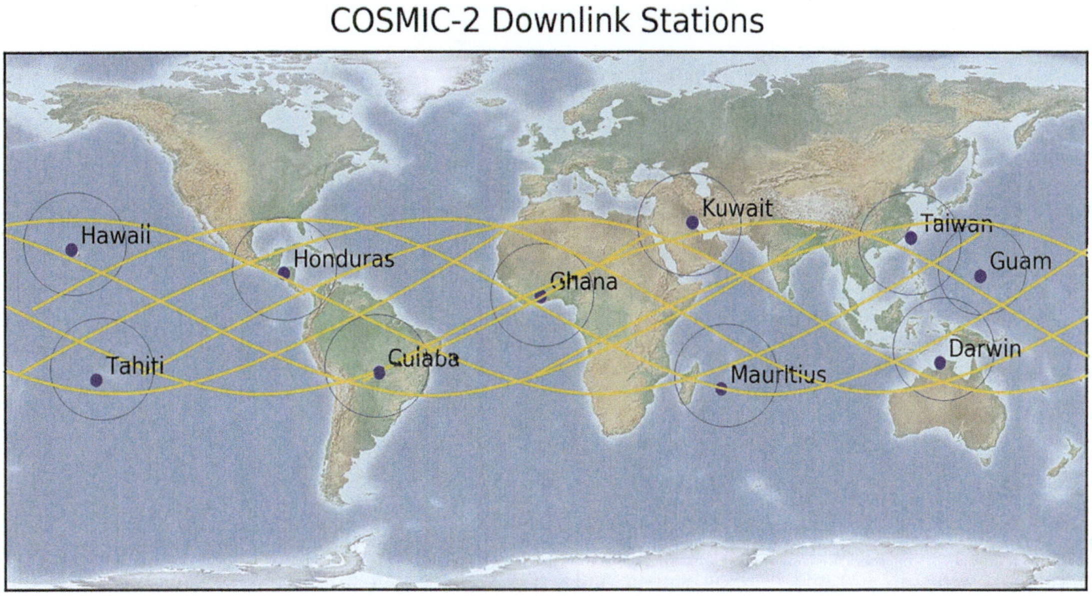

FIGURE 9.53 Map of COSMIC-2 ground stations. Map lines delineate study areas and do not necessarily depict accepted national boundaries. Image credit courtesy UCAR.

FIGURE 9.54 Dong Fang Hong 1 satellite. (https://en.wikipedia.org/wiki/Dong_Fang_Hong_1).

hurricanes, typhoons and other severe storms. The ability to monitor the effects of solar flares and other space weather variables increases the lead time for detecting events that could put our electric grids at risk.

9.6 Chinese satellites

The first orbiting Chinese satellite was Dong Fang Hong 1 (DFH-1; Fig. 9.54), which was successfully launched on April 24, 1970.

The satellite carried a radio transmitter which broadcast the then de facto national anthem of the same name. The satellite lasted for 20 days. It was followed by Dong Fang Hong 2, launched on April 8, 1984 into geostationary orbit and became the primary television broadcast satellite used by China. It was developed by the Chinese Academy of Space Technology (CAST) with a design life of 4.5 years. It was followed by a series of communication satellites up to 1991.

China's first satellite (The East is Red), made China the fifth space faring nation in the world. This 173 kg satellite was the heaviest first satellite ever placed in orbit by any country and exceeded the combined masses of the first four satellites launched by the superpowers. DFH-1 was a spherical polyhedron one meter in diameter and made of fiberglass-wound on aluminum alloy. Its surface was covered with insulation material designed to protect the instruments from the extreme heat

FIGURE 9.55 CBERS-1 satellite (https://en.wikipedia.org/wiki/CBERS-1).

and cold of Earth orbit. A sealed instrument module contained five silver zinc batteries to provide power to the satellite for 20 days. The instruments were an infrared horizon sensor and a solar angle gauge. On November 26, 1975 China successfully launched its first recoverable satellite and returned it to Earth. Between 1986 and 2000, China launched 27 foreign-made satellites during which time they experienced some failures that were resolved with help from the US science community.

The China-Brazil Earth Resources Satellite (CBERS-1, Fig. 9.55) was launched by China on October 14, 1999.

This satellite was instrumented with multiple systems:

- A Wide Field Imagery (WFI) with visible and near-infrared channels, with a spatial resolution of 260 m and a swath width of 890 km.
- A high-resolution CCD camera for multispectral imaging with a 20-m resolution and a swath width of 113 km.
- An Infrared Multispectral Scanner (IMS) with a spatial resolution of 20 m and a swath width of 120 km. This was followed by a series of four additional CBERS satellites up to CBERS-4, which carried four instruments:
- MUXCam a multispectral camera;
- PanMUX is a panchromatic imager;
- The Infrared Medium Resolution Scanner (IRSCAM);
- WFICAM a wide-field imaging camera.

China began its geostationary satellite program, called Feng Yun 2 (FY-2), in 1980, and the satellites (Fig. 9.56) were built by the Shanghai Institute of Satellite Engineering.

FY-2 was spin-stabilized with a rotational speed of 100 RPM. FY-2 data is open and available to international users who can receive S-VISSR high-resolution digital data and WEFAX low-resolution analogue images. The first FY-2 satellite was to be launched on April 2, 1994 when a fire broke out, and the associated explosion destroyed the launch vehicle and the satellite, killing one worker and injuring 20 more. The Chinese Meteorological Administration (CMA) launched the replacement FY-2A satellite on June 10, 1997 into geostationary orbit at 105°E longitude. On April 8, 1998 FY-2A ceased transmission of images due to a failure of the S-band communications antenna. FY-2B was launched on June 25, 200, and FY-2C was launched on October 19, 2004. FY-2D, FY-2E, FY-2F, and FY-2G followed in 2006, 2008, 2012, and 2014, respectively. The final satellite in the series, FY-2H, was launched in June 2018.

The FY-3 series of meteorological satellites ushered in the use of three-axis stabilized satellites (Fig. 9.57) in geostationary orbit. This series was developed collaboratively by CMA and the China National Space Administration (CNSA) and consisted of five spacecraft, FY-3A, FY-3B, FY-3C, FY-3D, and FY-3E, which were launched in May 2008, November 2010, September 2013, November 2017, and July 2021, respectively. FY-3A was decommissioned in 2018, and an additional five satellites (FY-3F, FY-3GK, FY-3H, FY-3I, and FY-3J) are scheduled to be launched between 2022 and 2025. The series objective is to provide global measurements of 3D temperature and moisture soundings, cloud and precipitation parameters, meteorological and hydrological events, biosphere anomalies and geophysical parameters in support of global change and climate monitoring. This series did not become operational until the launch of FY-3C in 2013.

FIGURE 9.56 Feng Yun 2B (https://en.wikipedia.org/wiki/Fengyun).

The instruments to be carried are SES, PR MWHS-1, MERSI, TOU/SBUS, ERM-1, ASI, ERM-2, MWTS-2, MWTS-1, SWS/SEM/IMS, VIRR, MERSI-2, MERSI-3, SWS/Tri-IPM, SSIM, MVIRS, GNOS, WindRAD, SIM-2, MERSI-LL, OMS, SIM, GAS, IRAS, HIRAS, SEM, SWS/SEM/HEPD, WAI. To help unravel this alphabet soup, here are some descriptions of these instruments.

SES satellite communications in geosynchronous orbit
PR precipitation radar
MWHS-1 microwave humidity sounder
MWTS-2 microwave temperature sounder
MERSI 2,3 medium resolution spectral imager
TOU/SBUS total column ozone mapper/solar backscatter ultraviolet sounder
ERM-1,2 broad-band radiometer that measures the Earth's radiation budget
ASI atmospheric structure instrument
GNOS global navigation satellite system occultation sounder
WindRAD wind radar
SIM-2 solar irradiance monitor
MERSI-L medium resolution spectral radiometer-low light
SWS/SEM/IMS space environment monitor
VIRR visible infrared radiometer
SWS/Tri-IPM space weather suite
SSIM spectral solar irradiance in UV/VIS/NIR/SCWIR
MVIRS multichannel visible infrared scanning radiometer
OMS ozone monitoring suite
SIM spectral irradiance monitor
GAS greenhouse assessing spectrometer
IRAS infrared astronomical telescope
HIRAS hyperspectral infrared atmospheric sounder
SEM space environment monitor (flux of energetic particles)
SWS/SEM/HEPD space weather suite/space environment monitor/high energy particle detector
WAI wide field auroral imager

It should be noted that this list covers the instruments carried by both geostationary and polar orbiting weather satellites.

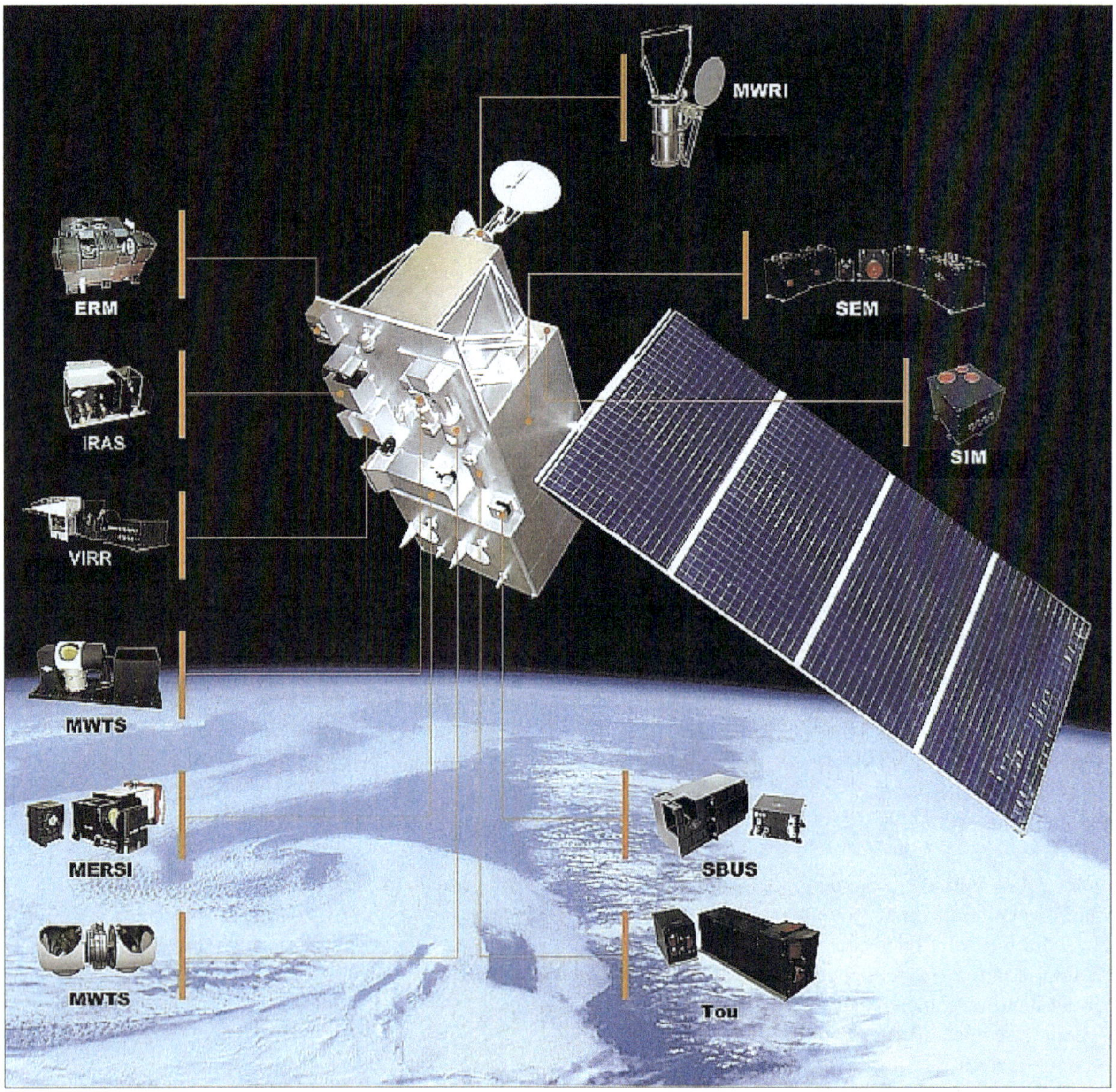

FIGURE 9.57 Illustration of the FY-3A spacecraft and instruments.
Image credit courtesy CMA/NSMC.

There was also a FY-3 series of polar-orbiting weather satellites (Fig. 9.58), which was again a collaborative effort between CMA and CNSA. The instruments listed above flew on both FY-3 geostationary and polar orbiting satellites. The first of the polar orbiters was also known as FY-3A, and it was launched on 27 May 2008 into a sun-synchronous near-circular orbit, at an average altitude of 836.4 km, an inclination of 98.75 degrees, and an orbital period of 101.49 minutes. The repeat period is about 6 days. This was considered an experimental satellite, as was its twin FY-3B, launched on November 4, 2010.

First, with FY-3C, launched on 23 September, the data were considered operational. There are three modes of FY-3 data:

- Delayed Picture Transmission (DPT), which is a direct playback mode. All the stored science and engineering data onboard (except the MERSI data) is transmitted at a high data rate of 110 Mbits/sec to the ground stations (Beijing, Guangzhou, Urumqi, Jamusi, and Kiruna) whenever the satellites pass over these stations.

FIGURE 9.58 Illustration of the FY-3 polar orbiting satellite.
Image credit courtesy CMA/NSMC.

- Mission Picture Transmission (MPT), which provides direct broadcast in X-band. The main purpose of this mode is for real-time broadcasting of the science and engineering data of the MERSI instrument with a data rate of 18.7 Mbits/sec to any available ground station.
- Advanced High-Resolution Picture Transmission (AHRPT), also known as HRPT.

FY-3D was launched on November 14, 2017, and FY-3E was launched on July 4, 2021. The CMA wants to have both a morning and an afternoon satellite in the FY-3 polar orbit. FY-3G went up on April 16, 2023, while FY-3F was launched on August 4, 2023. All of the satellites since FY-3B are considered operational, particularly for weather forecasting. FY-3D replaces FY-3B, which has operated for 8 years in geostationary orbit.

The next satellite, FY-3E, was launched on 4 July 2021 into a sun-synchronous near-circular orbit with an altitude of ~800 km, and an inclination of 98.7 degrees. This satellite crosses the equator in the early morning local time, making it the first Chinese civilian satellite to be launched directly into the early morning orbit. Fy-3E joins two other Chinese polar-orbiting weather observation satellites, FY-3C and FY-3D, flying in mid-morning and afternoon orbits. FY-3E carries 11 instruments, 3 of which are for the first time.

FY-3G was launched on April 16, 2023 into a drifting orbit of 407 km altitude and inclination of 50 degrees. Following it, FY 3F was launched on August 4, 2023 into a sun-synchronous orbit at 836 km altitude and an inclination of 98.8 degrees. The next satellite is FY-3H, planned for launch in 2025 into an early morning polar orbit. China made a commitment in 2014 that it would cover the early morning polar orbiting weather satellite operation.

The first visible light image (Fig. 9.59) from FY-3D (true color image of 3 channels on the MERSI-2 instrument) showing both ocean and land areas of the South China Sea (image credit CMA).

The MERSI-2 instrument that produced Fig. 9.49 is the Medium Resolution Spectral Imager, and it has 20 bands covering the spectrum (Table 9.12) with spatial resolutions from 250 m up to 1.0 km. This instrument was intended to cover the cloud and SST needs as well as ocean color sensing.

While this is a very ambitious new sensor, it has been very successful in achieving most of its observation goals of agriculture (crop and forest management), geological exploration, metal mapping, oil detection, soil sensing, and a wide range of vegetation parameters. As mentioned, it was also designed to compute ocean color from its radiance measurements. This is the primary imager carried by all the polar orbiting spacecraft.

In addition to weather satellites, China also has a very active commercial remote sensing sector. The Hongtu-1 (also called PIESAT-1) is a Chinese X-band Interferometric SAR (InSAR) satellite constellation in sun-synchronous orbit. The objective is to map nonpolar regions at a scale of 1:50,000 m to produce high-precision digital terrain models for commercial, scientific, and military applications. The structure is for 12 PIESATs that have been launched between March 30, 2023 and December 17, 2024. The long-term goal is a constellation of 38 satellite, 28 of which are InSAR satellites, and the remaining 10 are optical imaging satellites, including panchromatic and multispectral sensors.

China plans to launch eleven Hunjing satellites for disaster and environmental monitoring (huanjing is Chinese for environment). The first two satellites Huanjing-1A and 1B were launched September 6, 2008, and Huanjing-1C was launched November 16, 2012. They are all equipped with an S-band SAR instrument manufactured in Russia. The follow-on Huanjing-2A and 2B were launched September 27, 2020. All of these satellites continue to operate.

China has also built, launched, and operated a satellite in cooperation with Venezuela known as VRSS-1. It was launched on September 29, 2012 and carries two different resolution cameras; one with a spatial resolution of 2.5 m in

FIGURE 9.59 First visible light image from FY-3D (true color image from 3 channels of the MERSI-2 instrument), showing land and ocean areas of the South China Sea.
Image credit courtesy CMA.

panchromatic mode and 10 m in multispectral mode. The lower resolution camera has a resolution of 16 m. The satellites provide 350 images per day. The satellite flies in LEO. In addition, China operates military satellites that have classified data streams and operate both in the optical and SAR domains.

9.7 Satellites of India

The Indian space program was institutionalized in November of 1969 with the formation of the Indian Space Research Organization (ISRO). The Indian government established the Department of Space (DOS) and brought ISRO into it. DOS has the following responsibilities:

- Indian National Satellite (INSAT) program for telecommunications, TV broadcasting, meteorology, and developmental education.
- Remote sensing program for the application of satellite imagery for various developmental purposes.
- Indigenous capability for the design and development of spacecraft and associated technologies for communications, resources survey, and space sciences.
- Design and development of launch vehicles with indigenous technology for access to space and orbiting INSAT and IRS spacecraft and space science missions.
- R&D in space sciences and technologies, as well as application programs for national development.

ISRO built India's first satellite, Aryabhata, which was launched by the Soviet space agency in 1975. In 1980, India launched the satellite RS-1 onboard the SLV-3 launch vehicle, making India the seventh country to build and launch its own satellites. It then took ISRO 7 years to develop the Satellite Launch Vehicle (SLV) capable of putting a 40 kg satellite into a 400 km LEO. Their first attempt in 1979 carried a Rohini technology payload but could not inject the satellite into the desired orbit. It was followed by a successful launch in 1980 carrying a Rohini-1 satellite; RS-1 was the third Indian satellite to reach orbit. They then developed another solid-fueled rocket, the Augmented Satellite Launch Vehicle (ASLV), to carry satellites into geostationary orbit. The ASLV had numerous failures and was discontinued for the development of liquid launch vehicles. The year 1977 saw the launch of Bhaskara-1, which was India's first remote sensing satellite. It was equipped with two TV cameras, an MWR operating at 19 and 22 GHz, and an X-ray sky monitor. Bhaskara-2 was similar and was launched on November 20, 1981. One of its cameras ended early in the mission. Rohini-1 was the first

TABLE 9.12 MERSI-2 Spectral bands.

Band	CW (μm)	BW (μm)	IFOV (m)	NEΔ*T* (%) or NEΔ*T* (K@300K)	Dynamic range (maximum *ρ* or T)
1	0.470	0.05	250	0.45	100%
2	0.550	0.05	250	0.4	100%
3	0.650	0.05	250	0.4	100%
4	0.865	0.05	250	0.45	100%
5	11.25	2.5	250	0.54 K	330 K
6	1.640	0.05	1000	0.08	90%
7	2.130	0.05	1000	0.07	90%
8	0.412	0.02	1000	0.1	80%
9	0.443	0.02	1000	0.1	80%
10	0.490	0.02	1000	0.05	80%
11	0.520	0.02	1000	0.05	80%
12	0.565	0.02	1000	0.05	80%
13	0.650	0.02	1000	0.05	80%
14	0.685	0.02	1000	0.05	80%
15	0.765	0.02	1000	0.05	80%
16	0.865	0.02	1000	0.05	80%
17	0.905	0.02	1000	0.10	90%
18	0.940	0.02	1000	0.10	90%
19	0.980	0.02	1000	0.10	90%
20	1.030	0.02	1000	0.10	90

CW, Central wavelengths; *BW*, bandwidths; *IFOV*, instantaneous field of view (nadir).

geostationary satellite to be launched with an Indian launch vehicle. The satellite was spin-stabilized and was placed in geostationary orbit to serve as a weather satellite.

The PSLV arrived in the early 1990's. Its first flight carried a satellite into polar orbit in 1994 followed by a couple of failures. After that, the PSLV had more than 50 successful flights. India launched INSAT-1D in 1990 which replaced the defunct INASAT-1A and it provided telecommunications, meteorological imagery and broadcasting services. On October 15, 1994, the PSLV carried the IRS-P2 satellite into a sun-synchronous polar orbit of 800 km altitude. The satellite carried the Linear Imaging Self-Scanning Sensor-2M (LISS-2M) with two solid state push broom cameras using CCDs and were capable of providing imagery in four spectral bands in the visible and near-infrared with a 32-m resolution.

On May 26, 1999 ISRO launched its first commercial satellites (IRS-P4, KITSAT-3, and DLR-TUBSAT) aboard the PSLV-C2 launch vehicle. IRS-P4 was also known as Oceansat-1, which operated in a sun-synchronous orbit. It carried two instruments:

- Multifrequency Scanning Microwave Radiometer (MSmR) that operated at 6.6 GHz, ideal for monitoring SST.
- Ocean Color Monitor (OCM) is designed to monitor the color of the ocean with eight spectral channels ranging from 400 to 885 nm.

It had a spatial resolution of 350 m and a swath width of 1420 km.

In India space activities come under the ISRO, which comes under the DOS of the Indian Government. ISRO was founded on August 15, 1969, and its headquarters are in Bengaluru.

FIGURE 9.60 Bhaskara-1 satellite.
Image credit courtesy ISRO.

The first Indian environmental satellite was Bhaskara 1 (Fig. 9.50) launched on June 7, 1979 by ISRO carried two television cameras operating in the visible (0.6 μm) and near-infrared (0.8 μm). It also carried a microwave radionmeter MWR (SAMIR) operating at 19 and 22 GHz for the study of ocean state, water vapor, and liquid water in the atmosphere. The final instrument was an X-ray sky monitor operating in the 2–10-keV energy range to detect transient X-ray sources and monitor their variations. It was placed in LEO at an altitude of about 400 km and an inclination of 50.7 degrees. It was followed by an identical satellite Bhaskara-2 launched on November 20, 1981 (Fig. 9.60).

The next satellite was Rohini satellite RS-D1 which was an experimental spin stabilized satellite to test out some Earth observing instruments. This first launch, however, did not attain the intended orbit and the satellite fell out of orbit in 9 days. The second satellite in this series launched April 17, 1983, attained its orbit of 371×861 km and performed observations in both the visible and infrared portions of the spectrum. It operated for 7 years collecting over 2,500 images of the Earth.

India embarked on a mission to provide a unique capability in geostationary orbit of a communications and meteorological satellite. The first of the Indian National Satellite (INSAT) four satellites were strictly communications satellites, and only with INSAT-2A do we have a truly multipurpose satellite handling both communications traffic and meteorological observations. Like the earlier satellites, INSAT-3A was built by Ford Aerospace, and it was launched on Ariane 4 in August 1992. The radiometer it carried had a resolution of 2 km in the visible band and 8 km in the infrared channel. This was the world's first operational three-axis stabilized geostationary weather satellite. Its follow-on ONSAT-2B was also launched on Ariane July 23, 1993. These satellites also carried a DCS and a Satellite Aided and Rescue (SASAR) system. The Very High Resolution Radiometer (VHRR) was capable of operating in three scan modes, namely, full disc scan, normal scan, and sector scan mode.

INSAT-2E followed the earlier geostationary weather satellites and is positioned over 83 degrees East, providing both communications and meteorological data for Asia and Australia. It carried the VHRR as earlier but added a CCD camera capable of returning images with a spatial resolution of 1 km. INSAT-3A (Fig. 9.61) continues this series, but the satellites are now built in India.

The launch continued on Ariane-5 on April 9, 2003. It was eventually placed in geostationary orbit at 93.5 degrees East. The VHRR continues the earlier imaging capability with a visible band (0.55–0.75 μm), an infrared band (10.5–12.5 μm), and an atmospheric water vapor channel (5.7–7.1 μm). The visible, near infrared, and shortwave infrared all have a spatial resolution of 1 km. As earlier satellites, this satellite carries DCS and SARP payloads. The follow-on in this series is INSAT-3D launched on July 26, 2013, and the satellite incorporates new features such as star sensors and a new bus management system. It still carries a 6-channel multispectral imager and a 19-channel atmospheric sounder. It ceased operation in September 2020. The next in this series is INSAT-3DS, which was launched on February 17, 2024. It carries the same meteorological payload as previous satellites.

The actual first Indian indigenous remote sensing satellite was the Indian Remote Sensing satellite 1A (IRS-1A) which was launched on March 17, 1988 from the Soviet Cosmodrome at Baikonur. It was inserted into a sun-synchronous polar

FIGURE 9.61 INSAT-3A satellite.
Image credit courtesy ISRO.

orbit of approximately 900 km altitude. It carried two linear imaging self-scanning sensors LISS-1 and LISS-2 with spatial resolutions of 772 and 36 m, respectively. These were actually "push broom" scanners. It operated for 4 years. It was followed by a series of IRS satellites with the last one launched in 1997.

India also had an Oceansat-1 (also IRS-P4) that was launched on May 26, 1999 from India. It carried the Ocean Color Monitor (OCM) and the Multifrequency Scanning Microwave Ratiometer (MSMR). It operated in a sun-synchronous orbit of about 700 km altitude. The MSMR instrument collected radiation in the 6.6 GHz band. The OCM is a solid-state camera designed to monitor the color of the ocean in eight bands between 400 and 885 nm. It has a 1420-km swath width and a spatial resolution of 350 m. It completed its mission on August 8, 2010 after 11 years of operation.

The first strictly meteorological satellite was Kalapana-1, named after the NASA astronaut Kalapana Chawla who was lost in the Space Shuttle Columbia disaster. It was launched into geostationary orbit on 12 September 2002 and originally known as MetSat-1. On February 5, 2003, it was renamed Kalapana-1 to honor the astronaut. The satellite carried the VHRR scanning radiometer with channels in the visible, near infrared, and thermal infrared with a resolution of 2 km. It ceased operation in September of 2017.

The Indian Remote Sensing (IRS) satellites were a series of optical imaging satellites operated by the ISRO that started in 1988. The first satellite (Fig. 9.62) was launched on March 17, 1988 and continued to operate until January 1, 1994.

Part of the integrated LEO element of India's National Natural Resources Management System (NNRMS) had the objective to observe and enable the management of the country's natural resources (applications in agriculture, hydrology, geology, drought and flood monitoring, marine studies, snow studies, and land use). IRS-1A flew in a sun-synchronous

FIGURE 9.62 IRS satellite.
(image credit courtesy ISRO).

TABLE 9.13 Characteristics of the LISS 1,2 Sensors.

Parameter	LISS-1	LISS-II A/B
FOV/IFOV	9.4 degrees, 80 µrad	4.7+4.7 degrees, 40 µrad
Spectral bands	0.46–0.52 µm (blue)	0.46–0.52 µm (blue)
	0.52–0.59 µm (green)	0.52–0.59 µm (green)
	0.62–0.68 µm (red)	0.62–0.68 µm (red)
	0.77–0.86 µm (NIR)	0.77– 0.86 µm (NIR)
Integration time (ms)	11.2	5.6
Ground resolution	72.5 m all bands	36.25 m all bands
Swath width	148 km	148 km
Radiometric resolution	7 bit	7 bit
Detector elements	CCD 2038	CCD 2048

orbit with a nominal altitude of 904 km, an inclination of 99.048 degrees a period of 103.2 minutes, a repeat cycle of 22 days, and an equatorial crossing time of 10:26 hours on its descending node.

The satellite carried the LISS-1,2 (Linear Imaging Self-scanning Sensor), which are two multispectral cameras, each with different spatial resolutions. The characteristics of these instruments are given here in Table 9.13.

A follow-on satellite, IRS-P2, was launched October 15, 1994, in a sun-synchronous circular orbit with an altitude of 817 km, inclination of 98.7 degrees, period of 101.35 minutes, and a repeat cycle of 24 days. This satellite discontinued operation on September 1997.

Following the 2008 Mumbai attacks by Islamist terrorists ISRO decided to have its own radar imaging satellite which they procured from Israel Aerospace Industries and was known as RISAT and it was launched on 20 April 2009. It carried a C-band SAR developed by ISRO for the mission. It was intended to give all weather monitoring capability including tracking ships that are potentially hostile to India. It was actively used to locate the wreckage of a helicopter crash that claimed the life of Y.S. Rahasekhara Reddy, the chief minister of Andhra Pradesh.

The IRS series, IRS-P5, was known as Cartosat-1, which was the first in a series of Earth mapping satellites. It was launched on May 5, 2005 into a sun-synchronous orbit of about 620 km. Built and launched by ISRO it is operated by the National Technical Research Organization (NTRO) with a payload of two state-of-the-art panchromatic cameras that collect black and white stereoscopic images of the Earth in the visible band. The spatial resolution of these cameras is 2.5 m with a swath width of 30 km. The stereo images are collected with one camera pointed forward at +26 degrees and one aft at –5 degrees to yield nearly instantaneous stereo imagery. The time difference in the collection of the two scenes is about 52 seconds. The spacecraft is steerable to compensate for Earth's rotation and to force both the fore and aft cameras to view the same spot on the ground. The cameras are mounted on the spacecraft so that near simultaneous imaging of the same area from two different angles is possible, enabling the 3D reconstruction.

ScanSat-1 was India's first scatterometer satellite developed by ISRO with a payload developed by the Space Applications Center in Ahmedabad. This is a Ku-scatterometer similar to that carried by Oceansat-2. This mini satellite was launched September 26, 2016, and generated data used by meteorological agencies around the world. The scatterometer (OSCAT) operated in LEO at an altitude of approximately 740 km, an inclination of 98.1 degrees, and a period of 98.2 minutes. This mission came to a close after instrument failure on February 28, 2021.

HySIS was a hyperspectral imaging satellite launched on November 29, 2018 into a sun-synchronous polar orbit at about 645 km altitude. It carried two different instruments: (1) the visible near infrared radiometer (VNIR) with 60 spectral bands covering the range from 0.4 to 0.85 μm. (2) the Shortwave Infrared Range (SWIR) covering the spectral range of 0.85–2.4 μm with 10-nm bandwidths and 256 contiguous spectral bands. The instruments have spatial resolutions of 30 m and a swath width of 30 km.

Chapter 10

Commercial remote sensing

10.1 SPOT image

In recent years, commercial remote sensing has really exploded, and the advent of small satellites has enabled companies to collect optical and radar imaging data from constellations of small satellites. One of the first commercialization of space efforts was by the French company SPOT image (Satellite pour l'Observation de la Terre), which grew out of the French space agency (CNES) and became a public limited company in 1982. While SPOT was covered earlier in this book, we repeat it here for completeness. SPOT-1 (Fig. 10.1) was launched on 22 February 1986 into a syn-synchronous, circular, phased orbit with an inclination of 98.7 degrees at an altitude of 832 km. The satellite was equipped with a high resolution visible (HRV) imaging instrument that could operate either in panchromatic or multispectral modes. The panchromatic mode had a spatial resolution of 10 m while the multispectral mode (G, R near infrared [NIR]) had a resolution of 20 m. The scene size was 3,600 $km^{2,}$ and the revisit time varied from 1 to 4 days depending on the latitude. SPOT 2 was identical and launched on January 22, 1990, followed by another identical satellite SPOT-3, launched on September 26, 1993.

With SPOT-4, a new sensor was introduced, the high-resolution visible and infrared (HRVIR) radiometer, which had the same geometry as the earlier satellites but added a new shortwave infrared spectral band (SWIR). SPOT-4 carried two identical HRVIR instruments that could be operated together or individually. The HRVIR has an oblique viewing capability of 27 degrees on each side of nadir. The swath width was 117 $km^{2,}$ and if both sensors were used simultaneously, there was a 3 km^2 overlap.

SPOT-5 was launched on May 4, 2002 and carried two new high resolution geometrical instruments that were derivatives of the HRVIR on SPOT-4. They provided a higher spatial resolution of 2.5–5 m in the panchromatic mode or 10 m in the multispectral mode (20 m for the SWIR from 1.58 to 1.75 μm). This satellite also carried a high resolution sounder (HRS) imaging instrument operating in the panchromatic mode. This instrument can point forward and aft to make it possible to collect stereo images used to make a digital elevation map.

SPOT-6 was launched by India's Polar Satellite Launch Vehicle on September 9, 2012, and SPOT-7 followed on June 30, 2014. Together they form a constellation of Earth-imaging satellites to provide high-resolution Earth imaging for the following decade. The architecture followed the Pleiades satellites (Fig. 10.2) built under a joint French–Italian Optical and Radar Federated Earth Observation program. SPOT image is the official and exclusive worldwide distributor of Pleiades products and services.

SPOT 6,7 are phased in the same orbit as Pleiades 1A and 1B at an altitude of 694 km, forming a constellation of satellites 90 degrees apart from each other. The cameras carried by all of these satellites had both panchromatic (450–745 nm) and multichannel (blue 450–525 nm; green 530–590 nm; red 625–695 nm; and near infrared 760–890 nm) capabilities with varying spatial resolutions (panchromatic 1.5 m; color merge 1.5 m; multispectral 6 m).

In the US SPOT, data is available through the SPOT Image Corp. of the United States. All of the historical archive holdings are also maintained and preserved by the USGS/Earth Resources Observation and Science center in Sioux Falls, South Dakota. Thus, there are numerous ways to access SPOT data online.

10.2 Landsat

As discussed earlier, the transfer of Landsat operations to the private sector was not successful. In 1983 NOAA took over the operation of Landsat from NASA, and in anticipation of privatization, NOAA increased the cost of Landsat imagery, resulting in a dramatic decrease in the use of Landsat imagery. At the same time, there was pressure to privatize NOAA's meteorological satellites and their data. This was met with strong resistance by the weather community and by foreign operators of meteorological satellites who had been sharing their data. As a result of this strong resistance, the effort to move the weather satellites and their data to the private sector failed to win approval within Congress and was dropped.

Satellite Remote Sensing. DOI: https://doi.org/10.1016/B978-0-443-22086-9.00012-5

FIGURE 10.1 SPOT satellite (based on https://en.wikipedia.org/wiki/SPOT_(satellite)).

FIGURE 10.2 Pleiades and SPOT-6,7 satellite (based on https://en.wikipedia.org/wiki/Pléiades_(satellite)).

With no strong opposition within Congress to moving Landsat into the private sector, the Reagan administration pushed ahead with its plans to privatize Landsat. However, neither the small market for Landsat products nor the negative results of three studies commissioned by the Department of Commerce to review the matter provided support for commercializing the Landsat system. In spite of these conditions, Congress held a series of hearings on the issue and decided Landsat was ready for a phased transfer to private sector development and operation. On January 3, 1984, the Department of Commerce issued a request for proposal for a private firm to take over operational control of the Landsat system, market the data, and prepare to take full responsibility for the system in a few years. The competition was won by a new company, EOSAT, formed by RCA and Hughes Aircraft. According to the plan, NOAA would work with EOSAT to develop and build Landsat 6 and 7, which they would then operate and market the data. It was expected that Landsat-6 would be ready for launch in 1990 or 1991. Congress would later have problems with continued Landsat funding, and the result was that EOSAT agreed to develop only Landsat-6, and President Bush directed the National Space Council and the Office of Management and Budget to examine ways of continuing the program after the launch of Landsat-6. Part of the problem was that no agency wanted to provide funds to continue the Landsat program. It was estimated that operating costs were between $25 million and $40 million, of which only about $10 million could be recovered in data sales. This did not recognize the fact that the increase in the cost of Landsat imagery put it out of the reach of science users of the data.

At this time, competition arose with the advent of SPOT and other international land remote sensing satellites. By the early 1990's pressure had built up to return the Landsat system to government operation. Landsat and SPOT imagery had proved critical in the 1992 Gulf Conflict. In particular, the imagery was useful for mapping activities in the Persian Gulf. Moreover, global change researchers found an increased need for Landsat data, which was now priced out of their reach. John Knauss, NOAA Undersecretary for Oceans and Atmosphere, stated, "our experience with the Landsat program convinces us that commercialization of Landsat as previously envisioned, is not possible." As a result of these pressures, the control of the Landsat program transferred the control of Landsat from NOAA and EOSAT to the Department of Defense (DoD) and NASA. This was made official in the Land Remote Sensing Policy Act of 1992, which was signed into law in October.

EOSAT expected to continue supplying Landsat data through Landsat-6, which would carry an Enhanced Thematic Mapper (ETM) with, in addition to having better radiometric calibration, also involved additional pan-sharpening that would approach the capability of SPOT. Initially, DoD and NASA planned for a Landsat-7 that would look very much like Landsat-6 but later decided to replace the ETM with a High Resolution Multispectral Stereo Imager (HRMSI), which would collect 5 m resolution images. This addition would add $400 million to the budget not only in spacecraft costs but also in modifications of ground systems.

In September of 1993 Landsat-6 was launched but failed to reach orbit. NASA concluded that the costs of operating Landsat-7 with the HRMSI would be too great, while at the same time, DoD decided not to fund the satellite and pulled out of the agreement with NASA. In May 1994 the Clinton Administration resolved the issue of Landsat by assigning the system to NASA, NOAA, and the Department of the Interior (DOI). Under this plan, NASA will procure the satellite and instruments, NOAA will manage and operate the spacecraft and ground systems, and DOI will archive and distribute the data.

10.3 DigitalGlobe and Maxar

This left the commercialization of Earth remote sensing to the private sector, as clearly set out in the 1992 Remote Sensing Act. Shortly after the act was signed, a company called WorldView Imaging Corp. applied for a license, to operate a commercial remote sensing system capable of providing 3 m resolution panchromatic images in stereo. The company later became DigitalGlobe Corp., which later became the Maxar Corp. It received its first license for commercial Earth observation in 1993 and launched its first satellite, Ikonos, in 1990, providing 80 cm imagery of the Earth. It was followed by QuickBird in 2001 launched from Vandenberg AFB in California. In 2003 DigitalGlobe was awarded contracts from the National Geospatial Agency (NGA) in excess of $500 million.

Its first WorldView-1 satellite was launched on September 18, 2007 and was built by Ball Aerospace in Boulder, Colorado. It was equipped with an off-axis camera similar to Quickbird. The camera is a panchromatic imager capable of 50 cm spatial resolution. It had a revisit time of 1.7 days. It operated in low Earth orbit at a nominal altitude of 500 km, an inclination of 97.87 degrees and a period of 94.49 min. It was followed by WorldView-2 that was launched on October 8, 2009 becoming DigitalGlobe's third satellite in orbit. It operated in low Earth orbit with an altitude of approximately 770 km, inclination of 98.40 degrees and a period of 100.16 min. The sensor is capable of panchromatic images at 46 cm resolution or eight-band multispectral imagery with a 1.84 m resolution.

DigitalGlobe purchased GeoEye in 2013 and took over the operation of the GeoEye-1 satellite, which was launched on September 6, 2008 after three launch postponements. The satellite is in a sun-synchronous orbit of 681 km altitude, with an inclination of 98 degrees and a 10:30 am equator crossing time. At the time of its launch, GeoEye-1 was capable of the highest spatial resolution of 41 cm of all the commercial remote sensing satellites. That resolution, however, was only available to the US government. The GeoEye-1 satellite experienced an anomaly in its downlink antenna. This irregularity limits the range of movement of GeoEye-1's downlink antenna, which can influence the satellite's ability to image and downlink data at the same time. The National Geospatial Intelligence Agency (NGA) and Google paid a combined $502 million for upgrades to the satellite's four ground stations.

DigitalGlobe created the world's largest imagery archive with 1 billion square kilometers of Earth imagery, giving it the industry's largest commercial satellite imagery archive. In 2010 DigitalGlobe secured a $3.55 billion EnhancedView agreement with NGA. On August 13, 2014 WorldView-3 was launched, joining Ikonos, Quickbird, WorldView-1, GeoEye-1, and WorldView-2 satellites (Fig 10.3). The sensor is now capable of 31 cm panchromatic imagery, along with a 1.24 m multispectral imagery and 3.7 m SWIR imagery. It also carries a new instrument known as CAVIS (clouds, aerosols, vapors, ice, and snow) with a 30 m resolution.

In 2016 DigitalGlobe acquired Radiant Technologies with their advanced analytical capabilities and extended their customer base across the US Intelligence community. In 2007 DigitalGlobe itself was acquired by MacDonald Dettwiler and Associates (MDA), who then formed Maxar, composed of four business units: MDA, SSL, DigitalGlobe, and Radiant Solutions. SSL was originally Space Systems/Loral, LLC, founded originally as the Western Development Labs of Philco-Ford. In 2020 Maxar sells MDA to Northern Private Capital and acquires Vricon, adding 3D technology to its portfolio. In 2022 Maxar is awarded a 10-year contract by the US National Reconnaissance Office (NRO). In the same year, Maxar acquired Wovenware to bolster its machine learning and 3D capabilities. In 2023 the US private equity firm Advent International, alongside minority investor British Columbia Investment Management Corp., completed its acquisition of Maxar, which remains a US controlled, owned, and operated company. Moreover, in this year, Maxar Intelligence and Maxar Space Systems begin operating as independent business entities.

Most recently, Maxar began a series of new satellites known as WorldView Legion (Fig 10.4), with the first two satellites launched on May 2, 2024. Each satellite is equipped with a 30 cm resolution optical imager that can be pan-sharpened to 15 cm. Each sensor has bands from the coastal color wavelengths (400–500 nm), to the NIR (780–920 nm) with a panchromatic band from 450 to 800 nm. A complete list of the channels is given here in Table 10.1. Six sensors on orbit will cover five million square kilometers of the Earth's surface every day. The six satellites fly in mid-inclination low Earth orbits at 450 km altitude. The orbit covers the most densely populated areas around the mid-latitudes of the Earth and allows for 15 revisits per day. The orbit provides a 15-degree sun angle and a ground sample distance between 0.3 and 1.3 m at a latitude of 36 degrees north. Maxar believes they had four driving requirements. First, they had to assure continuity and offer the highest spatial resolution and most accurate calibration of any commercial satellite imagery in the industry, with collection and delivery characteristics that meet their customer needs and to begin new launches in 2021. Second, the land surface is not uniformly covered with paying customers, so they must image that part of the globe that will generate the most revenue. Third, not everything happens at 10:30 a.m., so they must have the ability to image from dawn to dusk. Fourth, since they are in business to be profitable, they have to do this for less than their previous constellation cost. Maxar plans to utilize

FIGURE 10.3 Illustration of the Maxar 500 series satellite bus (based on https://www.eoportal.org/satellite-missions/worldview-legion#eop-quick-facts-section).

FIGURE 10.4 Artist conception of all the operating Maxar satellites when Legion is fully operational (based on https://geoawesome.com/how-maxars-constellation-is-setting-new-standards-in-precision-change-detection-and-revisit-capabilities/#:~:text=Ideas%20%23Insights%20%23Space-,How%20Maxar%20constellation%20is%20setting%20new%20standards%20in,change%20detection%2C%20and%20revisit%20capabilities&text=With%20all%20six%20WorldView%20Legion,data%20captured%20by%20WorldView%20Legion).

TABLE 10.1 WorldView Legion sensor spectral bands.

Spectral band	Wavelength range (nm)
Panchromatic	450–800
Coastal	400–500
Blue	450–510
Green	510–580
Yellow	585–625
Red	630–690
Red edge 1	695–715
Red edge 2	730–750
NIR	780–920

NIR, Near infrared.

WorldView Legion's rapid revisit capacity, agility, and stereo abilities to create a living digital model of the Earth in 3D, which will provide accurate, reliable, and up-to-date geospatial data around the globe.

The WorldView Legion satellites (Fig. 10.3) are the first to use Maxar's 500 series spacecraft buses, which are constructed at the company's manufacturing facilities in Palo Alto and San Jose, California. The 500 series is branded as a versatile and mid-sized satellite platform with high stability, agility, and pointing accuracy. Each satellite has a 10-year design lifetime

The constellation plan is to have eight Legion satellites on orbit. When all of the satellites are operating, it will give Maxar an unparalleled suite (Fig. 10.4) of high spatial resolution satellite sensors.

The Legion 1 satellite was launched on May 2, 2024, along with the Legion-2 satellite. Legions 3 and 4 were launched on August 15, 2024, followed by Legions 5 and 6 on 4 February 2025. Legions 7 and 8 are scheduled for some time in 2026. All satellites fly in mid-inclination low Earth orbits at 450 km altitude, which enables up to 15 revisits per day of the most densely populated areas around the mid-latitudes of the Earth.

WorldView Legion will use Maxar's existing global network of receiving stations, which are being added to in order to provide better coverage in mid-latitude regions. The company's ground network is connected to a cloud environment, which feeds online platforms like SecureWatch and Global EGD Enhanced Global Delivery (EGD), which in turn serves high-priority imagery to customers in as little as 10 min. Some customers who require secure satellite service outside the internet are able to directly uplink and downlink Maxar data from the Maxar satellites.

It should be noted that Maxar has recently been acquired by Advent International, who subsequently split the company into two parts: (1) Vantor that is focused on data/software and (2) Lantis, which manufactures, launches, and operates satellites. Vantor provides information based on the satellites operated by Lantis.

10.4 Planet labs

Planet Labs public benefits coporation (PBC) is a publicly traded American Earth Imaging company based in San Francisco, CA. Their goal is to image the entire Earth on a daily basis from low Earth orbit with small satellites. They use the 3U-C CubeSat miniature satellites called DOVEs that are generally delivered into orbit as secondary payloads on other rocket launches. Each Dove satellite is equipped with a high-powered telescope and camera that collect optical images over the globe. Planet Labs was founded in 2010 as Cosmoogia by former NASA scientists Chris Boshuizen, Will Marshall, and Robbie Schingler, who then teamed up with John Kuolt in 2011, a student at Oxford University. Mr. Kuolt's Entrepreneurial Project at Oxford became the original business plan for Planet Labs. This group considered the main problem with satellite observations of the Earth was the large and complex nature of most satellites, prompting them to build small and inexpensive satellites that could be easily manufactured in large quantities. Called CubeSats, their first satellite was built in a California garage.

In April 2013 Planet Labs launched two demonstration CubeSats, Dove-1 and Dove-2. Dove-1 was launched on an Antares 110 rocket and Dove-2 went up on a Soyuz Rocket. They were both placed in a sun-synchronous orbit. Dove 3 and 4 were launched in November 2013. In June 2013 they announced plans for a Flock-1 constellation of 18 Earth-observing satellites. The Flock-1 satellites were carried to the International Space Station in January 2014 and deployed into orbit via the NanoRacks CubeSat Deployer in mid-February. The company planned to launch a total of 131 satellites by mid-2015.

As of May 2015 Planet Labs secured a total amount of $183 million in venture capital funding. In July 2015 Planet Labs acquired BlackBridge and its RapidEye satellite constellation. On April 18, 2017, Google completed the satellite of Terra Bella and its SkySat satellite constellation and delivered them to Planet Labs. As part of the sale, Google acquired an equity stake in Planet and entered into a multiyear agreement to purchase SkySat imaging data. On January 21, 2018 a Dove Pioneer CubeSat was part of the payload on a Rocket Lab Electron rocket launched from a privately owned and operated spaceport on Mahia Peninsula in New Zealand.

In September 2018 the company had launched a total of 298 satellites, 150 of which were still functioning. On December 18, 2018 Planet acquired the St. Louis company, Boundless Spatial, a geospatial software company. A few weeks later, Planet's attorneys discovered that Boundless had failed to disclose information concerning material customer contracts, and the acquisition was renegotiated down from $40 million to $16 million. The executives of Boundless had failed to disclose details concerning future work with the NGA. On July 3 the news said that the company had more than 120 active satellites providing daily imaging coverage over the entire Earth. In August of 2020 Planet completed its SkySat Constellation of 21 satellites by launching the final three SkySat satellites on a SpaceX Falcon 9 rocket. In August 2023 Planet cited restructuring as the reason for laying off 10% of its employees. In June 2024 Planet Labs announced a further reduction of nearly 17% of its global workforce. The company stated that the layoffs were necessary to align resources with market opportunities, to improve operational efficiency and support long term growth and profitability.

The Dove satellites are CubeSats that weigh only 4 kg and are 10 X 10 X 30 cm (Fig. 10.5).

In its deployed configuration, the solar panels deploy out away from the spacecraft, and a communications antenna is also deployed (Fig. 10.6).

The satellites carry either the ID PS2 or the ID PS2.5D telescopes. The PS2 has a detector frame with 6,600 pixels across and 4,400 lines down. The detector has a Bayer pattern filter that separates the incoming light into blue, green, and red channels. On top of the Bayer pattern filter, there is a 2-stripe filter where the top half blocks the NIR wavelengths, thus allowing only the RGB light to pass through, and the bottom half is restricted to the NIR wavelengths (Fig. 10.7).

Thus, each frame consists of an RGB image in the top half and a NIR image in the bottom half. The RGB half of each frame is then combined with the NIR half to generate the resulting 4-band image.

In late 2018 Planet launched 25 next-generation Dove satellites with a new instrument, the ID PS2.5D. The PS2.5D instrument uses the same telescope as the PS2 instrument and the same 2D frame detector as in the PS2, but the Bayer pattern filter and bandpass filters in the PS2 have been replaced with a high-performance butcher-block filter. This filter is made up of 4 individual bandpass filters that separate the light into each of the red, green, blue, and NIR channels (Fig. 10.8). The choice of the bandpass filters for PS2.5D matches closely with those of the Sentinel-2 satellite.

FIGURE 10.5 Dove Satellites (based on https://www.satimagingcorp.com/satellite-sensors/other-satellite-sensors/dove-3m/).

FIGURE 10.6 A Dove satellite with solar panels deployed (based on https://www.satimagingcorp.com/satellite-sensors/other-satellite-sensors/dove-3m/).

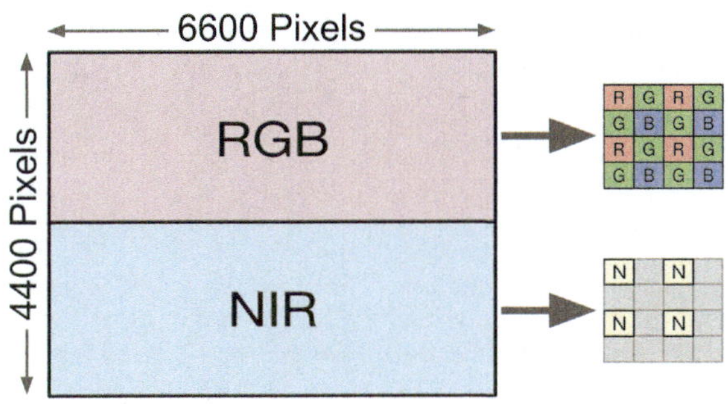

FIGURE 10.7 Image frame for the PS2 instrument (Apollo Mapping LLC). *NIR*, Near-infrared.

Thus, each frame of the PS2.5D instrument consists of four stripes that need to be combined to give the final image (Fig. 10.9). Each image has a spatial resolution of 3 m.

Plans are for a new PSB.5D with a PSBlue telescope with a larger 47 megapixel sensor and the same filter response as PS2.5D in the RGB and NIR bands. This extends the earlier capability by adding additional bands. These are red edge, green I, yellow, and coastal blue (Table 10.2).

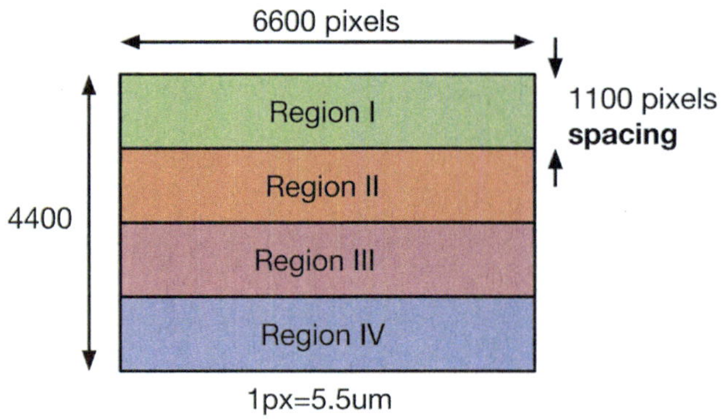

FIGURE 10.8 Frame of the PS2.5D instrument (Apollo Mapping LLC).

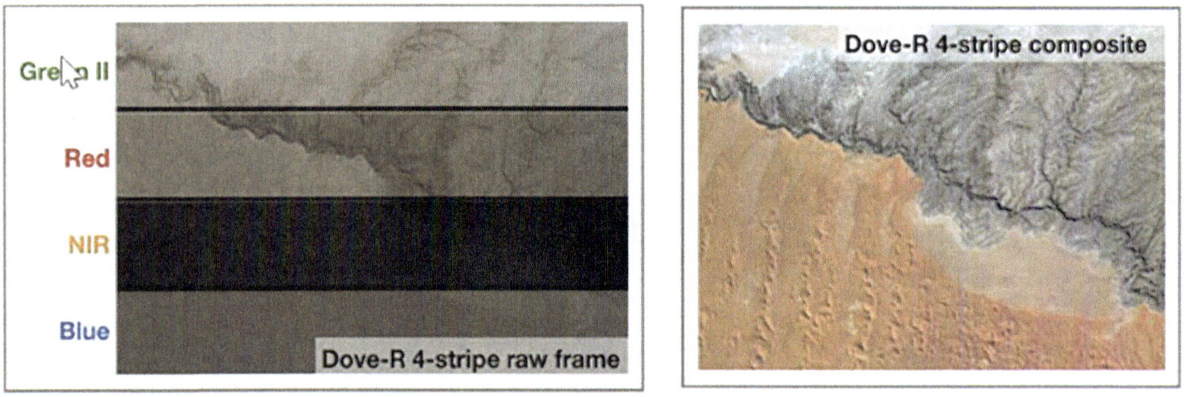

FIGURE 10.9 The four stripes of a PS2.5D image (*left*) and the combined final image (*right*) (all based on https://docs.planet.com/data/imagery/planetscope/).

TABLE 10.2 Spectral bands of the PSB.5D sensor (Apollo Mapping LLC).

Band	Name	Wavelenght (fwhm)	Interoperable with Sentinel-2
1	Coastal blue	443 (20)	Yes—with Sentinel-2 band 1
2	Blue	490 (50)	Yes—with Sentinel-2 band 2
3	Green I	531 (36)	No equivalent with Sentinel-2
4	Green	565 (36)	Yes—with Sentinel-2 band 3
5	Yellow	610 (20)	No equivalent with Sentinel-2
6	Red	665 (31)	Yes—with Sentinel-2 band 4
7	Red Edge	705 (15)	Yes—with Sentinel-2 band 5
8	NIR	865 (40)	Yes—with Sentinel-2 band 8a

NIR, Near infrared.

Within the Planet Organization, the RapidEye satellites constituted a five-satellite constellation producing 5 m resolution imagery. The satellites were built by Surrey Satellite Technology Ltd, and they were launched on a Dnepr rocket from Baikonur cosmodrome in Kazakhstan. Each carried a Jena-Optronik Spaceborne Scanner, which was a multispectral push broom scanner. They collected data in the blue, green, red, red-edge, and NIR.

Planet also assumed the operation of the SkySat constellation of satellites, collecting submeter resolution Earth imagery. The satellites are based on inexpensive automotive grade electronics and fast, commercially available processors. The satellites are roughly the size of a minifridge. The first SkySat satellite was launched on a Dnepr rocket from Baikonur cosmodrome in Kazakhstan on November 21, 2013, with the second going up on July 8, 2014. Four more were launched on September 16, 2016 on the Vega rocket from Kourou, and six more were launched along with four Dove CubeSats on a Minotaur-C from Vandenberg AFB on 31 October 2017. In 2020 Planet lowered its constellation of 15 SkySats from an altitude of 500 to 450 km to improve the resolution of its orthorectified imagery from 80 to 50 cm. On June 13, 2020 SpaceX's Falcon 9 rocket launched SkySats 16, 17, and 18, followed by a Falcon 9 launch of 19, 20, and 21 on August 18, 2020. SkySats carry a multispectral and panchromatic imager and a video sensor. It has a resolution of 90 cm in its panchromatic band. The multispectral sensor collects RGB and NIR imagery with a 2 m resolution. The Pelican constellation is planned to succeed the SkySat constellation. These satellites are planned to have a 30 cm spatial resolution and a revisit time of 10 times per day for most of the globe and 30 times per day for mid-latitudes. The first demonstration satellite was launched on November 11, 2003.

10.5 Satellogic

The Satellogic company was founded in 2010 by Emilliano Kargieman and Gerado Richarte. They launched their first constellation of NuSat satellites in May 2016. On December 19, 2019 Satellogic announced that they had received US $50 million in funding, and then in January 2022, the company went public through a merger with a special-purpose acquisition company (SPAC) (CF Acquisition Corp.).

Emiliano Kargieman worked for a while at the NASA Ames Research Center in Mountain View, California, where he realized the potential to bring the experience he had gained working in Information Technology to the satellite service business. Together with his friend and colleague Gerado Richarte they started Satellogic.

Satellogic built the first two nanosatellites, CubeBug-1, 2 that were launched in China on 26 April 2013. Their third satellite, BugSat 1, launched in June 2014. All of these satellites served as technology testbeds for the NuSat series of satellites. Satellogic plans to have 202 satellites on orbit by 2025 and an expected revenue of $480 million in 2025. Former Secretary of the Treasury Steven Mnuchin and Cantor Fitzgerald CEO Howard Lutnick are major investors in Satellogic. As of June 2024 Satellogc had 26 satellites in operation and a staff of about 140 people. After being appointed as the US Secretary of Commerce, Lutnick resigned from the Satellogic Board of Directors.

The NuSat is a small, light, and inexpensive satellite (Fig. 10.10) that can be mass produced. Each satellite carries three payloads: one for high-resolution multispectral imaging and another for a hyperspectral camera with a ground resolution

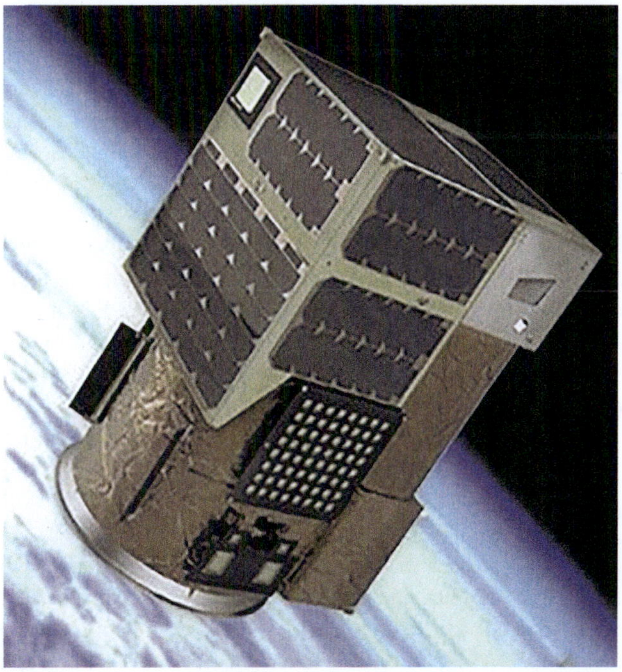

FIGURE 10.10 NuSat Satellogic satellite (based on https://satellogic.com).

FIGURE 10.11 One-meter multispectral image from Satellogic satellite based on https://satellogic.com. Image credit Satellogic.

of 30 m and a 150 km swath and a general-purpose HD camera. The satellite can acquire imagery in three different modes: strip mode, oblique strip mode, and spotlight mode.

The multispectral sensor has the traditional four channels (red, green, blue, and NIR) in the spectral range of 450–900 nm and a 0.7 m spatial resolution and swath of 5 km (Fig. 10.11). The hyperspectral camera is limited to select customers only. It collects imagery with a spatial resolution of 30 m over 600 spectral bands from the visible to the near-infrared (460–830 nm) with a swath of 150 km.

Data are all downlinked at the receiving station at Svalbard, Norway, which includes more than 10 passes per day. The ground station has two antennas to track and communicate with two polar orbiting satellites at the same time. The Satellogic satellites (in the Aleph-1 constellation) also broadcast their data to an S-band receiving station in Cordoba, Argentina, that supports six passes per day.

10.6 ÑuSat (NewSat)

NuSAT is an Argentinian commercial Earth observation company. They created the Aleph-1 constellation of satellites. The satellites (Fig. 10.12) are identical 51× 57 × 82 cm spacecraft that weighs 38.5 kg. The instrument operates in the visible and infrared wavelengths. The satellites are designed, built, and operated by Satellogic. Their first satellite was named BugSat 1d, which was a proof of concept satellite that weighed only 22 kg and carried only amateur radio capabilities.

The Aleph-1 constellation will consist of more than 300 satellites. The first two satellites were launched piggyback on a Chinese Long March 4B rocket in May 2016 into a 500 km sun-synchronous orbit with an inclination of 97.5 degrees. The third satellite was launched in June 2017, and the fourth and fifth were launched in February 2018. The sixth satellite was launched in September 2020, while seventh and eighth went up in January 2020, and the 9–18 were launched on November 6, 2020.

The NuSat imager characteristics are presented here in Table 10.3.

The instrument uses a push-broom scanner, and the constellation provides a 1.2-h revisit time.

10.7 BlackSky technology

BlackSky technology was founded in 2014 as the merger of Osprey Technologies and BlackSky holdings, originally part of Spaceflight Services. After the merger, BlackSky became a public company with an initial startup funding of $283 million. Each BlackSky microsatellite carries the same SV-24 (Spaceview-24) imaging telescope built by the Harris Corp's Exekus subsidiary. This imager has a spatial resolution of 1 m and the satellites orbit at an altitude of 500 km. The imager has four channels with three visible (RGB) and one panchromatic. The BlackSky plan is to have their satellites in a drifting orbit at 430 km altitude, an inclination of 42 degrees and periods of 90 min. Their first two satellites, Pathfinder 1

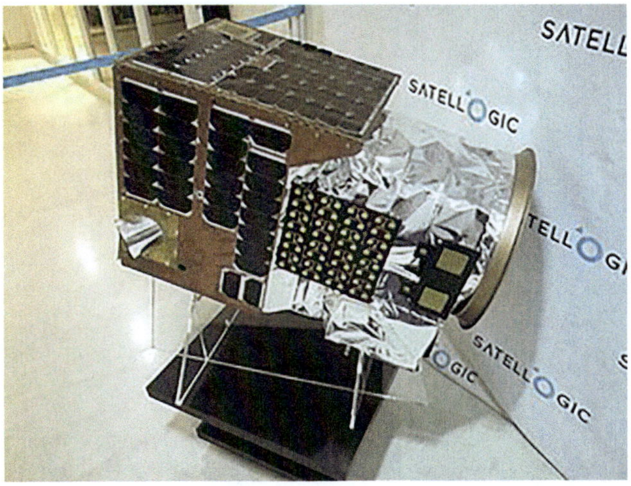

FIGURE 10.12 NuSat based on https://en.wikipedia.org/wiki/ÑuSat.

TABLE 10.3 NuSat image characteristics.

Spectral mode	Panchromatic	Multispectral	Hyperspectral	Thermal infrared
Resolution	1 m	1 m	30 m	90 m
Swath	5 km	5 km	150 km	92 km
Spectral bands	400–900 nm	400–690 nm	400–900 nm	8–14 μm
		400–510 nm	M	0.01 K accuracy
		510–580 nm	5 nm FWHM	
		580–690 nm		
		750–900 nm		

and 2, which were the precursors to the BlackSky satellites, of which 22 have been launched in mid-2024. BlackSky believes its advantage lies in its rapid monitoring capability with satellites able to capture images of the same location up to 15 times per day.

Following their earlier success, BlackSky started a new series of satellites known as Gen-3 (Fig. 10.13), which combines an hourly revisit time with very high resolution and AI-enabled outputs to their customers. This new instrument has a 35 cm spatial resolution (NIIRS-5+), multispectral measurements including the shortwave infrared, AI-enabled analytics, low-latency space communications for last-minute tasking capabilities, and a highly agile attitude control system for maximum efficiency.

The first Gen-3 satellite was launched on an Electron rocket by Rocket Lab in New Zealand on February 19, 2025. It was deployed into a circular orbit at 470 km altitude. BlackSky plans a constellation of 60 satellites in the future. BlackSky won a more than $100 million contract from a long-term strategic partner in the international defense sector for real-time, space-based monitoring capabilities. As stated by BlackSky, "This contract validates the market demand for the capabilities that BlackSky delivers. Customers rely on BlackSky as a trusted partner, delivering real-time decision quality data at mission speed."

The instrument carried by BlackSky Gen-3 satellites is known as SpaceView 24 (built by Harris Corp.'s Excelis division), which is an imaging telescope. It is a pushbroom scanner that is steerable in both the along-track and cross-track directions. It is a five-channel radiometer with one panchromatic band and three visible bands (RGB), including the shortwave infrared. It has a 35 cm resolution and uses the SWIR imagery to improve sensing in low-light and conditions such as clouds, smoke of haze. Image analysis uses AI to automatically understand objects in the image.

FIGURE 10.13 BlackSky Gen-3 satellite (based on https://ir.blacksky.com/overview/default.aspx).

10.8 Commercial synthetic aperture radar

While all these optical systems offer imagery with excellent spatial resolution, the imagery is compromised by cloud cover, fog, and haze. There is a need for an all-weather source of satellite imagery, and the solution is synthetic aperture radar (SAR). Both airborne and spaceborne experiments have clearly demonstrated the value of SAR imagery with its high spatial resolution and its all-weather sensing capability. Commercial SAR companies formed to provide this type of remote sensing imagery to complement and augment the many optical systems that were already available. It is interesting to note that while there were many European and Canadian sources of SAR data in the years between Seasat and 2022, there were no civilian US SAR satellites in operation. At the same time, it was clear that the Air Force was operating SAR satellites for which the data was strictly classified. As of the writing of this book, the United States finally has a civilian SAR satellite, which was developed between NASA and the Indian Satellite Research Organization (ISRO) and recently successfully launched a SAR satellite that was a cooperation between NASA and ISRO. This satellite was launched July 30, 2025, on an Indian launch vehicle. Called NISAR (NASA-ISRO SAR), the satellite is intended to observe Earth's land and ice surfaces with a dual frequency (L-band and S-band) radar system scanning the Earth surface twice every 12 days.

10.8.1 Capella space

In 2019 the US National Reconnaissance Office (NRO) announced its first-ever contract with a US commercial SAR firm, Capella Space, to study and evaluate the ability of a commercial SAR provider to integrate SAR into the national security enterprise. In 2020 the US Department of Commerce further relaxed restrictions on SAR satellites and streamlined the processing of licensing for commercial SAR satellites. Capella Space was founded in 2016 by Payam Banazadeh, a former engineer at NASA's Jet Propulsion Lab (JPL), together with William Walter Woods. Capella designs, manufactures, and operates a fleet of SAR satellites to provide high-resolution, all-weather satellite imagery to the US government and commercial customers. Their first satellite, Sequoia, was launched in August of 2020, followed by the launch of six second-generation Whitney satellites between January 2021 and January 2022. All of the satellites operate in a polar, sun-synchronous orbit. Capella has raised significant amounts of venture capital to continue their operation and data delivery. All of these satellites carry an X-Band SAR instrument.

In August of 2022 the company announced the development of their new generation of SAR satellites called "Acadia." The satellites (Fig. 10.14) carry an X-band (9.4–9.9 GHz) SAR instrument with a bandwidth of 700 MHz and single polarization (HH or VV). Ground resolution and swath width vary with look angle (Fig. 10.15). In general, the resolution is 25 cm for a swath of 5X5 km. They operate in a variety of sun-synchronous orbits at 575 km altitude and 97 degrees inclination, which makes possible a 2–5 hr revisit time with a constellation of eight satellites. Capella now has nine or

FIGURE 10.14 Capella Acadia synthetic aperture radar Satellite.
Image credit courtesy Capella Space.

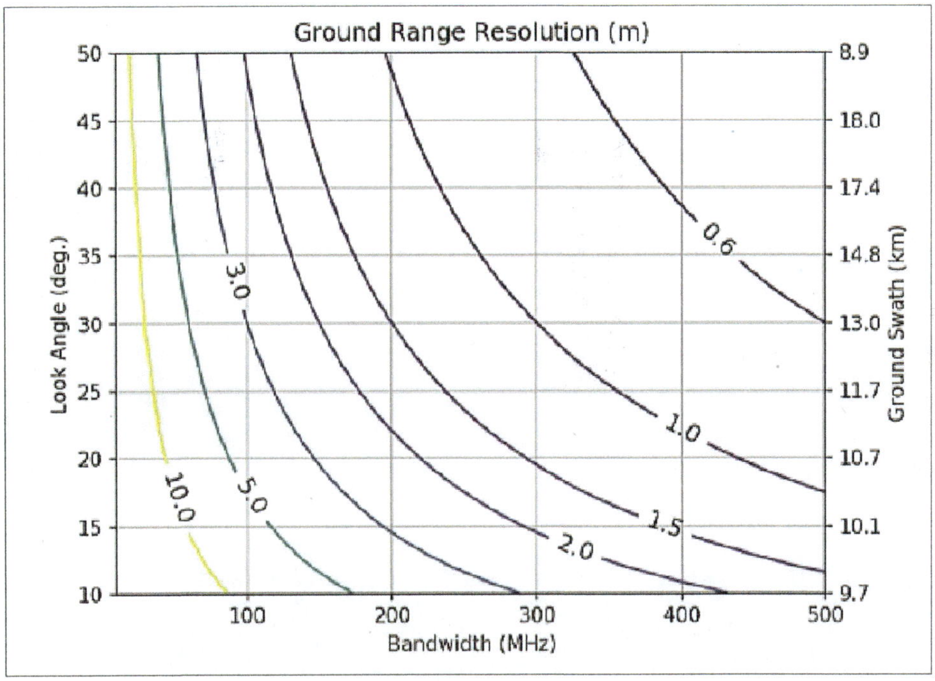

FIGURE 10.15 Ground range resolution as a function of transmit bandwidth and look angle.
Image credit courtesy Capella Space.

more satellites in operation, improving their revisit time. The X-band SAR operates at 500–700 MHz bandwidth in stripmap, spotlight, and sliding spotlight modes. Each satellite has a 3-year design life and by launching 12 satellites per year they will be able to maintain their constellation indefinitely.

In the spotlight mode, the SAR beam is focused on a single point on the Earth throughout the acquisition. In this mode, the azimuth resolution is limited by the duration of the antenna beam on the target. The resultant multilook images are acquired using a dwell time that results in pixels that have a finer resolution than 25 cm. The raw SAR data are processed to full resolution, and then pixels in the amplitude-only image are averaged over the multilooks to produce very high-resolution, speckle-free images that Capella is allowed to sell.

In the sliding spotlight mode, the instrument performs as a standard SAR where the swath is limited by a combination of the look angle, the radar beamwidth, and the pulse repetition frequency (PRF). The PRF is selected to either provide maximum swath with reduced azimuth ambiguity sidelobe performance or a reduced swath with better azimuth ambiguity sidelobe performance. The minimum slant range resolution is 30 cm.

The stripmap and multiswath stripmap modes are standard SAR modes where the swath is limited by a combination of the look angle, the radar beamwidth, and the PRF. The PRF is selected in the sliding spotlight mode, and the minimum slant range resolution is 30 cm.

The Capella spacecraft are not microsatellites as shown here in Fig. 10.16. It shows the difference between the small satellite used early on and the current satellites.

Once on orbit, the satellite unfurls the antennas made of a flexible material, and when deployed, they cover 8 m^2. The Acadia-5 satellite was launched on 16 August 2024 into the 97 degrees sun-synchronous orbit at 510–600 km altitude. The previous Acadia-2 was launched on August 11, 2024, in the 53 degrees orbit at 615 km altitude. The next earlier spacecraft Acadia-4, went up on April 7, 2024 into a 45.4 degrees orbit at 590 km altitude.

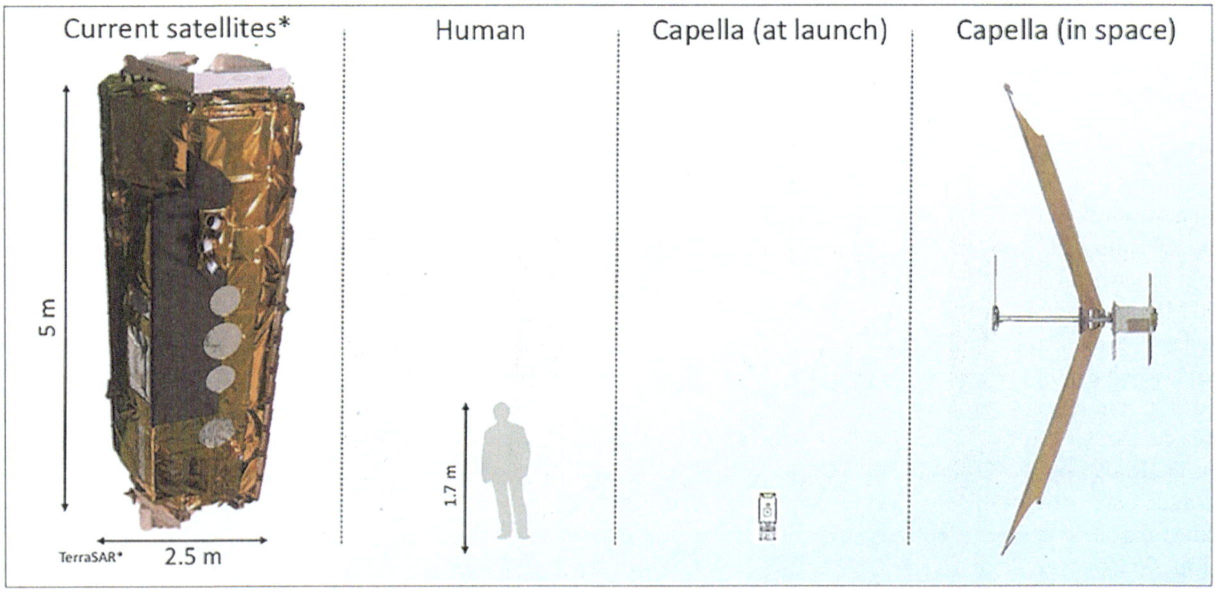

FIGURE 10.16 Capella satellite then and now, as compared with a human.
Image credit: Courtesy Capella Space.

FIGURE 10.17 Synthetic aperture radar image of the December 2021 Marshall Fire in Boulder County, Colorado.
Image credit: Courtesy Capella Space.

A SAR image of the Marshall Fire in Boulder County, Colorado, is presented here in Fig. 10.17. It is important to note that this image was collected in spite of smoke, clouds, and heavy snowfall. This fire destroyed 1084 homes in the Superior and Louisville parts of Boulder County.

Capella Space is headquartered in San Francisco with offices in Washington, D.C., and Louisville, Colorado. They design, build, and operate their own fleet of SAR satellites to provide all-weather satellite imagery to the US government and commercial customers.

10.8.2 Umbra

Umbra Space is a US-based commercial space company that develops and operates a SAR constellation, which began with the launch of the Umbra-01 microsatellite (Fig. 10.18) in June 2021. This is a 65 kg satellite with an X-band SAR with a spatial resolution of 25 cm over approximately 42 square kilometers. This first mission was to test new technologies and validate the performance of the instrument.

The technologies to be tested are as follows:

- Deployable high-gain mesh antenna newly developed by Umbra
- Radar and radio electronics developed by Umbra
- Novel thermal management systems
- Electrical and power systems
- Mission concept-of-operations
- Function of the integrated ground system
- SAR image formation using the full bandwidth allocation of 1,200 MHz.

This satellite was followed by Umbra-02 on January 13, 2022 and Umbra-03 on May 25, 2022. Launches of Umbra satellites continued up to Umbra-10, launched on 16 August 2024 along with Umbra-09. The total constellation is planned to be 24 satellites in sun-synchronous orbits at 500 km altitude. Each satellite is equipped with an X-band SAR instrument operating with a 1,200 MHz bandwidth capable of a 25 cm spatial resolution. As with other SARs, their instrument is capable of four different modes as presented in Table 10.4. The spatial resolution depends a lot on viewing options. Spotlight and Extended Dwell modes can achieve 25 cm ground spatial distance. For a 12 satellite constellation, there is a 15 min repeat time for polar latitudes and a 30 min repeat time for mid-latitudes, which becomes 45 min for the tropics.

Umbra Space has signed an agreement on February 14, 2025, with Maxar Technologies to add SAR imagery to complement the Maxar optical image portfolio. The partnership allows Maxar to directly task Umbra's satellites in order to integrate SAR data with the optical data imaging. Maxar has other agreements with SAR providers, but chose Umbra for a dedicated partnership since their comprehensive set of high-resolution SAR images best complements the Maxar portfolio of high-resolution optical images. This agreement is expected to extend over several years.

In August 2024 Umbra Space announced a new business segment called Mission Solutions aimed at the United States and other governments that seek satellite components, instruments, software, and mission capabilities in an effort to market their

FIGURE 10.18 The Umbra-01 Satellite (https://www.futurespaceflight.com/satellites/umbra-sar.html).

TABLE 10.4 Modes of the Umbra-synthetic aperture radar (SAR) instrument.

Operation mode	Resolution (m)	Swath width (km)	Polarization
Spotlight	0.25–4	4×4	HH or VV
Stripmap	3	5–20×50	HH or VV
Extended Dwell	0.25–2	4×4	HH or VV
Scanning	10	100×100	HH or VV

skills developed creating their own SAR constellation. Jason Mallare, Umbra's vice president of global solution said, "We see a growing demand for complete satellite systems as well as other mission capabilities." He said that Umbra could provide customers with dedicated access to Umbra's SAR data while their own systems are being built by Umbra.

Umbra is based in Santa Barbara, California, and feels that their competitors are also offering complete systems to collect SAR data, such as Capella Space. This foray into delivering SAR systems is a logical next step for the company.

10.8.3 ICEYE

ICEYE is a Finnish microsatellite manufacturer and operator. It was founded in 2014 and has a rather large constellation of SAR satellites. It is based in Espoo, Finland, and formed as a spin-off of Aalto University's Radio Technology Department. It now has offices in Poland, Spain, the United Kingdom, and the United States and over 700 employees worldwide. In September 2024 the company reported revenues of $100 million and expanded its funding to $438 million. The CEO and cofounder of ICEYE is Rafal Modzewski, together with Pekka Laurila. Their main offices are in Espoo, Finland, Irvine, California, Warsaw, Poland, and Bordon, UK.

On January 12, 2018 a PSLV-XL rocket carried ICEYE-X1 into orbit from the Satish Dhawan Space Centre, located in Sriharikota, Tirupati district, Andhra Pradesh, India. ICEYE-X2 was launched on December 3 on a Falcon 9 Block five rocket. The third ICEYE-X3 was launched on May 5, 2019 on the Electron rocket from the Rocket Lab Launch Complex in New Zealand. The fourth and fifth satellites were launched on July 5, 2019 by a Soyuz-2-1b rocket from the Vostochny Cosmodrome. The cooperation between SpaceX and ICEYE continued with six more satellites launched on Falcon 9 Block five rockets, four on June 30, 2021, and two on January 13, 2022. Another five have been launched on May 25, 2022 and three on January 3, 2023. Unfortunately, out of the last three satellites, only two are currently on orbit as one satellite failed to detach from the launch vehicle. Four more satellites were launched on June 12, 2023 and another four on November 11, 2023. Three more were launched on March 4, 2024, bringing the total number of satellites to 34. These satellites are in sun-synchronous, low Earth orbits with an altitude of 560 km and an inclination of 97.7 degrees. Each satellites makes 15 orbits per day with a 22 day ground repeat cycle. In the constellation, the mean revisit time at the equator is 20 h.

All satellites carry an X-band SAR operating at 9.8 GHz with multiple polarization and variable swath/resolution (Table 10.5).

A demonstration of the different ICEYE imaging modes is presented in Fig. 10.19.

The observational objectives of all ICEYE measurements are to provide all-weather imagery for oceans, land, and ice. Specifically, stated objectives are the fraction of vegetated land, glacier motion, glacier topography, land cover, land surface topography, soil moisture at the surface, soil type, and vegetation type. With 48 satellites as of January 2025, the company has the largest constellation of SAR satellites operating today. With this constellation, the company has the capability of revisiting any spot of the Earth within hours. People can task the satellites to collect data over a particular spot on the Earth at a particular time.

In addition to its customer base of SAR imagery, the company offers dedicated satellite launches and operations to fulfill a particular user need. Each satellite (Fig. 10.20) was dedicated to the X-band SAR imaging instrument called the ICEYE X1, which operates at 9.65 GHz with a resolution of 10 mX10 m, a swath width equal to or greater than 80 kmX40 km. The satellites are in sun-synchronous orbit with an altitude of 505 km and are VV polarized.

Originally intended to measure sea ice changes and glacial and sea ice tracking, the data have been used to monitor oil spills, and to prevent illegal fishing, along with many other parameters.

TABLE 10.5 X-band synthetic aperture radar (SAR) characteristics on ICEYE satellites.

Operation mode	Resolution (m)	Swath (km)	Field of regard	Polarization
Spotlight	1	5	600 km right/left	VV
Stripmap	3	30	600 km right/left	VV
ScanSAR	155	100	600 km right/left	VV

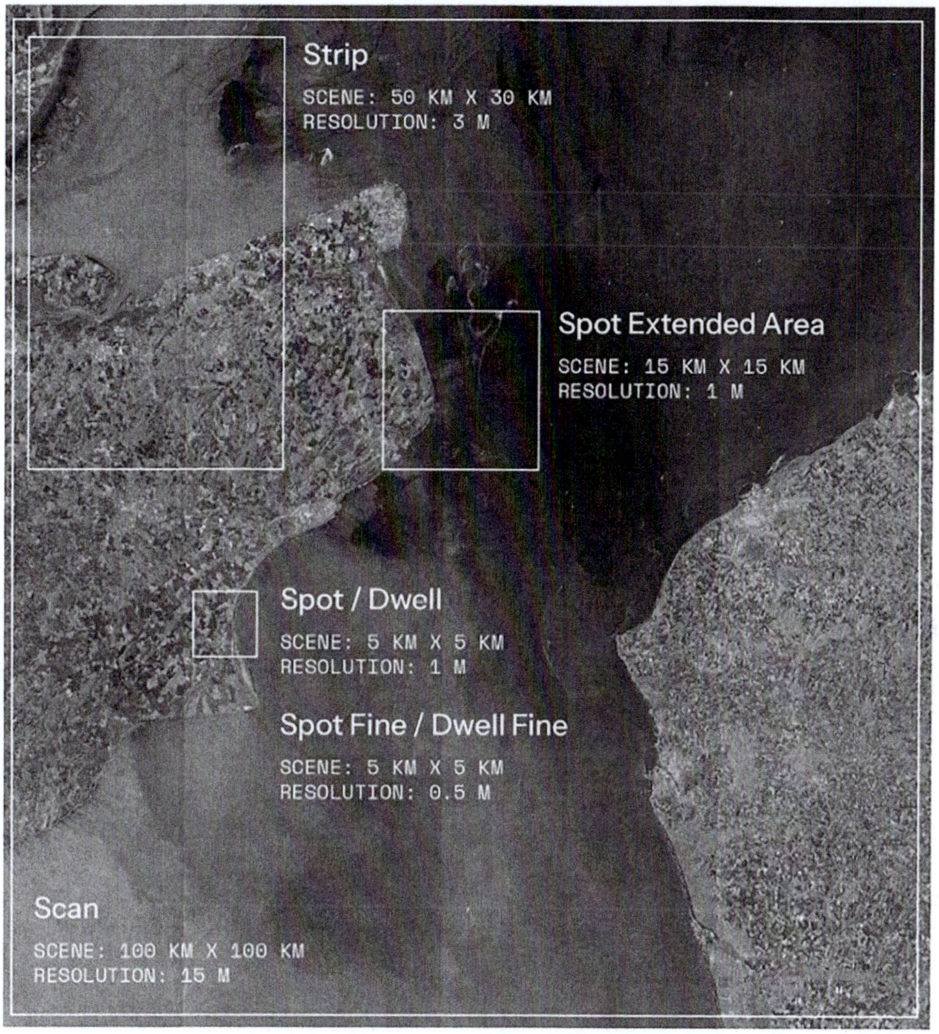

FIGURE 10.19 ICEYE image processed for the various imaging modes. Note that the scan and strip modes can be extended as much as 840 km along the ground track direction. A dwell mode was added to the Spotlight mode (https://www.iceye.com/sar-data/imaging-modes).

10.8.4 IMSAR

Another commercial SAR company, IMSAR, is not satellite-based but focuses more on airborne applications. Founded in 2004 as an outgrowth of Brigham Young University, IMSAR designs, builds, and supports security and intelligence products. IMSAR is dedicated to rapid technology innovation and is committed to decreasing the size and increasing the capabilities of its radars. They are involved with the US DoD and the US Department of Homeland Security (DHS).

FIGURE 10.20 Artist's rendition of the deployed ICEYE-X1 (https://www.eoportal.org/satellite-missions/iceye-x1#eop-quick-facts-section).

IMSAR multimode radar systems produce high-resolution SAR imagery; detect subsurface explosive hazards and other objects of interest beneath foliage; detect, identifie, and track moving targets of interest; and connect people, vehicles, and places with events, all from a small UAV.

10.9 Spire and atmospheric profiling

10.9.1 Spire global (file:///Users/williamjemery/Desktop/RS%20past%20pres%20futu/background%20off%20the%20web/SPIRE/Spire%20Global%20-%20Wikipedia.webarchive)

Spire was originally known as NanoSatisfi Inc, which was founded in June 2012 in San Francisco by International Space University graduates Peter Platzer, Jeroen Cappaert and Joel Spark as part of ArduSat, a project aiming to "democratize access to space." Early efforts used high-altitude balloons to test instruments. Their efforts were partly financed through "crowdfunding" through the "Kickstarter Program" that raised over $106,000 for Spire. Later, they signed an agreement with "NanoRacks" for the deployment of two satellites that were the first commercial satellites deployed from the International Space Station (ISS).

To raise the funding needed to build these satellites, Spire cooperated with Lemos Labs and raised $1.5 million in seed funding from Shasta Ventures, Lemos Labs, Emerge, Grishin Robotics, and Beamonte Investments in February 2013. A year after signing with NanoRacks, their first two satellites, ArduSat-1 and ArduSat-X (1U CubeSats), were successfully released into orbit by the ISS. On the basis of this success, Spire announced in July a follow-up $25 million Series A funding round led by RRE Ventures, backed by Emerge, Mitsui&Co Global Investment, and Mousse Partners. The following month, Spire announced that ArduSat would be spun off from the company and would concentrate on educational technology in partnership with US High schools. Soon thereafter, Spire opened its Singapore office and started to increase its network of ground stations.

To fund their next step, the company secured a $40 million Series B led by Promus Ventures with participation from Bessemer Venture Partners and Jump Capital. This funding was needed to build their first batch of Lemur satellites, the first of which, Lemur-2, was launched in September 2015 with the Polar Satellite Launch Vehicle XL, making Spire the first US based group to launch from India. This launch inaugurated the Spire tradition of naming each satellite after company employees, with first 3 Lemurs named Joel, Peter, and Jeoren after the company's cofounders.

Due to the need to mass produce satellites and their limited space in San Francisco Spire opened an office in Glasgow, Scotland, in February 2015. They initially used the facilities of Clyde Space Co before opening their own full-fledged cleanroom and manufacturing facilities in December 2015. The location was chosen to take advantage of what is widely considered the leading European ecosystem in small satellite production. These facilities enabled Spire to quickly produce its first batch of four nanosatellites that were launched in September. They then built eight Lemur satellites for an Atlas V launch in March 2016. Thus Spire had 10 simultaneously operating satellites in space as of June 2016. Two additional launches that year brought this total to 16.

Spire opened a second office in Boulder, Colorado in January 2016 and hired Dave Ector, the program manager for NASA's COSMIC satellites, and Alexander (Sandy) MacDonald, the retired director of NOAA's Earth Systems Research Laboratory, who brought a new emphasis on atmospheric sensing using the principle of Global Positioning System (GPS) occultations to profile the atmosphere. This had been the core mission of the COSMIC program, and Spire greatly expanded this effort by equipping all of its satellites with GPS receivers. Sandy MacDonald was interested in using these many atmospheric profiles to improve atmospheric numerical models and, therefore, weather forecasting. Spire participated in the inaugural Commercial Weather Data Pilot program of NOAA in September 2016. This cooperation increased in 2018, and Spire became a private company that sourced atmospheric profiles from radio occultation to the weather forecasting community to improve their weather models.

In 2017 Spire launched new satellites, bringing their total to 36. Unfortunately, the failure of a Soyuz launch vehicle destroyed 10 Lemur satellites. In that same year, Spire secured $70 million Series C funding led by the Government of Luxembourg and Spire opened its second European campus in that city in the same year. This round of funding put the Spire total at $140.5 Million.

In early 2018 satellites were launched by Rocket Lab's Electron rocket and also by Arianespce Vega as part of their proof of concept project, adding 28 new satellites to their constellation. At this time, Spire developed its own ADS-B payload, which tracked the movement of equipped aircraft, which was important in areas that conventional radars cannot cover. This capability was quickly adopted by many aircraft control systems.

In 2019 Spire created a new business unit called "Spire Marine," which was based in Luxembourg and tracked ships equipped with AIS systems (which are the majority of ships today). In that same year, Spire launched its 100th Lemur satellite on April 1st On March 1, 2021, the company went public via a merger with the SPAC Navsight. The merger was completed late in 2021 and valued the company at $1.6 billion. On August 17, 2021, Spire had about $265 million in cash on its books, while in the preceding year they had revenues of $36 million. They had 110 satellites in orbit and ground stations in 16 countries. At that time, Spire did not expect their constellation to grow much larger as it did not expect greater demand. Spire continued to build and launch satellites, as the Lemur satellites are refreshed on a three-year replacement cycle.

Due to missing some projected revenue targets and rising equipment losses, Spire's market value started decreasing after its first quarter report. As of April 30, 2022, the company had a market cap of $231 million, less than half of the $557 million in capital invested in the company and less than the cash that they had on hand. On November 13, 2024, Spire Global announced it was selling off its ship-tracking (AIS tracking) business to Kpler of Belgium for $241 million. Kpler gets the exclusive rights to the ship tracking data from the Spire satellites, but Spire retains ownership of the satellites, the infrastructure, and the technology. US government clients of Spire's maritime business remain clients of Spire and do not become clients of the Belgium company. The proceeds of this trade are intended to cover Spires's $100 million debt. In February 2025 the Canadian Space Agency awards a $72 million contract to Spire Global Canada to design a WildFireSat mission. Spire's satellites are all built in their Glasgow factory (Fig. 10.21).

These satellites are flexible platforms intended to carry a variety of in-house Spire payloads or other outside payloads. Spire hopes to further their "space-as-a-service" program with both aerospace and defense contractors. Spire designs, builds, and tests all of its satellites at its Glasgow Lab. The company uses a lot of COTS (commercial off the shelf)

FIGURE 10.21 Spire engineers assemble a batch of Lemur satellites (file:///Users/williamjemery/Desktop/RS%20past%20pres%20futu/background%20off%20the%20web/SPIRE/Spire%20Global%20-%20Wikipedia.webarchive).

FIGURE 10.22 An individual Lemur satellite (file:///Users/williamjemery/Desktop/RS%20past%20pres%20futu/background%20off%20the%20web/SPIRE/Spire%20Global%20-%20Wikipedia.webarchive).

electronics to reduce cost. Spire adheres to internationally recognized standards for the deposal of their no longer functioning satellites.

The GNSS-RO (global navigation satellite system-radio occultation) system profiles atmospheric temperature and moisture using the GPS satellite signals as its primary source of radio broadcasts. The GPS receivers are called "stratos" and are advanced science-grade GNSS receivers that were developed at Spire. These occultations rely on the changes in a radio signal that are caused by varying conditions in the atmosphere through which it is traveling (which is called occultation of the radio signal by the atmosphere). As the radio signal is refracted (or bent), the magnitude of this refraction depends on the density gradient of the atmosphere. This bending is calculated using the Doppler shift of the signal, given a certain geometry of the emitter and receiver. Thus, using inverse theory, these measurements can be converted to atmospheric temperature, pressure, and water vapor content, the very information that is needed by numerical weather prediction numerical models.

The individual Lemur satellites (Fig. 10.22) are CubeSats that are in sun-synchronous orbits with a 500–550 km altitude. The orbit period is 95 min and Spire operates 110 satellites. Each satellite weighs 4.6 kg and has dimensions of 10 cmX10 cmX30 cm. They are designed with a 2–14 year lifetime. At 110 satellites, Spire claims that they operate the largest constellation of satellites.

Chapter 11

Other individual satellite programs

11.1 Gravity Recovery and Climate Experiment

Gravity Recovery and Climate Experiment (GRACE) is an international cooperative US-German dual minisatellite (Fig. 11.1) with a satellite to satellite tracking geodetic mission having the overall objective to obtain long-term data with unprecedented accuracy for global, high-resolution models of the mean and time-variable components of the Earth's gravity field (a new model of the Earth's gravity field every 30 days for 5 years). Individual science objectives are as follows:

- To enable a better understanding of ocean surface currents and ocean heat transfer.
- To measure changes in the sea-floor pressure.
- To study ocean mass changes.
- To measure the mass balance of ice sheets and glaciers.
- To monitor changes in the storage of water and snow on the continents.

The concept of the GRACE mission takes advantage of the intersatellite range changes and their derivatives between two coplanar satellites (in low altitude, polar orbits), using a microwave tracking system. The orbits of the two separately flying spacecraft are perturbed differently in the Earth's gravity field, leading to intersatellite range variations. Each satellite carries a high-accuracy GPS receiver along with accelerometers to enable accurate orbit determination, spatial registration of gravity data, and the estimation of gravity field models.

These fluctuations in the strength of the Earth's gravity field reflect, in turn, changes in the distribution of mass in the ocean, atmosphere, and solid Earth, along with the storage of water, snow, and ice on land. Since ocean bottom pressure is the total column integral of the mass of the atmosphere and the ocean, GRACE measurements measure these ocean bottom pressure changes from space.

The GRACE satellite (Fig. 11.2) is equipped with three GPS antennas linked to a receiver of geodetic quality along with high-accuracy accelerometers to enable accurate orbit determination, spatial registration of gravity data, and the estimation of gravity field models. The GRACE mission is led by Byron Tapley (PI) of the University of Texas at Austin and by Frank Flechtner (CO-PI) of the GeoForschungsZentrum (GFZ) in Potsdam, Germany. NASA/JPL has led the spacecraft development in partnership with EADS Astrium GmbH. Satellite operations are carried out by DLR/GSOC with tracking stations at Weilheim and Neustrelitz. Germany also provided the Eurockot launch vehicle.

The satellites are identical as required for the gravity measurements. The leading satellite is between 170 and 270 km ahead of the trailing satellite, and the pair of satellites was named Tom and Jerry. The orbit is a circular orbit of 485 km altitude after launch (near solar maximum), decaying to about 300 km (near a solar minimum) after 5 years. The orbit was inclined at 89 degrees. There is an onboard cold-gas propulsion system used to maintain the separation of the two satellites. Orbit maneuvers are needed about every 50 days to maintain the separation. The rather low orbital altitude was selected to obtain the best possible gravity measurements (note that the gravity signal of any central body decays with the square of the orbital distance from the center of mass). The spacecraft is in a 30-day repeat orbit, and a new gravity field is measured each month. The GRACE system is sufficiently accurate to determine a change in mass equivalent to a volume of water with a depth of 1 cm over a radius of about 400 km.

The RF communications use S-band deployable booms beneath each satellite. The daily science data volume is about 50 Mbytes, including gravity data and GPS occultation data. All of the links are shown in Fig. 11.3. One can see the GPS connections that contribute to the accurate orbit calculations, along with the intersatellite communications and the data links to the ground. In addition to the two ground stations mentioned earlier. GFZ installed an automatic payload data acquisition station in Svalbard (Spitzbergen) to speed up the data processing and distribution chain. The polar location of

Satellite Remote Sensing. DOI: https://doi.org/10.1016/B978-0-443-22086-9.00013-7

FIGURE 11.1 Gravity Recovery and Climate Experiment satellites.
Image credit courtesy: NASA JPL.

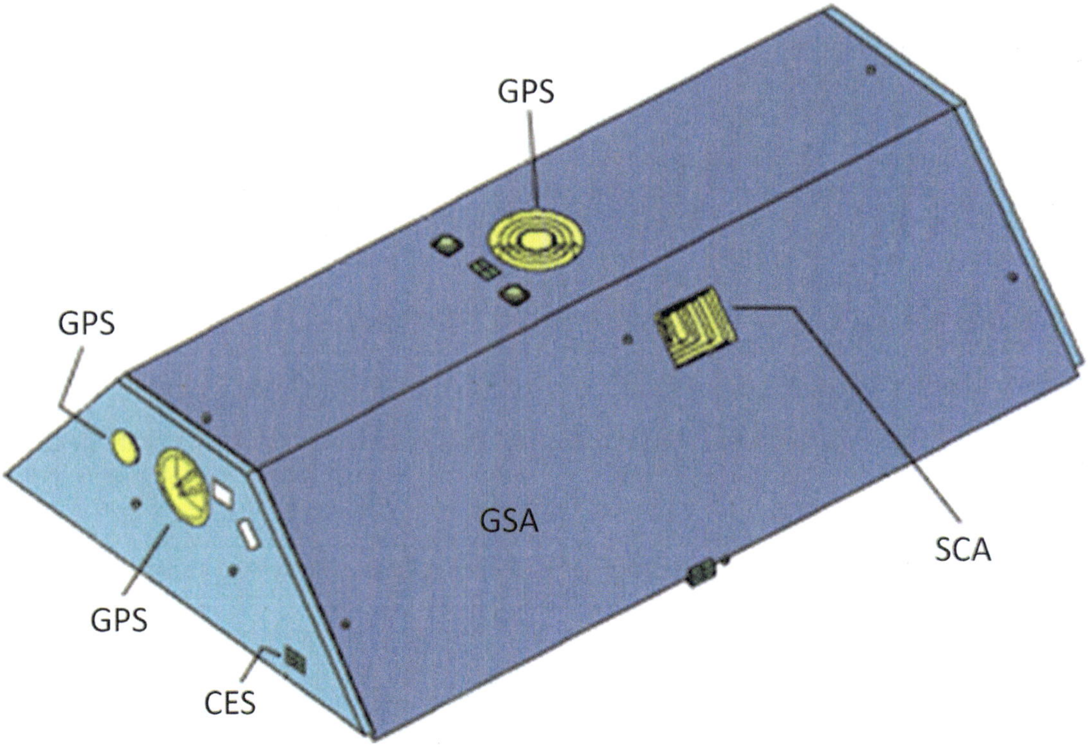

FIGURE 11.2 Top view of the Gravity Recovery and Climate Experiment (*GRACE*) spacecraft.
Image credit courtesy: GFZ Potsdam, Germany.

Svalbard makes it possible to have access to the data on almost all of the orbits. GRACE operated from March 17, 2002 till October 27, 2017.

Each GRACE satellite carried the science instrument system (SIS) and a K/Ka-Band ranging (KBR) instrument. SIS included all of the elements of the intersatellite ranging system, the GPS receivers required for precise orbit determination, and occultation experiments, along with associated sensors such as the star camera assembly. The SIS also coordinated all of the sensors, ensuring their compatibility with each other and the satellite. The KBR detected very small changes in the distance between the two satellites and used these data to derive gravitational measurements with a level of precision that had never been possible before.

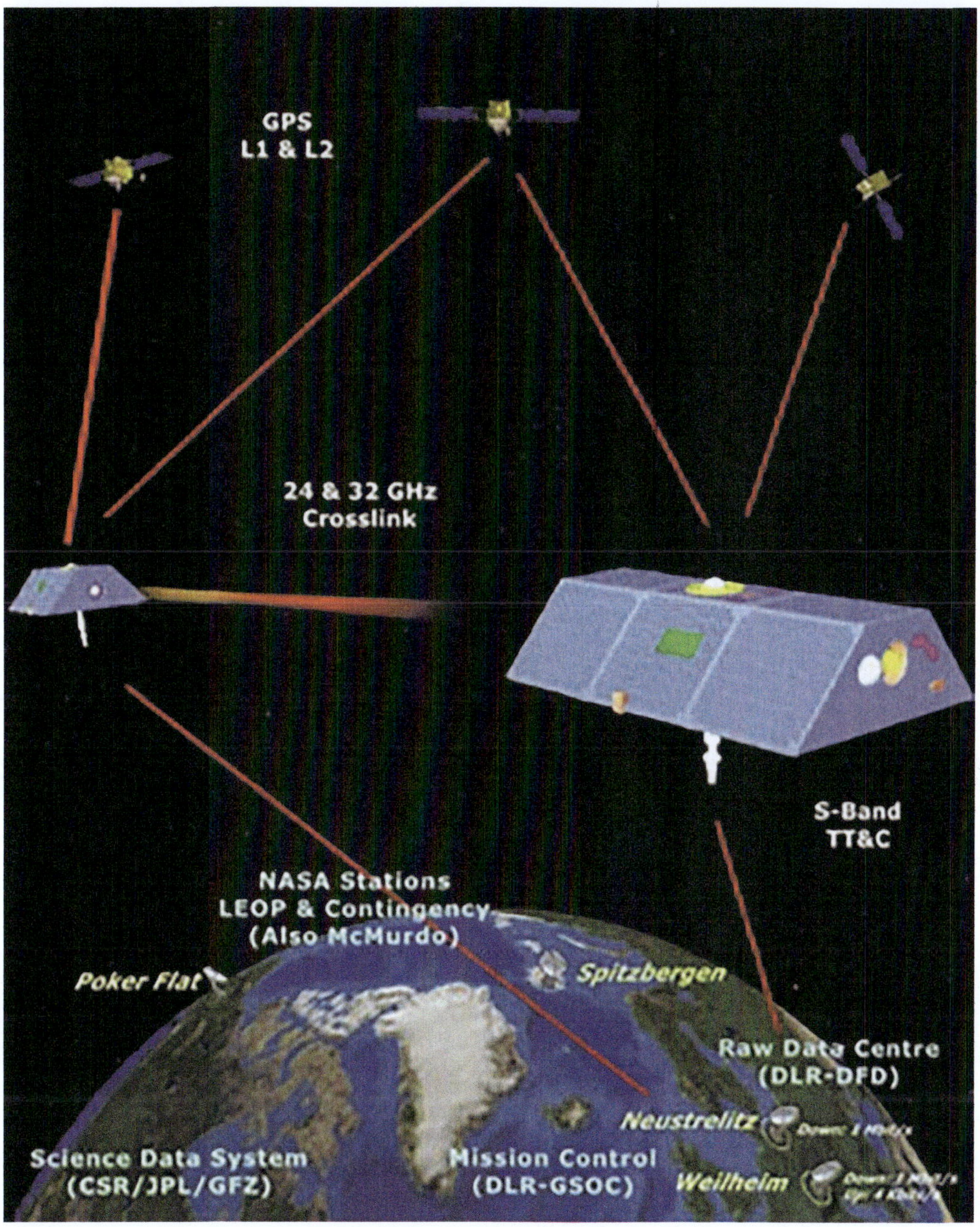

FIGURE 11.3 Flight configuration and ground support for a gravity mission.
Image credit courtesy NASA.

In 2019 a paper with authors led by Byron Tapley, the US principal investigator, titled "Contributions of GRACE to understanding climate change," helped to establish the utility of GRACE observations for the study of climate variability. GRACE and the GRACE follow-on have made it possible for scientists to create monthly global maps of mass changes, greatly improving our understanding of the impact of global climate change on water distribution in the ocean, ice, and continents. GRACE provided direct measurements of the ice-mass loss from ice sheets and glaciers, indicators of global warming, and a measure of water added to the ocean. The consequence of this added water is global sea level rise, which continues unabated. The GRACE follow-on mission continues these important measurements.

The GRACE follow-on (GRACE-FO) continues the partnership between NASA and the GFZ. GRACE-FO was launched on May 22, 2018, with the intention of continuing the time series started with the first GRACE mission. The satellites and their sensors are copies of the earlier GRACE satellites, and the orbital characteristics are the same.

11.2 Geostationary ocean color imager (https://www.earthdata.nasa.gov/data/instruments/goci)

Geostationary ocean color imager (GOCI) is the world's first geostationary satellite (Fig. 11.4) dedicated to the measurement of ocean color. This satellite was developed by the Korean Aerospace Research Institute and ASTRIUM (AIRBUS) and is operated by the Korean Ocean Satellite Center of the Korean Institute of Ocean Science & Technology (KIOST). It was launched June 27, 2010, into geostationary orbit at an altitude of 35,786 km over the nadir location at 128.2°E.

The GOCI instrument was installed on the Communication, Ocean and Meteorological Satellite. It was operated by KIOST.

The data products to be derived from the GOCI measurements were primarily chlorophyll concentration, the optical diffuse attenuation coefficients, the concentration of dissolved organic material or yellow substance (also called

FIGURE 11.4 Korea's geostationary ocean color imager satellite.
Image credit courtesy KIOST.

TABLE 11.1 Geostationary ocean color imager (GOCI) specifications (Korean Institute of Ocean Science & Technology [KIOST]).

Description	Specification
Detector	CMOS (2D, 1415×1432 pixels)
Radiometric calibration	Solar diffuser and Diffuser Aging Monitoring Device
Resolution (GSD)	500 m
Total FOV	16 slots, 5300×5300 Pixels
Coverage	2500 km×2500 km
Longitude	128.2°E
Altitude	35,786 km
SNR	> 1000
MTF	>0.3
No. of bands	8 (6-visible, 2-NIR)
Spectral coverage	400–900 nm

TABLE 11.2 Geostationary ocean color imager (GOCI) bands (Korean Institute of Ocean Science & Technology [KIOST]).

Band (#)	Central wavelengths (nm)	Bandwidth (nm)	SNR	Type	Primary application
1	412	20	1000	Visible	Yellow substance and turbidity
2	443	20	1090	Visible	Chlorophyll absorption maximum
3	490	20	1170	Visible	Chlorophyll and other pigments
4	555	20	1070	Visible	Turbidity, suspended sediment
5	660	20	1010	Visible	Atmospheric correction for turbid water, Baseline of fluorescence signal, Chlorophyll, suspended sediment
6	680	10	870	Visible	Fluorescence signal
7	745	20	860	NIR	Atmospheric correction and baseline of fluorescence signal
8	865	40	750	NIR	Atmospheric correction, vegetation, water vapor reference over the ocean

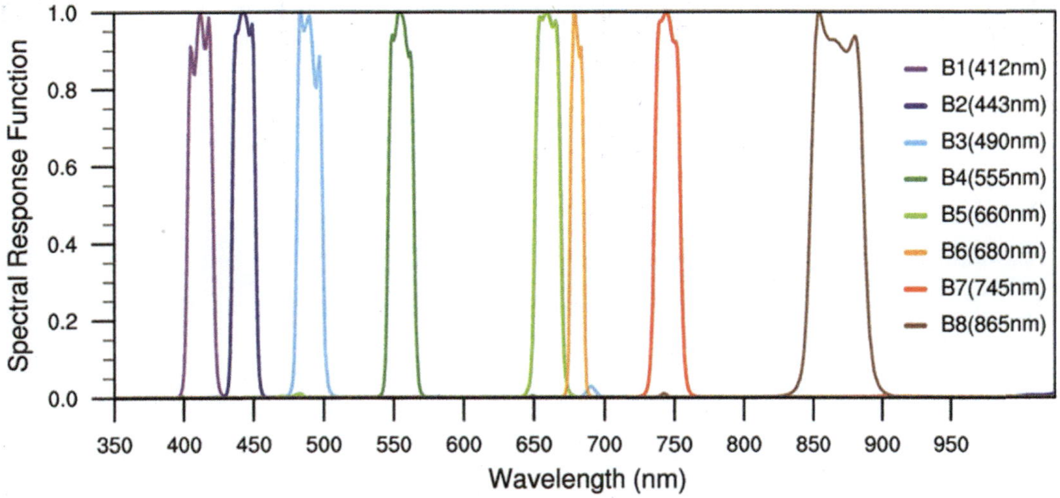

FIGURE 11.5 The geostationary ocean color imager spectral bands.
Image credit courtesy KIOST.

gelbstoff), and the concentration of suspended particles in the near-surface layer of the ocean. The goals of the project were to:

- detect, monitor, and predict short-term biophysical phenomena
- studies on biogeochemical variables and cycle
- detect, monitor, and predict noxious or toxic algal blooms
- monitor the health of the marine ecosystem
- coastal zone and resource management
- produce improved marine fisheries information to the fishing communities
- derive yellow dust and land classification information

The specifications of the GOCI instrument are given here in Table 11.1.

The spatial resolution of 500 m from geostationary orbit is a real achievement and primarily reflects the fact that this is a visible instrument. The calibration of these visible channels is a noted step forward for this instrument. The eight bands are spelled out in Table 11.2. These are well-known wavelength bands to measure ocean color. Still, this was a brave step forward to measure ocean color from geostationary orbit. The primary advantage of geostationary orbit is the rapid refresh made possible in this orbit versus the lower temporal resampling provided in polar orbit.

Located over 128.2°E, this instrument provided South Korea with rapid sampling of ocean color conditions around the Korean peninsula. Typical events studied are tidal changes around South Korea, forest fires, a typhoon in August 2011, sea fog, and yellow dust moving from Mongolia and northern China over the Korean peninsula.

These bands are shown as a graph in Fig. 11.5.

Chapter 12

The future of Earth remote sensing

Trying to predict the future of satellite remote sensing is very difficult, as it can be expected to move in so many different directions. It is fairly certain that geostationary weather satellites will continue to be launched and operated as part of the national weather forecasting systems for North America, Europe, India, and Asia. It is very likely that the polar orbiting weather satellites will continue in some fashion, although it is likely that they will carry a number of new sensors. The future role of smaller satellites is difficult to anticipate as they have been shown to be more capable that earlier expected.

If we look at an earlier forecast made by NASA in the early 1990s (Remote Sensing and the US future in space), we can see that this report missed a lot of the changes that would take place in the near and distant future. The report did not anticipate the capabilities of the geostationary satellites now in operation, and also predicted that the NPOESS system would take over the collection of polar orbiting weather satellites. They did not predict the total failure of the NPOESS program and its replacement by the much scaled down JPSS program. This report also predicted the success of the NASA EOS program, which was also dramatically scaled back along with its planned data system. At this earlier date, there is no anticipation of the role of commercial remote sensing and the explosion of the use of small CubeSats and microsats.

A more recent forecast was written by Humboldt State University in 2019 regarding the future of remote sensing. They conclude that the future will bring better, faster, and smaller technology to remote sensing. The trends that they expect to continue are the decreasing size of electronics and sensors, increases in computing power, increases in transmitting power for active systems and increasing tunability of systems. The consequences of these continued improvements will be improvements in the spatial, spectral, radiometric and temporal characteristics of satellite sensor observations that will result in more accurate and precise data. Smaller sensors can be mounted on a variety of platforms, allowing for greater flexibility in data collection.

This report goes on to say that newer satellites are capable of producing extremely high-resolution imagery. They refer to DigitalGlobe's Worldview-3 satellite (Maxar) capability to collect extremely high spatial resolution data, noting the high expense to acquire these data. They note the trends that smaller satellites are being launched into lower orbits. They note that the lower orbits compensate for the smaller size of the satellite and sensors, making it possible for the small satellites to still collect relatively high spatial resolution imagery. They also report on the use of Unmanned Aerial Systems (UAS) as an emerging platform for the collection of high spatial resolution imagery at a fraction of the cost of obtaining satellite data from Maxar satellites. Most UAS fly at altitudes less than 400 ft, which makes it possible to collect very high spatial resolution optical data.

This report also discusses improvements in spectral resolution that have given birth to the field known as hyperspectral imaging. Hyperspectral imagers are capable of collecting images with 100–200-narrow spectral bands. An example in their report is reproduced here as a "data cube" from the Airborne Visible Infrared Imaging Spectrometer (AVIRIS) built and operated by NASA Jet Propulsion Lab (Fig. 12.1).

Hyperspectral remote sensing is sometimes called "imaging spectroscopy." It has been used a lot in the detection and identification of minerals, terrestrial vegetation, and man-made materials. Data is typically collected as a data cube, and analysis requires a substantial amount of computer power. The AVIRIS instrument that collected the data cube in Fig. 12.1 collects visible reflectance data in 225 spectral bands ranging from 400 to 2500 nm. More information about AVIRIS can be found at the NASA/JPL AVIRIS website.

This report also discusses future improvements in temporal resolution, which they feel will be greatly enhanced by UAS and constellations of small satellites. UAS systems allow the user to specify the repeat sampling time, while the small satellite constellations provide shorter repeat time by the sampling of multiple satellites. They define a number of different small satellites as presented here in Table 12.1. They report that in 2013, 92 micro/nano satellites were launched. The latest numbers are 4,600 small satellites (180–500 kg), 900 minisatellites (100–180 kg), 1200 microsatellites (10–100 kg), and 500 nanosatellites (1–10 kg). These numbers include the many small communication satellites, which are not remote sensing platforms. In general, a small satellite is classed as any satellite weighing less than 500 kg.

Satellite Remote Sensing. DOI: https://doi.org/10.1016/B978-0-443-22086-9.00002-2

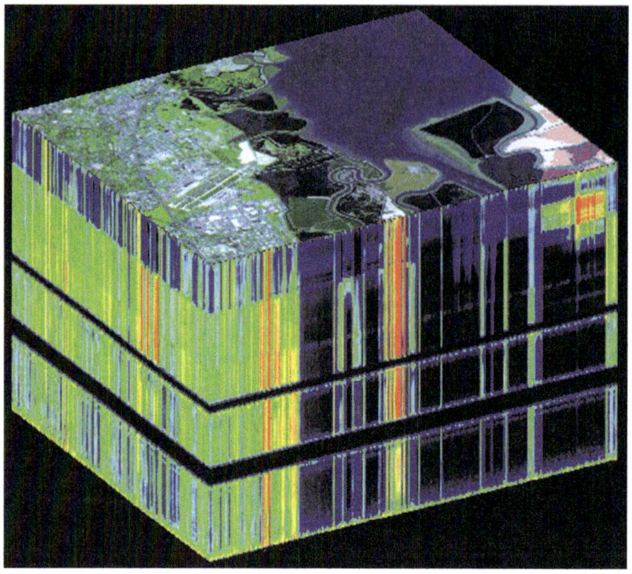

FIGURE 12.1 Airborne Visible Infrared Imaging Spectrometer Data cube of Jasper Ridge, CA.
Image credit courtesy JPL, NASA.

TABLE 12.1 Types of small satellites.

Satellite	Size (mass)
Femtosatellite	10–100 g
Picosatellite	<1 kg
Nanosatellite	1–50 kg
Microsatellite	10–100 kg
Small satellite	100–500 kg

NASA has a program to determine if small satellites can be built using smartphone technology. The first PhoneSat cube satellites (Fig. 12.2) cost approximately $3500, and these satellites have been launched and have successfully transmitted imagery to Earth.

A more recent paper (Zhang et al., 2022) addresses the need for intelligent hyperspectral remote sensing satellite systems to provide real-time, accurate and individually personalized remote sensing information services. This chapter reviewed the current developments in this area of remote sensing. Zhang et al. (2022) observe that the continued progress in computer chip technologies has greatly enabled remote sensing by smaller satellites. These developments have also made possible the intelligent processing of satellite data, increasing the spatial, spectral, and temporal sampling abilities. These developments present new problems, such as adequate communications bandwidth to get all of this data relayed to

FIGURE 12.2 PhoneSat 1 during a balloon test (https://www.researchgate.net/figure/PhoneSat 10 high altitude balloon test_fig6_289527065).

the ground. It also presents the satellite image analyst with a challenge to properly squeeze all of the information out of the images that are available. Fortunately, tools, such as artificial intelligence (AI) and deep learning, are available to tackle this challenge. This constitutes intelligent remote sensing.

According to Zhang et al. (2022) developing these intelligent remote sensing systems is a priority for China, The United States and the European Union. Satellites such as TacSat-3 (United States, 2009), TET-1 (Germany, 2012) are new technology demonstration satellites, while satellite G (China, 2020) along with a high-resolution multimode satellite (China, 2020) are all examples of satellites with intelligent systems onboard (Fig. 12.3). The improvement of solid state memory storage has eliminated the need for data compression and preprocessing. Satellite information services are evolving from large-scale ground stations to onboard processing, small receiving systems, and mobile terminals. Thus, it is important when the satellite is planned to consider the nature of the corresponding ground data system.

These intelligent satellites are capable of on-orbit target detection and identification. As shown in Fig. 12.3, the new intelligence level makes it possible for a single satellite to be used for multiple applications. These satellites may also be employed in a constellation to provide increased coverage in time and space. Indeed, intelligent satellite systems make it possible for users to order very specific products from these satellites. As a consequence, satellite information services are moving from large-scale ground systems to onboard satellite and hand-held terminals for receiving and processing the satellite imagery. This means that the satellite platforms must be flexible and the imagery made to conform to user demands. This also requires adaptable communication systems to deliver the desired image product to the interested user.

Improvements over the past decade signal the kinds of improvements that we may expect in the next decade. One important development is the transition from one or two large, dedicated satellites to a constellation of smaller but capable satellites. This is as true for optical remote sensing as it is for radar (synthetic aperture radar [SAR]). China's High-Resolution Earth Observation System (CHEOS) is a good example. CHEOS started with the launch of Gaofen-1 in 2013 and has progressed to a constellation of 24 satellites with 14 different types of satellites. The Beijing-2 satellite constellation started in 2015 and now includes three high-resolution satellites returning daily observations. The Jilin-1 constellation, launched and operated by the Chang Guang Satellite Company (a commercial satellite system), is expected to complete its network of 138 satellites by 2030.

The operation of constellations of satellites to achieve specific goals has resulted in the need for control software and satellite communications that would not have been previously possible or needed. Intersatellite laser communication has attracted increasing attention due to its large bandwidth, small size, light weight, low power consumption, reliability, and lack of spectral limitations. A number of tests have been completed, such as the 2015 laser communication terminal carried by the European Data Relay Satellite (EDRS), which communicated between a low Earth orbit (LEO) satellite and a geostationary satellite using a Binary Shift Keying modulation method. This was a communication distance of 45,000 km at a transmission rate of 1.8 Gb/s. The ILLUMA-T project in the United States plans to establish a two-way communication link between. GEO and LEO satellites with a communication rate of 2.88 Gb/s.

Regarding satellite to ground communication China's Tianlian Queqiao and Tiantong series backbone network constellation Ka-band ground communication can reach up to 300 Mb/s while Japan's Data Relay Test Satellite system is

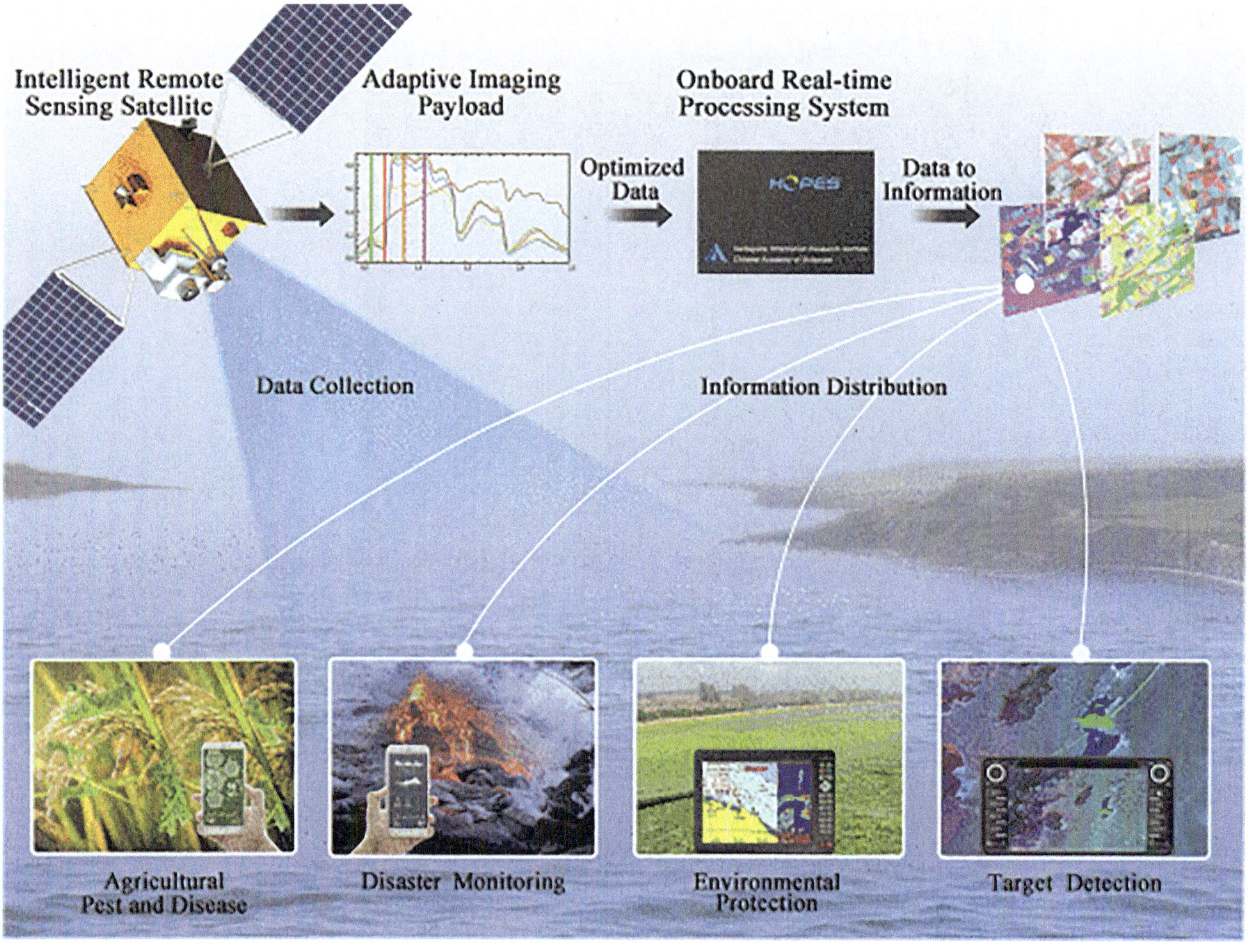

FIGURE 12.3 A depiction of intelligent remote sensing satellite systems (Zhang et al., 2022).

capable of speeds up to 240 Mb/s in the Ka-band while their EDRS system can attain Ka-band ground communications up to 300 Mb/s. The US tracking and data relay satellite and Starlink can attain ground downlink speeds of 800 Mb/s. Direct satellite-ground laser communication could easily double the present ground data transmission rates when they are available. This is significant because the radio frequencies are getting increasingly used and crowded.

In China, the rapid development of 5G and 6G communication technologies in the BeiDou Navigation Satellite System enabled the use of various types of miniaturized terrestrial data receivers, such as ships, automobiles, and handheld terminals. Thus, these portable devices make it possible to view and act on the satellite data while in the field. Hence intelligent satellite remote sensing will greatly enable the user community in various related applications.

In the development of these intelligent satellites, there are a few limits that must be respected. Spectral and spatial resolutions have an inverse relationship. When the spatial resolution is high (small spatial element resolution), the spectral resolution is, as a result, rather broad. If the spectral resolution is narrowed, it results in poorer spatial resolution. In any case, any such intelligent satellite system must strive to keep the signal-to-noise ratio to a minimum, as it will dictate the accuracy of the image target resolution. To address this condition, a load adaptive imaging mode has been proposed for intelligent remote sensing observations.

Onboard real-time data processing is one of the most important advancements in intelligent remote sensing satellite systems. Unlike the usual onboard data compression approach, intelligent onboard satellite systems seek to directly generate the desired ground target information from the satellite measurements. A lot of this onboard processing capability has been made possible by the rapid development of integrated circuit technology that provides significant processing power in relatively small packages that can be used on intelligent satellite systems. In addition, the low total ionization in LEO limits the impacts on these chips and makes it possible to replace high-cost aerospace-grade chips with regular industrial-grade chips for these onboard computing systems. The PhiSat-1, launched by ESA in 2020, is equipped with a Movidiua Myriad 2 visual processor. It uses this capability for cloud detection, and cloudy pixels are filtered out. China's

Jilin-1 spectral satellite uses GPU-based on-orbit processing. Processor chips with a large-scale graphics processor GPU as their core are becoming increasingly used in remote sensing.

The key to converting satellite radiance to ground target information relies on the fundamentals of calibration, geolocation, and atmospheric filtering. The algorithms needed to make this transformation need to be developed before launch and made available in the satellite's computer system. An example is the Pleiades satellite, which was launched after the end of the SPOT satellite program to provide a new type of high-resolution satellite imagery. The satellite carried out radiation correction, geometric correction (geolocation), image compression, and other processing functions using an MVP modular processor with an FPGA as the core.

China has also conducted onboard processing experiments with many different satellites. The Pujiang-1 satellite, built by the Shanghai Aerospace Technology Research Institute has the capability of performing image fusion onboard the satellite. It can distinguish, extract and locate anomalous target areas from orbit. A high-resolution multimodal imaging satellite launched by the Chinese Academy of Space Technology has been able to carry out target extraction and radiance correction and geometric correction to generate Level 2 image products of target regions and quickly transmit them to interested parties on the ground. The Luojia-01 satellite developed by Wuhan University has accomplished on-orbit geometric correction and radiometric calibration. Chang Guang Satellite Company's Jimlin-1 Spectrum satellites have sensors that can detect and locate forest fires and ships at sea.

Often, the onboard processing systems are merely a repeat of the algorithms used on the ground to convert satellite radiances into ground information. These algorithms are generally not optimized for use onboard the satellite, and there is a need to tailor the onboard software to meet the needs of the ground user while minimizing the demands on the onboard processing system. The key to a real breakthrough in the performance of onboard processing systems is the development of new computing architectures that satisfy the needs of people on the ground while enabling the reduced computing demands of the onboard processing system.

A primary motivation for onboard processing is the increasing need to relay information derived from satellite radiances to the ground users in a timely fashion. This is true of various emergencies such as forest fires, sea and land search/rescue, terrestrial law enforcement, landslides, and earthquakes. If the satellite radiances can be converted onboard to informational products that can then be directly transmitted to the application users, it will greatly facilitate the use of the satellite data for these various applications. This ignores the military applications, which are many and would also benefit greatly from onboard processing and dissemination. All of these developments need a set of communication standards to enable the transmission of the onboard-generated products to the ground. Such intelligent satellite systems raise the question of individual rights as to the use of data and products involving them.

We all know that AI will play a big role in the future, and it is very true of satellite remote sensing as well. One example is the use of AI by the University of Hawaii researchers to study coral reef halos. Features known as grazing halos or sand halos consist of ring-like patterns of bare sand that occur around coral patch reefs. These halos are readily visible in satellite images that have sufficient resolution. These "reef halos" may be important indicators of the vitality of coral reefs, but in the past, their observation has been difficult due to the difficult and time consuming observation process. Using satellite data, we can accurately identify and measure reef halos on a global scale, said Simone Franceschini, a postdoctoral research fellow at the university (Fig. 12.4).

"We're aiming to develop a freely available remote sensing tool to monitor ecological processes over large scales to improve the understanding and management of coral reef systems," said Elizabeth Madin, the study's lead author and an associate research professor at UH. She went on to say, "our current research shows that reef halos may represent an emerging opportunity to monitor reef ecosystems function at large scales including remote and otherwise inaccessible areas." She also said "reef halos are sometimes very clear in satellite imagery, with distinct edges and high contrast with background vegetation, but sometimes they are quite faint and hard to distinguish even by a highly trained observer. In the end, our team was able to develop a set of AI algorithms capable of taking into account the diversity of these patterns globally and identify and measure halos with surprising accuracy. It is hugely satisfying for us to now have built something that can accurately identify more than 90% of halos in some parts of the world."

Turning to a land application Nestle is using Pleiades Neo satellite imagery in its reforestation efforts. We are all familiar with how satellite imagery has been used to study deforestation, but the same can be done with reforestation by analyzing historical trends, which can be used to predict future changes in forest cover. Nestle uses high-resolution images to ensure that the coffee trees it has planted continue to thrive over the long term. This also helps Nestle demonstrate the amount of carbon it is removing from the atmosphere by its Global Reforestation Program, a key component of its 2050 net-zero emissions ambition.

Nestle will first apply this satellite technology in the Ranong and Chumphon provinces of Southeast Thailand. The Pleiades Neo satellites will monitor more than 150,000 shade trees on farms from where Nestle sources its coffee, over a

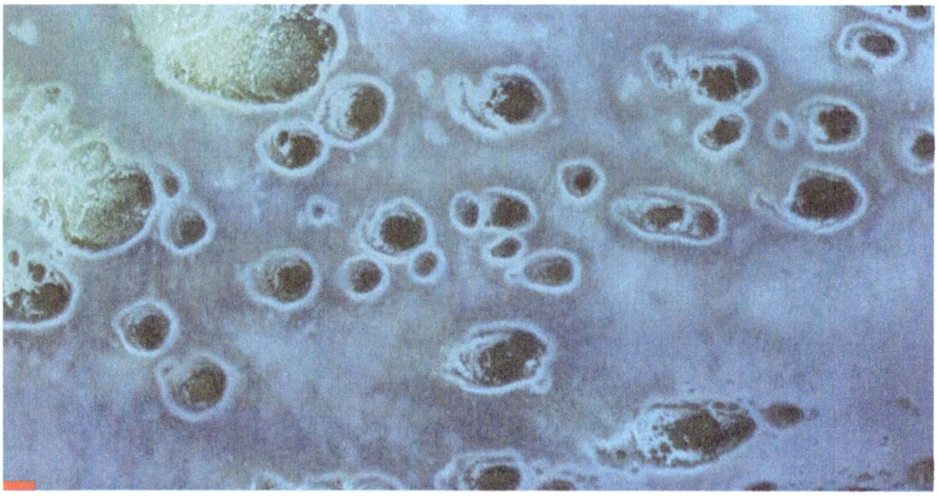

FIGURE 12.4 Example of reef halos around the Hawaiian Islands.
Image credit courtesy CNES/Airbus; DigitalGlobe.

20 year period. Shade trees help prevent coffee's overexposure to the sun, which increases yield and productivity in the long term while also removing carbon from the atmosphere. Based on their experience, Nestle will decide whether to expand this approach to other locations around the world.

The Pleiades Neo satellite constellation consists of four identical Airbus constructed satellites, each equipped with an imager capable of a ground resolution of 30 cm. This sensor also has advanced spectral capabilities enabling images across a wide range of wavelengths. Pleiades Neo is a commercial satellite program operated by Airbus. It has a capacity of collecting imagery over 1 million km^2 per day and the possibility of a daily revisit anywhere on the globe. The wavelengths are the traditional visible channels, including Deep Blue and Red Edge. Acquisitions can be mono, stereo, and tri-stereo as selected by the user. Plans are for the Pleiades Neo Next satellites to have a 20 cm resolution capability.

An interesting study that addresses the topic of "rewilding" (Pettorelli and Schulte to Bühne, 2022) speaks to the need for satellite remote sensing to fill the gaps in rewilding science. This science depends on the reality that our ability to tackle current societal challenges is closely linked to our ability to retain and increase biodiversity (Strassburg et al., 2020). Such a realization has recently led to an increase in calls for large-scale ecological restoration projects as exemplified by the Bonn Challenge, which aims to restore 350 million hectares of degraded and deforested lands by 2030 and the launch of the United Nations Decade on Restoration, which aims to motivate action to prevent, halt, and reverse the degradation of ecosystems on every continent and in every ocean.

Rewilding is broadly defined as the recovery of a self-organizing and self-sustaining ecosystems controlled by natural processes in equilibrium. It is increasingly being considered as a cost-effective environmental management option to curb the loss of biodiversity and enhance both biodiversity and ecosystems. As a conservation approach, it has been used increasingly in recent years. For example, the Union for the Conservation of Nature (IUCN) launched an intercommission working group on rewilding, along with a rewilding theme group under its Commission on Ecosystem Management.

Satellite remote sensing has the ability to map ecosystems and their change over time. Previous studies have demonstrated the use of satellite data in mapping changes in ecosystems and habitat distribution (Pettorelli, 2019), while other studies have used satellite imagery to monitor the changes in biodiversity as a consequence of rewilding efforts. Tracking changes in ecosystems includes providing information on the distribution of structural components (water bodies, etc.). Much as the use of satellite data to map and monitor land cover changes, the tracking of primary productivity dynamics has also been the topic of satellite measurements, particularly since the beginning of the Landsat program. The well-known normalized difference vegetation index (NDVI) introduced earlier provides a metric to monitor the health of vegetation on the ground. Nevertheless, the NDVI has seldom been used to monitor and evaluate rewilding efforts.

What are the barriers to using satellite data, such as the NDVI, for rewilding studies. Pettorelli and Schulte to Bühne (2022) introduce some possibilities in Fig. 12.5.

FIGURE 12.5 Barriers to using satellite data in rewilding studies (Pettorelli, 2019).

They break the barriers down into three categories: First, the need to convince people of the importance of monitoring studies and the capability of satellites to provide those measurements. Second are concerns about the complexities of processing satellite data into useful information. This involves the accuracy, resolution, geometric correction and derived relevant quantities such as the NDVI. Last are the trust issues between the ground scientists and the remote sensing community that only will come with a level of mutual understanding and trust.

They conclude that satellite remote sensing can provide information about changes in ecosystems on a scale and frequency available with no other method. This is particularly useful for rewilding studies where monitoring changes in the ecosystems of interest is critical. The barriers just discussed are not technological but rather social and can be overcome with concerted effort. The perception that satellite remote sensing will replace existing monitoring methods used in rewilding studies results in increased distrust in satellite methods. Since both field and satellite measurements are needed to properly classify an area, this mistrust needs to be overcome so that rewilding estimates can be made most efficiently. The ability of satellite imagery to supply a truly synoptic measurement of landscape changes that reflect rewilding is critical to future evaluation of the success of these efforts.

In a related study, Mairota et al. (2015) discuss the "challenges and opportunities in harnessing satellite remote sensing for biodiversity monitoring." They note that monitoring biodiversity changes is a major concern for many governmental agencies. Remote sensing can provide high-resolution synoptic images of habitats. Remote sensing is now able to rapidly deliver data on habitat quantity (amount and configuration) and quality (structure, distribution on individual plant species, habitat types) across a wide range of spatial, spectral, and temporal resolutions. These data, however, require a significant number of important processing steps to be useful. They must be accurately geolocated, properly calibrated and related to nearly coincident in situ measurements of relevant quantities. Expert knowledge and experience are needed to develop these relationships to enable satellite imagery to provide the information needed.

Specifically, Mairota et al. (2015) note that very high resolution sensors can be used to:

a. Advance theoretical and applied research areas such as landscape ecology and conservation biology.
b. Measure biophysical properties of surface environments (e.g., vegetation, water, and soil) that are relevant to conservation studies.
c. Measure changes in the state of land cover that dictate habitat quantity/quality to generate thematic maps. Such data can be used with field data to test hypotheses regarding factors causing biodiversity change.

All of these applications require accurate geolocation and radiometric calibration of the satellite image data. High spatial resolution of the satellite imagery is desired in the biophysical studies, but this higher spatial resolution exerts demands on the geometric correction accuracy. Another challenge in this geolocation is to account for vertical relief in the target ground structure. If a digital elevation map is available for the study region, it may be used to improve the satellite image of the target area. Calibration of the satellite reflectance is carried out prelaunch with optical methods, and vicarious calibration of onboard reference elements (solar diffuser) and celestial targets such as the moon can be used to compensate for any drift that the satellite sensor may experience.

There have been dramatic improvements in satellite image classification using convolutional neural networks (also called deep learning) to better resolve important features in the image. As these classification methods are matched up with field measurements of the relevant variables, it will be possible to develop models that best characterize the target areas in terms of their habitat and biodiversity. It is important in these analyzes that the analyst chooses the proper space and time scales for the problem being studied. The scales to be considered are the fundamental spatial scales of the landscape and the time/space scales of the different organisms being studied. The multiplicity of these scales presents a challenge both for the remote sensing scientist and their partner, providing the ground observations.

Mairota et al. (2015) report that "the combined use of remote sensing derived categorical and continuous products offers the opportunity to test the hypothesis that local habitat units are not interchangeable in terms of the internal quality, but instead their condition is dependent on variations in surrounding land cover composition or configuration at both patch and landscape levels." Moreover, they used linear mixed-effects models to test the influence of the surrounding patch and landscape structures. Such statistical models have the ability to explicitly consider different sources of space/time variability. Such models can also be examined in terms of the sampling scheme employed and the covariance structure of the data. They also note that "in addition to the estimated direct effects of patch or landscape predictors on the response variable, such as when a landscape variable alone influenced local habitat quality, the power of this approach lies in its ability to test for the influence of indirect effects when the landscape patch attributes have a conjoined effect on the response variable."

Mairota et al. (2015) conclude that remote sensing has the ability to increase the availability of information on changes in habitat quality and their impact on multiple aggregate components of biodiversity at various spatial scales. Further developments require that remote sensing scientists work more with ecologists to better merge the satellite data with direct ground measurements.

In a recent paper (Dubovik et al., 2021) titled "grand challenges in satellite remote sensing," the authors focus on the atmosphere and the upper ocean. They state that "the overall grand challenge in the advancement of satellite remote sensing is discovering innovative and affordable technologies and measurement concepts to solve new problems that have come to light in the first half of this century." These satellite programs have been spelled out earlier in this book, so the reader should be familiar with them. These authors first identify the need for increased spatial and temporal coverage to be able to sense and monitor targets on the ground/ocean, or in the atmosphere. They correctly point out that there is a trade-off between spatial resolution and temporal resolution when a single satellite is considered. Polar orbiting spacecraft have higher spatial resolution, while geostationary orbiting satellites have better temporal resolution that is limited in some wavelengths by the much higher orbit altitude. This problem has been overcome recently by having a constellation of smaller satellites that together are able to overcome both the need for higher spatial resolution and better temporal sampling. Another alternative has been implemented by Maxar, whose satellites are able to point in various directions to provide increased sampling of an area.

In this paper, they also recognize that newer sensors with additional capabilities are being proposed and created. As an example, they mention the multiangular polarimeters, which provide the most appropriate data for characterizing columnar properties of atmospheric aerosols and clouds. Several advanced polarimetric missions are planned to be launched in the coming few years by European and US Space agencies. Moreover, the China National Space Administration (CNSA) has recently launched several polarimetric remote sensing instruments. Similarly, the number of satellite-based LIDARs and radars is expected to increase. For example, NASA launched the Geosciences Laser Altimeter System (GLAS) onboard the ICESat satellite in 2003. Plans are for a wind LIDAR onboard the ESA Aeolus satellite as part of the Living Planet Program. LIDAR is expected to be an essential part of any Earth observation program.

Dubovik et al. (2021) also recognize the increased capability offered by a constellation of satellites. This multiplicity of sampling by various satellites in the constellation makes it possible to have the high spatial resolution desired while still having the rapid sampling to resolve features that change rapidly. The improvements in electronics have made it possible to have very capable small satellites that can be operated in a constellation to provide the most representative sampling of any target on the Earth. This constellation approach requires accurate geometric correction and sensor calibration to be able to merge the images provided by the different satellites in the constellation. This will demand satellite-to-satellite communication and increased satellite-to-ground data relay. Still, this opens a world of new possibilities of coincident high spatial and high temporal resolution. Improvements in sensor electronics will make it possible for the constellation satellites to sample over a range of wavelengths or frequencies, depending on if the target of interest is best sampled in the optical, passive microwave, or SAR domains.

It is recognized that AI and machine learning techniques will be needed to best process the volume of data that will be generated by this constellation of satellites. Another important thing is that the satellite data must be analyzed in concert with the direct objects measured in situ. The ultimate goal is to convert the satellite observations into representative ground measurements. This is only possible if there is a concerted effort to analyze the satellite measurements in conjunction with those from ground studies. Again, the need for remote sensing scientists to work with oceanographers and terrestrial ecologists is tantamount. In addition, for atmospheric studies, the remote sensing scientist must work with the more traditional meteorologists. This is perhaps less difficult since meteorologists have been using satellite data routinely almost from its inception. As pointed out in previous sections of this book, the early satellite sensors and programs were all focused on providing data for the weather forecaster.

In all the new developments, it is important to recognize that the time series of the satellite data may be essential to the science being studied. Thus while sensors and satellites will improve, it is important that these improvements do not introduce changes that make it difficult to relate the new data to earlier similar measurements. In this respect, satellite-to-satellite calibration is important to keep the time series viable for future studies. The absolute calibration of each instrument and the intercalibration of multiple relevant sensors are imperative for the success of any and all satellite programs. It should be noted that this chapter was highly biased toward atmospheric studies in general and atmospheric profiling in particular.

An article entitled "what are the most innovative trends in remote sensing" lists: cloud computing, hyperspectral imaging, small satellites, drone mapping, and AI (file:///Users/williamjemery/Desktop/RS%20past%20pres%20futu/background%20off%20the%20web/future/What%20are%20the%20most%20innovative%20trends%20in%20remote%20sensing%3 F.webarchive). They point out that one of the most significant shifts is the move from desktop-based computing to cloud-based processing and storage of large volumes of data. This has the advantages of scalability, accessibility, cost-effectiveness, and collaboration. Cloud platforms such as Google Earth Engine, Amazon Web Services and Microsoft Azure make it possible for users to access and analyze remote sensing data from various sources and formats. They can use advanced algorithms in their study and apply machine learning tools far beyond their own expertise. Cloud computing also enables the integration of satellite data with other geospatial data and services.

We have already discussed the advances offered with hyperspectral imaging and have spelled out the many advantages of small satellites and their constellations. We have not yet emphasized the use of drones or unmanned aerial vehicles (UAVs) in remote sensing applications. Drones are able to not only carry cameras but also can be equipped with other sensors, such as LIDARs and thermal infrared sensors. They fly at relatively low altitudes, eliminating the need for atmospheric corrections. Drones can also be used to sense hazardous targets such as volcanoes and forest fires. They are being used heavily in combat operations both for reconnaissance and for attack. There are a great many other applications, such as agriculture, forestry, archeology, and urban studies.

In an Earth Online article (https://earth.esa.int/eogateway/news/meet-the-future-of-earth-observation-super-resolution-data), they discuss what they refer to as the future of Earth Observations "superresolution data." Superresolution (SR) studies have been around for a while (Baldwin et al., 1998), and this article reports on an ESA/ESRIN SR workshop held in Frascati, Italy on July 11, 2024, called SUREDOS24 (SR and downscaling for Earth observation and Earth science). They report that ESA is taking steps to support SR techniques to enhance remote sensing data, while ensuring their validity and reliability for scientific and commercial use.

FIGURE 12.6 A WorldView-3 image of a city with 30 cm resolution on the left and 15 cm resolution on the right. Image credit Maxar.

FIGURE 12.7 A Pleiades Neo image: top 30 cm resolution, bottom 15 cm SR resolution.
Image credit Maxar.

SR techniques can be applied to many different data types, such as visible, hyperspectral, thermal infrared, SAR, and scatterometer data. The central focus is to reconstruct from lower resolution sources a higher resolution product, thereby enhancing the ability of the lower resolution data to respond to image targets that would be missed by the lower resolution original image. At the workshop, Peggy Fischer, ESA's third party mission (TPM) program manager, made a presentation

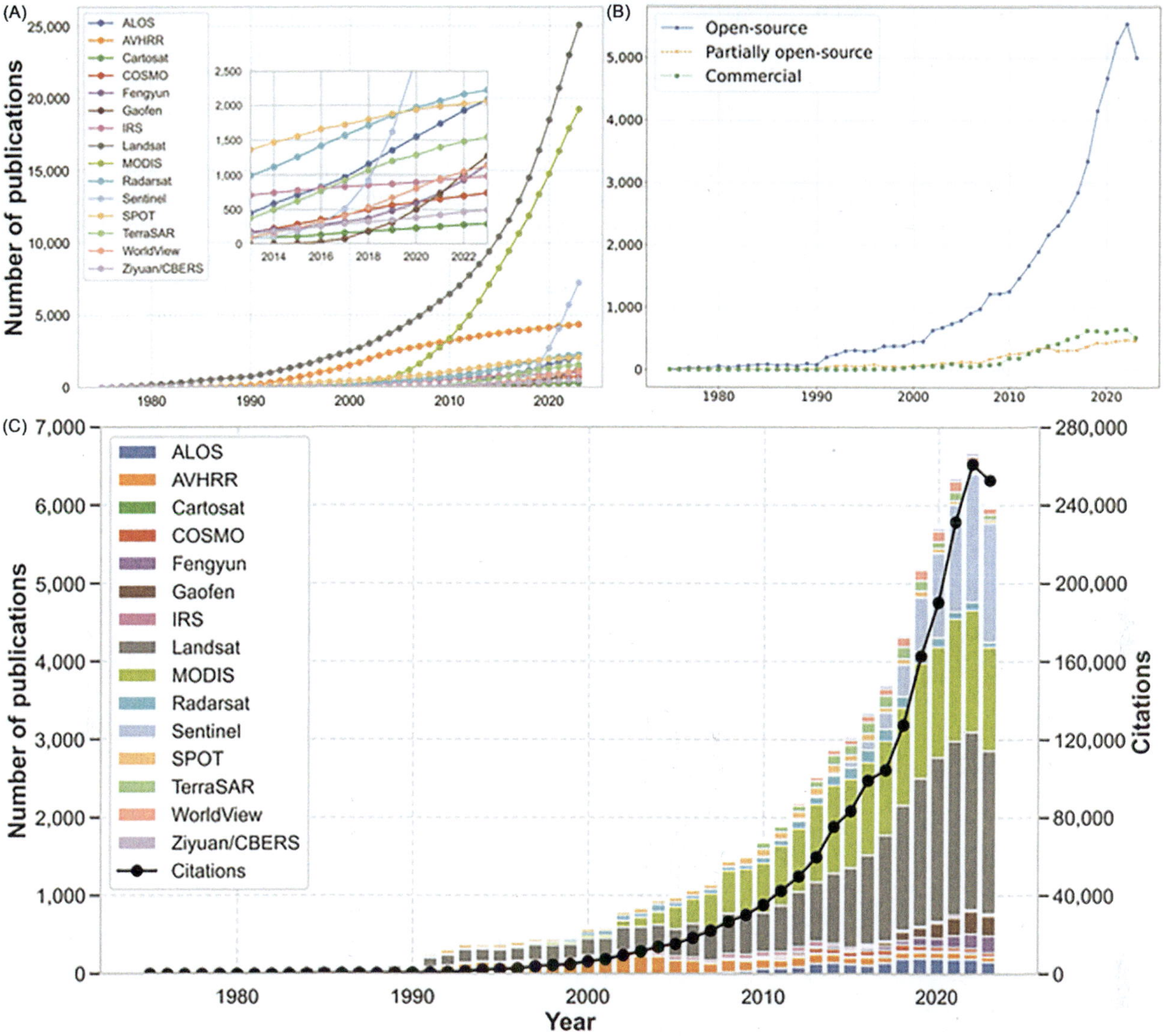

FIGURE 12.8 Number of (A) publications, (B) publications by open access, and (C) publications related to Earth Observation satellites, including Landsat, MODIS, Sentinel, AVHRR, Advanced Land Observing Satellite (*ALOS*), Fengyun, Indian Remote Sensing Satellite (*IRS*), SPOT, and Ziyuan (*ZY*) China-Brazil Earth Resources Satellite (*CBERS*), Cartosat (constellation of small satellites for Mediterranean basin observations), and the synthetic aperture radar (*SAR*) satellites COSMO-Skymed, Radarsat, TerraSAR, and WorldView optical satellites (these latter satellites are primarily commercial) (https://www.researchgate.net/figure/Number-of-A-publications-B-publications-by-data-open-access-level-and-C_fig1_387686331).

on how the TPM supports SR. She demonstrated how Maxar was able to generate 15 cm imagery from 30 cm images (Fig. 12.6). Here, the image on the left is the standard WorldView 30 cm image, and on the right is the same image using SR to get a 15 cm spatial resolution. You can clearly see the improvement in the depiction of ground elements. The 15 cm image is clearer, and image elements are more clearly displayed. The SR technique also increases the number of pixels in the image, which explains the added detail in the 15 cm image.

Pleiades Neo is another TPM mission follow-on to Pleiades, which offers 30 cm imagery. Fischer presented the images in Fig. 12.7 to show how SR can enhance an image to show the detailed structure of a particular element. Again, the SR version of the image better depicts the details of the structure. The increase in the number of SR pixels made it necessary to crop the image to show the tower more clearly.

One indicator of the growth in the use of satellite remote sensing is the dramatic increase in the number of science papers published that use remote sensing. In a paper by Zhao and Yu (2025), they track the number of publications associated with specific remote sensing missions (Fig. 12.8).

The dramatic increase from the year 2000 to 2023 is almost exponential. MODIS, Sentinel and Landsat account for a large part of this increase. There has been a simultaneous increase in the number of open access publications during this same time. There are economic reasons for some of the changes, and it also seems like AVHRR is definitely underestimated. Perhaps there is a bias toward a particular discipline (land) while AVHRR cuts across a number of disciplines.

To conclude this chapter on the future of satellite remote sensing, the basics are clear. Large government programs, particularly those involving weather forecasting, will continue and will gradually improve as sensors improve and satellites become smaller and less expensive. How the emergence of commercial remote sensing will have an impact on many of the government programs remains to be seen. It is clear that the commercial sector appears to be more efficient at creating new and innovative systems that can address old and fundamental problems. The science community will continue to improve the existing system while discovering new methods to use satellites to solve important societal issues. There will likely be greater coupling between commercial entities and research science as they together plan to attack new challenges in remote sensing. There will be lots of synergistic studies where drone (UAV) measurements will be coupled with satellite and ground observations. Ground truth measurements will continue to be essential in all satellite and UAV observations, as they need to be used to connect the satellite data to relevant ground information.

Glossary

AATSR	advanced along-track scanning radiometer
ABI	advanced baseline imager
AFB	air force base
AHI	advanced Himawari imager
AIRS	atmospheric infrared sounder
ALT	radar altimeter (on Seasat)
AMSR-E	advanced microwave scanning radiometer–EOS
AMSU	advanced microwave sounding unit
APL	Applied Physics Lab (John Hopkins University)
APT	automatic picture transmission for low resolution imagery
Argos	data collection system
ARS	angular rate sensors
ASAR	advanced synthetic aperture radar
ASCAT	Advanced Scatterometer
ASGC	along scan gain control
ASI	Italian Space Agency
ASOS	Automated Surface Observing System
ASS	atmospheric sounding system
ATS	Applied Technology Satellite
AVCS	Advanced Vidicon Camera System
AVHRR	advanced very high resolution radiometer
AVNIR	advanced visible and near-infrared radiometer
BRDF	bidirectional reflectance distribution function
CAS	Canadian forest system
CAVIS	clouds, aerosols, vapors, ice, and snow
CBERS	China–Brazil Earth Resources Satellite
CCD	charged coupled device
CCOR	compact coronagraph
CCRS	Canada Center for Remote Sensing
C CSAT	Satellite Chlorophyll-A
CDR	climate data record
CERES	clouds and radiant energy system
CHEOS	China's high resolution earth observation system
CG	cloud-to-ground lightning
CLASS	Comprehensive Large Array-data Stewardship System (repository for AVHRR data)
CLAVR	clouds from AVHRR
CLW	cloud liquid water
CMA	Chinese Meteorological Administration
CNSA	China National Space Administration
COMS	Communications, Ocean, and Meteorological Satellite
CONUS	North American Coverage (GOES)
CSAT	Satellite Chlorophyll-A
CSU	cross-strap unit
CW	cottonwood–willow
CZCS	coastal zone color scanner
DoD	Department of Defense
DOI	Department of the Interior

DAAC	Distributed Active Archive Center
DCPLS	data collection and platform location system
DEM	digital elevation model
DFH	Dong Fang Hong
DMSP	Defense Meteorological Satellite Program
DN	digital number
DORIS	doppler orbitography and radiopositioning by satellite
DOS	Department of Space
DPM	Damage Proxy Map
DPR	dual-frequency precipitation radar
DSP	digital signal processing (software created by the University of Miami)
DU	Dobson units (ozone)
DWSS	defense weather satellite system (US Air Force, cancelled)
ECMWF	European Center for Medium-Range Weather Forecasts
EDRS	European Data Relay Satellite
ENVISAT	environmental satellite
EOSAT	Earth Observation Satellite Company
EOSDIS	Earth Observing System Data and Information System
ERB	Earth radiation budget
EROS	Earth Resources Observation Satellite
ERS	Earth Research Satellite (Europe)
ESA	European Space Agency
ESMR	electrically scanning microwave radiometer
ESRI	Environmental Systems Research Institute (makers of ArcGIS)
ESSA	Environmental Science Services Administration
EUVS	extreme ultraviolet sensor
EV	essential variable
EXIS	extreme ultraviolet and X-ray irradiance sensors (EUVS+XRS)
FCI	Flexible Combined Imager
FD	function of the day of the year
FMC	fuel moisture content
FOV	field of view
FPA	focal plane assembly
FPGA	field programmable gate array
FPR	flat plate radiometer
GAC	global area coverage (of the AVHRR)
GAO	General Accounting Office
GDEM	global digital elevation model
GEOS	Geodetic Experimental Ocean Satellite
GEOSAT	US Navy Geodetic Satellite
GEOS&R	geostationary Earth radiation budget
GERB	geostationary Earth radiation budget
GFZ	GeoForschungsZentrum Potsdam, Germany
GHRSST	group for high-resolution sea-surface temperature
GIS	geographic information system
GLM	GOES lightning mapper
GMI	GPM microwave imager
GMS	Geostationary Meteorological Satellite (Japan, Himawari)
GMS-I	Geostationary Meteorological Satellite
GMT	Greenwich mean time
GNSS-RO	global navigation satellite system–radio occultation
GOCI	geostationary ocean color imager (South Korea)
GOES	Geostationary Operational Environmental Satellite
GOFC	global observations of forest cover
GOME	global ozone monitoring experiment
GOMOS	global ozone monitoring by the occultation of stars
GPM	Global Precipitation Measurement
GPS	global positioning system

GPU	graphic processor unit
GRACE	gravity recovery and climate experiment
GSFC	NASA's Goddard Space Flight Center
GSOC	German Space Operations Center
GTS	global transmission system
HEPAT	high-energy proton alpha telescope
HES	hyperspectral environmental suite
HIRS	high-resolution infrared sounder
HRG	high resolution geometrical
HRIR	high resolution infrared radiometer
HRPT	high resolution picture transmission (direct readout of the AVHRR)
HRWS-SAR	high resolution wide swath SAR
HSB	humidity sounder for Brazil
HSC	Hughes Space and Communication Company
IASI	infrared atmospheric sounding interferometer
IC	cloud-to-cloud lightning
ICT	internal calibration target
IGBP	International Geosphere–Biosphere Program (DIS data and information system)
IMS	infrared multispectral scanner
INSAT	Indian National Satellite
IORD	Integrated Operational Requirements Document
IRSCAM	infrared medium resolution scanner
IRU	inertial reference unit
ISRO	Indian Space Research Organization
ITOS	improved TIROS satellite
ITPR	infrared temperature profile radiometer
IVM	ion velocity meter
JAXA	National Space Development Agency of Japan
JMA	Japan Meteorological Administration
JPL	NASA's Jet Propulsion Lab
JPSS	Joint Polar Satellite System
KIOST	Korean Institute of Ocean Science & Technology
KITSAT	South Korean minisatellite
KMA	Korean Meteorological Administration
LAI	leaf area index
LASP	Laboratory for Atmospheric and Space Physics (University of Colorado at Boulder)
LEO	low Earth orbit
LIDAR	light detection and ranging
LISS	Linear Imaging Self-scanning Sensor
LIS	Lightning Imaging Sensor
LMSC	Lockheed Missiles and Space Company
LRIR	limb radiance inversion radiometer
LRR	laser reflector
LST	land surface temperature
MAB	mid-Atlantic Bight
MAG	magnetometer
MARMAP	multiyear study by NMFS of the MAB
MDA	MacDonald, Dettwiler, and Associates
MEIS	multispectral Earth imaging system
MEM	main electronics module (MODIS)
MEPED	medium energy proton and electron detector
MERIS	medium resolution imaging spectrometer
MESSR	multispectral electronic self-scanning radiometer (Japan)
METI	Japanese Ministry of Economy, Trade, and Industry
MIPAS	Michelson interferometer for passive atmospheric sounding
MIR	manipulated information rate processor
MISR	multiangle imaging spectrometer
MLS	microwave limb sounder

MND	metallic neutral density
MOBY	marine optical buoy
MODIS	moderate resolution imaging spectrometer
MODTRAN	moderate resolution atmospheric transmission
MOS	marine observation satellite (Japan)
MSD	mean square displacement
MSG	Meteosat Second Generation (SG)
MSMR	multifrequency scanning microwave radiometer
MSR	microwave scanning radiometer
MSS	multispectral scanner
MSU	microwave sounding unit
MTG	Meteosat Third Generation
MTSAT	multifunctional transport satellites
MVIRI	Meteosat visible and infrared imager
MVP	single chip multiprocessor of multimedia
MWR	microwave radiometer
MZM	monthly zonal means
NASA	National Aeronautics and Space Administration
NCAR	National Center for Atmospheric Research
NCDC	National Climate Data Center
NCEP	National Center for Environmental Prediction
NDBC	National Data Buoy Office
NDSI	normalized difference snow index
NDVI	normalized difference vegetation index
NEMS	Nimbus E Microwave Spectrometer
NESDIS	National Environmental Satellite Data and Information Service
NGDC	National Geophysical Data Center
NIR	near infrared
NIST	National Institute of Science and Technology
NMFS	National Marine Fisheries Service
NPOESS	National Polar-orbiting Operational Environmental Satellite System
NPP	NASA NPOESS preparatory platform
NRL.	Naval Research Lab
NRO	National Reconnaissance Office
NSF	National Science Foundation
NSSA	Navy Space Systems Activity
NSTC	National Science and Technology Council
NOAA	National Oceanic and Atmospheric Administration
OCM	Ocean Color Monitor
OKB	Russian Experimental Design Bureau
OLS	operational line scanner
OMI	UV grating spectrometer
OSC	Orbital Sciences Corporation
OSCAT	Indian scatterometer
PACUS	Pacific US (GOES West)
PATMOS	NOAA/NESDIS Pathfinder Atmosphere data set
PDF	probability distribution function
PMR	pressure modulated radiometer
PMT	photomultiplier tube
POES	polar-orbiting environmental satellite
POLDER	polarization and directionality of the Earth's reflectance
PR	photo reconnaissance
PR	precipitation radar
PRARE	precision range and range-rate experiment
PSLV	Polar Satellite Launch Vehicle (India)
QC	quality control
RAF	Royal Air Force
RAOB	Radiation Observing Balloon

RBV	Return Beam Vidicon (camera)
RCA	Radio Corporation of America
RCCM	radiometric camera-by-camera cloud mask (MISR)
RIP	riparian vegetation
RMI	Royal Meteorological Institute (Belgium)
RTE	radiative transfer equation
SAB	South Atlantic Bight
SAM-II	Stratospheric Aerosol Measurement II
SAMIR	microwave radiometer (India)
SAR	synthetic aperture radar
SARR	Search and Rescue Processor
SASAR	satellite sided and rescue system
SASS	satellite scatterometer (wind vector measurement)
SBEM	DoD space-based environmental monitoring
SBUV	Solar Backscatter Ultraviolet Spectrometer
SCAMS	scanning microwave radiometer
SCMR	surface composition mapping radiometer
SCR	selective chopper radiometer
SEA	Southeast Atlantic, NEA Northeast Atlantic
SEISS	collection of six instruments (EHIS energetic heavy ion sensor; MPS-HI and MPSS-LO magnetospheric particle sensors high and low; SGPS solar and galactic proton sensor; DPU data processing unit)
SCIAMACHY	scanning imaging absorption spectrometer for atmospheric cartography
SCT	solar calibration target
SEM	space environment monitor
SENSE	space environment scientific experiment
SEVIRI	spinning enhanced visible and infared imager (MSG)
SMMR	scanning multichannel microwave radiometer
SMR	snow melt runoff
SNR	signal-to-noise ratio
SPM	solar proton monitor (space weather instrument)
SPOT	satellite for observation of Earth
SR	scanning radiometer
SRTM	Shuttle Radar Topography Mission
SSM/I	special sensor microwave imager
SSMIS	special sensor microwave imager sounder
SSM/T	special sensor microwave temperature
SSL	space systems loral
SSP	sub satellite point (nadir)
SST	sea surface temperature
SSU	stratospheric sounding unit
ST	surface temperature (land and in situ)
SUVI	solar ultraviolet imager
SVC	spatial visible coherence
SWIR	short wave infrared
TAO	TOGA atmosphere ocean
TCWV	total column water vapor
T+DRE	tracking and data relay experiment
TDRSS	tracking and data relay satellite
TED	total energy detector
TGRS	tri-global navigation system radio occultation receiver
THIR	temperature humidity infrared radiometer
TIP	TIROS information processor
TIROS	Television Infrared Observation Satellite
TM	thematic mapper (Landsat)
TMA	three mirror anastigmat (telescope)
TMI	TRMM microwave imager
TOGA	Tropical Ocean Global Atmosphere
TOMS	Total Ozone Mapping Spectrometer

TOVS	TIROS operational vertical sounder (combines other instruments, a combination of the HIRS, the MSU, and the SSU)
TPW	total precipitable water
TRMM	Tropical Rainfall Mapping Mission
TWERLE	Tropical Wind Energy Conversion and Reference Level Experiment
UAV	unmanned aerial vehicle
UCAR	University Corporation for Atmospheric Research
USDA	US Department of Agriculture
UHI	urban heat island
USGS	US Geological Survey
UTC	coordinated universal time
UVN	ultraviolet visible near-infrared sounder
VAS	VISSR atmospheric sounder
VDGA	variable digital gain amplifier
VHF	very high frequency
VHRR	very high resolution radiometer
VI	vegetation index
VIRS	visible infrared radiometer
VISSR	visible infrared spin scan radiometer
VNIR	visible near infrared radiometer
VTPR	vertical temperature profile radiometer
WFI	wide field imagery
WFICAM	wide field imaging camera
WMO	World Meteorological Organization
WMONEX	winter monsoon experiment
WSF-M	weather system follow-on—microwave
XRS	X-ray sensor

Bibliography

REFERENCES

Allen, T.F.H., Starr, T.B., 1982. Hierarchy: Perspectives for Ecological Complexity. University of Chicago Press, Chicago.

Allen, T.F.H., O'Neill, R.V., Hoekstra, T.W., 1984. Inter- level relations in ecological research and management: some working principles from hierarchy theory. USDA Forest Ser- ViceGeneral Technical Report RM-10, 11pp. Rocky Mt. For. and Range Exp. Station, Ft. Collins, Colo.

Alonso, M., Camarasa, A., Chuvieco, E., Cocero, D., Kyun, I., Mart´ın, M.P., et al., 1996. Estimating temporal dynamics of fuel moisture content of Mediterranean species from NOAA–AVHRR data. EARSEL Adv. Remote. Sens.remote sensing 4 (4), 9–24.

Arino, O., Mellinotte, J.M., 1998. The 1993 Africa fire map. Int. J. Remote. Sens. 19, 2,019–2,023.

Asam, S., Eisfelder, C., Hirner, A., Reiners, P., Holzwarth, S., Bachman, M., 2023. AVHRR NDVI compositing method comparison and generation of multi-decadal time series—a timeline thematic processor. Remote. Sens. 15. Available at https://doi.org/10.3390/rs15061631.

Austin, P.M., Geotis, S.G., 1979. Raindrop sizes and related parameters for GATE. J. Appl. Meteor. 18, 569–592.

Baldwin, D., Emery, W., Cheeseman, P., 1998. Higher resolution earth surface features from repeat moderate resolution satellite imagery. IEEE Geosci. Rem. Sens. 36, 244–255.

Baumgartner, M.F., G. Apfl, Holzer, T., 1994. Monitoring alpine snow cover variations using NOAA-AVHRR data. In: Proceedings of IGARSS '94 - 1994 IEEE International Geoscience and Remote Sensing Symposium, Pasadena, CA, USA, pp. 2087-2089, vol. 4. Available at https://doi.org/10.1109/IGARSS.1994.399661.

Belward, A.S., Kennedy, P.J., Grégoire, J.-M., 1994. The limitation and potential of AVHRR GAC data for continental scale fire studies. Int. J. Remote. Sens. 15 (2), 215.

Bernstein, R.L., 1982. Sea surface temperature mapping with the Seasat microwave radiometer. J. Geophys. Res. 87, 7865–7872.

Bessho, K., Date, K., Hayashi, M., Ikeda, A., Imai, T., Inoue, H., et al., 2016. An introduction to Himawari-9, 9, Japan's new generation geostationary meteorological satellites. J. Met. Soc. Jpn. 94, 151–183. Available at https://doi.org/10.2151/jmsj.2016-009.

Bhartia, P.K., Klenk, K.F., Wong, C.K., Gordon, D., Fleig, A.J., 1984. Intercomparison of the NIMBUS 7 SBUV/TOMS total ozone data sets with Dobson and M83 results. J. Geophys. Res. 89, 5239–5247.

Bhartia, P.K., McPeters, R.D., Flynn, L.E., Taylor, S., Kramarova, N.A., Frith, S., et al., 2013. Solar backscatter UV (SBUV) total ozone and profile algorithm. Atmos. Meas. Tech. 6, 2533–2548. Available at https://doi.org/10.5194/amt-6-2533-2013.

Bontemps, S., Defourny, P., van Bogaert, E., Arino, O., Kalogirou, V., Perez, J.R., 2011. GLOBCOVER 2009: Products Description and Validation Report, European Space Agency, Paris, France.

Bowman, K.P., Krueger, A.J., 1985. A global climatology of total ozone from the Nimbus 7 total mapping spectrometer. J. Geophys. Res. 90, 7967–7976.

Boyer, P., Dawson, F.M., 2004. Sensitivity of spectral reflectance to variation is live fuel moisture content at leaf and canopy level. Rem. Sens. Environ. 92, 297–3008.

Brest, C.L., Rossow, W.B., 1992. Radiometric calibration and monitoring of NOAA AVHRR data for ISCCP. Int. J. Remote Sens. 13, 235–273.

Brown, O.B., Evans, R.H., Brown, J.W., Gordon, H.R., Smith, R.C., Baker, K.S., 1985. Phytoplankton blooming off the US East Coast: a satellite description. Science 229, 163–167.

Burke, H.K., Bussey, A.J., Jones, K.C., Trippsand, N.K., Ho, J., 1980. Inversion algorithm for environmental parameter extraction of the SSM/I program. Environmental Research and Technology, Inc., February.

Caetano, M.S., Mertes, L.A.K., Pereira, J.M.C., 1994. Using spectral mixture analysis for fire severity mapping. In: D.X. Viegas (Ed.), 2nd Int. Conf. Forest Fire Research. Portugal7 Coimbra, pp. 667–677.

Cahoon et al., 1991, 1994.

Cardone, V.J., 1969. Specification of the wind field distribution in the marine boundary layer for wave forecasting, Report TR 69-1, Geophys. Sci. Lab., New York University.

Chang, A.T.C., Wilheit, T., 1979. Remote sensing of atmospheric water vapor, liquid water, and wind speed at the ocean surface by passive microwave techniques from the Nimbus 5 satellite. Radio. Sci. 145, 793–802.

Cheney, R.E., Miller, L., 1988. Mapping the 1986-1987 El Nino with GEOSAT altimeter data. EOS. Trams/ Amer. Geophys. Union. 69, 754–755.

Chladil, M.A., Nunez, M., 1995. Assessing grassland moisture and biomass in Tasmania. The application of remote sensing and empirical models for a cloudy environment. Int. J. Wildland Fire 5, 165–171.

Chuvieco, E., Congalton, R.G., 1988. Using cluster analysis to improve the selection of training statistics in classifying remotely sensed data. Photogrammetric Eng. & Remote. Sens. 54, 1275–1281.

Chuvieco, E., Deshayes, M., Stach, N., Cocero, D., Rian˜o, D., 1999. Short-term fire risk: foliage moisture content estimation from satellite data. In: Chuvieco, E. (Ed.), Remote sensing of large wildfires in the European Mediterranean Basin. Springer-Verlag, Berlin, pp. 17–38.

Chuvieco, E., Rian˜o, D., Aguado, I., Cocero, D., 2002. Estimation of fuel moisture content from multitemporal analysis of Landsat Thematic Mapper reflectance data: applications in fire danger assessment. Inter- Natl J. Remote. Sens. 23 (11), 2145–2162.

Chuvieco, E., Cocero, D., Riano, D., Martin, P., Vega, J.M., de la Riva, J., et al., 2004. Combining NDVI and surface temperature for the estimation of live fuel moisture content in forest fire danger rating. Rem. Sens. Environ. 82, 322–331.

Clark, D., Murphy, M., Yarbrough, M., Feinholz, M., Flora, S., Broenkow, W., et al., 2003, MOBY, a radiometric buoy for performance monitoring and vicarious calibration of satellite ocean color sensors: measurement and data analysis protocols, SeaWiFs Postlaunch Technical Report Series (Accessed January 12, 2026).

Coll, C., Caselles, V., 1997. A global split-window algorithm for land surface temperature from AVHRR data: validation and algorithm com- parison. J. Geophys. Res. 102B14, 16697–16713.

Cracknell, A.P., 1997. The Advanced Very High Resolution Radiometer. CRC Press, Taylor and Francis, London, UK.

Croft, T.A., 1973. Burning waste gas in oil fields. Nature 245, 375–376.

Croft, T.A., 1978. Nighttime images of the earth from space. Sci. Am. 239, 86–98.

Croft, T.A. (1979), The brightness of lights on Earth at night, digitally recorded by DMSP satellite, Stanford Research Institute Final Report, U.S. Geological Survey, Washington, DC.

Crone, L.J., McMillin, L.M., Crosby, D.S., 1996. Constrained regression in satellite meteorology. J. Appl. Meteor. 35 (11), 2024–2035.

Csiszar, I., Abdelgadir, A., Li, Z., Jin, J.-Z., Fraser, R., Hao, W.-M., 2003. Interannual changes of active fire detectability in North America from long-term records of the advanced very high resolution radiometer. J. Geophys. Res. 108 (D2). Available at https://doi.org/10.1029/2001JD001373.

Diner, D., Beckert, J.C., Reilly, T.H., Bruegge, C.J., Conel, J.E., Kahn, R.A., et al., 1998. Multi-angle imaging spectro-radiometer (MISR) instrument description and experiment overview. IEEE Trans. Geosci. Rem. Sens. 36, 1072–1087.

Donlon, C., Berruti, B., Buongiorno, A., Ferreira, M.H., Femenias, P., Frerick, J., et al., 2012. The global monitoring for environment and security (GMES) Sentinel-3 mission. Rem. Sens. Environ. 120, 37–57. Available at https://doi.org/10.1016/j.rse.2011.07.024.

Dozier, J., 1981. A method for satellite identification of surface temperature fields of subpixel resolution. Remote. Sens. Environ. 11, 221–229.

Dubovik, O., Schuster, G.L., Xu, F., Hu, Y., Bösch, H., Landgraf, J., et al., 2021. Grand challenges in satellite remote sensing. Front. Rem. Sens. 2. Available at 10.3389/frsen.2021.619818.

Duda, R.O., Hart, P.E., 1973. Pattern Classification and Scene Analysis. Wiley Interscience, New York, pp. 482.

Durand, Y., Hallibert, P., Wilson, M., Ledouara, M., Grabarnik, S., Aminou, D., et al., 2015. The flexible combined imager on board MTG: from design to calibration. Sensors, Sys, Next-Gen Satsm Proc, SPIE 9639. Available at https://doi.org/10.1117/12.2196644.

Dwyer, E., Gregoire, J.-M., Malingreau, J.-P., 1998. A global analysis of vegetation fires using satellite images: spatial and temporal dynamics. Ambio 27, 175–181.

Eidenshink, J.C., 1992. The 1990 conterminous U.S. AVHRR data set. Photogrammetric Eng. Remote. Sens. 58, 809–813.

Elliott, C.C.H., 1979. The harvest time method as a means of avoiding quelea damage to irrigated rice in Chad/Cameroun. J. Appl. Ecol. 16, 23–35.

Elvidge, C.D., Baugh, K.E., Kihn, E.A., 1997. Relation between satellite observed visible-near infrared emissions, population, economic activity and electric power consumption. Int. J. Remote. Sens. 18, 1373–1379. Available at https://doi.org/10.1080/014311697218485.

Elvidge, C.D., Baugh, K.E., Dietz, J.B., 1999. Radiance calibration of DMSP-OLS low-light imaging data of human settlements. Remote. Sens. & Environ. 68, 77–88. Available at https://doi.org/10.1016/S0034-4257(98)00098-4.

Embury, O., Merchant, C.J., 2012. A reprocessing for climate of sea surface temperature from the along-track scanning radiometer: a new retrieval scheme. Rem. Sens. Env. 116, 47–61.

Embury, O., Merchant, C.J., Filipiak, M.J., 2012. A reprocessing for climate of sea surface temperature from the Alon-Track Scanning Radiometers: basis in radiative transfer. Rem. Sens. Environ. 115, 32–46.

Eva, H., Flasse, S., 1996. Contextual and multiple-threshold algorithms for regional active fire detection with AVHRR data. Remote. Sens. Rev. 14, 333–351.

Frederick, J.E., Cebula, R.P., Heath, D.F., 1986. Instrument characterization for the detection of long-term changes in stratospheric ozone: an analysis of the SBUV/2 radiometer. J. At-Mos. Ocean. Tech. 3, 472–480.

Frey, C., Kuenzer, C., Dech, S., 2017. Assessment of mono- and split-window approaches for time series processing of LST from AVHRR—A TIMELINE round robin. Remote. Sens. 9, 72.

Flasse, S.P., Ceccato, P., 1996. A contextual algorithm for AVHRR fire detection. Int. J. Remote. Sens. 17, 419–424.

Foster, J.L., 1983. Observations of the Earth using nighttime visible imagery. Int. J. Remote. Sens. 4, 785–791.

Franca, J.R., Brustet, J.M., Fontan, J., 1995. Multispectral remote sensing of biomass burning in West Africa. J. Atmos. Chem. 22, 81–110.

Fuglister, F.C., 1955. Alternative analyses of current surveys. Deep-Sea Res. 2, 213–229. Available at https://doi.org/10.1016/0146-6313(55)90026-5.

Giglio, L., Kendall, J.D., Justice, C.O., 1999. Evaluation of global fire detection using simulated AVHRR infrared data. Int. J. Remote. Sens. 20, 1947–1985.

Gitas, I.Z., Mitri, G.H., Ventura, G., 2004. Object-based image classification for burned area mapp;ing of Creus Cape, Spain, using NOAA-AVHRR imagery. Rem. Sens. Environ. 92, 409–413.

Gloersen, P., Cavalieri, D.J., Chang, A.T.C., Wilheit, T.T., Campbell, W.J., Johannessen, O.M., et al., 1984. A summary of results from the first Nimbus-7 SMMR observations. J. Geophys. Res. 89, 5335–5344.

Goldberg, M.D., Reale, A., Kratz, G., 1998. The use of pattern recognition to derive SSM/T2 moisture retrievals. Adv. Space Res. 21, 385–388.

Goldstein, H., 1951. Attenuation by Condensed Water in Propagation of Short Radio Waves, MIT Radiation Laboratory Series 13. McGraw-Hill, New York.

Goode, J.P., 1925. The Homolosine projection: a new device for portraying the Earth's surface entire. Assoc. Am. Geogr. Ann. 115, 119–125.

Gordon, H.R., 1987. Calibration requirements and methodology for remote sensors viewing the ocean in the visible. Remote. Sens. Environ. 22, 103–126. Available at https://doi.org/10.1016/0034-4257(87)90029-0.

Goward, S.N., Markham, B.L., Dye, D.G., Dulaney, W., Yang, J., 1991. Normalized difference vegetation index measurements from the Advanced Very High Resolution Radiometer. Remote. Sens. Environ. 35, 257–277.

Goward, S.N., Dye, D.G., Turner, S., Yang, J., 1993. Objective assessment of the NOAA global vegetation index data. Int. J. Remote. Sens. 14, 3365–3394.

Gregoire, J.-M., Cahoon, D.R., Stroppiana, D., Li, Z., Pinnock, S., Eva, H., et al., 2001. Forest fire monitoring and map- ping for GOFC: current products and information networks based on NOAA-AVHRR, ERS-ATSR, and SPOT-VGT systems. In: Ahern, F., Goldammer, J., Justice, C.O. (Eds.), Global and Regional Fire Monitoring From Space: Planning a Coordinated International Effort. SPB Academic, The Hague, pp. 105–124.

Grody, N.C., 1991. Classification of snow cover and precipitation using the special sensor microwave imager. J. Geophys. Rev. 96, 7423–7435.

Hall, D.K., Riggs, G.A., Salomenson, V.V., DiGirolamo, N.E., Bayr, K.J., 2002. MODIS snow-cover products. Rem. Sens. Environ. 83, 181–194.

Hardy, C.C., Burgan, R.E., 1999. Evaluation of NDVI for monitoring live moisture in three vegetation types of the Western U.S. Photogram- Metr. Eng. Remote. Sens. 65, 603–610.

Heath, D.F., Krueger, A.J., Roeder, H.A., Henderson, B.D., 1975. Solar backscatter ultraviolet and total ozone mapping spectrometer (SBUV-TOMS) for Nimbus G. Opt. Eng. 14, 323–331.

Hofer, R., Njoku, E.G., Waters, J.W., 1981. Microwave radiometric measurements of sea surface temperature from the Seasat satellite: first results. Science 212, 1385–1387.

Holben, B.N., 1986. Characteristics of maximum-value composite images from temporal AVHRR data. Int. J. Remote. Sens. 7, 1417–1434.

Houze Jr., R.A., Betts, A.K., 1981. Convection is GATE. Rev. Geophy. Space Phys. 19, 541–576.

Houze Jr., R.A., Hobbs, P.V., 1982. Organization and structure of precipitating cloud systems. Adv. Geophys. 24, 225–315.

Houze, R.A., Geotis, S.G., West, A.K., 1981. Winter monsoon convection in the vicinity of North Borneo. Part 1: sructure and time variation of the clouds and precipitation. Mon. Wea. Rev. 109, 1595–1614.

Hsu, F.-C., Baugh, K.E., Ghosh, T., Zjizhin, M., Elvidge, C.D., 2015. DMSP-OLS radiance calibrated nighttime lights time series with intercalibration. Remote. Sens. 7, 1855–1876. Available at https://doi.org/10.3390/rs70201855.

Huemmrich, K.F., Goward, S.N., 1992. Spectral vegetation indexes and the remote sensing of biophysical parameters. In: Proc. Int. Geosci. Rem. Sens. Sym, Houston TX, vol. 2, pp 10171019.

Huete, A.R., Tucker, C.J., 1991. Investigation of soil influences in AVHRR red and near-infrared vegetation index imagery. Int. J. Rem. Sens. 12, 1223–1242.

Huete, A., Dida, K., Miura, T., Rodriguez, E.P., Gao, X., Ferreira, L.G., 2002. Overview of the radiometric and bio- physical performance of the MODIS vegetation indices. Remote. Sens. Environ. 83, 195–213.

Inoue, T., 1987. A cloud type classification with NOAA-7 split-window measurements. J. Geophys. Res. 92, 3,991–4,000.

James, M., Kalluri, S., 1994. The pathfinder AVHRR land dataset: an improved coarse resolution dataset for terrestrial monitoring. Int. J. Remote. Sens. Available at https://doi.org/10.1080/01431169408954335.

Johnson, M.J., Durst, S.L., Calvo, C.M., 2008. Yellow-billed cuckoo distribution, abundance, and habitat use along the lower Colorado River and its tributaries. 2007 Annual Report U.S; Geological Survey Open-file Report 2008-1177, Reston, VA, U.S. Department of the Interior, U.S. Geological Survey, 274 p.

Justice, C.O., Malingreau, J.P., (Eds.), 1993. The IGBP-DIS satellite fire detection algorithm workshop technical report, IGBP-DIS Working Paper 9, NASA/GSFC, Greenbelt, MD, February.

Justice, C., Malingreau, J.P., (Eds.), 1996. Report of the IGBP-DIS fire algorithm workshop 2, IGBP-DIS Working Paper 14, Ispra, Italy, October 1995.

Kaufman, Y.J., Holben, B.N., 1993. Calibration of the AVHRR visible and near-IR bands by atmospheric scattering, ocean glint and desert reflection. Int. J. Remote. Sens. 14, 21–52.

Kaufman, Y.J., Holben, B.N., 1995. Calibration of the AVHRR visible and near-IR bands by atmospheric scattering, ocean glint and desert reflection. J. Appl. Met.

Kaufman, Y.J., Tucker, C.J., Fung, I., 1990. Remote sensing of biomass burning in the tropics. J. Geophys. Res. 95, 9,927-9,939.

Kaufman, Y.J., Justice, C., Flynn, L., Kendall, J., Prins, E., Ward, D.E., et al., 1998a. Potential global fire monitoring from EOS-MODIS. J. Geophys. Res. 103, 32,215–32,238.

Kaufman, Y.J., Kleidman, R.G., King, M.D., 1998b. SCAR-B fires in the tropics: properties and their remote sensing from EOS-MODIS. J. Geophys. Res. 103, 31,955–31,969.

Kawanishi, T., Sezai, R., Ito, Y., Imaoka, K., Takeshima, T., Ishido, Y., et al., 2003. The Advanced Microwave Scanning Radiometer for the Earth Observing System (AMSR-E), NASDA's contribution to the EOS for global energy and water cycle studies. IEEE Trans. Geosci. Remote. Sens. 41 (2), 184–194. Available at https://doi.org/10.1109/TGRS.2002.808331.

Kennedy, P.J., Belward, A.S., Grégoire, J.-M., 1994. An improved approach to fire monitoring in West Africa using AVHRR data. Int. J. Remote. Sens. 15, 2,235–2,255.

Klein, A.G., Hall, D.K., Riggs, G.A., 1998. Improving snow cover mapping in forests through the use of a canopy reflectance model. Hydrol. Process. 12, 1723–1744.

Kunkee, D.B., Poe, G.A., Boucher, D.J., Swadley, S.D., Yong, Y., Wessel, J.E., 2008. Design and evaluation of the first special sensor microwave imager/sounder. Trans. Geosci. Rem. Sens. 46, 863–883. Available at https://doi.org/10.1109/TGRS.2008.917980.

Langaas, S., 1992. Temporal and spatial distribution of Savannah fires in Senegal and the Gambia, West Africa, 1989-90, derived from multi-temporal AVHRR night images. Int. J. Wildland Fire 2, 21–36.

Le Traon, P.Y., October 15e20, 2007. Satellite altimetry, ESA-most dragon programme, 2nd Advanced Training Course in Ocean Remote Sensing, Hangzou, P.R. China slide #5. https://www.yumpu.com/en/document/view/21972852/principles-of-satellite-radar-altimetry/5.

Le Traon, P.Y., Ogor, F., 1998. ERS-1/2 orbit improvement using T/P: The 2 cm. challenge. J. Geophys. Res. 103, 8045–8057.

Laymon, S.A. ,Halterman, M.D., 1987. A proposed habitat management plan for yellow-billed cuckoos in California. USDA Forest Service Gen. Tech. Rep. PSW-110.

Laws, J.O., Parson, D.A., 1943. The relation of raindrop size to intensity. Trans. Am. Geophys. Union. 24, 452.

Lee, T.M., Tag, P.M., 1990. Improved detection of hotspots using the AVHRR 3.7 um channel. Bull. Am. Meteorol. Soc. 71, 1,722–1,730.

Legeckis, R., 1978. A survey of worldwide sea surface temperature fronts detected by environmental satellites. J. Geophys. Res. 83, 4501–4522. Available at https://doi.org/10.1029/JC083iC09p04501.

Levin, S.A., 1976. Population dynamic models in heterogeneous environments. Ann. Rev. Ecol. Syst. 7, 287–310.

Li, Z., Cihlar, J., Moreau, L., Huang, F., Lee, B., 1997. Monitoring fire activities in the boreal ecosystem. J. Geophys. Res. 102, 29,611–29,624.

Li, Z., Khananian, A., Fraser, R., 2000a. Detecting smoke from boreal forest fires using neural network and threshold approaches applied to AVHRR imagery, IEEE Tran. Geoscience and Re- mote Sensing, revised.

Li, Z., Nadon, S., Cihlar, J., Stocks, B., 2000b. Satellite mapping of Canadian boreal forest fires: evaluation and comparison of algorithms. Int. J. Remote. Sens. 21 (3), 71–3082.

Li, Z., Nadon, S., Cihlar, J., 2000c. Satellite detection of Canadian boreal forest fires: development and application of an algorithm. Int. J. Remote. Sens. 21, 3,057–3,069.

Li, Z., Kaufman, Y., Ichoku, C., Fraser, R., Trishchenko, A., Giglio, L., et al., 2001. A review of AVHRR-based active fire detection algorithms: principles, Limitations, and recommendations. In: Global and Regional Vegetation Fire Monitoring from Space: Planning a Coordinated International Effort, pp 199–225. Ed F.J. Ahera, J.G. Goldammer and C.O.

Liebe, H.J., 1985. An updated model for millimeter wave propagation in moist air. Radio. Sci. 20 1069-10.

Liu, X., Tang, B.-H., Yan, G., Li, Z.-L., Liang, S., 2019. Retrieval of global orbit drift corrected land surface temperature from long-term AVHRR data. Remote. Sens. 11, 2843.

London, J., Oltmans, S.J., 1978/79. The global distribution of long-term total ozone variations during the period 1957–1975. Pure Appl. Geophys. 117, 345–354.

Malingreau, J.P., 1990. The contribution of remote sensing to the global monitoring of fires in tropical and sub-tropical ecosystem.

Malingreau, J.P., Justice, C.O., (Eds.), 1997. The IGBP-DIS satellite fire detection algorithm workshop technical report, IGBP-DIS Working Paper 17, NASA/GSFC, Greenbelt, Maryland, USA, February.

Malingreau, J.P., Tucker, C.J., 1988. Large scale deforestation in the south-eastern Amazon basin. Ambio 17, 49–55.

Mairota, P., Cafarelli, B., Didham, R.K., Lovergine, F.P., Lucas, R.M., Nagendra, H., et al., 2015. Challenges and opportunities in harnessing satellite remote sensing for biodiversity monitoring. Ecolog. Infomat 30, 207–214. Available at https://doi.org/10.1016/j.ecoinf.2015.08.006.

Marra, J., Houghton, R.W., Garside, C., 1990. Phytoplankton growth at the shelf-break front in the Middle Atlantic Bight. J. Mar. Res. 48, 851–868.

Martin, P., Ceccato, P., Flasse, S., Downey, I., 1999. Fire detection and fire growth monitoring using satellite data. In: Chuvieco, E. (Ed.), Remote Sensing of Large Wildfires: in the European Mediterranean Basin. Springer-Verlag Berlin/Heidelberg, pp. 212 3-540-65767-3, XI.

Martinec, J., Rango, A., Roberts, R., 1994. Snowmelt runoff model (SRM) User's Manuarl (Ed M.F. Baumgartner). Geographica Bermensia P-29, Dept. of Geography, Univ of Berne, Switzerland.

Mateer, C.L., Heath, D.F., Krueger, A.J., 1971. Estimation of total ozone from satellite measurements of backscattered ultraviolet earth radiance. J. Atmos. Sci. 28, 1307–1311.

McClain, E.P., 1980. Passive radiometry of the ocean from space-an overview. Bound.-Layer Meteorol. 18, 7–24. Available at https://doi.org/10.1007/bf00117908.

McClain, E.P., Pichel, W.G., Walton, C.C., 1985. Comparative performance of AVHRR-based multichannel sea surface temperatures. J. Geophys. Res. 90, 11587–11601.

McMillin, L., 1975. Estimations of sea surface temperature from two infrared window measurements with different absorption. J. Geophys. Res. Available at https://doi.org/10.1029/JC080i036p05113.

McMillin, L.M., Crone, L.J., Goldberg, M.D., Kleespies, T.J., 1995. Atmospheric transmittance of an absorbing gas 4: optran, a computationally fast and accurate transmittance model for absorbing gases with fixed and variable mixing ratios at variable viewing angles. Appl. Opt. 34 (27), 6269–6274.

Merchant, C.J., Embury, O., Rayner, N.A., Berry, D.I., Corlett, G.K., Learn, K., et al., 2012. A 20 year independent record of sea surface temperature for climate from Along-Track Scanning Radiometers. J. Geophys. Res. 117. Available at https://doi.org/10.1029/2012JC008400.

Milne, A.K., 1986. The use of remote sensing in mapping and monitoring vegetational change associated with bushfire events in eastern Australia. Geocarto Int. 1, 25–34.

Minnett, P.J., Alvera-Azcarate, N.A., Chin, T.M., Cortlett, G.K., Gentemann, C.L., Karagali, I., et al., 2019. Half a century of satellite remote sensing of sea-surface temperature. Rem. Sens. Environ. 233. Available at https://doi.org/10.1016/j.rse.2019.111366.

Mitri, G., Gitas, I., 2002. The development of an object-oriented classification model for operational burned area mapping on the Mediterranean island of Thasos using Landsat-TM images. In: Viegas, D.X. (Ed.), Forest fire research and wildland fire safety. Rotterdam7 Mill press, pp. 79.

Monaldo, F.M., 1988. Expected differences between buoy and radar altimeter estimates of wind speed and significant wave height and their implications on buoy-altimeter comparisons. J. Geophys. Res. 93, 2285–2302.

Moody, A., Johnson, D.M., 2001. Land-surface phenologies from AVHRR using the discrete Fourier transform. Remote. Sens. Environ. 75, 305 3.

Muirhead, K., Cracknell, A., 1985. Straw burning over Great Britain detected by AVHRR. Int. J. Remote. Sens. 6, 827–833.

Njoku, E.G., Swanson, L., 1983. Global measurements of sea surface temperature, wind speed and atmospheric water content from satellite microwave radiometry. Month. Weath. Rev. 111, 1977–1987.

Paltridge, G.W., Barber, J., 1988. Monitoring grassland dryness and fire potential in Australia with NOAA/AVHRR data. Remote. Sens. Environ. 25, 381–394.

Peixoto, J.P., Oort, .H., 1992. Physics of Climate, Am. Inst. of Phys., College Park, Md.

Pereira, J.M.C., 1992. Burned area mapping with conventional and selective principal component analysis. Finisterra 27, 63–78.

Pereira, M.C., Setzer, A.W., 1993. Spectral characteristics of deforestation fires in NOAA-AVHRR images. Int. J. Remote. Sens. 14, 583–597.

Pereira, M.C., Chuvieco, E., Beaudoin, A., Desbois, N., 1997. Remote Sensing of Burned Areas. A Review of Remote Sensing Methods for the Study of Large Wildland Fires. Alcala′ de Henares7 Universidad de Alcala′, pp. 127–183.

Pettorelli, N., 2019. Satellite Remote Sensing and the Management of Natural Resources. Oxford University Press, Oxford, UK.

Pettorelli, N., Schulte to Bühne, H., 2022. Current and future opportunities for satellite remote sensing to inform rewilding. Rem. Sens. Ecol. Cons. Available at https://doi.org/10.1002/rse2.32.

Poe, G., St Germain, K., Bobak, J., Swadley, S., Wessel, J., Thomas, B., 2001. DMSP calibration/validation plan for the Special Sensor Microwave Imager Sounder (SSMIS). Nav. Res. Lab. 32.

Press, H., 1970. Introduction to the NIMBUS meteorological satellite program. IEEE Trans. Geoscience Electronics, GE-8, No. 4.

Pribble, J.R., Walsh, J.J., Dieterle, D.A., 1994. A numerical analysis of shipboard and coastal zone color scanner time series of new production within GulfStream cyclonic eddies in the South. Atl. Bight. J. Geophys. Res. 99, 7513–7538.

Price, J.C., 1987. Calibration of satellite radiometers and the comparison of vegetation indices. Remote. Sens. Environ. 21, 15–27.

Price, J.C., 1988. An update on visible and near-infrared calibration of satellite instruments. Remote. Sens. Envion. 24, 419–422.

Price, J.C., 1992. Estimating vegetation amount from visible and near infrared reflectances. Remote. Sens. Environ. 41, 29–34.

Reiners, P., Asam, S., Frey, C., Holzwarth, S., Bachmann, M., Sobrino, J., et al., 2021. Validation of AVHRR land surface temperature with MODIS and in situ LST—A TIMELINE thematic processor. Rem. Sens. 13. Available at https://doi.org/10.3390/rs13173473.

Rexer, M., Hirt, C., 2014. Comparison of free high-resolution digital elevation data sets (ASTER GDEM2, SRTM v2.1/v4.1) and validation against accurate heights from the Australian National Gravity database. Australian J. Earth Sci. 61. Available at https://doi.org/10.1080/08120099.2014.884983.

Riley, G.A., 1946. Factors controlling phytoplankton populations on Georges Bank. J. Mar. Res. 6, 54–73.

Riley, G.A., 1947. Seasonal fluctuations of the phytoplankton population in New England coastal waters. J. Mar. Res. 6, 114–125.

Robinson, J.M., 1991. Fire from space: global fire evaluation using infrared remote sensing. Int. J. Rem. Sens. 12, 2–24.

Roderick, M.L., 1993. Methods for calculating solar position and day length including computer programs and subroutine, Division of Resource Management Technical Report No. 137, Western Australia Department of Agriculture, Perth, WA, 22 pp.

Roderick, M.L., 1994. Satellite derived vegetation indices for monitoring seasonal vegetation conditions in Western Australia, Ph.D. thesis, School of Surveying and Land Information, Curtin Univ. Tech., Pert, Western Australia, 349 pp.

Roderick, M., Smith, R., Cridland, S., 1996. The precision of the NDVI derived from AVHRR observations. Rem. Sens. Environ. 56, 57–65.

Rössler, S., Dietz, A.J., 2002. Detection of snow cover from historical and recent AVHRR data-a thematic TIMELINE processor. Geomatics 2022 (2), 144–160. Available at https://doi.org/10.3390/geomatics2010009.

Ryan, J.P., Yoder, J.A., 1996. Long-term mean and event-related pigment distributions during the unstratified period in South Atlantic Bight outer margin and middle shelf waters. Continental Shelf Res. 16, 1165–1183.

Ryan, J.P., Yoder, J.A., Cornillon, P.C., 1999a. Enhanced chlorophyll at the shelfbreak of the Mid-Atlantic Bight and Georges Bank during the spring transition. Limnol. Oceanogr. 44, 1–11.

Ryan, J.P., Yoder, J.A., Barth, J.A., Cornillon, P.C., 1999b. Chlorophyll enhancement and mixing associated with meanders of the shelf break front in the Mid-Atlantic Bight. J. Geophys. Res. 104 23 479–23 493.

Ryther, J.H., Yentsch, C.S., 1958. Primary production of continental shelf waters off New York. Limnol. Oceanogr. 3, 327–335.

Salomonson, V.V., Appel, I., 2004. Estimating fractional snow cover from MODIS using the normalized difference snow index. Remote. Sens. Environ. 89, 351–360.

Salomonson, V.V., Appel, I., 2006. Development of the aqua MODIS NDSI fractional snow cover algorithm and validation results. IEEE Trans. Geosci. Remote. Sens. 44, 1747–1756.

Savage, R.C., 1976: The transfer of thermal microwaves through hydrometerws. PhD dissertation. Univ of Wisconsin, Madison, 147 pp.

Savage, R.C., Weinman, J.A., 1975. Preliminary calculates of the upwelling radiance from rainclouds at 37.0 and 19.35 GHz. Bull. Amer. Meteor. Soc. 56, 1272–1274.

Setzer, A.W., Malingreau, J.P., 1996. AVHRR monitoring of vegetation fires in the tropics: toward the development of a global product. In: LeVine, J.S. (Ed.), Biomass Burning and Global Change. MIT Press, Cambridge, MA, pp. 25–39.

Sokal, R.R., Rohlf, F.J., 1981. Biometry, second ed. W.H. Freeman and Co., New York.

Shea, D.J., K.E. Trenberth and R.W. Reynolds, 1990: A global monthly sea surface temperature climatology, NCAR Tech. Note 345, 167pp, Nat. Cent. For Atmos. Res., Boulder, CO.

Smith, G.L., Priestly, K.J., Loeb, N.G., Wielicki, B.A., Charlock, T.P., Minnis, P., et al., 2011. Clouds and Earth Radiant Energy System (CERES), a review: past, present and future. Adv. Space Res. 48, 254–263.

Soloviev, A.V., Lukas, R., 2014. The near-surface layer of the ocean: structure, dynamics and applications. Book second ed. Available at https://doi.org/10.1007/978-94-007-7621-0.

Southwood, T.R.E., 1977. Habitat, the template for ecological strategies? J. Amin. Ecol. 46, 337–365.

Spencer, R.W., 1986. A satellite passive 37 GHz scattering-based method for measuring oceanic rain rates. J. Clim. Appl. Meteor. 25, 754–766.

Spencer, R.W., Hinton, B.B., Olson, W.S., 1983. Nimbus-7 37 GHz radiances correlated with radar rain rates over the Gulf of Mexico. J. Clim. Appl. Meteor. 22, 2095–2099.

Steinwand, D.R., 1993. Mapping raster imagery into the interrupted Goode Homolosine Projection. Int. J. Remote. Sens. 15, 3463–3472.

Stogryn, A., 1970. The brightness temperature of a vertically structured medium. Radio. Sci. 5 (12), 1397–1406.

Strassburg, B.B.N., Iribarrem, A., Beyer, H.L., Cordeiro, C.L., Crouzeilles, R., Jakovac, C.C., et al., 2020. Global priority areas for ecosystem restoration. Nature 586, 724–729.

Tag, P.:M., Rankert, R.L., Brody, L.R., 2000. An AVHRR multiple cloud-type classification package. J. Appl. Meteor. 39. Available at https://doi.org/10.1175/1520-0450(2000)039<0125:AAMCTC>2.0.CO;2.

Tanaka, S., Kimura, H., Suga, Y., 1983. Preparation of a 1:25.000 LANDSAT map for assessment of burnt area on Etajima Island. Int. J. Remote. Sens. 4, 17–31.

Tanre, D., Holben, B.N., Kaufman, Y.J., 1992. Atmospheric correction algorithm for NOAA-AVHRR products: theory and application. Trans. Georsci. Rem. Sens. 30, 231–248.

Townshend, J.R.G., Justice, C.O., Gurney, C., McManus, J., 1992. The impact of misregistration on change detection. Trans. Geosci. Rem. Sens. 30, 1054–1060.

Tucker, C.J., Pinzon, J.E., Brown, M.E., Slayback, D.A., Pak, E.W., Mahoney, R., et al., 2005. An estimated AVHRR 8-km NDVI dataset compatible with MODIS and SPOT vegetation NDVI data. Int. J. Remote. Sens. 26, 4485–4498.

Urban, D.L., O'Neill, R.V., Shugart Jr., H.H., 1987. Landscape ecology: a hierarchical perspective can help scientists understand spatial patterns. Bioscience 37, 119127.

Verity, P.G., Lee, T.N., Yoder, J.A., Paffenhofer, G.-A., Blanton, J.O., Alexander, C.R., 1993. Processes on the outer continental shelf. Ocean Processes: US Southeast Continental Shelf, National Technical Information Service, Springfield, VA, DE93010744, 112 pp.

Vermote, E., 2019. AVHRR surface reflectance and normalized difference vegetation index—climate algorithm theoretical basis document, NOAA climate data record program CDRP-ATBD-0459 revsion 2. NOAA Central Library, Washington, DC, USA.

Wallace, C.S.A., 2002. Extracting temporal and spatial information from remotely sensed data for mapping wildlife habitat. PhD Thesis, University of Arizona, Tucson, AZ.

Waldrop, M.M., 1982. Imaging the Earth (II): the politics of Landsat. Science 216, 40–41.

Wallin, D., Tucker, C.J., Shugart, H.H., Wilhelmi, F., 1992. Satellite remote sensing of breeding habitat for an African weaverbird. Landsc. Ecol. Available at https://doi.org/10.1007/BF02418940.e.

Walsh, J.J., Whitledge, T.E., Barvenik, F.W., Wirick, C.D., Howe, S.O., Esaias, W.E., et al., 1978. Wind events and food chain dynamics within the New York Bight. Limnol. Ocanogr. 23, 659–683.

Walton, C.C., 2016. A review of differential absorption algorithms utilized at NOAA for measuring sea surface temperature with satellite radiometers. Rem. Sens. Environ. 187, 434–446. Available at https://doi.org/10.1016/j.rse.2016.10.011.

Walton, C.C., Pichel, W.G., Sapper, J.F., May, D.A., 1998. The development and operational application of nonlinear algorithms for the measurement of sea surface temperatures with the NOAA polar-orbiting environmental satellites. J. Geophys. Res. Available at https://doi.org/10.1029/98JC02370.

Wang, J., Dong, J., Liu, J., Huang, M., Li, G., Running, S.W., et al., 2014. Comparison of gross primary productivity derived from GIMMS NDVI3g, GIMMS, and MODIS in Southeast Asea. Remote. Sens. 6, 2108–2133.

Ward, P., 1971. The migration patterns of *Quelea quelea* in Africa. Ibis 113, 275–297.

Ward, P., 1973. Manual of techniques used in research on quelea birds. APG: RAF/67/087 UNDP/FAO, Rome.

Warnecke, G., Allison, L.J., McMillin, L.M., Szekielda, K.H., 1971. Remote sensing of ocean currents and sea surface temperature changes derived from the Nimbus 2 satellitei. J. Phys. Oceanogr. 1, 45–60. Available at https://doi.org/10.1175/1520-0485(1971)001<0045:Rsooca>2.0.Co;2.

Warner, M.D., Mass, C.F., Salathé, E.P., 2012. Wintertime extreme precipitation events along the Pacific Northwest coast. Climatology and synoptic evolution. Mon. Weath. Rev. 140, 2021–2043. Available at https://doi.org/10.1175/MWR-D-11-00197.1.

Welch, R., 1980. Monitoring urban population and energy utilization patterns from satellite data. Remote. Sens. Environ. 9, 1–9.

Wentz, F.J., 1975. A two-scale scattering model for foam-free sea microwave brightness temperatures. J. Geophys. Res. 80, 3441–3446.

Wentz, F.J., 1983. A model function for ocean microwave brightness temperatures. J,. Geophys. Res. 88, 1892–1908.

Wentz, F.J., 1992. Measurement of oceanic wind vector using satellite microwave radiometers. IEEE Trans. Geosci. Remote. Sens. 30, 960–972.

Wentz, F.J., 1997. A well-calibrated ocean algorithm for the special sensor microwave imager. J. Geophys. Res. 102, 8703–8718.

Wiens, J.A., Dyer, M.I., 1977. Assessing the potential impact of granivorous birds in ecosystems. In: Pinowski, J., Kendeigh, S.C. (Eds.), Granivorous Birds in Ecosystems. Cambridge University Press, Cambridge, pp. 205–266.

Wilheit, T.T., Chang, A.T.C., 1980. An algorithm for retrieval of ocean surface and atmospheric parameters from the observations of the Scanning Multichannel Microwave Radiometer(SMMR). Radio. Sci. 15, 525–544.

Wilheit, T.T., Greaves, J., Gatlin, J., Han, D., Krupp, B.M., Milman, A.S., 1984. Retrieval of ocean surface parameters from the Scanning Multichannel Microwave Radiometer on the Nimbus-7 satellite. Trans. Geosci. Rem. Sens. GE 22, 133–143.

Wong, E.W., Minnett, P.J., 2018. The response of the ocean thermal skin layer to variations in incident infrared radiation. J. Geophys. Res. 123. Available at https://doi.org/10.1002/2017JC013351.

Wu, R., Weinman, J.A., 1984. Microwave radiances from precipitating clouds containing aspherical ice, combined phase, and liquid hydrometers. J. Geophys. Res. 89, 7170–7178.

Yu, K., Zhu, S., Lan, L., Yang, B., 2023. High-resolution and wide-swatch SAR imaging with space-time coding array. Remote. Sens. 15. Available at https://doi.org/10.3390/rs15092465.

Zhang, B., Wu, Y., Zhao, B., Chanussot, J., Hong, D., Yao, J., 2022. Progress and challenges in intelligent remote sensing satellite systems. J. Sel. Top. Appl. Earth Obs. And. Rem. Sems. 15, 1814–1822. Available at https://doi.org/10.1109/JSTARS.2022.3148139.

Zhao, Q., Yu, L., 2025. Advancing sustainable development goals through Earth observation satellite data: current insights and future directions. J. Rem. Sens. 5. Available at https://doi.org/10.34133/remotesensing.0403.

Zhonghai, J., Simpson, J.J., 1999. Bidirectional anisotropic reflectance of snow and sea ice in AVHRR channel 1 and 2 spectral regions. I. Theoretical analysis. IEEE Trans. Geosci. Remote. Sens. 37, 543–554.

Further Reading

Bachmann, M., Tungalagsaikhan, P., Ruppert, T., Dech, S., 2015. Calibration and pre-processing of a multi-decadal AVHRR time series. In: Kuenzer, C., Dech, S., Wagner, W. (Eds.), Remote Sensing Time Series: Revealing Land Surface Dynamics. Springer, Cham, Switzerland, pp. 43–74 ISBN 978-3-319-15967-6.

Barnard, A., Boxx, E., Haëntjens, N., Orrico, C., Chamberlain, P., Frouin, R., et al., 2024. A float-based ocean color vicarious calibration program. Front. Rem. Sens. 5. Available at https://doi.org/10.3389/frsen.2024.1373540.

Barton, I.J., 1995. Satellite-derived sea surface temperatures: current status. J. Geophys. Res. 100, 8777–8790.

Becker, F., Li, Z.-L., 1990. Towards a local split window method over land surfaces. Int. J. Remote. Sens. 11, 369–393 1990.

Beer, T., 1980. P. Microwave sensing from satellites. Rem. Sens. Environ. 9, 65–86.

Bernstein, R.L., Morris, J.H., 1983. Tropical and mid-latitude North. Pacific sea surface temperature variability from the Seasat SMMR. J. Geophys. Res. 88, 1877–1891.

Burke, H.K., Jones, K.C., 1980. Models of environmental parameters for the SSM/I program, Environmental Research and Technology, Inc., February.

Choudhury, B.J., Schmugge, T.J., Cheng, A., Newton, R.W., 1979. Effect of surface roughness on the microwave emission from soils. J. Geophy. Res. 84 (C9), 5699.

Chuvieco, E., Aguado, I., Cocero, D., Riaño, D., 2003. Design of an empirical index to estimate fuel moisture content from NOAA– AVHRR analysis in forest fire danger studies. Int. J. Remote. Sens. 24 (8), 1621–1637.

Clark, D.K., 1980. Phytoplankton pigment algorithms for the Nimbus-7 CZCS. In: Gower, J.F.R. (Ed.), Oceanography from Space. Plenum Press, NewYork, pp. 227–237.

Cox, C., Munk, W., 1954. Measurement of the roughness of the sea surface from photographs of the sun's glitter. J. Optical Soc. Am. 44 (11) 1954.

Elliott, C.C.H., 1989. The pest status of the quelea. In: Bruggers, R.L., Elliott, C.C.H. (Eds.), *Quelea quelea*, Africa's Bird Pest. Oxford University Press, New York, pp. 17–34.

Franceschini, S., Meier, A.C., Suan, A., Stokes, K., Roy, S., Madin, E.M.P., 2023. A deep learning model for measuring coral reef halos globally from multispectral satellite imagery. Rem. Sens. Environ. 292. Available at https://doi.org/10.1016/j.rse.2023.113584.

Garcia-Moliner, G., Yoder, J.A., 1994. Variability in pigment concentration in warm-core rings as determined by coastal zone color scanner satellite imagery from the Mid-Atlantic Bight. J. Geophys. Res. 99 14 277–14 290.

Göttsche, F.M., Olesen, F.-S., 2001. Modeling of diurnal cycles of brightness temperature extracted from METEOSAT data. Remote. Sens. Environ. 337–348.

Gutman, G., Elvidge, C., Csiszar, I., Romanov, P., 2001. NOAA archives of data from meteorological satellites useful for fire products. In: Ahern, F., Goldammer, J., Justice, C.O. (Eds.), Global and Regional Fire Monitoring From Space: Planning a Coordinated Inter- national Effort. SPB Academic, The Hague, pp. 257–266.

Hall, R.C., 2001. The military polar-orbiting meteorological satellite program. Office of the historian, National Reconnissance Office.

Hilsenrath, E., Williams, D.E., Caffrey, R.T., Cebula, R.P., Hynes, S.J., 1993. Calibration and radiometric Stability of the Shuttle Solar Backscatter Ultraviolet (SSBUV) experiment. Metrologia 30, 243–248.

Hoogen, R., Rozanov, V.V., Burrows, J.P., 1999. Ozone profiles from GOME satellite data: algorithm description and first validation. J. Geophys. Res.-Atmos. 104, 8263–8280.

Huang, L.K., Cebula, R.P., Taylor, S.L., Deland, M.T., McPeters, R.D., Stolarski, R.S., 2003. Determination of NOAA- 11 SBUV/2 radiance sensitivity drift based on measurements of polar ice cap radiance. Int. J. Remote. Sens. 24, 305–314. Available at https://doi.org/10.1080/01431160304978.

Huang, F.T., Mayr, H.G., Russell, J.M., Mlynczak, M.G., 2010. Ozone diurnal variations in the stratosphere and lower mesosphere, based on measurements from SABER on TIMED. J. Geophys. Res.-Atmos. 115, D24308. Available at https://doi.org/10.1029/2010jd014484.

Jakubauskas, M.E., Legates, D.R., Kastens, J., 2002. Crop identification using harmonic analysis of time-series AV- HRR NDVI data. Computers Electron. Agriculture 37, 127–139.

Joiner, J., Bhartia, P.K., Cebula, R.P., Hilsenrath, E., McPeters, R.D., Park, H., 1995. Rotational raman-scattering (Ring Effect) in satellite backscatter ultraviolet measurements. Appl. Opt. 34, 4513–4525.

Justice, Academic Publishing by The Hague, The Netherlands.

Kareiva, P., Anderson, M., 1988. Spatial aspects of species interactions: the wedding of models and experiments. In: Hastings, A. (Ed.), Community Ecology. Springer-Verlag, New York, pp. 35–50.

Klenk, K.F., Bhartia, P.K., Fleig, A.J., Kaveeshwar, V.G., McPeters, R.D., Smith, P.M., 1982. Total ozone determination from the backscattered ultraviolet (BUV.) Experiment, J. Appl. Meteorol. 21, 1672–1684.

Kramarova, N.A., Bhartia, P.K., Frith, S.M., McPeters, R.D., Stolarski, R.S., 2013a. Interpreting SBUV smoothing errors: an example using the quasi-biennial oscillation. Atmos. Meas. Tech. 6, 2089–2099. Available at https://doi.org/10.5194/amt-6-2089-2013.

Kramarova, N.A., Frith, S.M., Bhartia, P.K., McPeters, R.D., Taylor, S.L., Fisher, B.L., et al., 2013b. Validation of ozone monthly zonal mean profiles obtained from the version 8.6 Solar Backscatter Ultraviolet algorithm. Atmos. Chem. Phys. 13, 6887–6905. Available at https://doi.org/10.5194/acp-13-6887-2013.

Kriegler, F.J., Malila, W.A., Nalepka, R.F., Richardson, W., 1969. Preprocessing transformations and their effect on multispectral recognition. Remote. Sens. Env. VI, 97–132.

Labow, G.J., McPeters, R.D., Bhartia, P.K., Kramarova, N., 2013. A comparison of 40 years of SBUV measurements of column ozone with data from the Dobson/Brewer network. J. Geophys. Res. Atmos. 118, 7370–7378. Available at https://doi.org/10.1002/jgrd.50503.

LaViolette, P.E., Chabot, P.L., 1968. Nimbus 2 satellite sea surface temperatures versus historical data in a selected region: a comparative study. Deep-Sea Res. 15, 617–622. Available at https://doi.org/10.1016/0011-7471(68)90068-5.

Liu, C.-T., 1982. Tropical Pacific sea surface temperatures measured by Seasat Microwave Radiometer and by ships. NOAA Rep. ERL PMEL-34, 160 pp.

Liu, X., Chance, K., Sioris, C.E., Spurr, R.J.D., Kurosu, T.P., Mar- tin, R.V., et al., 2005. Ozone profile and tropospheric ozone retrievals from the Global Ozone Monitoring Experiment: algorithm description and validation. J. Geophys. Res.-Atmos. 110, D20307. Available at https://doi.org/10.1029/2005JD006240.

Liu, X., Chance, K., Sioris, C.E., Kurosu, T.P., 2007. Impact of using different ozone cross sections on ozone profile retrievals from Global Ozone Monitoring Experiment (GOME) ultraviolet measurements. Atmos. Chem. Phys. 7, 3571–3578. Available at https://doi.org/10.5194/acp-7-3571-2007.

Liu, X., Bhartia, P.K., Chance, K., Spurr, R.J.D., Kurosu, T.P., 2010. Ozone profile retrievals from the Ozone Monitoring Instrument. Atmos. Chem. Phys. 10, 2521–2537. Available at https://doi.org/10.5194/acp-10-2521-2010.

Livesey, N., Read, W., Froidevaux, L., Lambert, A., Manney, G., Pumphrey, H., et al., 2013. EOS Aura MLS version 3.3 and 3.4 level 2 data quality and description document. Available at http://mls.jpl.nasa.gov/data/v3_data_quality_document.pdf (accessed 29.09.13).

Mclntosh, R.P., 1985. The Background of Ecology. Cambridge University Press, Cambridg.

Nordberg, W., Conaway, J., Ross, D.B., Wilheit, T., 1971. Measurements of microwave emissions from a foam covered wind driven sea. J. Atmos. Sci. 28, 429 1971.

O'Neill, R.V., DeAngelis, D.L., Waide, J.B., Allen, T.F.H., 1986. A Hierarchical Concept of Ecosystems. Princeton University Press, Princeton.

O'Reilly, J.E., Zetlin, C., 1998. Seasonal, horizontal, and vertical distribution of phytoplankton chlorophyll a in the northeast US continental shelf ecosystem. NOAA Technical Report NMFS 139, US Department of Commerce, Seattle.

Petty, G.W., Katsaros, K.B., 1990. Precipitation observed of the South China Sea by the nimbus-7 scanning multichannel microwave radiometer during winter MONEX. J. App. Meteotol. 29, 273–287.

Roderick, M.L., Smith, R., Lodwick, G., 1996. Calibrating long-term AVHRR derived NDVI imagery. Rem. Sens. Environ. 58, 1–12.

Rouse, J.W., Haas, R.H., Schell, J.A., Deering, D.W., 1973. Monitoring vegetation systems in the great plains with ERTS (Earth Resources Technology Satellite). In: Proceedings of 3rd Earth Resources Technology Satellite Symposium, Greenbelt, 10–14 December, SP-351, pp. 309–317.

Salomonson, V.V., Barnes, W., Xiong, J., Kempler, S., Masuoka, E., 2002. An overview of the Earth Observing System MODIS instrument and associated data system performance. In: IEEE, International Geoscience and Remote Sensing Symposium, Toronto, ON, Canada, pp. 1174–1176. Available at https://doi.org/10.1109/IGARSS.2002.1025812.

Schmitt, T.J., Lindstrom, S.S., Gerth, J.J., Gunshor, M.M., 2018. Applications of the 16 spectral bands on the Advanced Baseline Imager (ABI). J. Operational Metero. 6, 33–46.

Schulz, J., Schluessel, P., Grassl, H., 1995. Water vapor in the atmospheric boundary layer over oceans from SSM/I measurements. In,t. J. Remote. Sens. 14, 2773–2789.

Setzer, A.W., Pereria, M.C., 1991a. Operational detection of fires in Brazil with NOAA AVHRR. In: 24th Symposium on Remote Sensing of the Environment, ERIM, Rio de Janeiro, May.

Setzer, A.W., Pereria, M.C., 1991b. Amazonian biomass burnings in 1987 and an estimate of their tropospheric emission. Ambio. 20, 19–22.

Simmons, A.J., 1955. Phase shift by periodic loading of waveguide and its application to broad-band circular polarization. IRE Trans. Microw. Theory Tech. 3 (6), 18–21.

Siljestrfm, P., Moreno, A., 1995. Monitoring burnt areas by principal components analysis of multi-temporal TM data. Int. J. Remote. Sens. 16, 1577–1587.

Staelin, D.H., Kunzi, K.F., Perryjohn, R.L., Poon, R.K.I., Wilcox, R.W., 1976. Remote sensing of atmospheric water vapor and liquid water with the NIMBUS-5 microwave spectrometer. J. Appl. Met. 15, 1204–1214.

Stroppiana, D., Pinnock, S., Gregoire, J.-M., 2000. The global fire product: daily fire occurrence from April 1992 to December 1993 derived from NOAA AVHRR data. Int. J. Remote. Sens. 21, 1279–1288.

Stouffer, R.J., Manabe, S., 1999. Response of a coupled ocean-atmosphere model to increasing atmospheric carbon monoxide: sensitivity to the rate of increase. J. Clim. Available at https://doi.org/10.1175/1520-0442(1999)012<2224:ROACOA>2.0.CO;2.

Szekielda, K.H., Salomonson, V., Allison, L.J., Rapid variations of sea surface temperature in the Persian Gulf as recorded by Nimbus-2 HRIR. Limnol. Oceanogr, 17, pp. 307–309. Available at https://doi.org/10.4319/lo.1972.17.2.0307

Turner, M.G., 1987. Landscape heterogeneity and disturbance. Ecological Studies. Springer-Verlag, New York vol. 64.

Viegas, D.X., Viegas, T.P., Ferreira, A.D., 1992. Moisture content of fine forest fuels and fire occurrence in central Portugal. Int. J. Wildland Fire 2 (2), 69–85.

Vermote, E., NOAA CDR Program, 2019. NOAA Climate Data Record (CDR) of AVHRR Normalized Difference Vegetation Index (NDVI), version 5. [indicate subset used]. NOAA National Centers for Environmental Information. Available at https://doi.org/10.7289/V5ZG6QH9.

Wallace, C., van Riper, C., Villarreal, M., 2011. Mapping landscape phenology preference of yellow-billed cuckoo with AVHRR data. J. Biogeography. Available at https://www.researchgate.net/publication/258459849.

Ward, P., 1965. Feeding ecology of the black-faced dioch *Que- lea quelea* in Nigeria. Ibis 107, 173–214.

Wark, D.Q., 1993. Adjustment of TIROS operational vertical sounder data to a vertical view. NOAA Technical Report, NESDIS 64.

Index

Note: Page numbers followed by "*f*" indicate figures and "*t*" indicate tables.

T

U

V

W

Y

9780443220869